T0254716

.

Transition to Advanced Mathematics

Textbooks in Mathematics

Series editors:
Al Boggess, Kenneth H. Rosen

An Introduction to Analysis, Third Edition
James R. Kirkwood

Student Solutions Manual for Gallian's Contemporary Abstract Algebra, Tenth Edition
Joseph A. Gallian

Elementary Number Theory
Gove Effinger, Gary L. Mullen

Philosophy of Mathematics
Classic and Contemporary Studies
Ahmet Cevik

An Introduction to Complex Analysis and the Laplace Transform
Vladimir Eiderman

An Invitation to Abstract Algebra
Steven J. Rosenberg

Numerical Analysis and Scientific Computation
Jeffery J. Leader

Introduction to Linear Algebra
Computation, Application and Theory
Mark J. DeBonis

The Elements of Advanced Mathematics, Fifth Edition
Steven G. Krantz

Differential Equations
Theory, Technique, and Practice, Third Edition
Steven G. Krantz

Real Analysis and Foundations, Fifth Edition
Steven G. Krantz

Geometry and Its Applications, Third Edition
Walter J. Meyer

Transition to Advanced Mathematics
Danilo R. Diedrichs and Stephen Lovett

Modeling Change and Uncertainty
Machine Learning and Other Techniques
William P. Fox and Robert E. Burks

Abstract Algebra
A First Course, Second Edition
Stephen Lovett

Multiplicative Differential Calculus
Svetlin Georgiev, Khaled Zennir

Applied Differential Equations
The Primary Course
Vladimir A. Dobrushkin

Introduction to Computational Mathematics: An Outline
William C. Bauldry

Mathematical Modeling the Life Sciences
Numerical Recipes in Python and MATLAB™
N. G. Cogan

https://www.routledge.com/Textbooks-in-Mathematics/book-series/CANDHTEXBOOMTH

Transition to Advanced Mathematics

Danilo R. Diedrichs
Wheaton College, USA

Stephen Lovett
Wheaton College, USA

CRC Press
Taylor & Francis Group
Boca Raton London New York

CRC Press is an imprint of the
Taylor & Francis Group, an **informa** business

A CHAPMAN & HALL BOOK

First edition published 2022
by CRC Press
6000 Broken Sound Parkway NW, Suite 300, Boca Raton, FL 33487-2742

and by CRC Press
4 Park Square, Milton Park, Abingdon, Oxon, OX14 4RN

Library of Congress Cataloging-in-Publication Data

Names: Diedrichs, Danilo R., author. | Lovett, Stephen (Stephen T.), author.
Title: Transition to advanced mathematics / Danilo R. Diedrichs, Wheaton College, USA, Stephen Lovett, Wheaton College, USA.
Description: First edition. | Boca Raton : Chapman & Hall/CRC Press, 2022.|
Series: Textbooks in mathematics | Includes bibliographical references and index.
Identifiers: LCCN 2021055607 (print) | LCCN 2021055608 (ebook)
|ISBN 9780367494445 (hardback) | ISBN 9781032261003 (paperback)
|ISBN 9781003046202 (ebook)
Subjects: LCSH: Mathematics--Study and teaching.
Classification: LCC QA11.2 .D535 2022 (print) | LCC QA11.2 (ebook) | DDC 510--dc23/eng20220301
record available at https://lccn.loc.gov/2021055607
LC ebook record Δavailable at https://lccn.loc.gov/2021055608

ISBN: 978-0-367-49444-5 (hbk)
ISBN: 978-1-032-26100-3 (pbk)
ISBN: 978-1-003-04620-2 (ebk)

DOI: 10.1201/9781003046202

Typeset in CMR10 font
by KnowledgeWorks Global Ltd.

Contents

Preface

Purpose of this Book

This book prepares students for entrance into the discipline of mathematics, including its way of thinking and its way of communicating.

Every contemporary person with some formal education has encountered mathematical concepts. Surprisingly, when asked what mathematics is, few can offer a concise definition of mathematics. Some high school graduates might reply that mathematics is the study of properties of numbers and of space, but this answer is neither comprehensive nor cohesive. Some people contend that there is no generally accepted definition for mathematics. Some authors define mathematics as the science of patterns of numbers (arithmetic and algebra), of space (geometry), and so on. (See [24].) Though this definition offers a cohesive description of the topics usually accepted under the umbrella of mathematics, what unifies mathematics is the manner of thought. Mathematical thought is precise.

Perhaps because of its precision, for millennia mathematics has provided epistemological strength to countless areas such as agriculture, architecture, engineering, natural sciences, social sciences, forensics, and many more. In his famous 1960 article, "The Unreasonable Effectiveness of Mathematics in the Natural Sciences," Nobel Prize laureate physicist Eugene Wigner dubbed mathematics a "wonderful gift [77]." Despite the abstract nature of mathematics and the manner in which much of mathematics develops by working on problems internal to itself, interaction between mathematics and other disciplines often leads to new areas of investigation within mathematics and offers fruitful results in other sciences.

Another aspect about mathematics that surprises some outsiders is that it remains a very active field of inquiry. Mathematics does enjoy a timeless nature. What Euclid or Archimedes rigorously proved 2000 years ago remains valid today. And yet every day students and researchers prove new theorems and push the boundaries of what we know. The mathematics ArXiV (https://arxiv.org/archive/math) lists results produced by people today, right now. Anyone can engage in mathematical investigation alone, in research groups, and in any language, but stretching the boundaries of the discipline finds its full expression in the mathematical community. Though mathematical scientists regularly hope to develop some new area or prove a conjecture before anyone else, for the most part we view ourselves as collaborators, rather than competitors. After all, though mathematics boasts different branches, for the most part, the subject is not separated along schools of thought. In this context, the peer-review process involved in publication passes a value judgment as to whether a submission is a good fit for a given journal but also serves the role of a quality control check.

The book divides naturally into two parts that we label generally as the *content* and the *context* of advanced mathematics. Part I presents an introduction to proofs, where we introduce logic, various proof techniques, and certain fundamental structures that are ubiquitous in mathematics. Many examples foreshadow subsequent courses: algebra, analysis, statistics, and combinatorics, to name a few. Part II offers an overview of the culture, history and philosophy of the discipline; it outlines the dynamics of the research process; and it discusses current best practices for writing mathematics.

Proofs and Precision

The reader will notice that the approach in this text might feel like entering into a story in the middle. In elementary school, teachers introduce children to whole numbers, operations on whole numbers, fractions and so on. In high school, students encounter real numbers, complex numbers, Euclidean geometry, and much more. All of these topics rely on intuitive notions. Instructors teach proper habits of working with these mathematical objects and structures. However, from the perspective of advanced mathematics, the intuitive definitions do not carry enough precision to engage in deductive reasoning and therefore make rigorous claims.

Why does this precision matter? For hundreds, if not thousands, of years, scholars accepted Euclid's *Elements* as the pinnacle of precise reasoning.[1] Logical deduction applied to concepts with precise definitions provides certainty concerning our conclusions about abstract mathematical objects. This certainty also offers a foundation of confidence for those who use mathematical models to understand phenomena.

---

[1]United States Representative and Lincoln biographer, Henry Ketcham observes that, during his law career, Lincoln studied Euclid's *Elements* until he could reproduce the proofs of any proposition in the first six books, simply to understand the word "demonstrate [50]."

Despite Euclid's enduring legacy of proofs and precise thinking, some of Euclid's definitions fall short of standards of modern precision. For example, Euclid defines a "point" as "that which has no part," and he defines a line as a "breadthless length." Neither of these definitions carries much content; they are intuitive guides to how he manipulates them. As another example, Euclid uses without proof that a line segment connecting two points on either side of a line will intersect that line in a single point, which is an assumption of continuity. In the last 200 years, various mathematicians have proposed alternate sets of definitions and axioms that precisely define geometry in the sense that Euclid appears to have intended it. (See Hilbert's axioms and the School Mathematics Study Group (SMSG) axioms [70, appendices B and C].) Precision is essential in mathematical reasoning and equally important for defining objects.

Anyone who delves into advanced mathematics will observe a wide range of perspectives on the place of proofs in mathematical communication. One extreme, called the Satz-Beweis (German for "theorem-proof") style, emphasizes the definition of terms, the statement of theorems, and rigorous proofs, with the near exclusion of all else. Other writing styles help build intuition through numerous examples, motivate the definitions, dwell on the beauty of a theory, invite readers to "play" with ideas, underscore practical applications, emphasize the proper use of algorithms, and so on. Some authors put proofs front and center, some delay proofs to later sections or appendices, while in other instances, authors may simply cite a source for a proof. Despite all these variations of communication styles, mathematicians and statisticians of all types know that rigorous proofs lie at the foundation of all mathematical knowledge. Deductive reasoning of proofs is the hallmark of mathematical reasoning and the strength of its epistemology.

Context and Culture

The transition from elementary to more advanced courses requires more than merely adjusting to a different type of content or to more abstract concepts. Suddenly, students find they must express themselves with greater clarity and precision, read mathematical texts and articles written in a more formal style, collaborate with each other to approach complex problems, and write mathematical papers on a computer. These changes form a central part of the transition to advanced modern mathematics.

At the same time, students who transition into advanced mathematics courses should realize that the world of mathematics is much larger and more multifaceted than they previously thought. Many are surprised to learn that people are coming up with new mathematics still today, and many may have never heard of the rapidly-growing fields of topology, knot theory, operations research, and many others. The 21st century is indeed a new Golden Age for mathematics [25], a vibrant field that continues to grow and infiltrate more aspects of our daily life. Furthermore, at this stage in their studies, students often start making choices leading to a specific vocation and begin to wonder how their mathematical knowledge will play a role in their future career. The demand for mathematically-adept professionals is at an all-time high, but the student may not clearly see how to effectively leverage his or her mathematical abilities in today's economy.

The second part of this book guides students through the transition by addressing the aspects of the contemporary mathematical culture and the context in which people research and practice mathematics today. In Chapter 7, Mathematical Culture, Vocation, and Careers, we present the landscape of contemporary mathematics, highlighting its distinctive aspects, such as the role of computation, the mathematization of science, and mathematical associations. We also offer suggestions to effectively approach upper-level coursework and start planning a vocational pathway.

A proper understanding of contemporary mathematics, and indeed mathematics as a whole, requires delving into its historical roots. Chapter 8, titled History and Philosophy of Mathematics, provides a chronological bird's eye overview of the discipline, from its origins in ancient civilizations to modern times. Mathematics students may already be familiar with the names of certain famous mathematicians, such as Euclid, Newton, Euler, Gauss, and many others, as elementary mathematics textbooks often include historical vignettes about their lives and accomplishments; in this chapter, they appear in their historical and geographical context, alongside some lesser-known mathematicians. Along the way, we call attention to the influences of the Scientific Revolution, the Enlightenment, and the roles of the scientific and axiomatic methods. We also highlight connections to philosophical questions, some of which remain under debate today.

A student who transitions into advanced mathematics may have little to no experience reading mathematical journal articles and may feel intimidated by their formal and terse style. Chapter 9 on Reading and Researching Mathematics demystifies the process of scholarly communication so that the student becomes comfortable learning directly from primary sources. Reading research articles exposes the student to the dynamics of the research process and allows one to access the treasures of contemporary mathematics, most of which only exist in journal articles.

As students start reading advanced textbooks and primary literature, they gradually learn to express themselves, orally and in writing, with mathematical clarity and precision. Chapter 10, titled Writing and Presenting Mathematics, discusses the guidelines and best practices for mathematical style and syntax, from the terminology to the challenges of mathematical typesetting.

Prerequisites

Readers need the content of Calculus I and II, as examples and exercises sometimes pertain to continuous functions, derivatives or definite integrals. Furthermore, this book assumes that students would have completed or would be concurrently taking a first course in linear algebra. The examples and exercises in Part I that involve properties of vectors, matrices, or vector spaces arise in the same sequence as they appear in a linear algebra course.

Since students might not remember precisely some of the key definitions or theorems from calculus or linear algebra, Appendix B provides a list of the definitions and theorems that we use in Part I.

The Flow of Material

The authors have designed the content of this book to accompany two 2-credit courses offered at Wheaton College titled *Introduction to Proofs* and *Introduction to Upper-Level Mathematics*, whose content follows Parts I and II of this book, respectively. Both courses are required for students majoring in mathematics, and students typically take them in their sophomore year, after calculus and linear algebra but before any upper-level courses. Although in our program these are separate courses, we encourage mathematics majors to take them concurrently (and most students do). This way, students explore the content of the Introduction to Proofs course in its historical context, while simultaneously developing the tools and vocational motivation to approach upper-level coursework.

Therefore, an effective way to use this book is to progress through Parts I and II in parallel. Within each part, there is some flexibility in the sequencing of the sections. Part I generally follows a linear presentation. However, an instructor could choose to change the order of a few sections. For example, Section 4.5 on proofs by induction could be done any time after Section 4.1 on the definition of integers. An instructor could also organize sections to create a sort of "Topics in Set Theory" chapter as a conclusion to Part I by gathering together Sections 2.4, 3.3, 5.4, and 6.4 (generalized unions and intersections; choice functions; cardinality and countability; and quotient sets).

The four chapters of Part II are organized thematically (culture, history, reading, and writing), but the authors usually progress through all four in parallel throughout the semester. The sections within each of these chapters are designed to be covered in order, although there is some flexibility here, too. Section 7.1 on 21st century mathematics is a good starting point, as it introduces several historical and vocational aspects of modern mathematics discussed in subsequent sections. Another possible entry point is Section 7.3 (Studying Upper-Level Mathematics), which introduces students to the unique aspects of upper-level mathematics coursework in general. Although we recommend covering in order the subsections on the history and philosophy of mathematics, the instructor may want to introduce some of the content of Sections 8.3 (The Axiomatic Method), 8.4 (History of Modern Mathematics), and 8.5 (Philosophical Issues) early to give students some historical and philosophical context as they begin progressing through Part I.

As soon as students begin learning the fundamentals of mathematical writing and typesetting in Chapter 10, we assign exercises from Part I (for example, the simple proofs in Section 2.3) as "writing exercises" thereby allowing students to develop their writing and typesetting skills concurrently with their proof-writing skills. We also encourage students to apply their writing skills immediately to other mathematics courses they may be taking. At the end of the semester, the authors require that every student complete an expository research project, reading one or more contemporary research articles in depth, presenting the content orally, and writing a formal summary paper. This assignment allows students to demonstrate several skills they acquired through Chapters 9 (Reading and Researching Mathematics) and 10 (Writing and Presenting Mathematics).

Finally, the authors frequently accommodate guest presenters for several sections in Part II. For example, the instructor may invite other faculty members, career counselors, or professionals from the industry to highlight the content of Section 7.4 (Mathematical Vocations). Similarly, an academic librarian can provide additional insight into the aspects of researching primary and secondary sources (Section 9.4).

Acknowledgments

The authors would like to thank our colleagues and students for providing constructive feedback. We especially wish to thank Robert Brabenec, Darcie Delzell, Paul Isihara, Peter Jantsch, Terry Perciante, and Mary Vanderschoot, as well as our students Lucy Henneker, Daniel McKay, and Daniel Windus for their careful editing and insightful comments. We also would like to thank Wheaton College for granting sabbatical leaves to write this book.

Part I:

Introduction to Proofs

CHAPTER 1

Logic and Sets

Plato defined knowledge as "justified true belief." Every academic discipline that makes claims to knowledge possesses a notion of truth and a method of justification. In history, the area that analyzes historians' methods of justification is called historiography, the natural and social sciences use the scientific method, and knowledge claims in mathematics rely on proofs. Chapter 2 introduces the concept of a mathematical proof through logical deduction. This chapter, however, focuses on the notion of truth as it pertains to mathematics.

It does not take many thought examples before we recognize the polemical nature of the word "true." Each discipline involves its own sense of what it means by true, or true belief. Consider, for example, Newton's second law of motion, a fundamental principle in physics. A physicist does not hold this to be absolutely true at every scale in the same way that we hold that the area of a disk of radius R in the Euclidean plane is πR^2. Newton's second law of motion describes a model that is no longer applicable at the very small (quantum level) and at the very large scales (cosmological level). As an example from history, the statement that "the assassination of Archduke Ferdinand caused World War I," though often accepted as historical fact, carries considerable layers of context and interpretation; we do not hold it as true in the same way that we hold that $2 + 2 = 4$.

This chapter presents the notion of propositional logic, the fundamental way of expressing concepts mathematically. We also introduce the notion of a set, a concept which is closely related to propositional logic and consequently is ubiquitous throughout mathematics.

1.1 Logic and Propositions

1.1.1 Propositions

Consider the following strings of characters involving punctuation and letters from the English alphabet:

1) "Paris is the capital of France."

2) "What is the capital of Thailand?"

3) "The Burj Khalifa is 100 meters tall."

4) "It is raining."

5) "Open the door."

6) "Ghoudgh skjhdf neeert."

7) "Il Vesuvio scoppiò nel anno domini 79."

String (1) is an intelligible sentence in English. It is a statement of fact and it is true. It is also true by the definition of what capital means and because the French government, which has the authority to determine which city will serve as the country's capital, has declared it to be so. String (2) is also a sentence but a question, rather than a statement. String (3) is a statement but happens to be false; the roof height of the Burj Khalifa is 828 meters.

String (4) is a statement, but whether we consider it true depends critically on the context of the person speaking. As one reader reads the sentence, it might be true, while when another reads it, the statement might be false. This sentence also carries the definitional problem of how much precipitation must come down for us to declare that it is raining, e.g., does one drop of water falling per hour in one square foot count as raining? String (5) is still an intelligible sentence, but it is an imperative and not a statement.

Strings (6) and (7) are not intelligible English sentences. String (6) does not mean anything to anyone in any current natural language. On the other hand, string (7) is Italian. Translated into English it states, "Mount Vesuvius erupted in 79 AD," which is a true statement.

In the above examples, strings (1), (3), (4) and, when translated in English, string (7) are examples of propositions, the fundamental objects of study in logic.

Definition 1.1.1

> A *proposition* is a statement that is verifiably true or false. Whether a proposition is true or false is called the *truth value* of the proposition.

Philosophical debates may arise concerning what is meant about "true" or "false." However, as long as we are in a context in which the notions of "true" and "false" are incontrovertible, then the notion of propositions is well defined.

As we discussed with the above examples, it is common to allow a certain amount of context when deciding the truth value of a proposition. For example, when we stated that "Paris is the capital of France," we assumed that we meant now. That proposition is true now, but from 1682 to 1789 Versailles served as the capital of France.

The statement $2 + 2 = 4$ is an example of a propositions in mathematical symbols, as is $2^2 = 7$. The equals symbol is the verb of each sentence. The first proposition is true and the second one is false.

There are a few other ways in which a statement might not meet the standard of being a proposition. Consider the following examples.

"My computer is slow." Let us assume from context that we know who is speaking, when they are speaking, and to which computer they are referring. Even so, the adjective "slow" is too imprecise. In order to make the sentence a precise statement, we would need to define "slow." The statement "My computer is running at 20 gigaflops" is a proposition because it could be verified if it is currently running more or less than that benchmark of speed.

The mathematical phrase $x(x + 3) = 2$ also fails to be a proposition. If the variable x does not have a numerical value, the phrase may be true for some values of x and false for others.

In algebra, we regularly use variables like x and y to refer to unspecified real numbers. In calculus, we often use letters like f and g for unspecified functions, and we employ these letters in calculus identities. Similarly, in logic we also use variables to refer to propositions. We typically use the letters p, q, and r to represent an arbitrary proposition and we associate the symbol of $\mathbf{T}$ for the truth value of true and $\mathbf{F}$ for the truth value of false.

1.1.2 Boolean operators

We create *compound propositions* by combining propositions. When we do so, we may use a truth table to provide the truth value of the compound proposition based on the truth values of propositions that make it up. We often call *atomic propositions* the propositions that make up a compound proposition.

The *truth table* of a compound proposition is the table that lists the truth value of the compound proposition in terms of all the possible combinations of truth values of the atomic propositions.

The most basic compound propositions arise from combining two atomic propositions with a grammatical conjunction. In logic, these grammatical connectors are called *Boolean operators* or *logical operators*. We list the main ones here below.

Definition 1.1.2

Let p be a proposition. The *negation* of p, denoted by $\neg p$, is the proposition, "It is not the case that p."

Sometimes, English grammar (or the grammar of any natural language) allows us to shorten the negation statement. For example, let $p =$"The shirt I wore yesterday was white." By the above definition, $\neg p$ is the proposition, "It is not the case that the shirt I wore yesterday was white." This is correct but cumbersome. Instead, for $\neg p$ we usually say, "The shirt I wore yesterday was not white." However, simply putting "not" in front of the verb is not always correct or clear. For example, suppose that $p =$"The shirt I wore yesterday was white and black." Saying "The shirt I wore yesterday was not white and black" is not precise because we are unsure how the "not" applies to black. If ever there is possibility of confusion, using the longer phrase given in Definition 1.1.2 will always be correct.

The truth table for $\neg p$ is the following.

p	$\neg p$
T	F
F	T

Definition 1.1.3

Let p and q be propositions. The *conjunction* of p and q is the proposition "p and q." We denote the conjunction of p and q by $p \wedge q$. The conjunction of p and q is true only when p and q are both true.

Example 1.1.4. Let $p=$"The driver drove a red car." and $q=$"The driver wore a hat." The conjunction proposition $p \wedge q$ is, "The driver drove a red car and the driver wore a hat." In English, since the subject of each atomic proposition is the same, we often combine the two and say, "The driver drove a red car and wore a hat." Again, we need to take care when using these types of grammatical shorteners.△

The truth table for the conjunction is the following.

p	q	$p \wedge q$
T	T	T
T	F	F
F	T	F
F	F	F

Note that in this truth table, we write **T** and **F** such that the four rows run through all combinations of truth values for the atomic propositions p and q.

In English, and in many natural languages, we often use the grammatical conjunction "but." It is important to note that "but" has the same logical function as "and." For example, if we say, "Darius is short but is a fast runner," the word 'but' plays the same logical role as 'and' except that we use it to emphasize a contrast from what we might have expected given the first proposition.

Definition 1.1.5

Let p and q be propositions. The *disjunction* of p and q is the proposition "p or q." We denote the disjunction of p and q by $p \vee q$. The disjunction of p and q is false only when p and q are both false.

Example 1.1.6. Consider $p=$"Frank's bicycle has a flat tire" and $q=$"Frank is sick today." The disjunction of p and q is $p \vee q=$"Frank's bicycle has a flat tire or Frank is sick today." The only way this compound proposition can be false is if Frank's bicycle does not have a flat tire and if Frank is well. △

The truth table for the disjunction proposition is the following.

p	q	$p \vee q$
T	T	T
T	F	T
F	T	T
F	F	F

In English and in many other natural languages, we are often imprecise in our use of "or." Sometimes we use "or" in an exclusive sense. For example, a guest at a restaurant knows that the sentence "The meal comes with soup or salad" means that he or she can order soup or salad, but not both. Most natural languages make provisions to emphasize the use of an exclusive or. For example, in the sentence, "The meal comes with either soup or salad," the grammar of the word "either" makes precise that the "or" carries an exclusive sense.

Definition 1.1.7

The *exclusive or* of two propositions p and q, denoted by $p \oplus q$, is the compound proposition that is true exactly when one atomic proposition is true and the other is false.

The truth table for the exclusive or is the following.

p	q	$p \oplus q$
T	T	F
T	F	T
F	T	T
F	F	F

Note that in Example 1.1.6, since we assumed we were using the usual "or," the compound proposition of "Frank's bicycle has a flat tire or Frank is sick today" is true even if Frank's bicycle has a flat tire and Frank is sick today. The exclusive or proposition "Either Frank's bicycle has a flat tire or Frank is sick today" would be false if Frank's bicycle has a flat tire and Frank is sick today.

In all of mathematics, we always mean "or" in the inclusive way, as described in Definition 1.1.5. To refer to the exclusive or between p and q, we say "either p or q," we say "p or q but not both," or we say "p or q exclusive."

Definition 1.1.8

Let p and q be propositions. The *implication* statement $p \to q$ is the proposition that is false exactly when p is true and q is false.

The truth table for the implication proposition is the following.

p	q	$p \to q$
T	T	T
T	F	F
F	T	T
F	F	T

Since they form the core step in inferential reasoning, implication statements are extremely common in mathematics. In $p \to q$, the proposition p is called the *hypothesis* and q is called the *conclusion*. Because the implication statement is so common, there are many ways of expressing it. Here are a few: "if p then q"; "q if p"; "p implies q"; "p only if q"; "q whenever p."

We emphasize that if the hypothesis is false, the implication statement remains true regardless of the content of the conclusion. In this case when the hypothesis is false, we sometimes say that the implication statement is *vacuously true*.

Example 1.1.9. "If you score a goal, then the spectators cheer." This is written as $p \to q$ with the understanding that $p=$ "You score a goal" and $q=$ "the spectators cheer." This implication statement will only be false if you happen to score a goal, but the spectators do not cheer (a possible situation if the only spectators are rooting for the opposing team). This implication statement will remain true if you do not score, and the spectators cheer.   △

Example 1.1.10. "If $2+3 = 6$, then today is Friday." This sentence does not appear to make sense, but it is a valid compound proposition. Furthermore, it is true because the hypothesis is false.   △

Example 1.1.11. "I stay indoors whenever it rains." This is written in logic notation as $p \rightarrow q$ where p="It rains" and q="I stay indoors." △

The *converse* to the implication $p \rightarrow q$ is the compound proposition $q \rightarrow p$ and the *contrapositive* is the statement $(\neg q) \rightarrow (\neg p)$. For example, consider the implication statement, "If Allison comes to visit, then I will stay." The converse is, "If I will stay, then Allison comes to visit," whereas the contrapositive is, "If I do not stay then Allison does not come to visit." It is not hard to see that the implication and the converse do not mean the same thing: In the situation that Allison comes to visit and I do not stay, the original implication statement would be false, whereas the converse statement would be true. In Exercise 1.1.12, the reader is asked to verify an implication statement and its contrapositive have the same truth table, but that the converse does not.

Definition 1.1.12

Let p and q be propositions. The *biconditional* statement $p \leftrightarrow q$ is the proposition that is true exactly when p and q have the same truth value. This is usually spoken as "p if and only if q."

The truth table for the biconditional proposition is the following.

p	q	$p \leftrightarrow q$
T	T	T
T	F	F
F	T	F
F	F	T

Example 1.1.13. Suppose that a potential buyer says to the seller, "I will buy that chair only if the price of the chair is less than \$50." Note that this is the same as saying, "If I buy that chair, then the price of the chair is less than \$50." Assuming the buyer is truthful, this type of sentence still allows him or her the option not to purchase the chair if the seller offers the chair for less than \$50. Compare this statement to the biconditional proposition, "I will buy that chair if and only if the price of the chair is less than \$50." If the seller offers the chair for less than \$50, then the buyer must purchase the chair. △

1.1.3 Compound propositions

Just as in algebra we use parentheses to group algebraic expressions and impose an order of operations, we do the same with the parentheses for the operations we have defined on propositions. The study of properties of compound propositions is often called *propositional logic* or *Boolean logic*.[1]

Example 1.1.14. Consider the compound proposition $(\neg p) \vee q$. We point out in this example that when creating a truth table for a compound proposition, it is common practice to insert extra columns to show the truth values of the intermediate compound propositions as we build up to the full one. Consequently, when constructing a truth table for $\neg p \vee q$ we do the following.

p	q	$\neg p$	$\neg p \vee q$
T	T	F	T
T	F	F	F
F	T	T	T
F	F	T	T

We point out that $\neg p \vee q$ has the same truth table as $p \to q$. We say that these compound propositions are logically equivalent. This is the topic of Section 1.3.   △

In order to reduce parentheses, we institute one operator precedence. The negation $\neg$ is taken before any other connective operator. Thus, $\neg p \vee q$ means $(\neg p) \vee q$, but does not mean $\neg(p \vee q)$.

Example 1.1.15. Consider the compound proposition $(p \to q) \wedge (p \vee r)$. Since there are three atomic propositions, we will need $2^3 = 8$ rows to give us all possible combinations of truth value for p, q and r.

p	q	r	$p \to q$	$p \vee r$	$(p \to q) \wedge (p \vee r)$
T	T	T	T	T	T
T	T	F	T	T	T
T	F	T	F	T	F
T	F	F	F	T	F
F	T	T	T	T	T
F	T	F	T	F	F
F	F	T	T	T	T
F	F	F	T	F	F

Note that we use intermediate columns for $p \to q$ and $p \vee r$ to build this up.   △

Consider the statement, "I am 6 feet tall and I am not 6 feet tall." This compound proposition can never be true, no matter the truth value of the atomic proposition, "I am 6 feet tall." This is an example

---

[1]The term Boolean logic honors the work of George Boole (1815-1864) in *The Mathematical Analysis of Logic* (1848) and *The Laws of Thought* (1854).

of a contradiction. In contrast, the proposition, "I am 6 feet tall or I am not 6 feet tall," is always true.

Definition 1.1.16

A compound proposition is called a *tautology* (resp. a *contradiction*) if its truth value is **T** (resp. **F**) for all truth values of its atomic propositions. A compound proposition that is neither a tautology nor a contradiction is called a *contingency*.

As we will soon see, tautologies and contradictions play a fundamental role in developing propositional equivalences and rules of inference.

Example 1.1.17. As an example, we show that $(p \to q) \lor (q \to p)$ is a tautology. To do so, we create a truth table for the compound proposition.

p	q	$p \to q$	$q \to p$	$(p \to q) \lor (q \to p)$
T	T	T	T	T
T	F	F	T	T
F	T	T	F	T
F	F	T	T	T

Since **T** is the only truth value of the compound proposition, it is a tautology.   △

EXERCISES FOR SECTION 1.1

1. For each of the following sentences, decide whether it is a proposition and determine its truth value.
    (a) "The capital of Ohio is Toledo."
    (b) "Dogs are better than cats."
    (c) "Stop here."
    (d) "Leonhard Euler was born in 1707."
    (e) "$x^2 + 3x - 7 = 0$."

2. For each of the following sentences, decide whether it is a proposition and determine its truth value.
    (a) "Who wants to live forever?"
    (b) "$12 - 7 = 8$."
    (c) "France won the 2018 FIFA World Cup."
    (d) "New York City is a great tourist destination."
    (e) "$n \geq 50$."

3. Let p and q be the propositions $p =$ "I will garden today" and $q =$ "It will rain today." State each of the following compound propositions as an English sentence. (a) $\neg q \to p$; (b) $p \lor q$; (c) $\neg p \land q$; (d) $p \to q$.

4. Let p, q, and r be the propositions $p =$ "I earned less than \$15,000 last year", $q =$ "I have dependent children", and $r =$ "I can claim the EITC." (The EITC is the Earned Income Tax Credit.) State each of the following compound propositions as an English sentence. (a) $r \to (p \wedge q)$; (b) $r \vee (\neg p \wedge q)$; (c) $(\neg p \wedge r) \to r$; (d) $(p \wedge r) \vee (\neg p \wedge q)$.

5. This exercise shows a way in which natural grammar might fail to be precise. Let p, q and r be the propositions $p =$ "I earned an A on the final exam", $q =$ "I earned a B in the course", and $r =$ "I did not study for the final exam." Consider the sentence, "I earned a B in the course and I earned an A on the final exam or I did not study for the final exam." This could be interpreted in two ways.

   (a) Using compound propositions of p, q, and r, write the two ways in which this sentence could be interpreted.

   (b) Explain in English why these two compound propositions are not the same.

   (c) Give the truth tables of each of the two compound propositions from part (a).

6. Let p and q be the propositions $p =$ "My car is white" and $q =$ "My car is less than 5 years old." Express the following sentences as compound propositions.

   (a) My car is white and it is less than 5 years old.

   (b) If my car is less than 5 years old, then my car is white.

   (c) My car is less than 5 years old and is not white, or my car is white and is 5 years old or more.

7. Let p, q and r be the propositions about the current weather $p =$ "It is windy", $q =$ "It is cold", and $r =$ "It is sunny". Express the following sentences as compound propositions.

   (a) It is windy and cold, but it is sunny.

   (b) Whenever it is not sunny and windy, it is cold.

   (c) It is windy and cold, or it is not windy and sunny.

8. State the converse and the contrapositive to the following implication statements:

   (a) "If there is lightning, then the soccer game will be canceled."

   (b) "If the computer is more than 4 years old, then I will replace the computer."

9. State the converse and the contrapositive to the following implication statements:

   (a) "Whenever I have tea at night, I stay awake."

   (b) "Public works will cut down the tree if it has ash borers."

   (c) "If you touch the ball with your hands and you are in the penalty area, then the referee will call a penalty on you."

10. Construct a truth table for each compound proposition:

   (a) $(p \wedge q) \vee (\neg p \wedge \neg q)$ (c) $(p \to q) \wedge (\neg r \to p)$

   (b) $(p \wedge q) \to (p \vee q)$ (d) $\neg(p \to r) \vee (\neg p \oplus q)$

11. Construct a truth table for each compound proposition:
    (a) $(p \to q) \oplus (\neg p \lor q)$ (c) $(p \lor q) \oplus (\neg p \lor r)$
    (b) $\neg(p \to q) \lor p$ (d) $\neg p \land (\neg q \land r)$

12. Consider the implication statement $p \to q$.
    (a) Construct the truth table of the converse statement $q \to p$ and list how it differs from the truth table for $p \to q$.
    (b) Construct the truth table of the contrapositive $(\neg q) \to (\neg p)$ to verify that it has the same truth table as the original implication.

13. Construct a truth table for each compound proposition:
    (a) $(p \to q) \land (r \to s)$ (b) $((p \to q) \to \neg r) \to \neg s$

14. Show that each of the following is a tautology:
    (a) $(p \to q) \lor p$ (c) $(p \leftrightarrow q) \lor (p \oplus q)$
    (b) $(p \land q) \to p$ (d) $(p \to q) \lor (q \to p)$

15. Show that each of the following is a tautology:
    (a) $(p \land (q \land r)) \leftrightarrow ((p \land q) \land r)$ (c) $(p \land q) \to (p \leftrightarrow q)$
    (b) $(p \land q) \leftrightarrow (q \land p)$ (d) $p \to (p \lor q \lor \neg r)$

1.2 Sets

In mathematics, the concept of a set makes precise the notion of a collection of things. As broad as this concept appears, it is foundational for modern mathematics.

1.2.1 Definition and examples

Definition 1.2.1

1) A *set* is a collection of objects for which there is a clear rule to determine whether an object is included or excluded.

2) An object in a set is called an *element* of that set. We write $x \in A$ to mean "the object x is an element of the set A." We write $x \notin A$ if x is not an element of A.

Alternate expressions for $x \in A$ include "x is in A" or "A contains x." The essential criterion of a clear rule is that it carries the same precision as a logical proposition. In other words, if x is some object and A is some set, then the statement $x \in A$ is a logical proposition.

Some examples of sets include the members of the United Nations or the man-made structures above 800 feet tall. Both of these examples have clear rules that allow someone to clearly determine whether a given object is included in the collection or not. We can also refer to "the set of living people," but since people are born and people die every second, doing so requires us to implicitly think of the set of living people at a particular instant in time.

In natural languages, we regularly use terms or expressions that we treat as sets but in fact do not have a clear rule. For example, a person P cannot legitimately talk about the set of "my friends." The concept of friend does not carry enough precision: even at some fixed point in time, there are likely a number of people about whom P is not decided whether he or she considers them friends. In contrast, at some fixed point in time, the list of P's friends on a specific social networking site does form a set. As another nonexample of a set, consider the collection of all chairs. Whether this is a set is debatable. For instance, by some failure of construction, a piece of furniture intended as a chair may not be comfortable enough to sit on.

Some specific sets occur frequently in mathematics and so carry their own standard notation. Here are a few:

- Standard sets of numbers:

    - $\mathbb{N}$ is the set of whole numbers, including 0.
    - $\mathbb{Z}$ is the set of integers.
    - $\mathbb{Q}$ is the set of rational numbers.
    - $\mathbb{R}$ is the set of real numbers.
    - $\mathbb{C}$ is the set of complex numbers.

- Sometimes we use modifiers to the above sets. For example, $\mathbb{R}^+$ denotes the set of nonnegative reals and $\mathbb{R}^{<0}$ denotes the set of (strictly) negative reals.

- Following many authors, in this book we denote by $\mathbb{N}^*$, $\mathbb{Z}^*$, etc. any number set excluding 0. In particular, $\mathbb{N}^*$ denotes the set of positive integers.

- $\emptyset$, called the *empty set*, is the set that contains no elements.

- Intervals of real numbers:

    - $[a, b]$ is the interval of reals between a and b inclusive;
    - $[a, b)$ is the interval of reals between a and b, including a but not b;
    - $[a, \infty)$ is the interval of all reals greater than or equal to a;
    - Other self-explanatory combinations are possible such as $(a, b]$; $(a, b]$; (a, ∞); $(-\infty, b]$; and $(-\infty, b)$.

- In calculus, we often work with functions that are continuous or perhaps differentiable over some interval I. In analysis, the field that provides the theoretical underpinning of calculus, we denote by:

- $C^0(I)$ the set of functions that are continuous over I;
- $C^n(I)$, where n is a positive integer, the set of functions whose nth derivative exists and is continuous over I;
- $C^\infty(I)$ the set of functions such that all of its derivatives exist and are continuous over I.

- In linear algebra, we encounter:
  - $\mathbb{R}^2$ (resp. $\mathbb{R}^3$, and so on) the set of ordered pairs (resp. triples, and so on) of real numbers;
  - $M_{m \times n}(\mathbb{R})$ the set of $m \times n$ matrices with coefficients in $\mathbb{R}$;
  - $M_n(\mathbb{R})$ the set of square $n \times n$ matrices with coefficients in $\mathbb{R}$;
  - $M_{m \times n}(S)$, where S is any set of numbers, the set of $m \times n$ matrices with entries in S (e.g., $M_{2 \times 3}(Z)$ is the set of 2×3 matrices with integer entries).

There are two common notations for defining sets. Both of them explicitly provide the clear rule as to whether an object is in or out. In either case, the parenthesis { marks the beginning of the defining rule and the parenthesis } marks the end.

1) **List the elements.** For example, writing $S = \{1, 3, 7\}$ means that the set S is comprised of the three integers 1, 3, and 7. Since all that matters is whether an a specific object is in the set or not, in this list notation, order does not matter and we do not list elements more than once. As another example, writing $F = \{\text{New York}, \text{Los Angeles}, \text{Chicago}, \text{Atlanta}\}$ we mean the set consisting of the four cities that we named.

2) **State a defining property.** For example,

$$\{x \mid x \text{ is a rational number with } x^2 < 2\}$$

means the set of all x such that x is a rational number such that $x^2 < 2$. Since we already have a set label for the rational numbers, we will usually rewrite this more concisely as

$$\{x \in \mathbb{Q} \mid x^2 < 2\}$$

and read it as, "the set of rational numbers x such that $x^2 < 2$." An alternate notation for this expression is $\{x \in \mathbb{Q} : x^2 < 2\}$.

Two sets A and B are considered equal when they contain exactly the same elements. We write $A = B$ to denote set equality.

Example 1.2.2. A common exercise in precalculus involves determining the domain of a function. Consider the following typical exercise: Find the domain of $f(x) = 1 + \sqrt{x + 2}$. Since the square root function is defined on nonnegative reals, the domain consists of all real x such that $x + 2 \geq 0$. Some students will answer the question by saying, "The domain is all real x with $x \geq -2$." The proper way to answer the question uses set theory: "The domain is $[-2, \infty)$." $\triangle$

In the above example, we talked about a function in a manner similar to that used in typical calculus and precalculus books. However, in Chapter 3, when we study the general notion of functions between sets, we introduce habits of expression that are more precise. In particular, in set theory, we do not even discuss a function unless we state the domain to begin with.

Example 1.2.3. Recall from linear algebra the concept of a span of vectors. As a particular example, suppose that $\mathbf{u}_1, \mathbf{u}_2 \in \mathbb{R}^3$ are two vectors. The span of $\{\mathbf{u}_1, \mathbf{u}_2\}$, denoted $\mathrm{Span}(\mathbf{u}_1, \mathbf{u}_2)$, is the set of vectors in $\mathbb{R}^3$ consisting of all linear combinations of $\{\mathbf{u}_1, \mathbf{u}_2\}$. We can write this succinctly in either of the following ways

$$\mathrm{Span}(\mathbf{u}_1, \mathbf{u}_2) = \{c_1\mathbf{u}_1 + c_2\mathbf{u}_2 \in \mathbb{R}^3 \mid c_1, c_2 \in \mathbb{R}\} \tag{1.1}$$

$$\mathrm{Span}(\mathbf{u}_1, \mathbf{u}_2) = \{\mathbf{v} \in \mathbb{R}^3 \mid \mathbf{v} = c_1\mathbf{u}_1 + c_2\mathbf{u}_2 \text{ for some } c_1, c_2 \in \mathbb{R}\}.$$

Though they are equivalent, we tend to use the first one since it is shorter. $\triangle$

Some discussion in logic is appropriate here. Set theory based on the concept of a "clear rule" is called *naive set theory*, see [42]. However, naive set theory ultimately can lead to contradictions. Suppose for example that we defined S as the set of all sets that do not contain themselves. If $S \in S$, then by definition $S \notin S$, a contradiction; and if $S \notin S$, then $S \in S$, again a contradiction. Then S leads to a form of Russell's Paradox, which we can illustrate by the statement, "This sentence is false." If this statement is true then it is false, and if it is false then it is true.

The Zermelo-Fraenkel axioms of set theory, denoted by **ZF**, offer more technical foundations and avoid these contradictions. (See [61] for a presentation of set theory with **ZF**. See [33] for a philosophical discussion of **ZF** axioms.) We choose not to delve into the Zermelo-Fraenkel axioms in order to keep our introduction accessible. However, we point out one consequence of **ZF**. In **ZF**, the *axiom of comprehension* affirms that if A is a set, then for any formula or clear rule $P(x)$, there exists a set $\{x \in A \mid P(x)\}$ consisting of all elements in A satisfying $P(x)$. If there existed a universal set U of all possible objects or a set of all sets, then we would reach Russell's Paradox by setting $S = \{x \in U \mid x \notin x\}$. Consequently, there cannot exist either a set of all sets, or a set of all objects.

The most widely utilized form of set theory adds one axiom to the standard **ZF**, the so-called Axiom of Choice, and the resulting set of axioms is denoted by **ZFC**. Occasionally, certain theorems emphasize when their proofs directly utilize the Axiom of Choice. The reason for this was primarily historical. In the context of **ZF**, the Axiom of Choice implies some statements that appear obvious and others that feel counterintuitive. Consequently, for a time, mathematicians used to make clear when a certain result (and all results that use it as a hypothesis) relied on the Axiom of Choice.

A thorough treatment of **ZF** and **ZFC**, though ultimately essential to mathematical foundations, would detract from the introductory nature of this textbook. Naive set theory will suffice for our purposes, but we do briefly discuss the Axiom of Choice in Section 3.3. Whenever we need a technical aspect of set theory, we provide appropriate references. The interested reader is encouraged to consult [31, 45, 79] for a deeper treatment of set theory.

1.2.2 Subsets

Definition 1.2.4

> If A and B are sets, then B is called a *subset* of A, and we write $B \subseteq A$, if every element in B is also in A. We call a subset B of A a *proper* (or *strict*) subset, and write $B \subsetneq A$, if $B \neq A$.

The symbol $\subseteq$ resembles a rounded inequality $\leq$ symbol. This notation supports some authors' habit of writing $\subset$ for proper subset. However, some other authors use the same symbol $B \subset A$ to denote that B is a subset of A, not necessarily strict. Because of this notational inconsistency pertaining to $\subset$, we like many other authors, avoid the confusion by not using the symbol $\subset$ at all.

Following the pattern of symbols $\neq$ and $\notin$, we write $B \nsubseteq A$ as the negation of $B \subseteq A$, that is to say that B is not a subset of A.

As an example, the intervals $[1, \infty)$ or $(3, 17]$ are subsets of $\mathbb{R}$. With the sets of numbers that we defined earlier, we can write

$$\mathbb{N} \subseteq \mathbb{Z} \subseteq \mathbb{Q} \subseteq \mathbb{R} \subseteq \mathbb{C}.$$

As another example, following up on Example 1.2.3, we point out that the span of two vectors in $\mathbb{R}^3$ is a subset of $\mathbb{R}^3$. In fact, usually the definition of the span contains the term "subset." The notation in (1.1) where we read "$\{\mathbf{v} \in \mathbb{R}^3 \,|\, \text{something}\}$" indicates that the set we define is a subset of $\mathbb{R}^3$.

We point out that the empty set $\emptyset$ is a subset of every set. If this feels unintuitive for some readers, we can analyze it as follows. For any object x in any context, $x \in \emptyset$ is false. Hence, for any given set S and for any object x, the implication statement $x \in \emptyset \to x \in S$ is true, vacuously. Hence, every element of $\emptyset$ is in S.

The following intuitive proposition[2] has a simple proof. However, it is an important result that provides a strategy to prove that two sets are equal.

Proposition 1.2.5

Let A and B be two sets. Then $A = B$ if and only if $A \subseteq B$ and $B \subseteq A$.

Proof. Suppose first that $A = B$. Since every element of A is in B, we know that $A \subseteq B$. Since every element of B is in A, as well, we also have $B \subseteq A$.

Conversely, suppose that $A \subseteq B$ and $B \subseteq A$. Since $A \subseteq B$, every element of A is in B and since $B \subseteq A$ every element of B is in A. Hence, A and B contain all the same elements and hence $A = B$. $\square$

Though we discuss proofs formally in the next chapter, the reader should recognize that proving an "if and only if" statement requires proving one direction of the "if" statement and then the converse.

Example 1.2.6. As an example that illustrates how this proposition can be used effectively, we prove the following result: Let S be the smallest subset of $\mathbb{Z}$ that contains 22 and 30, such that for all $a, b \in S$, both $a + b$ and $a - b$ are in S; we prove that S is the set of all even numbers.

Let E be the set of even integers. E contains both 22 and 30 and is closed under addition and subtraction. Since S is the smallest such set, then $S \subseteq E$. Since

$$30 + 30 + 30 - 22 - 22 - 22 - 22 = 90 - 88 = 2,$$

then $2 \in S$. Then by adding or subtracting 2 to itself a sufficient number of times, we conclude that $E \subseteq S$. Since $S \subseteq E$ and $E \subseteq S$, by Proposition 1.2.5, we conclude that $S = E$. $\triangle$

This example encourages a number of comments. First, the manner in which we defined S offers an example of *a recursively defined set*. We defined S as containing certain elements and then specified a rule that knowing one subset, namely $\{a, b\} \subseteq S$, we also know that $\{a + b, a - b\}$ is also a subset of S.

Secondly, in the above proof, the reader might concede that it was fairly straightforward to observe that $S \subseteq E$ but that proving the

---

[2]We point out here an unfortunate reality of mathematical terminology. When we say Proposition 1.2.5 as we did here, the term Proposition is equivalent to Theorem, Lemma or Corollary, i.e., a statement that has been proven true. In contrast, in logic the term "proposition" means any sentence that is verifiably true or false. In this book, we sometimes say "logical proposition" to emphasize that we mean the latter. See Section 8.3 for more on this issue.

converse required a slightly smarter idea: of first showing that $2 \in S$. This is an example where we needed to devise our own strategy to establish the proof. Perhaps by being clear that we were trying to prove $E \subseteq S$, we could tell that showing $2 \in S$ would suffice to easily prove the rest. However, in Section 4.3, Theorem 4.3.9 generalizes the result of this example. Consequently, for someone familiar with the Extended Euclidean Algorithm, discussed in that section, this example should feel obvious.

Definition 1.2.7

Let S be a set. The *power set* of S, denoted $\mathcal{P}(S)$, is the set of all subsets of S.

The power set gives us a method to construct a new set from an old one. As an example, if $S = \{1, 2, 3\}$, then

$$\mathcal{P}(S) = \{\emptyset, \{1\}, \{2\}, \{3\}, \{1, 2\}, \{1, 3\}, \{2, 3\}, \{1, 2, 3\}\}.$$

As we emphasized before, $\emptyset$ is a subset of every set, so it occurs as an element of $\mathcal{P}(S)$. In the power set of a set, the subsets of S become elements themselves. So to be clear, $\mathcal{P}(\{1, 2, 3\})$ contains 8 elements.

1.2.3   Cartesian product

Another way to construct new sets from old ones is the Cartesian product. As its name suggests, it is inspired from the Cartesian plane, viewed as consisting of ordered pairs of real numbers.

Definition 1.2.8

Let A and B be two sets. We call the *Cartesian product* of A and B, denoted $A \times B$, the set of ordered pairs of elements in A and B:

$$A \times B = \{(a, b) \mid a \in A \text{ and } b \in B\}.$$

Example 1.2.9. Let $A = \{1, 2, 3\}$ and let $B = \{e, f\}$. We write out the sets $A \times B$, $B \times B$, and $B \times A$ explicitly:

$$A \times B = \{(1, e), (1, f), (2, e), (2, f), (3, e), (3, f)\},$$
$$B \times B = \{(e, e), (e, f), (f, e), (f, f)\},$$
$$B \times A = \{(e, 1), (e, 2), (e, 3), (f, 1), (f, 2), (f, 3)\}.$$

Note that since $A \neq B$, we also have $A \times B \neq B \times A$.   △

Example 1.2.10. Let C be the set of cities (officially registered townships) in the United States and let N be the set of last names of all U.S. citizens. The set $C \times N$ represents all possible pairings of cities to last names. △

It is not hard to imagine a construction that involves ordered triples or ordered four-tuples, or even n-tuples, where n is any specified positive integer.

Definition 1.2.11

We define the Cartesian product $A \times B \times C$ of three sets A, B, and C as the set of ordered triples (a, b, c), with $a \in A$, $b \in B$ and $c \in C$.

More generally, if $A_1, A_2, \ldots, A_n$ are n sets, the Cartesian product $A_1 \times A_2 \times \cdots \times A_n$ is the set of ordered n-tuples $(a_1, a_2, \ldots, a_n)$ with $a_i \in A_i$ for $i = 1, 2, \ldots, n$,

$$A_1 \times A_2 \times \cdots \times A_n$$
$$\overset{\text{def}}{=} \{(a_1, a_2, \ldots, a_n) \mid a_i \in A_i, \text{ for } i = 1, 2, \ldots, n\}.$$

If we take the Cartesian product of the same set A, we write

$$A^n \overset{\text{def}}{=} \overbrace{A \times A \times \cdots \times A}^{n \text{ times}}.$$

This notation generalizes the notation introduced in linear algebra where $\mathbb{R}^2$, $\mathbb{R}^3$ and $\mathbb{R}^n$ represented ordered pairs, triples, and n-tuples of real numbers.

Example 1.2.12. As a follow-up to Example 1.2.9, we list the elements

$$B \times A \times B = \{(e, 1, e), (e, 1, f), (e, 2, e), (e, 2, f), (e, 3, e), (e, 3, f),$$
$$(f, 1, e), (f, 1, f), (f, 2, e), (f, 2, f), (f, 3, e), (f, 3, f)\}.$$

Since order matters within the triples, $A \times B \times B \neq B \times A \times B$. △

The last example in this section illustrates two things. First, we illustrate how set theory notations provide symbols to accurately describe any object or set that students have encountered previously. Secondly, we make precise a distinction about the Cartesian product.

Example 1.2.13 (Euclidean Geometry). Definition 1 in Book I of Euclid's *Elements* defines a point as "that which has no part." This definition is awkward by modern standards. Modern mathematicians understand that though Euclid did not possess the formalism of set theory, he understood the Euclidean plane as a set of points. Some authors denote this set as $\mathbb{E}^2$, with the superscript to distinguish the plane from Euclidean space, denoted by $\mathbb{E}^3$. Other authors avoid the notation $\mathbb{E}^2$ because the Euclidean plane is not defined as a Cartesian product $\mathbb{E} \times \mathbb{E}$ for some set $\mathbb{E}$. This book will use this notation since topologists and geometers employ it regularly.

Euclid implicitly viewed geometric figures such as lines, line segments, circles, triangles, disks and so on as subsets of $\mathbb{E}^2$. So, using the notation AB to mean the distance between the points[3] A and B, the subset expression

$$\{M \in \mathbb{E}^2 \,|\, PM = r\}. \tag{1.2}$$

is a set theoretic definition of the circle of radius r and center P.

The Cartesian coordinate system of a plane allows us to identify $\mathbb{E}^2$ with $\mathbb{R}^2$, where $\mathbb{R}^2 = \mathbb{R} \times \mathbb{R}$ is the Cartesian product of $\mathbb{R}$ with itself. Then, employing the Pythagorean Theorem, we can define the circle of center $P = (a, b)$ and radius r by

$$\{(x, y) \in \mathbb{R}^2 \,|\, (x - a)^2 + (y - b)^2 = r^2\}. \tag{1.3}$$

The careful student will recognize (1.2) as the definition of a circle, while (1.3) states a formula for the circle.

Though mathematical education presents this identification as a fundamental reality, it is important to understand the difference. Euclid penned his *Elements* around 300 BC, whereas René Descartes introduced the concept of coordinates in 1637 AD, nearly 2,000 years later. Furthermore, the identification between $\mathbb{E}^2$ and $\mathbb{R}^2$ via the use of coordinates requires a proof from within Euclidean geometry.

The two notations $\mathbb{E}^2$ and $\mathbb{R}^2$ represent the difference between *synthetic geometry* and *analytic geometry*. In synthetic geometry, we prove results in the same style as Euclid's *Elements*, from definitions, postulates, and any previously proven proposition. In analytic geometry, we use Cartesian coordinates and algebraic methods to prove results. Despite this common dichotomy, we caution the reader away from assuming these are the only two methods of proof in geometry. For example, it is possible to prove many theorems in geometry using the algebra of vectors between points, while avoiding coordinates; such methods are technically neither synthetic nor analytic. $\triangle$

---

[3]This text uses AB for the distance between two points A and B and $\overline{AB}$ for the segment between A and B. Some other notations for the distance are $m(AB)$ or mAB or $d(A, B)$ or even $\|\overrightarrow{AB}\|$.

EXERCISES FOR SECTION 1.2

1. Write the following sets in list form:
   (a) $\{n \in \mathbb{Z} \mid n^2 < 12\}$;
   (b) $\{n \in \mathbb{N} \mid n \leq 21 \text{ and } n \text{ is odd}\}$;
   (c) $\{n \in \mathbb{N} \mid n \text{ is prime and is even}\}$.

2. Write the following subsets of $\mathbb{R}$ in interval notation:
   (a) The domain of $\ln(2x + 17)$;
   (b) $\{x \in \mathbb{R} \mid x^2 + x - 6 < 0\}$.

3. As subsets of the reals, describe the differences between the sets $\{3, 5\}$, $[3, 5]$, and $(3, 5]$.

4. Give the list description of $\mathcal{P}(\{1, 2, 3, 4\})$.

5. Let A, B, C be sets. Explain why $A \times B \times C$ is not the same set as $A \times (B \times C)$.

6. Let A, B, C, D be sets. Explain why $A \times (B \times C) \times D$ is not the same set as $(A \times B) \times (C \times D)$.

7. Let $A = \{1, 2, 3, 4\}$ and $B = \{a, b\}$. Write out as a list (a) $A \times B$; (b) $A \times A$; (c) $B \times B \times A$.

8. Write in list form $\{1, 3\} \times \{2, 4\} \times \{3, 5\}$.

9. Write in list form $\{1\} \times \{1, 2\} \times \{1, 2, 3\}$.

10. Write in list form $\{(a, b) \in \mathbb{Z} \times \mathbb{Z} \mid a^2 + b^2 < 5\}$.

11. Justify the statement that $A \times \emptyset = \emptyset$ for all sets A.

12. Suppose that A and B are sets with $A \times B = \emptyset$. What can you conclude about A and B?

13. Describe in geometric terms and sketch the following subsets of $\mathbb{R}^2$:
    (a) $\{(x, y) \in \mathbb{R}^2 \mid y = (x - 3)^2\}$;
    (b) $\{(x, y) \in \mathbb{R}^2 \mid 2x + 3y = 6\}$;
    (c) $\{(x, y) \in \mathbb{R}^2 \mid y = \sin x \text{ and } y \geq 0\}$;
    (d) $\{(x, y) \in \mathbb{R}^2 \mid 1 \leq x^2 + y^2 < 4\}$.

14. Let $S = \{1, 2, 3, 4, 5, 6\}$. Express in list form the following sets:
    (a) $\{(s, t) \in S^2 \mid |s - t| = 2\}$;
    (b) $\{(s, t) \in S^2 \mid s + t = 6\}$;
    (c) $\{(s_1, s_2, s_3) \in S^3 \mid s_1 \leq s_2 \leq s_3 \text{ and } s_1 + s_2 + s_3 = 10\}$;
    (d) $\{(s_1, s_2, s_3) \in S^3 \mid s_1 + s_2 = s_3\}$.
    [Note: In (a), because of the awkward notation of the absolute value vertical bar immediately following the such that |, we sometimes use the alternate notation when confusion may arise: $\{(s, t) \in S \times S : |s - t| = 1\}$.]

15. We denote by $M_2(\{0, 1\})$ the set of 2×2 matrices with only 0 or 1 for entries.

(a) Determine with proof how many elements are in $M_2(\{0,1\})$.

(b) Express in list form $\{A \in M_2(\{0,1\}) \mid \det A = 1\}$.

(c) Express in list form $\{A \in M_2(\{0,1\}) \mid \operatorname{Tr} A = 1\}$, where by $\operatorname{Tr} A$ we mean the trace of A, namely the sum of the diagonal elements.

(d) Express as a list $\{(A,B) \in M_2(\{0,1\}) \times M_2(\{0,1\}) \mid AB = I\}$.

16. Prove two sets A and B are equal if and only if $\mathcal{P}(A) = \mathcal{P}(B)$.

17. Prove that if $A \subseteq B$ and $B \subseteq C$, then $A \subseteq C$.

1.3 Logical Equivalences

1.3.1 Introduction

In Section 1.1.2, we observed that the implication compound proposition $p \to q$ has the same truth table as its contrapositive $\neg q \to \neg p$. As sentences, an implication statement is not the same sentence as its contrapositive; however, they are equivalent from the perspective of their truth values. This is a central concept in logic.

> **Definition 1.3.1**
>
> Two compound propositions p and q are called *logically equivalent* if $p \leftrightarrow q$ is a tautology. If this is the case, we write $p \equiv q$.

Another way to define this concept is to say that two compound propositions are logically equivalent if they have the same truth value for all possible cases of their atomic propositions.

In Example 1.1.14, we saw that $\neg p \lor q$ and $p \to q$ have the same truth values, no matter the truth value of p or q. We write this fact as

$$p \to q \equiv \neg p \lor q. \tag{1.4}$$

As we will see, this particular logical equivalence is very important and we will refer to it regularly.

Example 1.3.2. As another example, we show that $\neg(p \lor q) \equiv \neg p \land \neg q$ by using a truth table.

p	q	$p \lor q$	$\neg(p \lor q)$	$\neg p$	$\neg q$	$\neg p \land \neg q$
T	T	T	F	F	F	F
T	F	T	F	F	T	F
F	T	T	F	T	F	F
F	F	F	T	T	T	T

Since $\neg(p \vee q)$ and $\neg p \wedge \neg q$ have the same truth table, they are logically equivalent. This important propositional equivalence is called De Morgan's law.[4]   △

Example 1.3.3. The typical way of reading the biconditional $p \leftrightarrow q$ is "p if and only if q." This latter expression is a compressed form for the compound proposition $(p \to q) \wedge (q \to p)$. If these are indeed equivalent, we need to verify this to be the case. To do so, we construct a truth table.

p	q	$p \leftrightarrow q$	$p \to q$	$q \to p$	$(p \to q) \wedge (q \to p)$
T	T	T	T	T	T
T	F	F	F	T	F
F	T	F	T	F	F
F	F	T	T	T	T

We summarize this via the logic notation:

$$p \leftrightarrow q \equiv (p \to q) \wedge (q \to p). \tag{1.5}$$

△

1.3.2   Logical equivalence laws

Not unlike the many identities in algebra that allow us to solve equations or simplify algebraic expressions, there are a number of logical equivalences that allow us to simplify or restate compound propositions. Just as in Example 1.3.2, every equivalence listed in Table 1.1 can be verified using truth tables.

In the identity laws and dominance laws, the truth values of **T** and **F** appear explicitly in a compound proposition. In the identity law $p \wedge \mathbf{T} \equiv p$, we can think of **T** as a compound proposition that is always true. This identity law is not unlike the identity in the algebra of real numbers where $x \cdot 1 = x$ for every real number x.

Because of associativity of addition or multiplication, it is commonplace in algebra to write $1 + 2 + 3$ instead of $(1 + 2) + 3$, and to write $2 \cdot 3 \cdot 4$ instead of $(2 \cdot 3) \cdot 4$. Similarly, because of associativity, we write $p \wedge q \wedge r$ instead of $(p \wedge q) \wedge r$ and and we write $p \vee q \vee r$ instead $(p \vee q) \vee r$. From associativity of the $\vee$ operator, there is no logical confusion in a sentence like "Frank's car is a stick shift or it is red or it has leather seats."

In contrast to associativity, the compound proposition form $p \wedge q \vee r$ is not well defined. For example, the statement, "Frank's car is a stick shift and it is red or it has leather seats," leads to confusion. It is not

---

[4]Augustus De Morgan (1806-1871) was a prolific English mathematician and logician who taught at University College London for most of his career.

Equivalence	Name
$p \wedge \mathbf{T} \equiv p$ $p \vee \mathbf{F} \equiv p$	Identity Laws
$p \wedge q \equiv q \wedge p$ $p \vee q \equiv q \vee p$	Commutative Laws
$(p \wedge q) \wedge r \equiv p \wedge (q \wedge r)$ $(p \vee q) \vee r \equiv p \vee (q \vee r)$	Associative Laws
$(p \wedge q) \vee r \equiv (p \vee r) \wedge (q \vee r)$ $(p \vee q) \wedge r \equiv (p \wedge r) \vee (q \wedge r)$	Distributive Laws
$p \wedge \neg p \equiv \mathbf{F}$ $p \vee \neg p \equiv \mathbf{T}$	Complement Laws
$p \wedge p \equiv p$ $p \vee p \equiv p$	Idempotent Laws
$p \vee \mathbf{T} \equiv \mathbf{T}$ $p \wedge \mathbf{F} \equiv \mathbf{F}$	Dominance Laws
$\neg(\neg p) \equiv p$	Double Negative Law
$p \wedge (p \vee q) \equiv p$ $p \vee (p \wedge q) \equiv p$	Absorption Laws
$\neg(p \wedge q) \equiv \neg p \vee \neg q$ $\neg(p \vee q) \equiv \neg p \wedge \neg q$	De Morgan's Laws

Table 1.1: Logical equivalences.

particularly clear whether we mean $(p \wedge q) \vee r$ or $p \wedge (q \vee r)$, where we are using the atomic propositions of $p =$ "Frank's car is a stick shift", $q=$ "Frank's car is red", and $r =$ "Frank's car has leather seats."

Example 1.3.4. As an example of De Morgan's law, consider the following definition of continuity of a function at a point, that we encounter in a first calculus course. Let $f(x)$ be a real function defined over an interval $[a, b]$ and let $a < c < b$. We say that $f(x)$ is continuous at c if and only if $\lim_{x \to c} f(x)$ exists and $\lim_{x \to c} f(x) = f(c)$. As an application of De Morgan's law, consider the negation of the above definition: The function $f(x)$ is not continuous at c if and only if $\lim_{x \to c} f(x)$ does not exist or $\lim_{x \to c} f(x) \neq f(c)$.   △

Applications of De Morgan's laws are ubiquitous in mathematics, but it is valuable to be precise even in regular English. For example, consider the proposition, "My car is white and is a stick shift." Using De Morgan's law, the negation of this proposition is, "My car is not white or it is an automatic." It is incorrect to say that the negation of the original sentence is, "My car is not white and it is an automatic."

As another point of grammar, we point out that the construction "neither p nor q" means $\neg p \wedge \neg q$. De Morgan's law

$$\neg(p \vee q) \equiv \neg p \wedge \neg q$$

underscores that the expression "neither p nor q" is the negation of "p or q."

The next three examples illustrate how to use the logical equivalences in Table 1.1 to simplify compound propositions.

Example 1.3.5. The negation of an implication statement stands as a particularly important logical equivalence. Every time we prove that an implication statement is false, we must use this.

$$
\begin{aligned}
\neg(p \to q) &\equiv \neg(\neg p \vee q) && \text{by (1.4)} \\
&\equiv \neg(\neg p) \wedge \neg q && \text{De Morgan's law} \\
&\equiv p \wedge \neg q && \text{double negative.}
\end{aligned}
$$

Since $\neg(p \to q) \equiv p \wedge \neg q$, in order to prove that $p \to q$ is false, we need to prove that p is true and q is false.   △

(It is common in mathematical writing to label with a number all definitions, theorems, equations, and other items, thereby allowing the author to refer to it. In this case, we can find Equation (1.4) on page 23.)

Example 1.3.6. We show that $(p \to q) \wedge (p \to r) \equiv p \to (q \wedge r)$.

$$
\begin{aligned}
(p \to q) \wedge (p \to r) &\equiv (\neg p \vee q) \wedge (\neg p \vee r) && \text{by (1.4)} \\
&\equiv \neg p \vee (q \wedge r) && \text{distributivity} \\
&\equiv p \to (q \wedge r) && \text{by (1.4).}
\end{aligned}
$$

We could state this result by saying that $\to$ left-distributes over $\wedge$.△

Example 1.3.7. We simplify the compound proposition $r \to (\neg p \to (r \wedge q))$ as much as possible and justify all the steps.

$$
\begin{aligned}
r \to (\neg p \to (r \wedge q)) \\
\equiv \neg r \vee (\neg(\neg p) \vee (r \wedge q)) && \text{by (1.4) twice} \\
\equiv \neg r \vee (p \vee (r \wedge q)) && \text{double negative} \\
\equiv (\neg r \vee p) \vee (r \wedge q) && \text{associativity} \\
\equiv (p \vee \neg r) \vee (r \wedge q) && \text{commutativity} \\
\equiv p \vee (\neg r \vee (r \wedge q)) && \text{associativity} \\
\equiv p \vee ((\neg r \vee r) \wedge (\neg r \vee q)) && \text{distributivity} \\
\equiv p \vee (\mathbf{T} \wedge (\neg r \vee q)) && \text{complement} \\
\equiv p \vee (\neg r \vee q) && \text{identity}
\end{aligned}
$$

We could stop here, but for aesthetic reasons, we could go further:

$$\equiv \neg r \wedge (p \vee q) \qquad\qquad \text{associativity and commutativity}$$
$$\equiv r \rightarrow (p \vee q) \qquad\qquad \text{by (1.4).}$$

We point out in the above sequence of equivalences that at some point we used associativity and commutativity at the same time. This one step consists in fact of three steps, but we are familiar with this combination of rules from the properties of associativity and commutativity of $+$ and $\times$ with real numbers. $\qquad\triangle$

We point out that if we use logical equivalences to prove that a compound proposition is a tautology (respectively a contradiction), our sequence of equivalences would end in a **T** (resp. a **F**).

Example 1.3.8. Suppose that we wish to prove that $(p \wedge q) \rightarrow p$ is a tautology. We could do

$$\begin{aligned}
(p \wedge q) \rightarrow p &\equiv \neg(p \wedge q) \vee p && \text{by (1.4)} \\
&\equiv (\neg p \vee \neg q) \vee p && \text{De Morgan's law} \\
&\equiv (p \vee \neg p) \vee \neg q && \text{associativity and commutativity} \\
&\equiv \mathbf{T} \vee \neg q && \text{complement} \\
&\equiv \mathbf{T} && \text{identity.}
\end{aligned}$$

Since the compound proposition is equivalent to **T**, then it is a tautology. $\qquad\triangle$

Even though it is useful to remember all the logical equivalences in Table 1.1, they are not independent of each other. Though each equivalence can be verified by an appropriate truth table, some of the equivalences imply the others. The following theorem makes this precise. The proof of this theorem serves the pedagogical purpose of offering more examples of reasoning using logical equivalences.

Theorem 1.3.9

From the identity, commutative, associative, distributive and complement laws, the remaining laws in Table 1.1 follow.

Proof. We first prove the idempotent laws. Let p be any logical proposition. Then

$$\begin{aligned}
p &\equiv p \vee \mathbf{F} && \text{identity} \\
&\equiv p \vee (p \wedge \neg p) && \text{complement} \\
&\equiv (p \vee p) \wedge (p \vee \neg p) && \text{distributivity} \\
&\equiv (p \vee p) \wedge \mathbf{T} && \text{complement} \\
&\equiv p \vee p && \text{identity.}
\end{aligned}$$

The other idempotent law is similar.

For the first dominance law, let p be any logical proposition. Then

$$\mathbf{F} \wedge p \equiv (\neg p \wedge p) \wedge p \equiv \neg p \wedge (p \wedge p) \equiv \neg p \wedge p \equiv \mathbf{F}.$$

The third equality follows from the idempotent laws, which we just proved. Again, the second dominance law is similar.

Next we show the double negative law. For any proposition p,

$$
\begin{aligned}
\neg(\neg p) &\equiv (\neg(\neg p)) \wedge \mathbf{T} && \text{identity} \\
&\equiv (\neg(\neg p)) \wedge (\neg p \vee p) && \text{complement} \\
&\equiv ((\neg(\neg p)) \wedge \neg p) \vee ((\neg(\neg p)) \wedge p) && \text{distributivity} \\
&\equiv (\neg(\neg p)) \wedge p && \text{complement; identity} \\
&\equiv ((\neg(\neg p)) \wedge p) \vee (p \wedge \neg p) && \text{identity; complement} \\
&\equiv p \vee ((\neg(\neg p)) \wedge \neg p) && \text{distributivity} \\
&\equiv p \vee \mathbf{F} && \text{complement} \\
&\equiv p && \text{identity.}
\end{aligned}
$$

For the absorption law, let p and q be propositions. Then

$$p \wedge (p \vee q) \equiv (p \vee \mathbf{F}) \wedge (p \vee q) \equiv p \vee (\mathbf{F} \wedge q) \equiv p \vee \mathbf{F} \equiv p.$$

The second absorption law is similar. (Though we do not name the law used in each step, the reader ideally should be able to recognize it. This is the same practice as in basic algebra.)

The proof of the De Morgan's laws is the most challenging. We first claim that for any propositions a and b, if $a \wedge \neg b \equiv \mathbf{F}$ and $a \vee \neg b \equiv \mathbf{T}$, then $a \equiv b$. (This does not mean that they are the same proposition but simply that they must have the same truth values.) We can see this as follows

$$
\begin{aligned}
b &\equiv b \wedge \mathbf{T} \equiv b \wedge (a \vee \neg b) \\
&\equiv (b \wedge a) \vee (b \wedge \neg b) \equiv (b \wedge a) \vee \mathbf{F} \equiv (b \wedge a) \vee (a \wedge \neg b) \\
&\equiv a \vee (b \wedge \neg b) \equiv a \vee \mathbf{F} \equiv a.
\end{aligned}
$$

Using this result, our strategy to prove $\neg(p \wedge q) \equiv \neg p \vee \neg q$ involves proving $a \wedge \neg b \equiv \mathbf{F}$ and $a \vee \neg b \equiv \mathbf{T}$ for $a \equiv \neg p \vee \neg q$ and $b \equiv \neg(p \wedge q)$. First,

$$
\begin{aligned}
(\neg p \vee \neg q) \wedge \neg(\neg(p \wedge q)) &\equiv (\neg p \vee \neg q) \wedge (p \wedge q) \\
&\equiv (\neg p \wedge (p \wedge q)) \vee ((p \wedge q) \wedge \neg q) \\
&\equiv ((\neg p \wedge p) \wedge q) \vee (p \wedge (q \wedge \neg q)) \equiv (\mathbf{F} \wedge q) \vee (p \wedge \mathbf{F}) \\
&\equiv \mathbf{F} \vee \mathbf{F} \equiv \mathbf{F}.
\end{aligned}
$$

Then second,

$$(\neg p \vee \neg q) \vee \neg(\neg(p \wedge q)) \equiv (\neg p \vee \neg q) \vee (p \wedge q)$$
$$\equiv ((\neg p \vee \neg q) \vee p) \wedge ((\neg p \vee \neg q) \vee q)$$
$$\equiv ((\neg p \vee p) \vee \neg q) \wedge (\neg p \vee (\neg q \vee q))$$
$$\equiv (\mathbf{T} \vee q) \wedge (\mathbf{T} \vee p) = \mathbf{T} \wedge \mathbf{T} = \mathbf{T}.$$

Using the claim, we conclude that $\neg(p \wedge q) \equiv \neg p \vee \neg q$. A similar argument establishes the second De Morgan's law.   □

1.3.3   Comment on other logics

The term *logic* refers to the study of methods of reasoning that assigns and analyzes the validity of statements. In this broad sense, logic forms a discipline of philosophy with a long tradition. With phrases like "that is logical" replete in daily conversation, it may come as a surprise to students that different types of logic exist. In fact, mathematicians may forget that many logic systems exist outside of formal mathematical logic.

In a group of works collectively entitled *Organon*, Aristotle developed an analysis of structured reasoning. The *Stanford Encyclopedia of Philosophy* says of Aristotelian logic, that it "has had an unparalleled influence on the history of Western thought." Despite this, "today, very few would try to maintain that it is adequate as a basis for understanding science, mathematics, or even everyday reasoning."[5]

Boolean logic refers to concept of proposition and system of operators we have introduced in Sections 1.1 and 1.3. Work by Frege (1848-1925) and Russell (1872-1970) established a formal system of logic based on the notions of Boolean logic and introducing "predicate calculus," which we will cover in Sections 1.5 and 1.6. Though logic remains an active field of investigation both from the philosophical and the mathematical perspectives, Frege's formal system stands as the foundation of modern mathematical logic.

In Section 8.3.3, we mention a constructivist perspective on the philosophy of mathematics, in which a mathematical object or idea is considered true only after a proof creates, or provides a method to create, the object. In the logic that follows from this perspective, called *intuitionistic logic*, the notion of truth differs from Boolean logic. Though this system possesses similar rules as given in Table 1.1, the Complement Laws and the Double Negative Law no longer hold. Most mathematicians use Boolean logic; those who use intuitionistic logic clearly signal that they do so.

---

[5] *Stanford Encyclopedia of Philosophy* [72].

EXERCISES FOR SECTION 1.3

1. Using De Morgan's law, state the negation of each of the following:
   (a) It is snowing, or it is not cold.
   (b) My friend Frank is Italian; therefore, he likes pizza. [Hint: Use (1.4).]

2. Using De Morgan's law, state the negation of each of the following:
   (a) My car is red, and it has less than 20,000 miles.
   (b) If the number 1 is even, then my dog can fly. [Hint: Use (1.4).]

3. In Calculus, we encounter the theorem: "If a function f is differentiable at a, then f is continuous at a."
   (a) State the contrapositive of this implication statement.
   (b) State the converse of this implication statement.
   (c) State the negation of the converse of this implication statement and then provide an example of a function f and a real number a to show that the converse is not always true.

4. Construct the truth tables to establish both absorption laws.

5. Construct the truth tables to establish both distributivity laws.

6. Use logical equivalences and label all the steps to prove the triple De Morgan's law
$$\neg(p \wedge q \wedge r) \equiv \neg p \vee \neg q \vee \neg r.$$

7. Prove the following by using logical equivalences, labeling all your steps.
   (a) $(p \wedge q) \vee (r \wedge s) \equiv (p \vee r) \wedge (p \vee s) \wedge (q \vee r) \wedge (q \vee s)$.
   (b) $(p \vee q) \wedge (r \vee s) \equiv (p \wedge r) \vee (p \wedge s) \vee (q \wedge r) \vee (q \wedge s)$.
   [Note: Both of these equivalences resemble the FOIL method in elementary algebra of $\times$ distributing over $+$.]

8. Prove that $(p \vee q) \rightarrow r \equiv (p \rightarrow r) \wedge (q \rightarrow r)$ in two ways: (a) Using a truth table. (b) Using logical equivalences.

9. Prove that $p \oplus q \equiv \neg(p \leftrightarrow q)$ in two ways: (a) Using a truth table. (b) Using logical equivalences. [Hint: Use the equivalence in Equation (1.5).]

10. Prove that $(p \wedge (p \rightarrow q)) \rightarrow q$ is a tautology in two ways: (a) Using a truth table. (b) Using logical equivalences.

11. Prove that $((p \rightarrow q) \wedge (q \rightarrow r)) \rightarrow (p \rightarrow r)$ is a tautology in two ways: (a) Using a truth table. (b) Using logical equivalences.

12. Show that $(p \rightarrow q) \rightarrow r$ and $p \rightarrow (q \rightarrow r)$ are not equivalent by any means.

13. Show that $(p \wedge q) \rightarrow r$ is not equivalent to $(p \rightarrow r) \wedge (q \rightarrow r)$ by any means.

14. Prove or disprove that $(p \oplus q) \oplus r$ is equivalent to $p \oplus (q \oplus r)$.

15. Find a compound proposition involving p, q and r that is true if and only if exactly one of p, q or r is **T**.

16. (*) Explain why every compound proposition of $p_1, p_2, \ldots, p_n$ can be expressed using only the logical operators $\neg$ and $\vee$.

1.4 Operations on Sets

1.4.1 Definition and examples

The logical operations of conjunction, disjunction, and so on applied to sets inspire operations on sets.

Definition 1.4.1

Let A and B be sets.

- The *union* of A and B is $A \cup B = \{x \mid x \in A \vee x \in B\}$.
- The *intersection* of A and B is $A \cap B = \{x \mid x \in A \wedge x \in B\}$.
- The *set difference* of A by B is $A \setminus B = \{x \mid x \in A \wedge x \notin B\}$. An alternate notation for the set difference is $A - B$.

With these operations, we can succinctly denote the set of irrational numbers by $\mathbb{R} \setminus \mathbb{Q}$. As another example, consider the function f defined over the reals by $f(x) = \sqrt{x+2}/(x-3)$. The domain of f consists of all reals $x \geq -2$, except for 3. Using set operations, this domain is $[-2, \infty) \setminus \{3\}$. As yet another example, the domain of $g(x) = \ln(x^2 - 9)$ requires $x^2 > 9$, which we write as the set $(-\infty, -3) \cup (3, \infty)$.

Though there does not exist a universal set of all objects, set operations allow us to define sets that we might consider unorthodox, such as the union of the set of all people living now with the set of triangles in the Euclidean plane. Furthermore, we cannot talk about an operation on sets that is tantamount to the logic operation of negation, i.e., if A is a set, we cannot talk about the set of all objects not in A. However, in many situations, we work with subsets of a specific context set U. Then we can talk about the operations on subsets of U. In this situation, we call U the *universal set*.

Definition 1.4.2

If U is any set and $A \subseteq U$, the *complement* of A is $A^c = \{x \in U \mid x \notin A\}$. Alternate notations for the complement of A include $\complement A$ and $\overline{A}$.

Another way to define the complement of a subset is simply $U \setminus A$. For this reason, some call the set difference of A by B as the *relative complement* of B in A as $A \setminus B$.

Example 1.4.3. Let $U = \{1, 2, \ldots, 10\}$ and consider the subsets $A = \{1, 3, 6, 7\}$, $B = \{1, 5, 6, 8, 9\}$, and $C = \{2, 3, 4, 5, 6\}$. We calculate the following operations.

1) $A \cap B = \{1, 6\}$.

2) $B \setminus (A \cup C) = B \setminus \{1, 2, 3, 4, 5, 6, 7\} = \{8, 9\}$.

3) $(B \cup C)^c = \{1, 2, 3, 4, 5, 6, 8, 9\}^c = \{7, 10\}$.

4) $(A \cap B) \cap C = \{1, 6\} \cap C = \{6\}$.

As with other contexts, parentheses determine the precedence of operators.   △

Definition 1.4.4

We say that two sets A and B are *disjoint* if $A \cap B = \emptyset$.

Example 1.4.5. Let $A = [1, 6]$ and $B = [2, \pi]$ be intervals. Then we can write $A \setminus B = [1, 2) \cup (\pi, 6]$, expressed as the union of two disjoint intervals.   △

1.4.2   Venn diagrams; membership tables

Like a truth table, we can use a membership table to determine which elements are contained in a given composition of operations on sets. A *membership table* is a table that, for all combinations of whether an element is in or not in the sets occurring in the operations on sets, lists whether an element is in or not in the given set operation. Keeping with the notation that we use for truth tables, we list **T** for when an element is in a set and **F** for when an element is not in the set.

Example 1.4.6. Let A, B, and C be sets and consider the operation of $(A \cup B) \setminus C$. This composition of operators involves three sets, so we use a table with eight rows. We remind the reader that an element is in $U \setminus V$ exactly when an element is in U (indicated by **T**) but not in V (indicated by **F**).

A	B	C	$A \cup B$	$(A \cup B) \setminus C$
T	**T**	**T**	**T**	**F**
T	**T**	**F**	**T**	**T**
T	**F**	**T**	**T**	**F**
T	**F**	**F**	**T**	**T**
F	**T**	**T**	**T**	**F**
F	**T**	**F**	**T**	**T**
F	**F**	**T**	**F**	**F**
F	**F**	**F**	**F**	**F**

The intermediate column gives us the membership table for $A \cup B$, but the last column gives the membership table for $(A \cup B) \setminus C$.   △

As we already discussed truth tables in previous sections, we do not provide more examples of membership tables. However, since there does not exist a universal set of anything we could conceive of as an object, the last row in the table of Example 1.4.6 might not represent a set. We could at best consider the elements in $A \cup B \cup C$. However, if the sets in question are subsets of given set U, then the row of the membership table is the set corresponding to $U \backslash (A \cup B \cup C)$.

Another common method of visualizing the relationship of set operations to the original sets involves depicting each set as a region in the plane and shading the regions corresponding to being in the set operation. We call this the *Venn diagram* of a set operation. For example, if A and B are subsets of a set S, the following pictures are the Venn diagrams for $A \cup B$ and $A \cap B$.

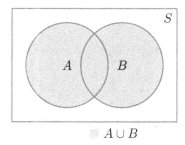
$A \cup B$

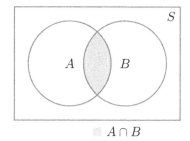
$A \cap B$

In the above diagrams, if A and B are not subsets of a context set S, then we do not write S in the diagram and ignore everything outside of $A \cup B$. The Venn diagrams for $A \backslash B$ and A^c are given below.

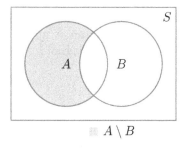
$A \backslash B$

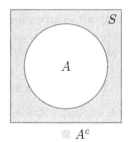
A^c

Using a Venn diagram to illustrate another set operation, we introduce the *symmetric difference* on sets. If A and B are sets, their symmetric difference is $A \triangle B = \{x \mid x \in A \oplus x \in B\}$. Because of the exclusive or occurring in the definition, an alternative notation is $A \oplus B$. Here is the Venn diagram.

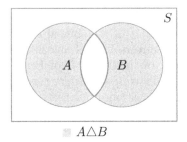

$A \triangle B$

Example 1.4.7. As another illustration of Venn diagrams but now with three sets, we draw the Venn diagram for the set operation in Example 1.4.6, $(A \cup B) \setminus C$.

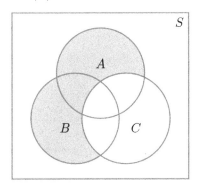

The shaded area of the diagram corresponds to $(A \cup B) \setminus C$. We notice from the Venn diagram that $(A \cup B) \setminus C$ is the union of three "undivided" regions, which we can write as

$$(A \cup B) \setminus C = (A \cap B^c \cap C^c) \cup (A \cap B \cap C^c) \cup (A^c \cap B \cap C^c).$$

Note that with three subsets A, B, and C of a given set S, any triple intersection like $A \cap B \cap C^c$, where any of the subsets X appear as X or X^c gives the eight "undivided" regions of the diagram. $\triangle$

1.4.3   Identities in set operations

Because of their definition from logic operators, set operators share many similar properties with them. Suppose that $A, B, C \in \mathcal{P}(S)$. Then the truth value of **T** corresponds to the whole context set S, while the truth value of **F** corresponds to $\emptyset$. Under these assumptions, set operators have the following properties, which are identical to those in Table 1.1.

We trust the reader will notice the strong similarity in how the operation $\wedge$, $\vee$ and $\neg$ behave on logical propositions and in how $\cap$, $\cup$ and ^{c} behave on subsets of a set. Each of these contexts is an

Equivalence	Name
$A \cap S = A$ $A \cup \emptyset = A$	Identity Laws
$A \cap B = B \cap A$ $A \cup B = B \cup A$	Commutative Laws
$(A \cap B) \cap C = A \cap (B \cap C)$ $(A \cup B) \cup C = A \cup (B \cup C)$	Associative Laws
$(A \cap B) \cup C = (A \cup C) \cap (B \cup C)$ $(A \cup B) \cap C = (A \cap C) \cup (B \cap C)$	Distributive Laws
$A \cap A^c = \emptyset$ $A \cup A^c = S$	Complement Laws
$A \cap A = A$ $A \cup A = A$	Idempotent Laws
$A \cup S = S$ $A \cap \emptyset = \emptyset$	Dominance Laws
$(A^c)^c = A$	Double Negative Law
$A \cap (A \cup B) = A$ $A \cup (A \cap B) = A$	Absorption Laws
$(A \cap B)^c = A^c \cup B^c$ $(A \cup B)^c = A^c \cap B^c$	De Morgan's Laws

Table 1.2: Identities on set operations.

example of what we call a *Boolean algebra*. We caution the reader again that if A, B and C are not subsets of a specific set S, then we cannot consider the set complement or any identity in Table 1.2 that involves S. On the other hand, we could implicitly set S to be the union of all sets involved.

Example 1.4.8. One particularly common identity involves an alternate expression for the set difference. We observe from the definition that

$$A \setminus B = A \cap B^c. \tag{1.6}$$

This comes from the definition $A \setminus B = \{x \mid x \in A \land x \notin B\}$ so we can write $A \setminus B = \{x \mid x \in A \land x \in B^c\} = A \cap B^c$.   △

Example 1.4.9. Let A, B and C be sets. We simplify the expression:

$$
\begin{aligned}
&(A \cup B) \setminus (A \cup C) \\
&= (A \cup B) \cap (A \cup C)^c &&\text{by (1.6)} \\
&= (A \cup B) \cap (A^c \cap C^c) &&\text{De Morgan's law} \\
&= (A \cap (A^c \cap C^c)) \cup (B \cap (A^c \cap C^c)) &&\text{distributivity.}
\end{aligned}
$$

Using a combination of associativity and commutativity, we have

$$(A \cup B) \setminus (A \cup C)$$
$$= ((A \cap A^c) \cap C^c) \cup (A^c \cap B \cap C^c)$$
$$= (\emptyset \cap C^c) \cup (B \cap A^c \cap C^c) \qquad \text{complement and commutati}$$
$$= \emptyset \cup (B \cap (A \cup C)^c) \qquad \text{dominance and De Morgan'}$$
$$= B \setminus (A \cup C) \qquad \text{identity and (1.6).}$$

(We point out that though we did not specify A, B and C as subsets of a context set S, we could assume here that $S = A \cup B \cup C$.)   △

EXERCISES FOR SECTION 1.4

1. Let $S = \{a, b, c, d, e, f, g, h, i, j\}$ be the set of the first ten letters of the English alphabet. Let $A = \{x \in S \mid x \text{ is a letter in "jade"}\}$, let $B = \{x \in S \mid x \text{ is a vowel}\}$, and let $C = \{d, e, f, g, h\}$. Determine the following sets: (a) $B \cup (A \cap C)$. (b) $(A \cup B \cup C)^c$. (c) $(A \cap B) \cup (A \cap C) \cup (B \cap C)$.

2. Let $S = \{1, 2, 3, 4, 5, 6, 7, 8\}$ and consider the three subsets $A = \{1, 2, 3, 4, 5, 6\}$, $B = \{2, 3, 6, 7\}$ and $C = \{1, 4, 8\}$.

   (a) Calculate $A \setminus (B \cup C)$.

   (b) Calculate $(A \setminus B) \cup (B \setminus C) \cup (C \setminus A)$.

   (c) Is it possible to obtain the subset $\{1\}$ as a sequence of operations involving A, B and C? If so, show how; if not, explain why.

   (d) Repeat the previous question with the subset $\{2\}$.

3. Write the domain of each of the following functions as an interval or a union of intervals of $\mathbb{R}$ or $\mathbb{R} \setminus A$ for some set A.

   (a) $f(x) = \sqrt{\dfrac{x - 3}{x + 1}}$.

   (b) $g(x) = \ln(x + 5) + \ln(3 - 2x)$.

   (c) $h(x) = \sqrt{(x^2 - 4)(7 - x^2)}$.

4. In $\mathbb{Z}$ consider the subsets $A = \{x \in \mathbb{Z} \mid 4 \text{ divides } x\}$ and $B = \{x \in \mathbb{Z} \mid 6 \text{ divides } x\}$. Determine $A \cup B$ and $A \cap B$.

5. Draw the Venn diagram and a membership table for each of the set operations: (a) $A \cup (B \cap C)$. (b) $(B \cup C)^c \cup (A \cup C)^c$.

6. Draw the Venn diagram and a membership table for each of the set operations: (a) $(A \cap B^c) \cup (B \cap C^c)$. (b) $(A \cup B) \setminus C$.

7. Let $a, b, c, d \in \mathbb{R}$. Let $A = [a, \infty) \times [b, \infty)$ and $B = [c, \infty) \times [d, \infty)$ be subsets of $\mathbb{R}^2$. Depending on the relative values of a, b, c, d, describe the shape of each of the following sets. (a) $A \cup B$. (b) $A \cap B$. (c) $A \setminus B$.

8. For this exercise, we will call up-quadrants subsets of $\mathbb{R}^2$ that have the form of A or B as described in the previous exercise. (a) Describe

how to obtain the rectangle $[a, b) \times [c, d)$ using a finite number of set operations of up-quadrants. (b) Prove or disprove whether it is possible to obtain the rectangle $(a, b) \times (c, d)$ using a finite number of set operations of up-quadrants.

9. Prove that the following are true for all sets A and B.
   (a) $A \cap B \subseteq A$.
   (b) $A \subseteq A \cup B$.

10. Let A and B be subsets of a set S.
    (a) Prove that $A \subseteq B$ if and only if $\mathcal{P}(A) \subseteq \mathcal{P}(B)$.
    (b) Prove that $\mathcal{P}(A \cap B) = \mathcal{P}(A) \cap \mathcal{P}(B)$.
    (c) Show that $\mathcal{P}(A \cup B) = \mathcal{P}(A) \cup \mathcal{P}(B)$ if and only if $A \subseteq B$ or $B \subseteq A$.

11. Prove that the following are equivalent: (i) $A \subseteq B$; (ii) $A \cap B = A$; (iii) $A \cup B = B$; (iv) $A \setminus B = \emptyset$.

12. In each of the following cases, what holds true about the sets A and B? (a) $A \setminus B = A$; (b) $A \cup B = A$; (c) $A \cap B = A$; (d) $A \triangle B = B$.

13. Let A and B be subsets of a set S. Prove that $A \subseteq B$ if and only if $B^c \subseteq A^c$.

14. Let A, B, and C be subsets of a set S.
    (a) Prove that $A \triangle B = \emptyset$ if and only if $A = B$.
    (b) Prove that $A \cap (B \triangle C) = (A \cap B) \triangle (A \cap C)$.

15. Use the set identities in Table 1.2 to simplify:
    (a) $(A \cup B)^c \cap (C \cup A^c)$;
    (b) $(A^c \cup B) \cap (B^c \cup C)^c$.

16. Use the set identities in Table 1.2 to simplify:
    (a) $((A \cup B \cup C) \cap (A \cup B)^c) \cup C$;
    (b) $((A \setminus B) \setminus (C \setminus A))$.

17. Let A, B, and C be subsets of a set S. In a Venn diagram for three sets, shade the region that corresponds to the subset D of S consisting of elements that are in exactly two of the sets A, B, or C. Give a disjunctive normal form expression of A, B, and C for the set D.

1.5 Predicates and Quantifiers

1.5.1 Predicates

In mathematics, we often encounter statements like "$x > 3$" or "$x^2 + x - 1 = 0$". In Section 1.1 we underscored that such statements are not logical propositions: when the variable does not carry a specific value, the statement is neither true nor false.

If we have set theory at our disposal, we might assume that the variable x is an element of some set. However, even if we are working in the context of logic prior to set theory, we will say that the variable x can refer to anything in some specified *universe of discourse*.

> **Definition 1.5.1**
>
> A statement $P(x)$ whose truth value depends on the object x is called a *predicate* or a *propositional function*. The universe of discourse for the variable is called the *domain* of the predicate.

If we let $P(x)$ be the predicate "$x > 3$," then $P(1)$ is the proposition $1 > 3$, whose truth value happens to be false, while $P(4)$ is the proposition $4 > 3$, whose truth value is true. We call the process of giving the variable a value *instantiating* the variable. This is the formal term for the colloquial expression "plugging in" a value for a variable.

Just like how we can combine real-valued functions by adding or multiplying them, given two predicates $P(x)$ and $Q(x)$, we can create a new predicate by combining them with any logical operator. For example, $P(x) \land Q(x)$, $P(x) \lor Q(x)$, $P(x) \to Q(x)$, $\neg P(x)$ and so on, are new predicates.

Example 1.5.2. Suppose that the universe of discourse for x consists of all people living now. Consider the predicates $P(x) =$"x is older than 60" and let $Q(x) =$"x works full time." Then $\neg P(x) \to Q(x)$ is a new predicate whose truth value still depends on x. Replacing x with any specific person, the predicate becomes a proposition. △

1.5.2 Quantifiers

There are other common ways of producing a proposition from a predicate besides instantiating the variable. These methods are central to all of mathematics.

> **Definition 1.5.3**
>
> The *universal quantification* of $P(x)$, written $\forall x\, P(x)$, read as "for all x, $P(x)$", is the proposition that is true exactly when $P(x)$ is true for all x in its universe of discourse.

If the universe of discourse of a variable x in a predicate $P(x)$ is a set S, then we may make this clear by writing $\forall x \in S\, (P(x))$. We call the symbol $\forall$ a quantifier.

Example 1.5.4. Consider the predicate $9x^2 - 9x + 2 \geq 0$. Then $\forall x \in \mathbb{R}\, (9x^2 - 9x + 2 \geq 0)$ is a proposition, which we can also express with more English as $9x^2 - 9x + 2 \geq 0$ for all real x. This proposition happens to be false because $9(1/2)^2 - 9(1/2) + 2 = -1/4$. Factoring, we have

$$9x^2 - 9x + 2 = (3x - 1)(3x - 2)$$

and by analyzing the signs, we know that $9x^2 - 9x + 2 \geq 0$ exactly when $x \in (-\infty, 1/3] \cup [2/3, \infty)$. Consequently, the proposition $\forall x \in \mathbb{Z}\,(9x^2 - 9x + 2 \geq 0)$ has a truth value of true. △

Example 1.5.5. If we revisit Example 1.5.2, assuming the universe of discourse is all living people, then we might consider the predicate $\forall x\,(Q(x) \to \neg P(x))$. Expressed in English, this proposition says, "Every person who works full times is 60 years old or younger." This is a proposition whose truth value is false because there do exist people for whom this predicate is not true. △

Definition 1.5.6

> The *existential quantification* of $P(x)$, written $\exists x\, P(x)$, read as "there exists an x such that $P(x)$," is the proposition that is true exactly when $P(x)$ is true for at least one x in the universe of discourse.

Often in mathematics we are concerned not just with the existence of an object with a certain property, but that only one such object exists. There is a quantifier for that situation.

Definition 1.5.7

> The unique existential quantification, written $\exists! x\, P(x)$, read as "there exists a unique x such that $P(x)$," is the proposition that is true exactly when $P(x)$ is true for exactly one x in the universe of discourse.

Example 1.5.8. Consider the algebraic equation $x^3 - x - 6 = 0$ over the set of real numbers. The equation is a predicate. To ask whether it has a solution amounts to determining the truth value of the logical statement $\exists x \in \mathbb{R}\,(x^3 - x - 6 = 0)$. The truth value of this proposition happens to be true since $x = 2$ solves the equation. △

Example 1.5.9. Consider once more Examples 1.5.2 and 1.5.5. Using the existential quantifier and our predicates, we can express the English sentence, "Someone over 60 is still working full time," by $\exists x\,(P(x) \wedge Q(x))$. Our proposition does not include the stress of the word "still," but it fully expresses the logic. △

Quantifications, both existential and universal, occur ubiquitously in natural languages. For example, the statement, "There is someone alive who is over 110 years old" is an existentially quantified statement. In English, the words "some" or "something" or "someone" indicate an existential quantification, while we use words like, "every" or "everything" or "everyone" to employ a universal quantifier. In mathematical writing, though we now have symbols for quantifiers, it is still quite common to use these natural linguistic constructions to express the quantifiers.

$\forall x\,(M(x) \to R(x))$	Every person I have met is right-handed.
$\forall x\,(R(x) \to M(x))$	I have met every right-handed person.
$\forall x\,(M(x) \land R(x))$	I have met every person, and they are all right-handed.
$\forall x\,(M(x) \lor R(x))$	Every person is right-handed, or I have met them.
$\forall x\,(M(x) \leftrightarrow R(x))$	Every person I have met is right-handed and vice versa.
$\exists x\,(M(x) \land R(x))$	I have met a right-handed person.
$\exists! x\,(M(x) \land R(x))$	I have met exactly one right-handed person.
$\exists x\,(M(x) \lor R(x))$	There is someone whom I have met or who is right-handed.
$\exists x\,(M(x) \to R(x))$	There is someone, whom if I met them, is right-handed.

Table 1.3: Examples of quantifiers to English

To illustrate the connection between quantifiers and grammar, Table 1.3 lists a number of possible combinations with quantifiers and expresses the proposition in natural English.

Our universe of discourse is the set of living people, and let $M(x) =$ "I have met x" and $R(x)=$ "x is right-handed." Not every English expression in Table 1.3 feels equally natural. Consider the last example in the table; it sounds awkward primarily because it is not an interesting sentence and so is not common.

Though this introduction to quantifiers might serve as a reader's first formal encounter with them, it is crucial to realize that they are always and have always been present, even if only implicitly. For example, if someone writes $x^2 \geq 0$ when talking about real numbers, we usually mean this in a universal sense. Technically, this statement is $\forall x \in \mathbb{R}\,(x^2 \geq 0)$ and its truth value is true. On the other hand, when we solve an equation, say $x^2 - 3x + 2 = 0$, the equation itself is not a logical proposition but a predicate. When we ask if an equation has a solution, we are asking an existential question.

Example 1.5.10. We consider two definitions from linear algebra and emphasize the role of quantifiers. Consider first the concept of linear independence.

> **Definition:** A subset of vectors $\{\mathbf{v}_1, \mathbf{v}_2, \ldots, \mathbf{v}_k\} \subseteq \mathbb{R}^n$ is called *linearly independent* if $c_1\mathbf{v}_1 + c_2\mathbf{v}_2 + \cdots + c_k\mathbf{v}_k = \mathbf{0}$ implies $c_1 = c_2 = \cdots = c_k = 0$.

We first point out that the "if" in the above definition does not play the usual role of a conditional in compound proposition. Instead, it serves to define the new terms. Pertaining to quantifiers, the phrase after the "if" is a universal quantification, which we can rephrase in symbols as

$$\forall (c_1, c_2, \ldots, c_k) \in \mathbb{R}^k$$
$$(c_1\mathbf{v}_1 + c_2\mathbf{v}_2 + \cdots + c_k\mathbf{v}_k = \mathbf{0} \to (c_1, c_2, \ldots, c_k) = (0, 0, \ldots, 0)).$$

For an example with an existential quantifier, consider the concept of linear combinations.

> **Definition:** A vector $\mathbf{u} \in \mathbb{R}^n$ is said to be a linear combination of a set of vectors $\{\mathbf{v}_1, \mathbf{v}_2, \ldots, \mathbf{v}_k\} \subseteq \mathbb{R}^n$ if there exist real numbers $c_1, c_2, \ldots, c_k$ such that $\mathbf{u} = c_1\mathbf{v}_1 + c_2\mathbf{v}_2 + \cdots + c_k\mathbf{v}_k$.

We can express the phrase after the "if" using an existential quantifier by

$$\exists (c_1, c_2, \ldots, c_k) \in \mathbb{R}^k \, (\mathbf{u} = c_1\mathbf{v}_1 + c_2\mathbf{v}_2 + \cdots + c_k\mathbf{v}_k). \qquad \triangle$$

1.5.3 Negation of quantifiers

Suppose that we make a universal statement like $p =$"Every living person is shorter than 8 feet." To show that p is false, we simply need to find one example of a person who is 8 feet or taller. By this example, we notice that the negation of this universal statement is, "There is someone who is 8 feet or taller." (At the writing of this book, there are five living people who are 8 feet or taller.) This is the role of a counter-example. This turns out to be a generalized version of De Morgan's law.

> **Proposition 1.5.11 (De Morgan's Law)**
>
> Let S be a set and let $P(x)$ be a predicate about elements of S. Then
>
> $$\neg \forall x \in S \, P(x) \equiv \exists x \in S \, (\neg P(x)) \qquad (1.7)$$
> $$\neg \exists x \in S \, P(x) \equiv \forall x \in S \, (\neg P(x)) \qquad (1.8)$$

Observe that (1.8) follows from (1.7) by considering a double negative and replacing $\neg P(x)$ with $P(x)$.[6]

---

[6]Note that in (1.7) the negation does not affect the universe of discourse for x. We still have $x \in S$.

Since implication statements are ubiquitous in mathematics, it is important to be clear on the negation of a universally quantified implication statement. If $P(x)$ and $Q(x)$ are predicates, then we have the following chain of logical equivalences:

$$\neg \forall x \, (P(x) \to Q(x))$$
$$\equiv \exists x \, (\neg(P(x) \to Q(x))) \quad \text{negation of a quantifier}$$
$$\equiv \exists x \, (\neg(\neg P(x) \vee Q(x))) \quad \text{by (1.4)}$$
$$\equiv \exists x \, (P(x) \wedge \neg Q(x)). \quad \text{De Morgan; double negative}$$

We summarize this in a proposition.

Proposition 1.5.12

Let $P(x)$ and $Q(x)$ be predicates over any universe of discourse. Then $\neg \forall x \, (P(x) \to Q(x)) \equiv \exists x (P(x) \wedge \neg Q(x))$.

Example 1.5.13. As a specific example of Proposition 1.5.12, consider the number theory result, Proposition 4.4.8, which states that if $2^n - 1$ is a prime number, then n is a prime number. We can write this as, $\forall n \in \mathbb{N}(2^n - 1$ is prime $\to n$ is prime). However, the converse to Proposition 4.4.8 false. To prove that the converse is false, we must prove that the following is true:

$$\neg \forall n \in \mathbb{N} \, (n \text{ is prime} \to 2^n - 1 \text{ is prime}).$$

By Proposition 1.5.12, this is equivalent to

$$\exists n \in \mathbb{N} \, (n \text{ is prime} \wedge 2^n - 1 \text{ is not prime}).$$

So we must show that there exists an integer n that is prime and such that $2^n - 1$ is not prime. Finding one instance suffices. Here is one: 11 is prime and $2^{11} - 1 = 2047 = 23 \times 89$ is not prime. $\triangle$

Proposition 1.5.12 shows the role of a *counter-example*. When proving that a universally quantified statement is false, it suffices to find one example in the universe of discourse for which the predicate is false. Proposition 1.5.12 occurs so frequently in mathematical reasoning that authors expect the reader to recognize it.

EXERCISES FOR SECTION 1.5

1. The universe of discourse is $\mathbb{R}$. Let $P(x)$ be the predicate $e^x > x + 3$. Determine the truth values of (a) $P(1)$; (b) $P(2)$; (c) $P(-1)$.

2. The universe of discourse is $C^1(\mathbb{R})$. Let $P(f)$ be the predicate $f' + f = 0$ (as a function). Determine the truth values of (a) $P(f)$, where $f(x) = \cos x$; (b) $P(g)$, where $g(x) = 3e^{-x}$.

3. The universe of discourse is people in your math class. Let $M(x) =$ "x eats meat." Express in English each of the following propositions:
   (a) $\forall x\, M(x)$ (c) $\exists x\, M(x)$
   (b) $\exists x\, (\neg M(x))$ (d) $M(\text{Frank}) \rightarrow \neg M(\text{Corbin})$

4. The universe of discourse is all animals. Let $H(x) =$ "x has fur," $D(x) =$ "x is a dog," and $F(x) =$ "x can fly." Translate into a natural English sentence each of these quantified statements.
   (a) $\forall x\, (D(x) \rightarrow F(x))$ (c) $\exists x\, (H(x) \wedge \neg F(x) \wedge \neg D(x))$
   (b) $\forall x\, (H(x) \vee F(x))$ (d) $(\exists x\, (D(x) \wedge F(x))) \rightarrow (\forall x\, H(x))$

5. The universe of discourse is all cars (registered in some country). Let $S(x) =$ "x is a stick shift," let $W(x) =$ "x has a W16 engine," and let $C(x) =$ "x cost less than \$100,000." Express in natural English each of the following propositions.
   (a) $\exists x\, ((\neg S(x)) \wedge W(x))$;
   (b) $\forall x\, (W(x) \rightarrow C(x))$;
   (c) $(\exists x\, (W(x) \wedge C(x))) \wedge (\forall x\, ((W(x) \wedge \neg S(x)) \rightarrow \neg C(x)))$.

6. Let B be the set of buildings (assuming this is clear). Consider the following predicates: $T(x) =$ "x is over 40 meters tall," $S(x) =$ "x has a steel frame," and $E(x) =$ "x has an elevator." Express the following sentences exclusively using logical operators and quantifiers.
   (a) Every building over 40 meters tall has a steel frame.
   (b) There exists a building with a steel frame that does not have an elevator.
   (c) Every building with a steel frame and an elevator is over 40 meters tall.
   (d) Some building 40 meters or shorter has an elevator.

7. Let the universe of discourse be at your college or university. Translate each of these sentences into logical expressions using predicates, quantifiers, and logical operations.
   (a) Some math major knows Python.
   (b) Everyone who knows Python owns a bicycle.
   (c) There's a math major who owns a bicycle but does not know Python.
   (d) Everyone who owns a bicycle and knows Python is a math major.

8. Let $C(x) =$ "x is in my math class," $S(x) =$ "x is on the swim team," and $E(x) =$ "x exercises daily." Express the following sentences using logical operators and quantifiers. For each of the following statements, first suppose that the universe of discourse is everyone in your math class and repeat the question with the universe of discourse being all people.
   (a) Everyone in my math class is on the swim team.
   (b) Someone in my math class exercises daily but is not on the swim team.

   (c) Everyone in my math class who exercises daily is on the swim
       team.

   (d) No one in my classes is either on the swim team or exercises
       daily.

9. State the negation in natural English (in which you effect De Morgan's law) to the statement, "Every country I visited is in the northern hemisphere and requires driving on the left side of the road."

10. State the negation in natural English (in which you effect De Morgan's law) to the statement, "Someone in my class juggles and does not eat red meat."

11. State the negation in natural English (in which you effect De Morgan's law) to the statement, "Every person I have met has traveled to five other countries or hates traveling."

12. Find a counterexample, if possible, to these universally quantified statements. If instead the statement is true, then prove it.
    (a) $\forall x \in \mathbb{R}\,(x^2 + 2 > 3x)$.
    (b) $\forall x \in \mathbb{R}\,((x > 0) \wedge (x = 1))$.
    (c) $\forall x \in \mathbb{R}\,(|x| + |x - 1| > 0)$.

13. For each of the following propositions, state and prove the truth value.
    (a) $\exists x \in \mathbb{R}\,(x^2 + x + 1 = 0)$.
    (b) $\exists n \in \mathbb{Z}\,((n - 1)/(n + 3) = 5)$.
    (c) $\exists (x, y) \in \mathbb{R}^2\,((x^2 + y^2 < 4) \wedge (y < 1 - (x - 2)^2))$.

14. For each of the following propositions, state and prove the truth value.
    (a) $\forall x \in \mathbb{R}\,(2x \neq x^2)$.
    (b) $\exists n \in \mathbb{N}\,(\sqrt{n^2 + 1} \in \mathbb{N})$.
    (c) $\exists n \in \mathbb{Z}\,(n$ is the sum of two squares $)$.
    (d) $\exists a \in \mathbb{R}\,(x^2 + ax + a^2 = 0$ has a solution in $\mathbb{R})$.

15. In each of the following propositions, exhibit an element that shows the existential quantifier to be true.
    (a) $\exists f \in C^1(\mathbb{R})\,(f' = f)$.
    (b) $\exists f \in C^2(\mathbb{R})\,(f'' = -f)$.
    (c) $\exists f \in C^0(\mathbb{R})\,(\forall x \in \mathbb{R}\,|f(x)| < 0.5)$.

16. In each of the following propositions, exhibit an element that shows the existential quantifier to be true.
    (a) $\exists \mathbf{v} \in \mathbb{R}^3\,(\mathbf{v} \cdot \langle 1, 3, 5 \rangle = 3)$.
    (b) $\exists (c_1, c_2) \in \mathbb{R}^2\,(c_1 \langle 3, -1, 5 \rangle + c_2 \langle 9, 4, 10 \rangle = \langle 7, 0, 10 \rangle)$.
    (c) $\exists A \in M_2(\mathbb{R})\,(A^2 = 0 \wedge A \neq 0)$.

17. For each of the following propositions, decide with explanation if it is true or false.
    (a) $\forall \mathbf{a} \in \mathbb{R}^3\,(\|\mathbf{a} + \langle 2, 3, 6 \rangle\| \leq \|\mathbf{a}\| + 7)$.
    (b) $\forall t \in \mathbb{R}\,(\|\langle 2, 2, 3 \rangle + t\langle 1, -1, -4 \rangle\| > 3)$.

18. Let $P(x)$ and $Q(x)$ be predicates. Consider the two propositions:

$$\forall x\, (P(x) \to Q(x)) \qquad \text{and} \qquad (\forall x\, P(x)) \to (\forall x\, Q(x)).$$

Give a counterexample to show that these are not logically equivalent.

19. Let $P(x)$ and $Q(x)$ be predicates. Show that

$$\exists x (P(x) \vee Q(x)) \equiv (\exists x\, P(x)) \vee (\exists x\, Q(x)).$$

20. Let $P(x)$ and $Q(x)$ be predicates. Show that

$$\forall x (P(x) \wedge Q(x)) \equiv (\forall x\, P(x)) \wedge (\forall x\, Q(x)).$$

1.6 Nested Quantifiers

1.6.1 Multivariable predicates

In the previous section, we introduced predicates and showed how quantifiers applied to predicates create logical propositions. We only considered predicates involving a single variable. However, many mathematical statements and logical expressions involve predicates of more than one variable.

Consider the algebraic expression $2x + 3y = 20$. The reader should recognize this as an equation of a line in plane. We analyze the logic behind this equation. First, because of habits developed in elementary algebra, the reader will likely assume that the variables x and y represent real numbers. A priori x and y could represent anything for which the expression might makes sense, such as complex numbers or real-valued functions over an interval. Hence, the variables are understood to come from a context, usually a set. For this example, we also assume that x and y represent real numbers.

Secondly, the expression $2x + 3y = 20$ will be true for some values of x and y while it may be false for others. Consequently, in itself, the expression is not a proposition. It is a predicate, but a *multi-variable predicate*. If $P(x, y)$ is $2x + 3y = 20$, then, for example, $P(1, 1)$ is false, whereas $P(1, 6)$ is true. Both the expressions $P(1, 1)$ and $P(1, 6)$ are examples of instantiating the predicate.

> **Definition 1.6.1**
>
> A statement $P(x_1, x_2, \ldots, x_n)$ whose truth value depends on the objects $x_1, x_2, \ldots, x_n$ is called a *multi-variable predicate*.

As with single variable predicates, the universe of discourse for a given variable is called its domain. In a multi-variable predicate the domains of the variables may all be the same or all different.

Example 1.6.2. Let $A(x, y)$ be the statement "x has an account with y" where the domain of x is all people and the domain of y consists of all online services. This is a predicate. When Frank refers to a specific person, the instantiation $A(\text{Frank}, \text{Facebook})$ is a proposition with a truth value, either true or false.   △

1.6.2   Nested quantifiers

In Section 1.5, we saw that besides instantiating the predicate, we can create a proposition from a predicate by attaching quantifiers the variable. With multi-variable predicates, we can do the same thing. For example, assuming that x and y are real numbers, then $\forall x \forall y (2x + 3y = 20)$ is a false statement, while $\exists x \exists y (2x + 3y = 20)$ is true.

However, with multi-variable predicates, we can associate a different quantifier to each variable. Furthermore, the order of the quantifiers may matter for the meaning of the resulting logical proposition. Table 1.4 illustrates this using the predicate from Example 1.6.2 and restating in English all eight possible combinations of order and quantifiers associated to the two-variable predicate.

(1) $\forall x \forall y\, A(x, y)$	Every person has an account with every online service.
(2) $\forall x \exists y\, A(x, y)$	Every person has an account with some online service.
(3) $\exists x \forall y\, A(x, y)$	Someone has an account with every online service.
(4) $\exists x \exists y\, A(x, y)$	There exists someone who has an account with an online service.
(5) $\forall y \forall x\, A(x, y)$	For every online service, every person has an account with it.
(6) $\forall y \exists x\, A(x, y)$	For every online service, someone has an account with it.
(7) $\exists y \forall x\, A(x, y)$	There is an online service with which everyone has an account.
(8) $\exists y \exists x\, A(x, y)$	There is an online service with which someone has an account.

Table 1.4: Two nested quantifiers.

First of all, we can tell that (1) $\equiv$ (5) and that (4) $\equiv$ (8). Besides those two equivalences, Table 1.4 gives six different propositions. To illustrate the distinction in roles of contrasting quantifiers compare (2) to (3). Restricting the domain of x to adults in North America, statement (2) could be true; most modern adults have an account with some service with an online presence. On the other hand, statement (3) seems ludicrous. To see that (2) and (6) are different, (2) could be true and (6) false if everyone has an account with some online service, but that a new online service has not yet acquired members. If we compare (2) with (7), we notice that if (2) is false then (7) must also be false, but if (7) is false, it is still possible that (2) be true. Consequently, (2) and (7) are not equivalent either.

When negating a logical proposition that involves nested quantifiers, the negation must work its way all the way through each of the quantifiers. So for example,

$$\neg\,(\forall x \exists y \forall z\, P(x, y, z)) \equiv \exists x \neg\,(\exists y \forall z\, P(x, y, z))$$
$$\equiv \exists x \forall y \neg\,(\forall z\, P(x, y, z)) \equiv \exists x \forall y \exists z\, \neg P(x, y, z).$$

Example 1.6.3. Consider the predicate $T(x, y) =$ "x has taken y" where the domain of x is all students at your college and the domain of y is the set of classes offered at the same college. Let $F(x, z) =$ "x and z are friends" where the domain of z is the same domain as x. (We assume that we can put a clear rule on the concept of friends). We translate a few sentences into quantified statements.

- There is a class that everyone has taken: $\exists y \forall x\, T(x, y)$.

- If any two people are friends then they have taken a class together:
$$\forall x \forall z \exists y\, \big(F(x, z) \rightarrow \big(T(x, y) \wedge T(z, y)\big)\big).$$

- Every student has taken more than one class:
$$\forall x \exists y_1 \exists y_2\, \big(T(x, y_1) \wedge T(x, y_2) \wedge (y_1 \neq y_2)\big).$$

- There is class such that any person who has taken it has also had a friend take it:
$$\exists y \forall x \exists z (T(x, y) \rightarrow (F(x, z) \wedge T(z, y))). \qquad \triangle$$

The above example illustrates how many sentences in natural languages can be phrased precisely using nested quantifiers. As for their role in mathematics, it is impossible to underscore enough how ubiquitous are statements involving nested quantifies throughout mathematics. Even if a particular statement does not specifically use the symbols $\forall$, $\exists$, or $\exists!$, many if not most mathematical theorems involve quantification in some way.

For example, consider the role of quantifiers in various algebra properties. We can properly express the commutativity, associativity, existence of an identity, and existence of inverse of addition as follows. Assume that the domain of the variables x, y, and e is the set of real numbers, then

Commutativity	$\forall x \forall y\, (x + y = y + x)$
Associativity	$\forall x \forall y \forall z\, (x + (y + z) = (x + y) + z)$
Additive Identity	$\exists e \forall x\, (x + e = x)$
Additive Inverses	$\forall x \exists y\, (x + y = e).$

Obviously, the additive identity is $e = 0$. It is important to notice that the order of quantifiers is essential in the last two statements. In particular, if we switched the order in the last statement about additive inverses, the logical statement $\exists y \forall x \, (x + y = 0)$ is not true at all. That is because the additive inverse depends on the element x as opposed to the additive inverse being the same real number for all x.

For stylistic reasons, plenty of authors use a natural language to express these propositions. So for example, we may state the commutativity property by saying, "For all $x, y \in \mathbb{R}$, $x + y = y + x$." Then the words "for all" form the quantifier. As we pointed out earlier, $\forall x \forall y P(x, y) \equiv \forall y \forall x P(x, y)$ with any predicate $P(x, y)$. Consequently, by a slight abuse of the notation, we might also write $\forall x, y \in \mathbb{R}$, $x + y = y + x$. The benefit of this notation is that it includes the domain of the variables without needing to write out the domain in English ahead of the quantified statements.

The reader should now recognize that simply writing "$(a + b)^2 = a^2 + 2ab + b^2$" does not have a truth value; the statement is only a predicate. To properly assert the quadratic binomial formula, we should say

"For all $a, b \in \mathbb{R}$, $(a + b)^2 = a^2 + 2ab + b^2$"; or
"$(a + b)^2 = a^2 + 2ab + b^2$, for all real numbers a and b"; or
$\forall a, b \in \mathbb{R} \, ((a + b)^2 = a^2 + 2ab + b^2)$.

In contrast, consider an equation, such as $x^2 + x + 1 = 0$. As stated, the equation is a predicate; it may be true or false depending on the value of x. If we ask the question whether this equation has a solution, we are determining the truth value of the existential statement $\exists x \, (x^2 + x + 1 = 0)$. However, the careful reader may notice that we are still missing something: we must be clear about the universe of discourse of the variable. Indeed, $\exists x \in \mathbb{R} \, (x^2 + x + 1 = 0)$ is false, whereas $\exists x \in \mathbb{C} \, (x^2 + x + 1 = 0)$ is true.

We point that solving an equation involves more than determining the truth value of an existential statement. "Solve $x^2 + x + 1 = 0$," means in precise terms, "Write $\{x \mid x^2 + x + 1 = 0\}$ in list form." Again, to be precise, we must specify the universe of discourse for the variable x. The solution to $x^2 + x + 1 = 0$ over $\mathbb{R}$ is

$$\{x \in \mathbb{R} \mid x^2 + x + 1 = 0\} = \emptyset$$

whereas over $\mathbb{C}$ it is

$$\{x \in \mathbb{C} \mid x^2 + x + 1 = 0\} = \left\{ \frac{-1 - i\sqrt{3}}{2}, \frac{-1 + i\sqrt{3}}{2} \right\}.$$

The following examples take a closer look at the role of quantifiers in theorems or definitions from prior topics in mathematics.

1.6.3 More examples of quantification in mathematics

Example 1.6.4 (Vector Subspaces). Recall from linear algebra the concept of a subspace of a vector space. Here is the definition of a subspace.

> **Definition.** A nonempty subset W of $\mathbb{R}^n$ (or more generally of a vector space V) is called a *subspace* of $\mathbb{R}^n$ (or V) if (1) for all $\mathbf{u}, \mathbf{v} \in W$, $\mathbf{u} + \mathbf{v} \in W$ and (2) for all $\mathbf{u} \in W$ and all $c \in \mathbb{R}$, $c\mathbf{u} \in W$.

The quantifiers figure in an essential way in the definition. Using the symbols of quantifiers, we can restate (1) and (2) as follows: (1) $\forall \mathbf{u}, \mathbf{v} \in W \, (\mathbf{u} + \mathbf{v} \in W)$; and (2) $\forall \mathbf{u} \in W \, \forall c \in \mathbb{R}, (c\mathbf{u} \in W)$.

We consider two examples with concept of subspace, one involving the proof that some subset is a subspace, and another in which we must prove that a subset fails to be a subspace.

Let $\mathbf{a}$ be a specific vector in $\mathbb{R}^n$ and let $W = \{\mathbf{v} \in \mathbb{R}^n \,|\, \mathbf{a} \cdot \mathbf{v} = 0\}$. We show that W is a subspace of $\mathbb{R}^n$. Note that W is nonempty since $\mathbf{a} \cdot \mathbf{0} = 0$, so the zero vector $\mathbf{0}$ is in W. To prove (1), let $\mathbf{u}, \mathbf{v} \in W$. By definition of W, since $\mathbf{u} \in W$, then $\mathbf{a} \cdot \mathbf{u} = 0$ and since $\mathbf{v} \in W$ we also have $\mathbf{a} \cdot \mathbf{v} = 0$. Then

$$\mathbf{a} \cdot (\mathbf{u} + \mathbf{v}) = \mathbf{a} \cdot \mathbf{u} + \mathbf{a} \cdot \mathbf{v} = 0 + 0 = 0.$$

Thus, $\mathbf{u} + \mathbf{v} \in W$. Now, also let $c \in \mathbb{R}$. Then $\mathbf{a} \cdot (c\mathbf{u}) = c(\mathbf{a} \cdot \mathbf{u}) = c0 = 0$. Thus $c\mathbf{u} \in W$. This establishes that W is a subspace of $\mathbb{R}^n$.

As a second example, we will prove that some subset is not a subspace. To do so, we need to prove the negation of the definition. We can write the definition of a subspace very precisely as

$$(W \neq \emptyset) \wedge (\forall \mathbf{u}, \mathbf{v} \in W, \mathbf{u} + \mathbf{v} \in W) \wedge (\forall \mathbf{u} \in W \, \forall c \in \mathbb{R}, c\mathbf{u} \in W).$$

Using De Morgan's law repeatedly, the negation of the definition of a subspace is

$$(W = \emptyset) \vee (\exists \mathbf{u}, \mathbf{v} \in W, \mathbf{u} + \mathbf{v} \notin W) \vee (\exists \mathbf{u} \in W, \exists c \in \mathbb{R}, c\mathbf{u} \notin W). \tag{1.9}$$

In English, this means W is empty, or W fails to be closed under addition, or W fails to be closed under scalar multiplication.

Let $V = \mathbb{R}^3$ and let W be the solution set to $x^2 + y^2 = z^2$. We will prove this is not a subspace by finding a counter-example. Note that $W \neq \emptyset$ so the $(W = \emptyset)$ part of (1.9) is false. However, if we take $\mathbf{u} = \langle 1, 0, 1 \rangle$ and $\mathbf{v} = \langle 0, 1, 1 \rangle$, then $\mathbf{u}, \mathbf{v} \in W$ but $\mathbf{u} + \mathbf{v} = \langle 1, 1, 2 \rangle$ and $1^2 + 1^2 \neq 2^2$, so $\mathbf{u} + \mathbf{v} \notin W$. Thus these examples of vectors show that $(\exists \mathbf{u}, \mathbf{v} \in W, \mathbf{u} + \mathbf{v} \notin W)$ of (1.9) is true. This in turn makes all of (1.9) true, so we have found a counterexample to the definition of a subspace. Hence, W is not a subspace of $\mathbb{R}^3$. $\triangle$

The first proof in the above example began with the phrase, "Let $\mathbf{u}, \mathbf{v} \in W$." The simple expression "Let" preceding one or more variables means that we consider arbitrary elements from the given set. This simple phrase is a concise formulation often necessary for the proofs that establish the truth of a universally quantified statement.

Example 1.6.5 (Limits of Functions). A foundational concept in calculus is the limit of a function at a point c. A student's introduction to calculus may or may not emphasize or work with the precise definition of a limit. We present it here as an important application of quantifiers.

Let f be a real-valued function defined over an interval $[a, b]$ except possibly at a point $c \in [a, b]$. We say that L is the limit of f as x approaches c and we write

$$\lim_{x \to c} f(x) = L \qquad \text{if}$$

$$\forall \varepsilon > 0 \, \exists \delta > 0 \, \forall x \in [a, b] \, (0 < |x - c| < \delta \to |f(x) - L| < \varepsilon). \quad (1.10)$$

Intuitively, we think of ε as an error value. The definition formalizes the intent that no matter how small of an error (ε) value we set, if x is close enough to c (within δ), then all values $f(x)$ fall within the desired ε of the limit L. This formal definition does not carry any assumption of smallness for ε, simply that it remains positive. However, the point in requiring ε to remain positive is that it *can* get arbitrarily small. △

Example 1.6.6 (Extreme Value Theorem). In a first course in calculus, we encounter the Extreme Value Theorem. Along with the First Derivative Test, this theorem provides the theoretical justification behind optimization.

> **Theorem.** Let f be a continuous function defined over an interval $[a, b]$. Then f attains both a global maximum and a global minimum over the interval.

Using quantifiers and set theory notation, we can write the Extreme Value Theorem as

$$\forall (a, b) \in \{(x, y) \in \mathbb{R}^2 \mid x < y\} \, \forall f \in C^0([a, b])$$
$$(\exists m \in [a, b] \, \exists M \in [a, b] \, \forall x \in [a, b] \, (f(m) \leq f(x) \leq f(M))).$$

A new feature in this example is that the universe of discourse of a variable may depend on the value of the variable that is already bound by a quantifier. so for example, the universe of discourse of the function f depends on the pair (a, b). △

Most readers would agree that this way of stating the theorem feels much harder to parse than using the English. That is why many authors favor phrasing statements in linguistically natural ways. However, it is essential that we can track with the symbolic logic that undergirds mathematical statements.

Example 1.6.7 (Fermat's Last Theorem). According to Fermat's Last Theorem, no three positive integers a, b, and c satisfy the equation $a^n + b^n = c^n$ for any integer value of n greater than 2. Pierre de Fermat, first conjectured this result in 1637, but it remained an unsolved problem until proved by Andrew Wiles in 1994.

The statement of the theorem in English is perfectly accurate and displays a nice flow. We could express it entirely using symbols and quantifiers as

$$\forall n \in \mathbb{N} \setminus \{0, 1, 2\} \,\neg\, (\exists a, b, c \in \mathbb{N}^*, \, a^n + b^n = c^n)$$

or alternative using De Morgan's law,

$$\forall n \in \mathbb{N} \setminus \{0, 1, 2\} \,\forall a, b, c \in \mathbb{N}^* \,\neg (a^n + b^n = c^n).$$

Though either of these compact propositions fully expresses the theorem, it is much more common to express quantifiers with natural linguistic constructions. △

Example 1.6.8 (Euclidean Geometry). As a last example in this section we consider a proposition from Euclidean geometry to make a subtle but important point.

Consider Proposition 9 in Book I of Euclid's *Elements*: "It is possible to bisect a given rectilinear angle" (EE I.9). Referring to the Euclidean plane by $\mathbb{E}^2$ as we did in Example 1.2.13, we may be tempted to write this proposition with quantifiers as

$$\forall A, B, C \in \mathbb{E}^2 \,\exists P \in \mathbb{E}^2 \,(2\angle BAP = \angle BAC).$$

This does *not* express Proposition EE I.9 because this proposition in Euclid's *Elements* pertains to questions of constructibility using a compass and a straightedge, and not to a question of existence. For example, Euclid would probably not deny that given a rectilinear angle $\angle BAC$ there exists an angle whose measure is a third of $\angle BAC$; however, it can be proven that such an angle is not constructible, i.e., that it is impossible to construct such an angle with a compass and straightedge. △

1.6.4 Universal implication and biconditional

In Section 1.1, we already introduced the logical operators of implication $\longrightarrow$ and biconditional $\longleftrightarrow$. We conclude this section by introducing the common symbols of $\implies$ and $\iff$.

Definition 1.6.9

If $P(x, y, \ldots, z)$ and $Q(x, y, \ldots, z)$ are two multi-variable predicates, then

$P(x, y, \ldots, z) \implies Q(x, y, \ldots, z)$ means
$\quad \forall x \forall y \cdots \forall z (P(x, y, \ldots, z) \longrightarrow Q(x, y, \ldots, z))$ is true.
$P(x, y, \ldots, z) \iff Q(x, y, \ldots, z)$ means
$\quad \forall x \forall y \cdots \forall z (P(x, y, \ldots, z) \longleftrightarrow Q(x, y, \ldots, z))$ is true.

In other words, the symbols $\implies$ and $\iff$ respectively mean universally quantified conditional or biconditional statements.

For example, we could shorten the limit definition in (1.10) by writing

$$\forall \varepsilon > 0 \, \exists \delta > 0 \, (0 < |x - c| < \delta \implies |f(x) - L| < \varepsilon).$$

The $\implies$ symbol quantifies the one unbound variable x with a universal quantifier.

Example 1.6.10. We take the opportunity to underscore a difference between $\implies$ and $\iff$ using elementary algebra. We consider the following two algebraic identities involving real numbers:

$$ab = c \text{ and } b \neq 0 \iff a = \frac{c}{b} \tag{1.11}$$

$$a = b \implies a^2 = b^2. \tag{1.12}$$

We propose to solve the algebraic equation $(3x-5)/(x-3) = x+3$. Clearing denominators using (1.11) and then using other algebra rules gives us

$$\frac{3x - 5}{x - 3} = x + 3 \iff 3x - 5 = (x - 3)(x + 3) \text{ and } x - 3 \neq 0$$
$$\iff 3x - 5 = x^2 - 9 \text{ and } x \neq 3$$
$$\iff x^2 - 3x - 4 = 0 \text{ and } x \neq 3$$
$$\iff (x - 4)(x + 1) = 0 \text{ and } x \neq 3$$
$$\iff (x - 4 = 0 \vee x + 1 = 0) \wedge (x \neq 3)$$
$$\iff x = 4 \text{ or } x = -1.$$

We were able to reduce to the last line because if $x = 4$ or if $x = -1$, then $x \neq 3$, so the condition of $x \neq 3$ is redundant. Since we have shown that

$$\frac{3x - 5}{x - 3} = x + 3 \iff x = 4 \text{ or } x = -1,$$

then the equation holds (saying that the equation holds means that the predicate of the equation is true) if and only if $x = 4$ or $x = -1$. We conclude that the roots of the equation are 4 and -1.

In contrast, consider the equation $\sqrt{3x + 4} = x$. First using (1.12), we have

$$\sqrt{3x + 4} = x \implies 3x + 4 = x^2 \implies x^2 - 3x - 4 = 0$$
$$\implies (x - 4)(x + 1) = 0 \implies x - 4 = 0 \lor x + 1 = 0$$
$$\implies x = 4 \text{ or } x = -1.$$

Since at the first step we used an algebraic rule that is an implication and not a biconditional, we cannot immediately say at the conclusion of our reasoning that the roots of equation are 4 and -1. All that we have proved is that if x is a root of the equation, then it must be one of those two values. In this situation, we must check whether these values are in fact roots. If we plug $x = 4$ into the equation, we have $\sqrt{3 \times 4 + 4} = 4$, which is true. Hence, 4 is a root of the equation. If we plug $x = -1$ into the equation, we have $\sqrt{3 \times (-1) + 4} = -1$, which is not true, because the square root function always returns nonnegative values. We conclude that the only root of the equation is $x = 4$.

In (1.12), it would be incorrect to use the biconditional $\iff$ rather than the more restrictive implication $\implies$ because

$$a^2 = b^2 \iff a^2 - b^2 = 0 \iff (a - b)(a + b) = 0$$
$$\iff a - b = 0 \lor a + b = 0 \iff a = b \text{ or } a = -b. \quad \triangle$$

EXERCISES FOR SECTION 1.6

1. Let $R(x, y) =$ "x has read y" where the domain of x is the set of all people and the domain of y is the set of all published books. Express the following quantified statements into natural English.

   (a) $R(\text{Sarah}, \text{To Kill a Mockingbird})$.
   (b) $\forall y \, \neg R(\text{John}, y)$.
   (c) $\neg (\exists x \forall y \, R(x, y))$.
   (d) $\exists y \exists! x \, R(x, y)$.

2. Let $H(x, y) =$ "x said hi to y" and $C(x, y) =$ "x and y are in a class together" where x and y is the universe of discourse of all students at your university. Translate the following quantified statements into natural English.

   (a) $\forall y \, H(\text{Anne}, y)$.
   (b) $\exists x \, (C(x, \text{Corbin}) \land H(\text{David}, x))$.
   (c) $\exists x \forall y \, (C(x, \text{Henry}) \land H(x, y))$.
   (d) $\forall x \forall y \, (H(x, y) \to H(y, x))$.

    (e) $\exists x \forall y \, (C(x,y) \to H(x,y))$.

    (f) $\forall x \exists y \, (C(x,y) \wedge \neg H(y,x))$.

    (g) $\exists x \exists y \, ((x \neq y) \wedge C(x, \text{Zeke}) \wedge C(y, \text{Zeke}) \wedge H(x,y))$.

3. Let $C(x,y,z) = $ "x, y and z are collinear" and let $B(x,y,z) = $ "x is between y and z" where the universe of discourse for all the variables is the set of points in the Euclidean plane. Express the following statements in natural English and state the truth value of each statement.

    (a) $\forall x \forall y \exists z \, C(x,y,z)$.

    (b) $\forall x \forall y \forall z \, (B(x,y,z) \longrightarrow C(x,y,z))$.

    (c) $\exists x \exists y \forall z \, B(x,y,z)$.

4. Let $M(x,y) = $ "x has met y" where the domain of both x and y is the set of all people, and $S(x,y) = $ "x has sent a text to y" with the same domain. Express the following statements using quantifiers.

    (a) Frank has not sent a text to anyone.

    (b) Jasper met Janice but has not sent a text to her.

    (c) There is someone who has met everybody.

    (d) Anyone who has sent a text to someone has met them.

    (e) Someone has sent a text to someone they have not met.

    (f) Meeting someone is commutative.

    (g) Sending texts to someone is not commutative.

5. Let $H(x) = $ "x is happy" and $A(x,y) = $ "x is angry at y" be predicates whose variables are in the universe of discourse of people. Express the following statements using quantifiers.

    (a) Carla is angry at someone.

    (b) No one is angry at themselves.

    (c) Happy people are not angry at anybody.

    (d) There is a happy person who is angry at everybody.

    (e) If someone is happy, then at most one person is angry at them.

    (f) If somebody is angry at everybody, then not everybody is happy.

6. Suppose that the universe of discourse of all variables is the set of reals. Express the following statements with quantifiers and then state the truth value of each statement.

    (a) The product of any two positive numbers is again positive.

    (b) If the product of two numbers is 0, then one of the numbers must be 0.

    (c) It the product of two numbers is 1, then one of the numbers must be 1.

    (d) Every positive number has a power that is less than 1.

7. Determine the truth value of each of the following statements where the domains of the variables nonnegative integers.

    (a) $\exists m \exists n \, (mn = 6)$.

(b) $\forall m \exists n \, (mn = 6)$.

(c) $\exists m \exists n \exists p \, (m^2 + n^2 + p^2 = 30)$.

(d) $\forall m \exists n \, (m + n = 20)$.

8. State using quantifiers and predicates that there exist two distinct subsets of the set $\{1, 2, \ldots, 10\}$ whose entries sum to 43. Show whether this is true or false.

9. The Goldbach Conjecture, one of the oldest unsolved problems in number theory, states that every even integer greater than 2 can be written as the sum of two prime numbers.

   (a) Write the Goldbach Conjecture using quantifiers, clearly specifying universes of discourse for each of the variables.

   (b) State the negation of the Goldbach Conjecture (pushing the negative inside the quantifiers).

10. Lagrange's Four-Square Theorem states that every nonnegative integer can be expressed as the sum of four integer squares. State this theorem using quantifier symbols.

11. State the negation of the definition for a limit in (1.10).

12. The Intermediate Value Theorem states: "If f is a continuous function over an interval $[a, b]$, then it takes on any given value between $f(a)$ and $f(b)$ at some point within the interval." Express the statement of this theorem entirely using quantifiers (without any English words). [Hint: Start with $\forall (a, b) \in \{(x_1, x_2) \in \mathbb{R}^2 \mid x_1 < x_2\} \, \forall f \in C^0([a, b]) \ldots$ Also, you may wish to use $\min(u, v)$ (respectively $\max(u, v)$) as the minimum (resp. maximum) between u and v.]

13. The Mean Value Theorem states: "If f is a continuous function over the interval $[a, b]$ and differentiable over (a, b), then

$$\frac{f(b) - f(a)}{b - a} = f'(c)$$

for some c with $a < c < b$. Express the statement of this theorem entirely using quantifiers. [Hint: See the hint of the previous exercise.]

14. Let f be a function that is continuous over an interval $[a, b]$. Recall that f is said to have a global maximum at $c \in [a, b]$ if $f(c)$ is equal or larger to every other value of the function over $[a, b]$.

   (a) Rewrite the definition of a global maximum in the following way. "A function f has a global maximum at $c \in [a, b]$ if..." and complete with a statement that only involves symbols and quantifiers.

   (b) Complete the following sentence using quantifiers. "A function f does not have a global maximum at $c \in [a, b]$ is..."

15. Let f be a function that is continuous over an interval $[a, b]$. Recall that f is said to have a local maximum at $c \in [a, b]$ if there is some real $\varepsilon > 0$ such that $f(c) \geq f(x)$ for any x with $c - \varepsilon \leq x \leq c + \varepsilon$.

(a) Rewrite the definition of a local maximum in the following way. "A function f has a local maximum at $c \in [a, b]$ if..." and complete with a statement that only involves symbols and quantifiers.

(b) Complete the following sentence using quantifiers. "A function f does not have a local maximum at $c \in [a, b]$ is..."

16. A subset S of $\mathbb{R}$ is said to be *closed under addition* if for all $x, y \in S$, $x + y \in S$. Suppose that S and U are two subsets that are closed under addition. Prove that $S \cap U$ is also closed under addition.

17. Using only quantifiers and symbols, state the claim that if an $n \times n$ matrix commutes with every other $n \times n$ matrix, then it is equal to a scalar multiple of the identity matrix I_n. (Do not prove this.)

18. Recall the definition of a subspace of a vector space in Example 1.6.4. Let V be a vector space and let U and W be two subspaces. Prove that $U \cap W$ is also a subspace of V.

19. Let A be an $n \times n$ matrix. In linear algebra, we encounter the concept of an eigenvalue. A number λ is called an *eigenvalue* of A if there is a nonzero vector $\mathbf{v}$ such that $A\mathbf{v} = \lambda\mathbf{v}$. Restate this definition using quantifiers.

20. Let A be an $n \times n$ matrix. Referring to the previous exercise, express using only quantifiers and predicates that A does not have any real eigenvalues. Write this expression so that there is no $\neg$ symbol in front of any quantifiers.

CHAPTER 2

Arguments and Proofs

In a letter to a friend, Sofia Kovalevskaya (1850-1891) wrote, "Many who have never had an opportunity of knowing any more about mathematics confound it with arithmetic, and consider it an arid science. In reality, however, it is a science which requires a great amount of imagination." The beauty of mathematical imagination is that it rests on the solid foundation of logic.

The introduction to Chapter 1 cast mathematical work in light of Plato's classical definition of knowledge as "justified true belief." In mathematics, the justification always comes in the form of a proof. Less formally, proofs are how we know what we know, how we confirm the reality of our imagination.

This chapter begins with the rules of inference in Section 2.1. Using Boolean logic, these rules allow us to know that the truth of one compound proposition leads to the truth of another. With rules of inference at our disposal, we present a first few proof methods in Sections 2.2 and 2.3. Drawing on examples from familiar topics, we also offer tactics on how to find proofs, and discuss some common logical errors, known as fallacies. Finally, in Section 2.4 we discuss generalized unions and intersections and use proof techniques to prove equalities about such objects.

2.1 Constructing Valid Arguments

2.1.1 Introduction

At the heart of a mathematical proof lies an argument, which takes a collection of premises and establishes a conclusion. Though we will soon define our terms, we underscore a distinctive feature of mathematical thought at the outset.

Mathematical reasoning is primarily *deductive*. Logical deduction starts with a set of accepted premises and argues to specific consequences that follow with certainty from the premises. This type of reasoning contrasts with *inductive* reasoning, in which the conclusion of the argument holds only with some level of confidence. This aspect of mathematical thinking gives the discipline its epistemological strength. The empirical nature of the natural and social sciences by necessity involves inductive reasoning.

2.1.2 Common rules of inference

In Section 1.3, we considered proposition equivalences. Recall that a logical equivalence is a situation in which two compound propositions take on exactly the same truth values regardless of the various values of the constituent atomic propositions.

A more general situation occurs when one compound proposition implies another. This means that whenever the first is true, the second must also be true. The following definition makes this precise.

Definition 2.1.1

Given two compound propositions p and q, if $p \to q$ is a tautology, then we say that q is a logical implication of p. In this case, we write $p \vdash q$ or also $p \implies q$.

The notation $p \vdash q$ is particular to the branch of logic, whereas the universally quantified implication $\implies$ appears in other areas.

Obviously, given two compound propositions, we can always use a truth table to verify whether the implication from one to the other is a tautology. However, some specific tautologies model common steps of reasoning. We call these *rules of inference*. By sequencing rules of inference, we can infer conclusions from a given set of premises. When properly done, such a sequence is called a valid argument. This method of arriving at conclusion forms the core of all mathematical reasoning. Table 2.1 lists common rules of inference.

Logic Symbols	Name of Rule
$(p \wedge q) \vdash p$	Simplification
$p \vdash (p \vee q)$	Addition
$(p \wedge (p \to q)) \vdash q$	Modus Ponens
$(\neg q \wedge (p \to q)) \vdash \neg p$	Modus Tollens
$((p \to q) \wedge (q \to r)) \vdash (p \to r)$	Hypothetical syllogism
$((p \vee q) \wedge \neg p) \vdash q$	Disjunctive syllogism
$((p \vee q) \wedge (\neg p \vee r)) \vdash (q \vee r)$	Resolution

Table 2.1: Rules of inference.

For each rule of inference in Table 2.1, we can check that its corresponding implication statement is a tautology either by constructing the truth table or by using propositional equivalences. For example, the tautology associated to modus ponens is $(p \wedge (p \rightarrow q)) \rightarrow q$, and the truth table for this compound proposition is the following.

p	q	$p \rightarrow q$	$(p \wedge (p \rightarrow q))$	$(p \wedge (p \rightarrow q)) \rightarrow q$
T	T	T	T	T
T	F	F	F	T
F	T	T	F	T
F	F	T	F	T

Only using propositional equivalences, we can establish the same tautology as follows.

$$(p \wedge (p \rightarrow q)) \rightarrow q \equiv \neg(p \wedge (\neg p \vee q)) \vee q \equiv \neg((p \wedge \neg p) \vee (p \wedge q)) \vee q$$
$$\equiv \neg(\mathbf{F} \vee (p \wedge q)) \vee q \equiv \neg(p \wedge q) \vee q$$
$$\equiv \neg p \vee \neg q \vee q \equiv \neg p \vee \mathbf{T} \equiv \mathbf{T}.$$

Example 2.1.2. As a first example, suppose that we have the following premises: "If it snows, then it is overcast" and "It is snowing." Using modus ponens, we conclude that "It is overcast." $\triangle$

Example 2.1.3. Suppose that we begin with these two statements: "If it is sunny, then I will go jogging," and "If I go jogging, then I will get sweaty." Assuming these are true, we can conclude that "If it is sunny, then I will get sweaty." This follows from an application of the Hypothetical syllogism rule. To see this clearly, define $p =$"It is sunny," $q =$"I go jogging," and $r =$"I get sweaty." Our first statement is $p \rightarrow q$, while the second statement is $q \rightarrow r$. The hypothetical syllogism tells us that $((p \rightarrow q) \wedge (q \rightarrow r)) \vdash (p \rightarrow r)$, so assuming the validity of our initial statements, we conclude that $p \rightarrow r$, or that "If it is sunny, then I get sweaty." $\triangle$

Example 2.1.4. Any combination of truth values for p and q that makes the biconditional $p \leftrightarrow q$ true, also makes the implication $p \rightarrow q$ true. We can easily check this. Consequently, $(p \leftrightarrow q) \rightarrow (p \rightarrow q)$ is a tautology, and we can write

$$(p \leftrightarrow q) \vdash (p \rightarrow q).$$

This is a common step of reasoning in logic: if we know an if-and-only-if statement, we can conclude one of the if-then statements, and subsequently use that if-then statement for other rules of inference. In fact, this implication follows from the simplification rule of inference, as we show here:

$$p \leftrightarrow q \equiv (p \rightarrow q) \wedge (q \rightarrow p) \quad \text{by Equation 1.5}$$
$$\vdash (p \rightarrow q) \quad\quad\quad\quad\quad\quad \text{by simplification.2}$$

In this case, the propositions that arise in the simplification are not atomic, but rather compound propositions.   △

There are four rules of inference associated with quantifiers. They all involve the difference between an instantiated predicate and a quantification. Table 2.2 names and defines these rules. In these rules we understand that the element c comes from the universe of discourse for the variable in the quantification statement.

Logic	Name of Rule
$P(c)$ for some $c \vdash (\exists x \, P(x))$	Existential generalization
$(\exists x \, P(x)) \vdash P(c)$ for some c	Existential instantiation
$P(c)$ for arbitrary $c \vdash (\forall x \, P(x))$	Universal generalization
$(\forall x \, P(x)) \vdash P(c)$ for arbitrary c	Universal instantiation

Table 2.2: Rules of inference with quantifiers.

The existential generalization assumes the validity $P(c)$ of one instance of $P(x)$ for a specific c in the universe and concludes that the existential quantifier is true. If the universe of discourse consists of a finite set $\{c_1, c_2, \ldots, c_n\}$, then the existential generalization rule is tantamount to the fact that the following compound proposition is a tautology

$$P(c_1) \to (P(c_1) \vee P(c_2) \vee \cdots \vee P(c_n)).$$

This follows from associativity and commutativity of $\vee$ and a repeated application of the addition rule of inference.

In parallel, the universal instantiation assumes the validity of $\forall x \, P(x)$ and concludes that $P(c)$ is true for any specific c in the universe of discourse. If the universe of discourse were a finite set $\{c_1, c_2, \ldots, c_n\}$, then the universal instantiation rule is tantamount to the following compound proposition being a tautology

$$(P(c_1) \wedge P(c_2) \wedge \cdots \wedge P(c_n)) \to P(c_1).$$

This follows from associativity and commutativity of $\wedge$ and a repeated application of the simplification rule of inference.

The reader might wonder at the value of universal generalization. It seems too straightforward to be useful. However, as we will see, it is a key behind many proofs in mathematics. When we first begin to use it, we will point it out in the margin notes.

Example 2.1.5. Consider the statement, "All trees have bark." This is $\forall x \, B(x)$ where $B(x) =$ "x has bark" and the universe of discourse consists of all trees. We conclude B(Mr. Roger's oak tree).   △

2.1.3   Arguments and argument forms

Let us build on Example 2.1.3 by adding one more proposition, which we hold to be true: "I am not sweaty." Using the symbols from that example, this last proposition is $\neg r$. We can add $\neg r$ to $p \to r$, which we already obtained from a hypothetical syllogism and then use modus tollens to conclude that $\neg p$, i.e., "it is not sunny."

Analyzing how we arrived at our conclusion, we notice that we began with three propositions (1) $p \to q$, (2) $q \to r$, and (3) $\neg r$ and assumed that they were true. From this assumption, we used a few rules of inference to conclude that "it is not sunny." This example illustrates the style of reasoning in mathematics.

We formalize our terminology in order to offer a precise framework for our methods of reasoning.

Definition 2.1.6

> In logic, an *argument* is a finite sequence of propositions. The final proposition is called the *conclusion* of the argument, while all the other propositions are called *premises*. An argument is called *valid* if it is impossible for the conclusion to be false, while all the premises are true. An argument is called *sound* if it is valid and if all its premises are true.

As the reader has surely noticed, in order to analyze compound propositions, we regularly use propositional variables. So we make a distinction between the specific propositions and propositional variables with the following definition.

Definition 2.1.7

> An *argument form* is a sequence of compound propositions $p_1, p_2, \ldots, p_n$ involving propositional variables. An argument form is called *valid* if $(p_1 \wedge p_2 \wedge \cdots \wedge p_{n-1}) \to p_n$ is a tautology.

When creating a valid argument or a valid argument form, some premises follow from rules of inferences involving previous premises. We may call these intermediate conclusions. However, any premise that is not an intermediate conclusion is called a *hypothesis*. A statement of a hypothesis or the statement of a conclusion following from an application of a single rule of inference is called a *step* in an argument or argument form.

Example 2.1.8. We return once more to Example 2.1.3, with the expanded initial list of propositions. Suppose we have the hypotheses: "If it is sunny, then I go jogging," and "If I go jogging, then I get sweaty," and "I did not get sweaty." We can rephrase these with

variables by (1) $p \to q$, (2) $q \to r$, and (3) $\neg r$. We formalize the
process of building a valid argument as follows.

Step	Form	Proposition	Rule of Inference
1.	$p \to q$	If it is sunny, then I go jogging	Hypothesis
2.	$q \to r$	If I go jogging, then I get sweaty	Hypothesis
3.	$p \to r$	If it is sunny, then I get sweaty	Hypothetical syllogism on (1) and (2)
4.	$\neg r$	I did not get sweaty	Hypothesis
5.	$\neg p$	It is not sunny	Modus tollens on (3) and (4)

The argument consists of the list of propositions in the column
labeled Propositions, whereas the argument form is the sequence of
compound propositions in the column labeled Form. There are five
steps in the argument and each step is either a statement of a hypothesis or the result of a rule of inference.   △

It is important to point out a key property about the process ex-
hibited in this example. By adding a new premise only if it follows
as a conclusion of a rule of inference involving or a logical equivalence of previous premises, we know that the truth of the argument's
conclusion follows from the truth of the hypotheses.

When we apply rules of inference to construct a valid argument,
we write the statements out in English. However, for the purposes of
this section, using the propositional variables allows us to focus on the
argument form in order to analyze the validity of the construction.

Example 2.1.9. In logic broadly, the term *syllogism* refers to a form
of reasoning in which a conclusion is drawn from two premises, each
of which shares a term with the conclusion. Various philosophical
texts claim the most famous syllogism as:

> All men are mortal. Socrates is a man. Therefore, Socrates
> is mortal.

Though it is not clear where this argument first appeared, it has
become something of a meme. Here is how we establish the conclusion
using the rules of inference at our disposal. Note first that argument
involves two predicates $P(x) =$ "x is a man" and $M(x) =$ "x is mor-
tal." The hypotheses are $\forall x \, (P(x) \to M(x))$ and $P(\text{Socrates})$. The
following table provides the argument form.

Step	Premise	Rule of Inference
1.	$\forall x \, (P(x) \to M(x))$	Hypothesis
2.	$P(\text{Socrates})$	Hypothesis
3.	$P(\text{Socrates}) \to M(\text{Socrates})$	Universal instantiation on (1)
4.	$M(\text{Socrates})$	Modus ponens with (2) and (4)

Using rules of inference, the conclusion of this famous syllogism follows from two rules of inference. △

Example 2.1.10. None of the rules of inference in Table 2.1 involve the biconditional. However, many steps of reasoning involving the biconditional follow from a simplification step. Consider the following argument.

Step	Premise	Rule
1.	p	Hypothesis
2.	$p \leftrightarrow q$	Hypothesis
3.	$(p \to q) \land (q \to p)$	Equivalence (1.5)
4.	$(p \to q)$	Simplification on (3)
5.	q	Modus ponens on (1) and (4)

Using the symbols we have introduced so far, we could organize the same work in the following way:

$$p \land (p \leftrightarrow q) \equiv p \land ((p \to q) \land (q \to p)) \quad \text{by (1.5)}$$
$$\vdash p \land (p \to q) \quad \text{by simplification}$$
$$\vdash q \quad \text{by modus ponens.} \quad (2.1)$$

This shows the precise steps in extending a modus ponens rule to apply with the biconditional. △

Example 2.1.11. Suppose that we have the three hypotheses.

- "Everyone who studied at the college or came for a conference stayed in a dorm."
- "If someone stayed in the dorm, then the college hired custodial staff."
- "Ben studied at the college."

We can conclude that the college hired custodial staff. The following list of steps constructs a valid argument to support the conclusion.

Define the following predicates and propositions: $S(x) =$"x studied at the college", $C(x) =$"x came for a conference", $D(x) =$"x stayed in a dorm", and $p =$"the college hired custodial staff." Here are the steps for a valid argument.

Step	Premise	Rule
1.	$\forall x \, ((S(x) \lor C(x)) \to D(x))$	Hypothesis
2.	$(\exists x \, D(x)) \to p$	Hypothesis
3.	$S(\text{Ben})$	Hypothesis
4.	$S(\text{Ben}) \lor C(\text{Ben})$	Addition
5.	$(S(\text{Ben}) \lor C(\text{Ben})) \to D(\text{Ben})$	Universal instantiation of (1)
6.	$D(\text{Ben})$	Modus ponens (4) and (5)
7.	$\exists x \, D(x)$	Existential generalization on (6)
8.	p	Modus ponens on (7) and (2)

Note that we could have reordered the steps. We only used the hypothesis $(\exists x\, D(x)) \to p$ right before step (8) so we could have brought in that hypothesis as a step just before it. It is not necessary to list all the hypotheses first. $\triangle$

Mathematical arguments involve previously proven theorems. In such arguments, these theorems play the role of hypotheses. It is usually considered poor writing to list all of the definitions or theorems involved in a argument at the outset of the argument. Instead, we bring them in at the point that they are useful.

Example 2.1.12. Consider these two statements about real numbers: $x, y \in \mathbb{Q} \Rightarrow xy \in \mathbb{Q}$ and $x_0 y_0 \notin \mathbb{Q}$. We can rewrite the first one as $\forall x, y \in \mathbb{R}\, ((x \in \mathbb{Q} \land y \in \mathbb{Q}) \to xy \in \mathbb{Q})$. The second statement refers to some specific elements x_0 and $y_0 \in \mathbb{R}$ with the property that $x_0 y_0 \notin \mathbb{Q}$.

By universal instantiation with the first proposition, we have $((x_0 \in \mathbb{Q} \land y_0 \in \mathbb{Q}) \to x_0 y_0 \in \mathbb{Q})$. Using modus tollens on this with the second hypothesis, we conclude that $\neg(x_0 \in \mathbb{Q} \land y_0 \in \mathbb{Q}) \equiv x_0 \notin \mathbb{Q} \lor y_0 \notin \mathbb{Q}$. $\triangle$

The following example exhibits the use of various rules of inference with quantifiers.

Example 2.1.13. Suppose that (i) $\forall x(P(x) \lor Q(x))$, (ii) $\forall x(R(x) \to \neg Q(x))$, and (iii) $\exists x(\neg P(x))$ are true. We construct a valid argument form to conclude that $\exists x(\neg R(x))$ is true.

Step	Premise	Rule
1.	$\exists x\,(\neg P(x))$	Hypothesis
2.	$\neg P(c)$ for some c	Existential instantiation with (1)
3.	$\forall x(P(x) \lor Q(x))$	Hypothesis
4.	$P(c) \lor Q(c)$	Universal instantiation with (3)
5.	$Q(c)$	Disjunctive syllogism (4) and (2)
6.	$\forall x(R(x) \to \neg Q(x))$	Hypothesis
7.	$R(c) \to \neg Q(c)$	Universal instantiation with (6)
8.	$\neg(\neg Q(c))$	Double negative of (5)
9.	$\neg R(c)$	Modus tollens of (7) and (8)
10.	$\exists x(\neg R(x))$	Existential generalization with (9).

Note that we do not conclude $\forall x(\neg R(x))$ because c is a specific element and not an arbitrary element in the universe of discourse for the variable x. $\triangle$

In this last example, we first provide an argument form to establish a certain conclusion from a set of hypotheses but then show a few ways to analyze the argument's validity if we change the hypotheses.

Example 2.1.14. Consider the statements: "If you are not allowed to drive the car, then the car is not registered or the car is not insured," and "The car is insured but you are not allowed to drive it." We conclude that the car is not registered.

Define atomic propositions $a =$ "You are allowed to drive the car," $r =$ "The car is registered," and $i =$ "The car is insured." The following steps construct a valid argument to support the conclusion.

Step	Proposition	Rule
1.	$\neg a \to (\neg r \vee \neg i)$	Hypothesis
2.	$i \wedge \neg a$	Hypothesis
3.	$\neg a$	Simplification on (2)
4.	$\neg r \vee \neg i$	Modus ponens on (3) and (1)
5.	i	Simplification on (2)
6.	$(\neg r \vee \neg i) \wedge i$	Conjunction of (4) and (5)
7.	$(\neg r \wedge i) \vee (\neg i \wedge i)$	Distributivity on (6)
8.	$(\neg r \wedge i) \vee \mathbf{F}$	Negation law in (7)
9.	$\neg r \wedge i$	Identity law on (8)
10.	$\neg r$	Simplification on (9)

Note that in step (6), we simply stated the conjuction of two previous premises. This is sometimes useful. It is not so much a rule of inference but a matter of organization.

Just as in elementary algebra, we ultimately will involve more than one rule of inference or logical equivalence in one step. For example, in the above argument form, it should already feel natural to go from (7) to (9) in one step, combining two logical equivalences.

Now suppose that we had started with the two hypotheses, "If the car is not registered or the car is not insured, then you are not allowed to drive the car," and "The car is insured but you are not allowed to drive it." Note that the hypotheses are now $(\neg r \vee \neg i) \to \neg a$ and $i \wedge \neg a$. The conclusion of "The car is not registered" does *not* produce a valid argument. Let's see this in three ways.

1) Because these hypotheses might resonate with our regular experience, we can intuit that there may be other reasons as to why "you are not allowed to drive the car." However, this intuition remains vague and could be misleading.

2) If we started to construct a valid argument as in the above table but with the modified hypotheses, it is not readily apparent that a modus ponens would ultimately isolate the conclusion. This intuition might be correct but is still not incontrovertible.

3) In this moderately simple case, we can building a truth table for the proposition that the conclusion follows from the hypotheses, namely

$$((\neg r \vee \neg i) \to \neg a) \wedge (i \wedge \neg a) \to \neg r. \tag{2.2}$$

a	r	i	① $\neg r \vee \neg i$	② ① $\to \neg a$	③ $i \wedge \neg a$	④ ② $\wedge$ ③	④ $\to \neg r$
T	T	T	F	T	F	F	T
T	T	F	T	F	F	F	T
T	F	T	T	F	F	F	T
T	F	F	T	F	F	F	T
F	T	T	F	T	T	T	F
F	T	F	T	T	F	F	T
F	F	T	T	T	T	T	T
F	F	F	T	T	F	F	T

Since the last column is not all **T**, then (2.2) is not a tautology and so the argument form is not valid. In fact, this truth table confirms our intuition that the argument fails precisely when $\neg a \wedge r \wedge i$, which means when you are not allowed to drive the car and the car is registered and it is insured.   △

2.1.4   Proofs in mathematics

Definition 2.1.15

A *proof* of a mathematical statement is a sound argument that establishes the truth of the conclusion from the truth of the stated hypotheses.

At first pass, this definition might appear limited to conditional statements. For example, the statement "6 is an even number" is not a conditional statement. We point out that for any logical proposition p, we have $p \equiv (\mathbf{T} \to p)$. Hence, every statement can be viewed as a conditional. Many philosophers of mathematics call all mathematical truths contingent truths. We can already see this in this simple statement "6 is an even number," because part of the premises that will enter the proof involve the definition of integers, the concept of even, and the number 6.

The term *proof* appears in other areas of human inquiry, and it is important to recognize the differences in meaning based on context. Consider the concept of proof in law. When an attorney claims, "We will prove beyond reasonable doubt that [such and such]," he or she makes no claims to a deductive reasoning. Instead, their notion of proof involves an inductive argument that may involve varying levels of certainty. Furthermore, the concept of "reasonable doubt" pertains to the strength of conclusions that a thinking (reasonable) person might draw based on physical or circumstantial evidence.

Consider also the phrase: "The precession of Foucault's pendulum proves that the Earth rotates." The term "prove" does not involve a proof in the mathematical sense. The argument is inductive. It may involve some differential equations and vector algebra, but the complete argument from physics relies on the Newtonian theory of mechanics and various other assumptions, which are themselves supported by empirical evidence.

Within mathematics, a fundamental application of statistics is that of hypothesis testing. Without describing the theory, sometimes scientists using hypothesis testing might say that a "small p-value proves the null hypothesis." This is technically an abuse of language because a small value of the statistic p does not prove the null hypothesis but rather gives a measure to the certainty of the null hypothesis. So statisticians will say, "reject the null hypothesis" or instead "retain the null hypothesis." In fact, scientists may choose different values (usually 0.1, 0.05 or even 0.01) as what constitutes a small p-value, thereby shifting the threshold for rejection or retention.

2.1.5 Fallacies

We emphasized above that an argument will be valid if it involves only (a) hypotheses and (b) premises obtained by (i) conjunctions of previous premises, (ii) logical equivalences to previous premises, and (iii) rules of inference involving previous premises. The concept of "skipping steps" means to compress a few of these steps into one. After providing a valid argument to derive a conclusion from two hypotheses, Example 2.1.14 gives an example of a *fallacy*.

In logic, the term *fallacy* refers to faulty reasoning in the construction of an argument. We can subdivide fallacies into two types: formal and informal. A *formal fallacy* resembles a familiar rule of inference but is in fact not a tautology. The term *informal fallacy* applies to an error in reasoning that does not arise from an invalid argument.

We list a few common fallacies with rudimentary examples.

The argument that involves $p \to q$ and q as premises and concludes p is invalid. More precisely, $((p \to q) \land q) \to p$ is not a tautology. The error of treating this compound proposition as a tautology is called *affirming the conclusion*. For example, consider the following argument:

> If you are allowed to travel overseas, then you have a passport. You have a passport. Therefore, you are allowed to travel overseas.

This is not a valid argument because there could be other requirements for being allowed to travel overseas besides just having a passport. This erroneous reasoning is an example of affirming the conclusion. We could view it as a garbled modus ponens.

Similarly, consider the argument:

> If you are allowed to travel overseas, then you have a passport. You are not allowed to travel overseas. Therefore, you do not have a passport.

Again, this is an invalid argument because $((p \to q) \wedge \neg p) \to \neg q$ is not a tautology. To treat this compound proposition as a tautology is the fallacy of *denying the hypothesis*.

Both affirming the conclusion and denying the hypothesis are formal fallacies because they incorrectly treat a contingency as a tautology. Though we will consider these more in depth in later sections, we mention two informal fallacies that are particularly common in incorrect proofs.

Faulty generalization refers to assuming the truth of a universally quantified statement based on an incomplete sampling. From a purely logical perspective, we could label this fallacy as saying that $(\exists x \, P(x)) \to (\forall x \, P(x))$ is not a tautology. Though we will discuss prime numbers with precision in Section 4.4, we mention the following example already. Consider this statement: $\forall n \in \mathbb{N} \, (n^2 + n + 41$ is prime). If our instinct is to start trying a number of examples, we might suspect (conjecture) the statement to be true; it turns out that $n^2 + n + 41$ is prime for $n \in \{0, 1, 2 \dots, 40\}$. However, if $n = 41$, then $n^2 + n + 41$ is not prime. A preponderance of examples does not prove a universally quantified statement, unless we check every possible option.

Some unsound arguments arise from a fallacy called *circular reasoning*. This occurs when a premise in an argument is equivalent to the conclusion. By the definition of validity, an argument involving circular reasoning may still be valid. If our argument only involves hypotheses and premises obtained from rules of inference or logical equivalences, then circular reasoning is a situation where the conclusion is a hypothesis of the argument.

Consider the following erroneous proof that $\sqrt{2}$ is an irrational number.

> Since there do not exist integers $m, n \in \mathbb{N}^*$ such that $2 = m^2/n^2$, then by taking square roots, there do not exist integers m, n such that $\sqrt{2} = m/n$. Hence, $\sqrt{2}$ is irrational.

Though it is true that $\sqrt{2}$ is an irrational number, this is not a correct proof. The statement "since there do not exist integers $m, n \in \mathbb{N}^*$ such that $2 = m^2/n^2$" is equivalent to the conclusion that we are asked to prove. The reasoning is fallacious because it is circular.

Appendix A provides a rubric for assessing the quality of a proof. Within this appendix, after providing a rubric for scoring logic, Section A.1 lists many of the fallacies with some further problems in the logic of proof construction.

EXERCISES FOR SECTION 2.1

1. What rule of inference is involved in each of these arguments?
    (a) If Frank is over 40, he will buy a motorcycle. Frank did not buy a motorcycle. Hence, Frank is not over 40.
    (b) If April gets tired, then she will drink a cup of coffee. If April drinks a cup of coffee, then she gets the jitters. If April gets tired, then she gets the jitters.
    (c) Corbin is home or he is at school. Corbin is not home or he is out jogging. Hence, Corbin is at school or out jogging.

2. What rule of inference is involved in each of these arguments?
    (a) If someone read the book, then we can have a discussion. Someone read the book. Hence, we can have a discussion.
    (b) My car is white or is a stick shift. My car is not white. Hence, my car is a stick shift.
    (c) I am running a temperature and I have a stomach ache. Hence, I have a stomach ache.

3. From the set of stated hypotheses, draw a relevant conclusion and label all the rules of inference or logical equivalences used to arrive at the conclusion.
    (a) If Fred smells the back of his fridge, then he will pass out or throw up. If Fred throws up, then he will rush to the bathroom. Fred did not rush to the bathroom but he did smell the back of his fridge.
    (b) In order to get the car registered, I must get it inspected. In order to get the car inspected, I have to drive to a testing site. I did not drive to a testing site.
    (c) If I exercise or work outside, then I take a shower. I exercised or slept in. I worked outside or did not sleep in.

4. From the set of stated hypotheses, draw a relevant conclusion and label all the rules of inference or logical equivalences used to arrive at the conclusion.
    (a) If Jasmine drives 300 miles or gets tired, then she will stop driving. If Jasmine goes to visit Beth, then she will drive 300 miles. Jasmine did not stop driving.
    (b) If I spend over $30,000 on my car, then my car will have a spoiler and a racing stripe. If I do not spend $30,000 on my car, then my car won't have a spoiler and will be red. My car does not have a spoiler.
    (c) If Jack eats late or drinks coffee late, then he doesn't fall asleep quickly. If Jack has to work late, then he eats late. Jack feel asleep quickly.

5. From the set of stated hypotheses, draw a relevant conclusion and label all the rules of inference or logical equivalences used to arrive at the conclusion.

    (a) Every English major likes to read. Everyone in my class likes to read or likes to watch movies. No one in my class is an English major.

    (b) Joe only eats sweets. Sweets are not healthy. Pork rinds are not healthy. Carrots are healthy. Tandi does not like Joe.

    (c) Everyone who visits Paris sees the Eiffel Tower. Everyone who visits Rome sees the Colosseum. Everyone in the class has visited Paris or Rome. Pauline has not seen the Colosseum.

6. From the set of stated hypotheses, draw a relevant conclusion and label all the rules of inference or logical equivalences used to arrive at the conclusion.

    (a) Every full-time student likes at least one course they have taken. No full-time student likes the course Painful 101. Frank liked the course Painful 101.

    (b) Lester murdered Jester or Chester murdered Jester. People who are happy don't murder people. People who exercise every day are happy. Lester exercises every day.

    (c) Every bird has feathers. Every creature that can fly is an insect or a bird. A penguin has wings. Every insect has six legs. A mute swan does not have six legs but can fly.

7. Locate and describe the fallacy in drawing the conclusion from the hypotheses:

    (a) Hypotheses: If I have enough, then I will be happy. If I make a lot of money, then I will have enough. I do not make a lot of money. Conclusion: I will not be happy.

    (b) Hypotheses: If I study and do all the exercises, then I will pass the course. I did not do all the exercises. Conclusion: I will not pass the course.

    (c) Hypotheses: Every mathematics major has taken a linear algebra course. Anyone who understands eigenvalues has taken a linear algebra course. Joe is a mathematics major. Conclusion: Joe understands eigenvalues.

8. Suppose that we have the following premises: (1) $p \to (q \lor r)$; (2) $s \to \neg r$; and (3) $p \land s$. Provide an argument form to conclude that q is true.

9. Suppose that we have the following premises: (1) $p \to q$; (2) $q \to r$; (3) $r \to s$; (4) $s \to p$. Provide an argument form to conclude that $p \leftrightarrow q$.

10. Suppose that we have the following premises: (1) $(p \land q) \to (r \lor s)$; (2) $t \to (p \land \neg s \land u)$; (3) $u \to q$. Provide an argument form to conclude that $t \to r$.

11. Suppose that we have the following premises: (1) $\forall x \, (P(x) \lor \neg Q(x))$; (2) $\exists x \, R(x)$; and (3) $\forall x \, (R(x) \to Q(x))$. Conclude: $\exists x \, P(x)$.

12. From the given hypotheses, give an argument that arrives at the conclusion, and label all the rules of inference or logical equivalences used.

   (a) Hypotheses: Every integer greater than 1 is composite or prime. 42 is not prime. Conclusion: 42 is composite.

   (b) Hypotheses: For all reals x, y, and z, if $x \leq y$ and $y \leq z$, then $x \leq z$. $1 \leq 3$. The number a satisfies $a > 3$. Conclude: $a \geq 1$.

   (c) Hypotheses: A real number a is a critical point of a function f if and only if f is not defined at a or $f'(a) = 0$. If $f(a)$ is an extreme value of f, then a is a critical point of f. Furthermore, $f(a)$ is an extreme value of f and f is defined at a. Conclude: $f'(a) = 0$.

13. From the given hypotheses, give an argument that arrives at the conclusion, and label all the rules of inference or logical equivalences used.

   (a) Hypotheses: If f and g are any continuous functions over $\mathbb{R}$, then fg is continuous over $\mathbb{R}$. The function y with $y(x) = \sin x$ is continuous over $\mathbb{R}$. The function h with $h(x) = x^2$ is continuous over $\mathbb{R}$. Conclude: The function F with $F(x) = x^2 \sin x$ is continuous over $\mathbb{R}$.

   (b) Hypotheses: $\forall x, y, z \in \mathbb{R} \, (x \leq y \longrightarrow x + z \leq y + z)$. $\forall x, y \in \mathbb{R}, \forall c \in \mathbb{R}^{>0} \, (x \leq y \longrightarrow cx \leq cy)$. $2a + 3 \leq 15$. Conclude: $a \leq 6$.

   (c) Hypotheses: For any positive integer n, if A is a nilpotent $n \times n$ matrix, then $I + A$ is invertible. (This is true, but you are not asked to prove it.) For any positive integer n, if $B \in M_n(\mathbb{R})$ is invertible, then $\det(B) \neq 0$. Also $\det \begin{pmatrix} 4 & 5 \\ 1 & 2 \end{pmatrix}$. Conclude that $\begin{pmatrix} 3 & 5 \\ 1 & 1 \end{pmatrix}$ is not nilpotent.

14. Label the fallacy involved in each of the invalid arguments below. (It is not necessary to know anything about the mathematical objects in question.)

   (a) If n is a real number such that $n > 0$, then $n^2 > 0$. Suppose that a has $a^2 > 0$. Hence, $a > 0$.

   (b) $\forall y \in \mathbb{R} \exists x \in \mathbb{R} \, (x < y)$. Hence, $\exists x \in \mathbb{R} \, (x < 2)$. Therefore $2 < 2$ and so $\exists x \in \mathbb{R} \, (x < x)$.

15. Label the fallacy involved in each of the invalid arguments below.

   (a) If the series $\sum_{n=1}^{\infty} a_n$ converges, then $\lim_{n \to \infty} a_n = 0$. With the sequence $a_n = 1/n^2$ for $n \geq 1$ we have $\lim_{n \to \infty} \frac{1}{n^2} = 0$. Hence, $\sum_{n=1}^{\infty} \frac{1}{n^2}$ converges. [In this case, the conclusion is true but argument is not valid.]

   (b) If a matrix B satisfies $B = A^\top A$ for some matrix A, then B is symmetric. (Spectral Theorem:) If B is symmetric, then B is diagonalizable. $B \neq A^\top A$ for all matrices A. Hence, B is not diagonalizable.

2.2   First Proof Strategies

In the previous section, we presented the concept of a logically valid argument. We can now develop the concept of a proof. After defining a few terms related to proofs and theorems, we present a first few proof methods.

The labels we put on different proof methods or strategies serve as guides. They give examples of what approaches might avail themselves as effective. For many situations, more than one method will work. The student of mathematics might find some strategies easier to grasp or faster to apply in a given situation but should become adept with all of these approaches.

The examples in the next two sections draw primarily from calculus, linear algebra, and set theory. We offer here three precise definitions for familiar concepts that we will use occasionally for various elementary examples.

Definition 2.2.1

A integer n is called *even* if there exists an integer k such that $n = 2k$. An integer is called *odd* if there exists an integer k such that $n = 2k + 1$.

Definition 2.2.2

A real number α is called *rational* if $\alpha = p/q$, where p and q are integers with $q \neq 0$. A real number that is not rational is called *irrational*.

Definition 2.2.3

Let x be a real number. The *floor* of x, denoted $\lfloor x \rfloor$, is the largest integer n less than or equal to x. More precisely,

$$\lfloor x \rfloor = n \iff n \leq x < n + 1 \text{ where } n \in \mathbb{Z}.$$

2.2.1   Formal versus informal proofs

When writing a mathematical proof, labeling the rules of inference that we use at each stage is not unlike using training wheels to learn how to ride a bicycle. The manner in which we began to construct valid arguments in the previous section is called a *formal proof*. A formal proof clearly delineates the source of every premise, whether a hypothesis of the theorem or statement whose truth was already established prior, and it labels the rules of inference or logical equivalences.

In contrast, an *informal proof* might skip some steps, not explicitly cite every theorem used, or not label the rules of inference. In practice, the application of rules of inference from earlier premises usually hides under connective words like 'thus', 'therefore', 'consequently', 'hence' and many other common expressions.

The following example of a proof from calculus contrasts the typical informal proof with a formal proof.

Example 2.2.4. As an example of the principle we just mentioned, we propose to show that the sign function

$$\text{sign}(x) = \begin{cases} -1 & \text{if } x < 0 \\ 0 & \text{if } x = 0 \\ 1 & \text{if } x > 0 \end{cases}$$

is discontinuous at 0. We give a proof with its usual concision and then analyze the steps of logic.

Proof. As x approaches 0, the left- and right-hand limits are

$$\lim_{x \to 0^+} \text{sign}(x) = \lim_{x \to 0^+} 1 = 1 \text{ and } \lim_{x \to 0^-} \text{sign}(x) = \lim_{x \to 0^-} -1 = -1.$$

Since these two are not equal, then $\lim_{x \to 0} \text{sign}(x)$ does not exist. Consequently, $\text{sign}(x)$ is not continuous at 0. However, we know from a theorem in calculus that if a function f is differentiable at a point a, then f is continuous at a. Consequently, since $\text{sign}(x)$ is not continuous at 0, we deduce that $\text{sign}(x)$ is not differentiable at 0. □

This is a typical proof with a fairly standard efficiency. It involves some theorems and definitions implicitly and explicitly states others. The following list provides and labels all the steps for a formal proof. The reader should notice that every step involves a definition or theorem as a premise, a logical equivalence of a previous premise, or a rule of inference involving one or more previous premises.

1) Definition of function: We point out that $\lim_{x \to 0^+} \text{sign}(x) = \lim_{x \to 0^+} 1$, and expect the reader to recognize this because $f(x) = 1$ for all $x > 0$.

2) Theorem, not cited: $\lim_{x \to a^+} c = c$ and $\lim_{x \to a^-} c = c$.

3) Universal instantiation of (2) with $a = 0$ and $c = 1$ and simplification: $\lim_{x \to 0^+} 1 = 1$.

4) Transitivity of equality: $\lim_{x \to 0^+} \text{sign}(x) = 1$.

5) Definition of function: $\lim_{x \to 0^-} \text{sign}(x) = \lim_{x \to 0^-} -1$.

6) Universal instantiation of (2) and simplification: $\lim_{x \to 0^-} -1 = -1$.

7) Transitivity of equality: $\lim_{x \to 0^-} \text{sign}(x) = -1$.

8) Theorem, not cited: For all functions f and real numbers a, the limit $\lim_{x \to a} f(x)$ exists if and only if $\lim_{x \to a^+} f(x)$ and $\lim_{x \to a^-} f(x)$ exist and are equal.

9) Universal instantiation of (8) with $f = \text{sign}$ and $a = 0$: The limit $\lim_{x \to 0} \text{sign}(x)$ exists if and only if $\lim_{x \to 0^+} \text{sign}(x)$ exists and $\lim_{x \to 0^-} \text{sign}(x)$ exists and they are equal.

10) Simplification of (9) using (2.1): If $\lim_{x \to 0} \text{sign}(x)$ exists, then the limits $\lim_{x \to 0^+} \text{sign}(x)$ and $\lim_{x \to 0^-} \text{sign}(x)$ also exist, and $\lim_{x \to 0^+} \text{sign}(x) = \lim_{x \to 0^-} \text{sign}(x)$.

11) (4) and (7): $\lim_{x \to 0^+} \text{sign}(x) \neq \lim_{x \to 0^-} \text{sign}(x)$.

12) Addition on (10): We have $\lim_{x \to 0^+} \text{sign}(x) \neq \lim_{x \to 0^-} \text{sign}(x)$ or $\lim_{x \to 0^+} \text{sign}(x)$ does not exist or $\lim_{x \to 0^-} \text{sign}(x)$ does not exist.

13) De Morgan's law on (12): It is not true that $\lim_{x \to 0^+} \text{sign}(x)$ exists and $\lim_{x \to 0^-} \text{sign}(x)$ exists and that $\lim_{x \to 0^+} \text{sign}(x) = \lim_{x \to 0^-} \text{sign}(x)$.

14) Modus tollens on (10) with (13): $\lim_{x \to 0} \text{sign}(x)$ does not exist.

15) Definition, not stated: A function f is continuous at a if and only if the limit $\lim_{x \to a} f(x)$ exists and $\lim_{x \to a} f(x) = f(a)$.

16) Universal instantiation of (15): The function sign is continuous at 0 if and only if $\lim_{x \to 0} \text{sign}(x)$ exists and $\lim_{x \to 0} \text{sign}(x) = 0$.

17) Addition with (14): $\lim_{x \to 0} \text{sign}(x)$ does not exist or we have $\lim_{x \to 0} \text{sign}(x) \neq \text{sign}(0)$.

18) De Morgan's law on (17): It is not the case that, $\lim_{x \to 0} \text{sign}(x)$ exists and $\lim_{x \to 0} \text{sign}(x) = \text{sign}(0)$.

19) Modus tollens on (18) and (16): $\text{sign}(x)$ is not continuous at 0.

20) Theorem, stated: If a function f is differentiable at a point a, then f is continuous at a.

21) Universal instantiation of (20): If $\text{sign}(x)$ is differentiable at 0, then $\text{sign}(x)$ is continuous at 0.

22) Modus tollens on (21) and (19): $\text{sign}(x)$ is not differentiable at 0. $\triangle$

The ideal level of informality for a proof depends on the audience. For someone writing a proof in analysis, the above proof in paragraph form should suffice. For a mathematical journal read by professionals, the author might shorten the proof further. For someone interacting with or writing to an audience with less developed proof skills, the writer might wish to add a few more details, like restate the definition of continuity or cite the theorems used in steps (2) and (8).

2.2.2 Terms concerning mathematical statements

Mathematical writing is unique in that we often encapsulate statements with labels and names. The reader might have heard of the Intermediate Value Theorem, the Rank-Nullity Theorem, Fermat's Last Theorem, the Riemann Hypothesis, or the Collatz Conjecture. In most books or articles, the author will number the theorems, propositions, definitions and so on in order to refer to them easily as premises in later proofs, or to use them in examples.

Speaking technically, we call a *theorem* any mathematical statement whose truth has been established through a proof. In practice, however, when we give names to theorems, we tend to call something "Theorem a.b" or "Laplace's Theorem" if it is a singularly important result. Otherwise, we often label a theorem as "Proposition y" if it does not quite meet the level of importance of a Theorem. A *lemma* is a theorem that, serves primarily as a stepping stone to a Proposition or Theorem that involves a long proof. The term *corollary* refers to a theorem whose proof is usually short and follows fairly quickly from a Theorem or Proposition.

Commonly, people who work in an area begin to formulate guesses about what might be true before they can find a proof. A *conjecture* refers to a yet unproven mathematical statement for which there exists support in the form of numerical evidence or intuition from mathematicians who are well versed in the particular topic. Though the term hypothesis plays a specific role when describing a $p \to q$ statement, if we talk about "The So-and-So Hypothesis," the term *hypothesis* is synonymous with conjecture.

2.2.3 Direct proofs

In a *direct proof* of a conditional statement $p \to q$, we assume that p is true, use rules of inference, when needed include as premises other propositions already proved, and conclude that q must be true. This strategy establishes the first two rows of the truth table for $p \to q$, namely

p	q	$p \to q$
T	T	T
T	F	F

Implied in this strategy is that if the hypothesis p is false, then whether q is **T** or **F**, the conditional $p \to q$ is true.

Example 2.2.5. As a first example of a direct proof, consider the following proposition: The product of two odd integers is again an odd integer. Here is a proof: Let a and b be two odd integers. By Definition 2.2.1, there exist $m, n \in \mathbb{Z}$ such that $a = 2m + 1$ and

$b = 2n + 1$. Then

$$ab = (2m + 1)(2n + 1) = 4mn + 2m + 2n + 1$$
$$= 2(2mn + m + n) + 1.$$

Since $k = 2mn + m + n \in \mathbb{Z}$ and $ab = 2k + 1$, then by definition ab is an odd integer. $\triangle$

The statement of the proposition is a universally quantified statement. It is common to begin proofs of universal statements with the phrase, "Let a and b be [odd integers]" and replace [odd integers] with whatever is contextually appropriate. (For example, see below the first sentence in the proof of Proposition 2.2.6.) This phrase indicates the genericity of the object. Proving the desired result for these arbitrary odd integers gives us the universally quantified statement via universal generalization.

We also point out that we used the common phrase, "by definition." This phrase should only be used when restating for reference a specific definition. In this proof, the reader will notice that we first provided a reference for the definition we cited; the second time we used the phrase, we did not provide the citation because we had recently used that same definition. It is an essential part of writing proofs to know precise definitions of terms.

As a second example of a direct proof, we provide a proof for the familiar addition rule for derivatives.

Proposition 2.2.6 (Addition Rule)

If $f(x)$ and $g(x)$ are functions that are differentiable at c, then the addition function $(f + g)(x)$ is differentiable at c and $(f + g)'(c) = f'(c) + g'(c)$.

Proof. Let $f(x)$ and $g(x)$ be two functions that are differentiable at c. Then the limit rules give us

$$\lim_{h \to 0} \frac{(f + g)(c + h) - (f + g)(c)}{h}$$
$$= \lim_{h \to 0} \frac{f(c + h) + g(c + h) - (f(c) + g(c))}{h}$$
$$= \lim_{h \to 0} \left(\frac{f(c + h) - f(c)}{h} + \frac{g(c + h) - g(c)}{h} \right).$$

By the addition rule for limits, this becomes

$$\left(\lim_{h \to 0} \frac{f(c + h) - f(c)}{h} \right) + \left(\lim_{h \to 0} \frac{g(c + h) - g(c)}{h} \right) = f'(c) + g'(c),$$

where the last equality follows by the assumption that $f(x)$ and $g(x)$ are differentiable at c. Since this limit exists, $(f + g)(x)$ is differentiable at c and furthermore, $(f + g)'(c) = f'(c) + g'(c)$. □

As another example of a direct proof, we revisit a proof of the quadratic formula.

Proposition 2.2.7 (Quadratic Formula)

Given constants $a, b, c \in \mathbb{C}$ with $a \neq 0$, the roots of the equation $ax^2 + bx + c = 0$ are

$$x = \frac{-b \pm \sqrt{b^2 - 4ac}}{2a}.$$

Proof. Let $a, b, c \in \mathbb{C}$ with $a \neq 0$. For all $x \in \mathbb{C}$, we have

$ax^2 + bx + c = 0$ and $a \neq 0$

$\iff x^2 + \dfrac{b}{a}x + \dfrac{c}{a} = 0$ dividing by a

$\iff x^2 + \dfrac{b}{a}x + \dfrac{b^2}{4a^2} - \dfrac{b^2}{4a^2} + \dfrac{c}{a} = 0$ adding $\dfrac{b^2}{4a^2} - \dfrac{b^2}{4a^2} = 0$

$\iff \left(x + \dfrac{b}{2a}\right)^2 - \dfrac{b^2}{4a^2} + \dfrac{c}{a} = 0$ by the Binomial Formula

$\iff \left(x + \dfrac{b}{2a}\right)^2 - \dfrac{b^2 - 4ac}{4a^2} = 0$ adding fractions.

Factoring the difference of squares, this is equivalent to

$$\left(x + \frac{b}{2a} + \frac{\sqrt{b^2 - 4ac}}{2a}\right)\left(x + \frac{b}{2a} - \frac{\sqrt{b^2 - 4ac}}{2a}\right) = 0.$$

Finally, we know that for all $u, v \in \mathbb{C}$, we have $uv = 0$ if and only if $u = 0$ or $v = 0$. Therefore,

$$x + \frac{b}{2a} + \frac{\sqrt{b^2 - 4ac}}{2a} = 0 \text{ or } x + \frac{b}{2a} - \frac{\sqrt{b^2 - 4ac}}{2a} = 0,$$

from which we deduce that

$$x = -\frac{b}{2a} - \frac{\sqrt{b^2 - 4ac}}{2a} \text{ or } x = -\frac{b}{2a} + \frac{\sqrt{b^2 - 4ac}}{2a}.$$

Using the $\pm$ symbol, the proposition follows. □

Since the first part of the proof involves a sequence of biconditionals, one could be tempted to view that portion as a chain of (two-way) hypothetical syllogisms. However, the whole algebraic proof is a sequence of modus ponens. A formal proof would draw this out. In the lines that are explicitly connected by the $\Longleftrightarrow$ symbol, the algebra rules (theorems) stated to the right are brought in as premises in the argument. In the second half of the proof, we varied the format of presentation and mentioned the relevant algebra rules ahead of their use.

The next proposition is an exercise from linear algebra. Recall that an $n \times n$ matrix $A = (a_{ij})$ is upper triangular if $a_{ij} = 0$ whenever $i > j$.

Proposition 2.2.8

The product of two $n \times n$ upper triangular matrices is again upper triangular.

Before we give a proof, we point out that this statement is a conditional statement. We can write this as

$$\forall A, B \in M_n(\mathbb{R}) \, (A \text{ and } B \text{ are upper triangular}$$
$$\longrightarrow AB \text{ is upper triangular}).$$

Proof. Let A and B be two upper triangular $n \times n$ matrices. By definition, whenever $i > j$, we have $a_{ij} = 0$ and $b_{ij} = 0$. Let $C = AB$. Then by definition of multiplication, if $C = (c_{ij})$, we have

$$c_{ij} = \sum_{k=1}^{n} a_{ik} b_{kj}.$$

Suppose that $i > j$. Then,

$$c_{ij} = \sum_{k=1}^{i-1} a_{ik} b_{kj} + \sum_{k=i}^{n} a_{ik} b_{kj} \qquad (2.3)$$

For $1 \leq k \leq i - 1$, we have $i > k$ so $a_{ik} = 0$. For $i \leq k \leq n$, since $j < i$, we have $j < k$ so $b_{kj} = 0$. Thus, when $i > j$, we have

$$c_{ij} = \sum_{k=1}^{i-1} 0 \cdot b_{kj} + \sum_{k=i}^{n} a_{ik} \cdot 0 = 0.$$

Hence, we conclude that C is upper triangular. $\square$

We call a proof *indirect* if it is any type of proof that is not direct. The next two subsections present two styles of indirect proofs – proofs by contrapositive and proofs by contradiction.

2.2.4 Proofs by contrapositive

A proof by contrapositive also applies to proving a condition statement $p \to q$ but is a direct proof of the contrapositive $\neg q \to \neg p$. Since the contrapositive of a conditional statement is equivalent to the original conditional statement, proving $\neg q \to \neg p$ proves $p \to q$.

Our first example involving a proof by contrapositive is a simple result about factoring integers. This proposition plays an interesting role when we approach the topic of prime factorization in Section 4.4.

Proposition 2.2.9

Suppose that n is a positive integer such that $n = ab$ with $a, b \in \mathbb{Z}$ with $a, b > 1$, Then $a \leq \sqrt{n}$ or $b \leq \sqrt{n}$.

Proof. We prove the contrapositive. Let n be a positive integer and suppose that a and b are integers with $a > \sqrt{n}$ and $b > \sqrt{n}$. Then taking the product of these two inequalities of positive reals, $ab > n$, which implies that $ab \neq n$. The proposition follows. □

The above proof is concise. The signal "We prove the contrapositive" signals to the reader our proof strategy. A direct proof of the contrapositive involves supposing the negation of the conclusion, which, by De Morgan's law, is that $a > \sqrt{n}$ and $b > \sqrt{n}$. From this supposition, we give a direct proof of the negation of the hypothesis, i.e., that $n \neq ab$.

We draw the next example of a proof by contrapositive from linear algebra.

Example 2.2.10. Let $T : V \to W$ be a linear transformation from the vector space V to the vector space W. Let $\mathbf{v}_1, \mathbf{v}_2, \ldots, \mathbf{v}_k \in V$. We prove that if

$$\{T(\mathbf{v}_1), T(\mathbf{v}_2), \ldots, T(\mathbf{v}_k)\}$$

is a linearly independent set in W, then $\{\mathbf{v}_1, \mathbf{v}_2, \ldots, \mathbf{v}_k\}$ is also linearly independent.

We prove the contrapositive. Suppose that $\{\mathbf{v}_1, \mathbf{v}_2, \ldots, \mathbf{v}_k\}$ is linearly dependent. By definition, there exist $c_1, c_2, \ldots, c_k \in \mathbb{R}$ not all zero such that

$$c_1\mathbf{v}_1 + c_2\mathbf{v}_2 + \cdots + c_k\mathbf{v}_k = \mathbf{0}.$$

Applying the linear transformation to both sides and using the linear properties, we have

$$T(c_1\mathbf{v}_1 + c_2\mathbf{v}_2 + \cdots + c_k\mathbf{v}_k) = T(\mathbf{0})$$
$$\implies c_1T(\mathbf{v}_1) + c_2T(\mathbf{v}_2) + \cdots + c_kT(\mathbf{v}_k) = \mathbf{0}.$$

Consequently, $\{T(\mathbf{v}_1), T(\mathbf{v}_2), \ldots, T(\mathbf{v}_k)\}$ is linearly dependent. The result follows by the contrapositive. △

2.2.5   Proofs by contradiction

In a proof by contradiction of a statement p, we assume that p is false and then show that this leads to a contradictory statement. The logic behind this strategy is that we prove the truth of the compound statement $\neg p \rightarrow \mathbf{F}$. However, this compound statement is only true when $\neg p \equiv \mathbf{F}$, which means that $p \equiv \mathbf{T}$.

Some authors signal the beginning of a proof by contradiction by bluntly saying, "We prove this by contradiction." This may be helpful but is not necessary. In this textbook, the authors take the habit of signaling the beginning of a proof by contradiction by saying, "Assume that...". Anywhere the reader sees this beginning, he or she can expect that the proof will establish a contradiction.

Proposition 2.2.11

The number $\sqrt{2}$ is irrational.

Proof. Assume that $\sqrt{2}$ is rational. Then we can write $\sqrt{2} = p/q$, where the fraction p/q is expressed in least form. Squaring both sides and multiplying by q^2, we deduce that $2q^2 = p^2$. Since p^2 is 2 times an integer, then p^2 is even. We deduce that p also must be even because if p were odd, then by Example 2.2.5, p^2 would need to be odd. Since p is even, $p = 2m$ for some integer m. Hence, we have $2q^2 = 4m^2$. Dividing by 2, we deduce that $q^2 = 2m^2$.

By a similar reasoning, we deduce that q^2 and therefore that q is even. Hence $q = 2n$ for some integer n. Then we can write

$$\sqrt{2} = \frac{p}{q} = \frac{2m}{2n} = \frac{m}{n}.$$

But this contradicts the statement that the fraction p/q was expressed in least form. Hence, we conclude that $\sqrt{2}$ is irrational.   □

Our next example of a proof by contradiction is the Arithmetic Mean - Geometric Mean (AM-GM) Theorem. This innocuous theorem about nonnegative real numbers has profound consequences, applications and generalizations.

Theorem 2.2.12 (AM–GM Inequality)

For all nonnegative real x and y,

$$\sqrt{xy} \leq \frac{x+y}{2}.$$

Proof. Let x and y be nonnegative reals. Assume the contrary, namely that $\sqrt{xy} > (x+y)/2$. Then $2\sqrt{xy} > (x+y)$. Since x and y are nonnegative, squaring this expression keeps the inequality in the same direction. Then,

$$4xy > (x+y)^2 \implies 4xy > x^2 + 2xy + y^2$$
$$\implies 0 > x^2 - 2xy + y^2 \implies 0 > (x-y)^2.$$

This is a contradiction: The square of any real number is nonnegative. The theorem follows. □

Example 2.2.13. We draw a third example of a proof by contradiction from calculus. Suppose that $f(x)$ is a function defined and differentiable over some interval $[a, b]$ such that $|f'(x)| < 1$ for all $x \in [a, b]$. We prove that $|f(x_2) - f(x_1)| < |x_2 - x_1|$ for all $x_1, x_2 \in [a, b]$ with $x_1 \neq x_2$.

Assume the contrary, namely that there exist distinct $x_1, x_2 \in [a, b]$ with $|f(x_2) - f(x_1)| \geq |x_2 - x_1|$. By the Mean Value Theorem there exists $c \in [\min(x_1, x_2), \max(x_1, x_2)]$ such that

$$f'(c) = \frac{f(x_2) - f(x_1)}{x_2 - x_1}.$$

Since, $|f(x_2) - f(x_1)| \geq |x_2 - x_1|$, we deduce that $|f'(c)| \geq 1$. This contradicts the hypothesis that $|f'(x)| < 1$ for all $x \in [a, b]$. Hence, we conclude that $|f(x_2) - f(x_1)| < |x_2 - x_1|$ for all distinct $x_1, x_2 \in [a, b]$. △

2.2.6 Quod erat demonstrandum

As a matter of style, some authors prefer to make clear the beginning and the end of a proof. This habit serves a variety of roles. In particular, this form of bookending helps the reader distinguish between the expository writing that develops intuition, examples that illustrate a phenomenon, or the definitions themselves. (Subsection 9.2.3 discusses the history behind this habit.) As the reader will have noticed by now, in this book we end our proofs with a right-justified symbol □. Though less common among authors, this book also follows the habit of ending an example with the symbol △.

Exercises for Section 2.2

1. Use a direct proof to show that the product of two rational numbers is a rational number.

2. Prove that the sum of a rational number and an irrational number is irrational.

3. Prove that the product of two perfect square integers is again a perfect square.

4. Use a direct proof to show that if $x, y \in (0, 1)$ with $x < y$, then $\dfrac{x}{1-x} < \dfrac{y}{1-y}$.

5. Use a direct proof without calculus to show that for all $x, y \in \mathbb{R}$, if $x < y$ then $2^x < 2^y$.

6. (*) Suppose that m and n are two integers that can be written as the sum of two squares (of integers). Prove that mn can be written as the sum of two squares.

7. Suppose $n \in \mathbb{Z}$. Show that if $n^2 - 4n + 7$ is even, then n is odd. Use a proof by contrapositive.

8. Let $x, y \in \mathbb{R}$. Prove that if $x + y \geq 10$, then $x \geq 4$ or $y \geq 6$.

9. Let $a, b \in \mathbb{Z}$. Prove that if $a^3(b^2 - 2b)$ is odd, then a and b are odd.

10. Prove that if m and n are integers with $mn = 1$, then either $m = n = 1$ or $m = n = -1$.

11. Prove that there exist no integers a and b such that $14a + 30b = 1$.

12. Prove that $\sqrt[3]{2}$ is irrational.

13. Suppose that $a, b, c \in \mathbb{N}$ with $a^2 + b^2 = c^2$. Prove that a or b is even.

14. Prove that $(x + y)\left(\dfrac{1}{x} + \dfrac{1}{y}\right) > 4$ for all positive and distinct reals x, y.

15. The *harmonic mean* of two positive reals x_1 and x_2 is the number m such that
$$\frac{1}{m} = \frac{1}{2}\left(\frac{1}{x_1} + \frac{1}{x_2}\right).$$
Prove that for all positive real numbers x_1 and x_2, the harmonic mean is less than or equal to the geometric mean.

16. Prove that $|\tan^{-1} x - \tan^{-1} y| \leq |x - y|$ for all $x, y \in \mathbb{R}$.

17. Prove that $x^5 + 5x^3 + 15x + 10$ has exactly one real root. [Hint: Use a contradiction and Rolle's Theorem.]

18. Prove the following result form calculus using a proof by contradiction and the Mean Value Theorem. If f is a function defined on the interval (a, b) with $f'(x) = 0$ for all $x \in (a, b)$, then f is a constant function.

19. A nonzero matrix A is called a *zero divisor* if there is a nonzero matrix B such that $AB = 0$ or $BA = 0$. Prove by contradiction that a zero divisor matrix is not invertible.

20. Prove the claim in Exercise 1.6.17.

21. Show that matrix multiplication is distributive over addition. More precisely, let n be a positive integer and let $A, B, C \in M_{n \times n}(\mathbb{R})$. Prove that $(A + B)C = AC + BC$ and that $C(A + B) = CA + CB$.

22. Recall that a $n \times n$ matrix A is called *orthogonal* if $A^\top A = I$. Prove that the product of two orthogonal matrices is again orthogonal.

23. In geometry of the plane, a subset S of the plane is called *convex* if for all $p, q \in S$ the line segment $\overline{pq}$ connecting p and q is a subset of S. Prove that the intersection of two convex sets is a convex set.

24. (*) Prove that $|e^{-a^2} - e^{-b^2}| \leq \sqrt{\dfrac{2}{e}} |a - b|$ for all $a, b \in \mathbb{R}$. [Hint: Show that if $f(x) = e^{-x^2}$, then $|f'(x)| \leq \sqrt{2/e}$ for all $x \in \mathbb{R}$.]

25. (*) Prove that $2^x \geq x^2$ for all $x \geq 4$. [Hint: Study the shape of $f(x) = 2^x - x^2$ for $x \geq 4$.]

2.3 Proof Strategies

This section continues the introduction of proof methods begun in the previous section. In addition to some additional methods, we also mention a few standard time-saving strategies and conclude with typical ways to get ideas on how to find a proof.

2.3.1 Proof by cases

A *proof by cases* is a proof of a universal statement that subdivides the universal set into a small number of subsets (cases) and provides a proof for each case separately. The adjective "small" is a value judgment. A proof involving 20 cases, the proofs of which may seem similar, might be valid but, unless the approach were absolutely necessary, would seem inelegant by most mathematicians.

Example 2.3.1. We prove that $\lfloor 2x \rfloor = \lfloor x \rfloor + \lfloor x + 1/2 \rfloor$ for all $x \in \mathbb{R}$. Let $x \in \mathbb{R}$ and call $n = \lfloor x \rfloor$. We consider two cases:

Case 1: $n \leq x < n + 1/2$. Multiplying by 2, we have $2n \leq 2x < 2n + 1$. Hence, by definition of the floor function, $\lfloor 2x \rfloor = 2n$. Adding $1/2$ to the inequality, we have $n + 1/2 \leq x + 1/2 < n + 1$. Since this implies that $n \leq x + 1/2 < n + 1$, we deduce that $\lfloor x + 1/2 \rfloor = n$. Hence, $\lfloor x \rfloor + \lfloor x + 1/2 \rfloor = n + n = 2n = \lfloor 2x \rfloor$.

Case 2: $n + 1/2 \leq x < n + 1$. Multiplying by 2 we have $2n + 1 \leq 2x < 2n + 2$. Then by definition of the floor $\lfloor 2x \rfloor = 2n + 1$. Adding $1/2$ to the inequality, we have $n + 1 \leq x + 1/2 < n + 3/2$. Since this implies that $n + 1 \leq x + 1/2 < n + 2$, we deduce that $\lfloor x + 1/2 \rfloor = n + 1$. Hence, $\lfloor x \rfloor + \lfloor x + 1/2 \rfloor = n + n + 1 = 2n + 1 = \lfloor 2x \rfloor$.

The formula holds in both cases. Consequently, it holds universally. $\triangle$

The strategy of breaking a proof or a portion of a proof into cases is quite common. Writers do not always signal the cases as clearly as we have in the above example with the bold text. Nonetheless, quality exposition should make the cases clear enough.

2.3.2 Existence proofs

Some theorems in mathematics affirm the existence of some object with a specified property. Such theorems come in two different flavors: constructive and nonconstructive. A *constructive proof* either explicitly determines or provides a process with a finite number of steps that determines the object in question. In contrast, a *nonconstructive proof* establishes the existence of an object without providing a construction.

Example 2.3.2. Is there a matrix A such that $A^2 = \begin{pmatrix} 1 & 1 \\ 0 & 1 \end{pmatrix}$? Consider matrices of the form

$$B = \begin{pmatrix} 1 & x \\ 0 & 1 \end{pmatrix},$$

where x is any real number. We notice that

$$B^2 = \begin{pmatrix} 1 & x \\ 0 & 1 \end{pmatrix} \begin{pmatrix} 1 & x \\ 0 & 1 \end{pmatrix} = \begin{pmatrix} 1 & 2x \\ 0 & 1 \end{pmatrix}.$$

Hence, we deduce that

$$\begin{pmatrix} 1 & \frac{1}{2} \\ 0 & 1 \end{pmatrix}^2 = \begin{pmatrix} 1 & 1 \\ 0 & 1 \end{pmatrix}.$$

This is a constructive proof to the existence problem since we provided an actual solution.   △

Example 2.3.3. We say that a nonnegative integer n can be written as a sum of 2 squares in k different ways ignoring signs and order if there exists k pairs $(a, b) \in \mathbb{N}$ such that $a \leq b$ and $n = a^2 + b^2$. For example, 25 can be written as a sum of 2 squares in two different ways ignoring signs and order, namely $0^2 + 5^2$ and $3^2 + 4^2$. Consider the following question: Does there exist a nonnegative integer that is the sum of 2 squares in 4 different ways, ignoring signs and order?

Yes, there is. The integer 1105 works because

$$1105 = 4^2 + 33^2 = 9^2 + 32^2 = 12^2 + 31^2 = 23^2 + 24^2.$$

That's all we need for a proof. This is a constructive proof. The reader may (and should) ask, "How did we find this example?". In this case, a computer program running a for loop. From a theoretical perspective, this feels unsatisfactory because this approach does not exhibit any underlying patterns as to why 1105 might work.   △

As an example of a nonconstructive existence proof, consider the following application of the Intermediate Value Theorem (IVT).

Example 2.3.4. We show that the cubic equation $x^3 + x + 1 = 0$ has a root. By limit theorems, since the function $f(x) = x^3 + x + 1$ is a polynomial, it is continuous over $\mathbb{R}$. Clearly, $f(-1) = -1$ and $f(0) = 1$. By the Intermediate Value Theorem, since $-1 < 0 < 1$ there exists $c \in (-1, 0)$ such that $f(c) = 0$. This value c is a root of $x^3 + x + 1 = 0$.

This proof for the existence of a root to the cubic equation neither calculates explicitly nor offers a method to calculate the root. Consequently, the proof that we gave is a nonconstructive proof. The reader might object and point out that our proof involved citing the IVT, and speculate that maybe there might exist a constructive proof for the IVT. The usual proofs for the IVT given in analysis are not constructive, so that is not the case.   △

2.3.3   Uniqueness proofs

Some theorems in mathematics affirm that if an object with a given property exists, then there must exist only one such object. The strategy for a uniqueness proof is to assume the two objects with the desired property exist, and then show that they must be equal.

Example 2.3.5. In Example 2.3.4 we showed that $x^3 + x + 1 = 0$ has a solution over $\mathbb{R}$. In this example, we show that this solution is unique.

Suppose that x_1 and x_2 are real roots of the equation. Then we have

$$
\begin{aligned}
(x_1^3 + x_1 + 1) - (x_2^3 + x_2 + 1) &= 0 \\
\iff x_1^3 - x_2^3 + (x_1 - x_2) &= 0 \\
\iff (x_1 - x_2)(x_1^2 + x_1 x_2 + x_2^2) + (x_1 - x_2) &= 0 \\
\iff (x_1 - x_2)(x_1^2 + x_1 x_2 + x_2^2 + 1) &= 0. \qquad (2.4)
\end{aligned}
$$

This equation implies that $x_1 = x_2$ or $x_1^2 + x_1 x_2 + x_2^2 + 1 = 0$. Solving for x_1 in terms of x_2 in this quadratic equation we get

$$
x_1 = \frac{-x_2 \pm \sqrt{x_2^2 - 4(x_2^2 + 1)}}{2} = \frac{-x_2 \pm \sqrt{-3x_2^2 - 4}}{2}
$$

However, since $x_2 \in \mathbb{R}$, then $x_2^2 \geq 0$ so $-3x_2^2 \leq 0$ and $-3x^2 - 4 < 0$. Hence, there is no real solution for $x_1 \neq x_2$ if there is a solution for

x_2. Consequently, the only possibility for (2.4) is that $x_1 = x_2$. Thus, the real solution to $x^3 + x + 1 = 0$ is unique.   △

Example 2.3.6. We show that there exists a unique differentiable function $f(x)$ such that $f'(x) = 3x^2$ for all $x \in \mathbb{R}$ and also satisfying $f(1) = 17$. We notice that $f(x) = x^3 + 16$ fits the desired property. (This is the existence part.) Suppose that $f(x)$ and $g(x)$ both satisfy this property. Then $g'(x) - f'(x) = \frac{d}{dx}(g(x) - f(x)) = 0$. This implies that $g(x) - f(x)$ is a constant function. However, since $g(1) - f(1) = 0$, then $g(x) - f(x) = 0$ for all $x \in \mathbb{R}$. Hence, $f(x) = g(x)$ for all $x \in \mathbb{R}$. This shows that there exists a unique function with the desired property.

The statement, "This implies that $g(x) - f(x)$ is a constant function," should not have surprised anyone acquainted with calculus. As stated, it is sufficient for most readers. However, we could have chosen to be more explicit by saying, "By the Mean Value Theorem, this implies that $g(x) - f(x)$ is a constant function." We could provide even more detail by showing how the Mean Value Theorem leads to this result: "For all $a, b \in \mathbb{R}$, by the Mean Value Theorem,

$$\frac{(g(b) - f(b)) - (g(a) - f(a))}{b - a} = g'(c) - f'(c)$$

for some $c \in [a, b]$. Since $g'(c) - f'(c) = 0$, we deduce that $g(b) - f(b) = g(a) - f(a)$. Hence $g(x) - f(x)$ is a constant function."   △

2.3.4 Backward and forward

A proof must start somewhere. It is not always easy to find a good beginning point. Some of the methods and strategies that we have discussed so far offer ideas on how to start in certain cases.

As we work on a proof, we may discover some pieces of the puzzle. From a stylistic perspective, simply listing various observations without flow does not constitute a valid proof. On the other hand, a logical beginning may feel far from the issues at hand in the proposition we are proving. Consequently, it is not uncommon for a proof of a conditional statement $p \to q$ to work on results near the hypothesis, then work on results near the conclusion, and finally fill in the middle. A proof may therefore involve some backwards and forwards movements in the reasoning.

In a first course in calculus, we introduce limit theorems as the foundation for derivative rules. However, many calculus courses introduce the definition of the limit and give a few useful limit theorems, but never prove these theorems nor work directly with the limit definition. The following example involves a limit proof from the definition and illustrates backwards and forwards proof strategy.

Example 2.3.7. We prove that $\lim_{x\to 2} \frac{1}{x} = \frac{1}{2}$ from the limit definition (see Example 1.6.5). To do so means that we need to prove that for all $\varepsilon > 0$, there exists $\delta > 0$ such that $|x - 2| < \delta$ implies $|1/x - 1/2| < \varepsilon$.

Let $\varepsilon > 0$. (Since we are proving a statement with $\forall \varepsilon$, we begin with "Let $\varepsilon > 0$.") Note that $|x-2| < \delta$ if and only if $2-\delta < x < 2+\delta$. Assuming that $\delta < 2$ so that $2 - \delta > 0$, we can take the inverse of this inequality, which involves changing the order. This is equivalent to

$$\frac{1}{2+\delta} < \frac{1}{x} < \frac{1}{2-\delta} \iff \frac{1}{2+\delta} - \frac{1}{2} < \frac{1}{x} - \frac{1}{2} < \frac{1}{2-\delta} - \frac{1}{2}$$

$$\iff -\frac{\delta}{2(2+\delta)} < \frac{1}{x} - \frac{1}{2} < \frac{\delta}{2(2-\delta)}.$$

Since $0 < 2 - \delta < 2 + \delta$, we recognize that $1/(2+\delta) < 1/(2-\delta)$ and so $\delta/(2(2+\delta)) < \delta/(2(2-\delta))$. Hence, we have shown that

$$|x - 2| < \delta \implies \frac{-\delta}{2(2-\delta)} < \frac{1}{x} - \frac{1}{2} < \frac{\delta}{2(2-\delta)}$$

$$\implies \left|\frac{1}{x} - \frac{1}{2}\right| < \frac{\delta}{2(2-\delta)}.$$

So if it is possible to find δ such that $\delta/(2(2-\delta)) < \varepsilon$, then we are done. Note that since $1 + 2\varepsilon > 0$, we have

$$\delta/(2(2-\delta)) < \varepsilon \iff \delta < 4\varepsilon - 2\delta\varepsilon \iff \delta + 2\varepsilon\delta < 4\varepsilon$$

$$\iff \delta(1 + 2\varepsilon) < 4\varepsilon \iff \delta < \frac{4\varepsilon}{1 + 2\varepsilon}.$$

We deduce that for a δ satisfying if $0 < \delta < \min(2, 4\varepsilon/(1 + 2\varepsilon))$,

$$|x - 2| < \delta \implies \left|\frac{1}{x} - \frac{1}{2}\right| < \frac{\delta}{2(2-\delta)} \implies \left|\frac{1}{x} - \frac{1}{2}\right| < \varepsilon.$$

This proves the existence of a δ that satisfies the requirements of the limit definition. We conclude that $\lim_{x\to 2} \frac{1}{x} = \frac{1}{2}$.

The phrase "If [such and such] is true, then we are done" is not uncommon for a backward and forward strategy. We started from $|x - 2| < \delta$ and found an equivalent statement involving $|1/x - 1/2|$. Then we went to the end of the proof, recognizing it would be useful to have $\delta/(2(2-\delta)) < \varepsilon$, so we decided to prove it then. $\triangle$

2.3.5 A few proof shortcuts

Concision and efficiency are valued characteristics of a proof. There are a few acceptable ways to shorten a proof.

The phrase "without loss of generality," informally abbreviated to WLOG, signals a choice about an object or variable that is not the general case but could be easily modified and perhaps repeated to cover all situations. (Section 10.1.2 contrasts the informal and formal habits in mathematical style.)

Example 2.3.8. The limit

$$\lim_{x \to 0} \frac{\sin x}{x} = 1$$

plays a key role in proving the differentiation formulas for trigonometric functions. We will not repeat a proof of this limit, but only point out a useful application of the shortcut. A valid proof could start with the following. "Note that $f(x) = (\sin x)/x$ is an even function. Without loss of generality, suppose that $0 < x < \pi/2$." There are two things at work in this: (a) Since we are trying to calculate a limit as $x \to 0$, what happens far away from 0 is irrelevant. (b) Since $f(x)$ is an even function, $f(-x) = f(x)$ so $\lim_{x \to 0^-} f(x) = \lim_{x \to 0^+} f(x)$, which means we only need to calculate the limit from above 0.

This restriction is valuable because the standard proofs for this limit utilize the geometry of the unit circle in the first quadrant. When $0 < x < \pi/2$, the variable x can represent the angle of a ray in the first quadrant.   △

As another example of a proof involving the WLOG signal, we prove the following proposition from linear algebra:

Proposition 2.3.9

> A set of vectors $\{\mathbf{v}_1, \mathbf{v}_2, \dots, \mathbf{v}_k\}$ in a vector space V is linearly dependent if and only if one of the vectors is a linear combination of the others.

$(\Longrightarrow)$ *Proof.* Suppose[1] that $\{\mathbf{v}_1, \mathbf{v}_2, \dots, \mathbf{v}_k\}$ is a linearly dependent set. Hence, there exists $c_1, c_2, \dots, c_k \in \mathbb{R}$, not all zero, such that

$$c_1 \mathbf{v}_1 + c_2 \mathbf{v}_2 + \cdots + c_k \mathbf{v}_k = \mathbf{0}.$$

Without loss of generality, suppose that $c_1 \neq 0$. Then

$$c_1 \mathbf{v}_1 = -c_2 \mathbf{v}_2 - \cdots - c_k \mathbf{v}_k \implies \mathbf{v}_1 = -\frac{c_2}{c_1} \mathbf{v}_2 - \cdots - \frac{c_k}{c_1} \mathbf{v}_k.$$

So $\mathbf{v}_1$ is a linear combination of the others.

---

[1] We will occasionally use the symbol $(\Longrightarrow)$ in the margin as an aide to the reader to signal the start of the proof of the forward implication in the proof of a biconditional statement. The symbol $(\Longleftarrow)$ signals the start of the proof for the converse. Some authors include this in the text, but this habit is informal and the prose should be clear enough not to need these signals.

Conversely, suppose that one of the vectors is a linear combination $(\Longleftarrow)$
of the others. Without loss of generality, suppose that $\mathbf{v}_1$ is a linear
combination of $\{\mathbf{v}_2, \ldots, \mathbf{v}_k\}$. Then $\mathbf{v}_1 = d_2\mathbf{v}_2 + \cdots + d_k\mathbf{v}_k$. Hence

$$\mathbf{v}_1 - d_2\mathbf{v}_2 - \cdots - d_k\mathbf{v}_k = \mathbf{0}.$$

Since $(c_1, c_2, \ldots, c_k) = (1, -d_2, \ldots, -d_k)$, we have c_1 so not all c_i are
zero. Thus $\{\mathbf{v}_1, \mathbf{v}_2, \ldots, \mathbf{v}_k\}$ is a linearly dependent set.   □

In this proof, we used the WLOG signal once in each direction.
First, in the forward ($\Longrightarrow$) direction, we do have that not all c_i coefficients are zero. In general, c_1 could be 0. All we can really say
exactly is that $\exists i \in \{1, 2, \ldots, k\}$, $c_i \neq 0$. However, we could renumber
the vectors in the list so that the coefficient in front of $\mathbf{v}_1$ is nonzero.
Second, in the other direction, we again used the shortcut to our advantage. Here is how we would have to write the rest of the proof
without the shortcut: "Let $i \in \{1, 2, \ldots, k\}$ such that

$$\mathbf{v}_i = d_1\mathbf{v}_1 + \cdots + d_{i-1}\mathbf{v}_{i-1} + d_{i+1}\mathbf{v}_{i+1} + \cdots + d_k\mathbf{v}_k.$$

Then

$$d_1\mathbf{v}_1 + \cdots + d_{i-1}\mathbf{v}_{i-1} - \mathbf{v}_i + d_{i+1}\mathbf{v}_{i+1} + \cdots + d_k\mathbf{v}_k = \mathbf{0},$$

which gives a trivial linear combination of coefficients that are not
all zero. Hence, $\{\mathbf{v}_1, \mathbf{v}_2, \ldots, \mathbf{v}_k\}$ is linearly dependent." Admittedly,
this is not particularly difficult, but taking advantage of the WLOG
shortcut simplifies the presentation.

A second common shortcut involves the Latin phrase *mutatis mutandis* which means "having changed what needs to be changed."
Sometimes mathematicians use this phrase to shorten a proof when
it requires two cases, both of which are very similar. The author who
uses this shortcut will work out the first case completely and then
say, "*Mutatis mutandis* for the second case." The notion of similarity between cases is a judgment call on the author's part, but the
differences between the two cases should be obvious.

Example 2.3.10. A fundamental theorem in linear algebra is that
if a vector space V has a basis with n elements in it, then every basis
has n elements. Again, we do not give a full proof here, but we sketch
the steps of the proof:

1) Let $\mathcal{B}_1$ be a basis of V consisting of n elements and let $\mathcal{B}_2$ be
   another basis, consisting of m elements.

2) Assume that $m > n$. The creative part of the proof enters at
   this stage and leads to a contradiction. (We skip this.)

3) Assume that $m < n$. At this stage we only need to say, "*Mutatis
   mutandis* with the previous case, we again arrive at a contradiction." We can summarize this case so quickly because the

work done in the first case (step (2)) would be repeated with very minor changes to arrive to the contradiction.

4) Since $m > n$ and $m < n$ both lead to contradictions, conclude that $m = n$. $\triangle$

2.3.6   How to start finding proofs

There are certain areas of elementary mathematical work that primarily involve carefully following various rules. For example, computing derivatives of functions simply requires a precise application of differentiation rules. Similarly, using Gauss-Jordan elimination on the augmented matrix of a system of linear equations and then correctly interpreting the resulting reduced row echelon form simply requires a careful application of rules. These examples illustrate the algorithmic side of mathematical work.

There is no universal strategy to prove or disprove any given mathematical statement. One of the challenges of advanced mathematics is that the work becomes less and less algorithmic the further one advances. Proving mathematical statements requires creativity. Though practice improves one's intuition and experience in an area helps one's insight, it still takes some creativity to find valid proofs. Nonetheless, there are a few common things to try.

- Modify an Example or Other Proof: When taking a course, many exercises share some similarities with proofs or examples in the text. Consequently, studying examples or proofs may help with ideas for the exercises. This phenomenon may generalize to research in which some new results arise from modifying or extending earlier results.

- Try a Few Examples: Sometimes, running examples with the objects in question allows us to discern patterns. Studying these patterns may help us see the underlying phenomena. From these, we may be able to identify the general situation. The patterns generally do not constitute a proof, so simply claiming "This pattern continues" is a form of faulty generalization.

Though it may take considerable work to find a proof of a statement, we do not include in the proof all of the ideas, mental movements, dead ends, or examples that led up to our proof. For expository purposes and to aid reader comprehension, we may include an example before the statement of a theorem. However, to narrate one's thought process like "First we thought such and such because such and such, but then we realized that could not work..." has no place in a proof.

2.3.7 Other proof methods

There are many other methods of proof besides the ones discussed so far. In some ways, we are scratching the surface. In future chapters of this part of the book, we will continue to introduce set theory concepts, elementary number theory, and ideas from combinatorics. However, along the way, we deliberately introduce new proof techniques. Here is a list of some of the proof methods we will encounter in this book beyond this chapter.

Proof Method	Location in Book
Proofs using well-ordering	Section 4.1.2
Proofs by induction	Section 4.5
Combinatorial proof	Section 5.2
Pigeon-hole principle	Section 5.3
Invariance proofs	Exercises 2.3.24 through 2.3.26

When attempting to prove a conjecture, at the absolute least we need a valid argument. At best, we would like a proof that is concise, elucidates what is going on in the problem, and flows well. Mathematicians may call a proof elegant when it satisfies these ideals. Awareness and experience of multiple proof techniques allows us to draw on many possible avenues; some may help and some may not. Sometimes a combination of techniques helps. Occasionally, a mathematician may devise an entirely new proof strategy that can be applied in other situations.

EXERCISES FOR SECTION 2.3

1. Prove that $n(n + 1)$ is an even number for all integers n.

2. Prove the familiar algebra result $|xy| = |x||y|$ for all $x, y \in \mathbb{R}$ by cases.

3. Prove that $n^2 + 3n + 7$ is an odd number for all integers n.

4. Prove that $\lfloor 3x \rfloor = \lfloor x \rfloor + \left\lfloor x + \frac{1}{3} \right\rfloor + \left\lfloor x + \frac{2}{3} \right\rfloor$ for all $x \in \mathbb{R}$.

5. Prove that $\min(x, y) + \max(x, y) = x + y$ for all $x, y \in \mathbb{R}$.

6. Prove that there are no solutions in integers a and b to the equation $a^2 + b^2 = 7$.

7. Prove that there are no solutions in integers a and b to the equation $a^2 + 3b^2 = 14$.

8. Let n be a positive integer. Prove that the last two digits of 7^n can be 01, 07, 49, or 43. [Hint: Use cases depending on whether $n = 4k$, $n = 4k + 1$, $n = 4k + 2$, or $n = 4k + 3$.]

9. We call a subset S of $\mathbb{R}^2$ an open rectangle if $S = \{(x, y) \in \mathbb{R}^2 \mid a < x < b \text{ and } c < y < c\}$ for some $a, b, c, d \in \mathbb{R}$. Prove that the intersection of two open rectangles is again an open rectangle.

10. Prove that between any two rational numbers $x < y$, there exists a rational number z with $x < z < y$.

11. Prove that there exists 2020 consecutive positive integers, none of which are perfect squares.

12. Show that $\exists! x\, P(x)$ is equivalent to $\exists x P(x) \wedge (\forall x \forall y\, P(x) \wedge P(y) \to x = y)$.

13. Let n be a nonnegative integer. Prove that are $2n + 1$ nonnegative integers k such that $\lfloor \sqrt{k} \rfloor = n$.

14. Prove that every 3×3 matrix A with real coefficients has a real eigenvalue.

15. Let S be any set.
    (a) Prove that there exists a unique subset B of S such that $B \cap A = A$ for all $A \subseteq S$.
    (b) Prove that there exists a unique subset C of S such that $C \cup A = A$ for all $A \subseteq S$.

16. For sets A and B, prove that the following statements are equivalent: (a) A and B are disjoint; (b) $A \setminus B = A$; (c) $B \setminus A = B$.

17. Recall that the inverse of an $n \times n$ matrix A, is a matrix B such that $AB = BA = I$. Prove that if A has an inverse, then it is unique.

18. Consider the two functions $f(x) = x^2$ and $g(x) = x^2 - 4x + 10$.
    (a) Prove or disprove that the curves $y = f(x)$ and $y = g(x)$ share a unique point.
    (b) Prove or disprove that the curves $y = f(x)$ and $y = g(x)$ share a unique tangent line.

19. Let $(x, y) \in \mathbb{R}^2 \setminus \{(0, 0)\}$. Show that there is a unique pair $(r, \theta) \in \mathbb{R}^{>0} \times [0, 2\pi)$ such that $(x, y) = (r \cos \theta, r \sin \theta)$.

20. Prove that between any two rational numbers there exists an irrational number.

21. Modify Example 2.3.7 to prove that $\lim\limits_{x \to 3} \dfrac{1}{x} = \dfrac{1}{3}$.

22. Modify Example 2.3.7 to prove that $\lim\limits_{x \to 3} x^2 = 9$.

23. Consider the generic cubic function $f(x) = x^3 + px + q$.
    (a) Prove that $f(x) = 0$ has exactly one root if $p \geq 0$.
    (b) For this part and the next, suppose that $p < 0$. Let x_1 and x_2 be the two distinct roots of $f'(x) = 0$. Show that $f(x_1)f(x_2) = q^2 + \dfrac{4}{27}p^3$.
    (c) Use the Intermediate Value Theorem to show that $f(x) = 0$ has exactly 3 roots if $f(x_1)f(x_2) < 0$; exactly 2 roots if $f(x_1)f(x_2) = 0$; and exactly 1 root if $f(x_1)f(x_2) > 0$.

The remaining exercises involve the proof strategy called an invariance *proof. In a proof that involves invariance, the proof capitalizes on a property of a mathematical object that remains unchanged under transformations applied to that mathematical object. Sometimes invariance proofs require some creativity in identifying the invariant property.*

24. Let n be an odd positive integer. Suppose that someone does the following: first, writes the numbers $1, 2, \ldots, 2n$ on a sheet of paper; picks any two integers j and k, erases them from the list, and appends $|j - k|$ to the list; and continues until only one integer remains. Prove that the remaining integer is odd. [Hint: First show that the sum of integers starts out as odd. Then show that in the process described, the sum must change by an even number.]

25. Start with the set $\{3, 4, 12\}$. In each step, replace any two of the elements of the set a and b and replace them with $\frac{3}{5}a + \frac{4}{5}b$ and $\frac{4}{5}a - \frac{3}{5}b$. Is it possible to reach the set $\{2, 7, 10\}$ in a finite number of steps?

26. (*) Let $n \geq 3$. Start with integers placed evenly around a circle such that any two adjacent ones are distinct. At each stage, put in the midpoint of each arc, the average of the values on either side of that midpoint and erase the numbers at the previous stage. Show that after repeating this process long enough, none of the numbers are integers.

2.4 Generalized Unions and Intersections

We end this second chapter with a short section on generalized unions and intersections. These occur regularly in advanced mathematics. Though the topic falls squarely in set theory, we delayed it until now, after we introduced our first proof methods. From a broader pedagogical perspective, the reader will note that our examples offer applications of quantifiers and further illustrate how to prove the equality of sets.

It is very common in mathematics to consider collections of sets, i.e., sets of sets. When we do, we use another set that indexes all the sets in the collection.

For example, $L_n = \{(n, y) \in \mathbb{R}^2 \mid y \in \mathbb{R}\}$ describes the vertical line in $\mathbb{R}^2$ corresponding to $x = n$. Then $\{L_n\}_{n \in \mathbb{Z}}$ represents the set of all vertical lines in $\mathbb{R}^2$ with integer values for x. We could use the notation $\{L_n \mid n \in \mathbb{Z}\}$ but the notation $\{L_n\}_{n \in \mathbb{Z}}$ is more common in that it expresses the role of $\mathbb{Z}$ as an indexing set for this collection of lines.

Indexing sets can be finite. For example if $\mathcal{I} = \{1, 2, \ldots, 7\}$, then the collection $\{A_i\}_{i \in \mathcal{I}}$ is the same as $\{A_1, A_2, \ldots, A_7\}$. However, indexing sets can themselves be any set. For example, if $C_r = \{(x, y) \in \mathbb{R}^2 \mid x^2 + y^2 = r\}$, the notation $\{C_r\}_{r \in \mathbb{R}^{\geq 0}}$ refers to the set of all circles of center $(0, 0)$ and radius r, with $r \in \mathbb{R}^{\geq 0}$.

Definition 2.4.1

Let $\{A_i\}_{i\in\mathcal{I}}$ be a collection of sets, indexed by a set $\mathcal{I}$. The *generalized union* of this collection is

$$\bigcup_{i\in\mathcal{I}} A_i \overset{\text{def}}{=} \{x \,|\, \exists i \in \mathcal{I}\,(x \in A_i)\} = \{x \,|\, x \in A_i \text{ for some } i \in \mathcal{I}\}.$$

The *generalized intersection* of this collection is

$$\bigcap_{i\in\mathcal{I}} A_i \overset{\text{def}}{=} \{x \,|\, \forall i \in \mathcal{I}\,(x \in A_i)\} = \{x \,|\, x \in A_i \text{ for all } i \in \mathcal{I}\}.$$

In each of the above two definitions, we expressed the set descriptor in two ways: first using the quantifier symbol and second using natural linguistic quantification.

Example 2.4.2. Consider the collection of intervals $I_n = (0, 1+1/n)$ with $n \in \mathbb{N}^*$. We propose to determine the generalized intersection

$$A = \bigcap_{n=1}^{\infty} \left(0, 1 + \frac{1}{n}\right).$$

By thinking about it, we might guess that A is equal to $(0,1)$ or maybe $(0,1]$. We show that $A = (0,1]$.

First, suppose that $x \in (0,1]$. For all $n \in \mathbb{N}^*$, we have $1/n > 0$ so $1 + 1/n > 1$. Consequently, $x \in (0, 1 + 1/n)$ for all $n \in \mathbb{N}^*$, so $x \in A$. This shows that $(0,1] \subseteq A$.

Now suppose that $x \in A$. By definition, $x \in (0, 1 + 1/n)$ for all $n \in \mathbb{N}^*$. Clearly, $x > 0$. Suppose that y is a real number with $y > 1$. The condition $y > 1 + 1/n$ is equivalent to $y - 1 > 1/n$, which is equivalent to $n > 1/(y-1)$. So for any real $y > 1$, there exists some integer n larger than $1/(y-1)$ and then $y \notin A_n$. Hence, if $x \in A$, we must have $x \leq 1$. We have shown that $A \subseteq (0,1]$.

Since $A \subseteq (0,1]$ and $(0,1] \subseteq A$, we conclude that $A = (0,1]$. $\triangle$

We pause to look more carefully at the strategy in the above example. Our reasoning used Proposition 1.2.5, which states that two sets A and B are equal if and only if $A \subseteq B$ and $B \subseteq A$. Though the proposition might have appeared rather basic when first introduced, it often gives a valuable strategy to prove that two sets are equal. Digging into further detail, to prove that $A \subseteq B$, we must show $\forall x\,(x \in A \rightarrow x \in B)$. So, if we use a direct proof for this part, we will begin with a phrase like, "Suppose that $x \in A$," and then we will do work to establish that $x \in B$. If instead, we use a proof by contrapositive, we will begin with a phrase like, "Suppose that

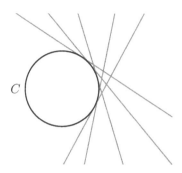

Figure 2.1: A few tangent lines to a circle.

$x \notin B$," and then we will try to establish that $x \notin A$. Then, we face the choice of strategy when we prove that $B \subseteq A$.

In the above example, we first made a guess as to the generalized intersection and then proceeded to prove it in two steps: $(0,1] \subseteq A$ and then $A \subseteq (0,1]$. The "Now" at the beginning of the third paragraph in the example emphasizes the switch from proving $(0,1] \subseteq A$ to proving that $A \subseteq (0,1]$.

Example 2.4.3. As a more complicated example, suppose we are working in the Euclidean plane. Let C be a circle and for each $p \in C$, call L_p the tangent line to C at p. Consider

$$B = \bigcup_{p \in C} L_p,$$

the union of all the tangent lines to the circle C. A sketch as in Figure 2.1 leads us to a guess that B is the set of points on or outside the circle C. We prove this by referring to theorems in Euclidean geometry, which the reader hopefully encountered in high school. We cite the theorems out of Euclid's *Elements*.

By Book III Proposition 1, we can construct the center of C. Call it O. By Book III Proposition 17, for each point M outside of the circle, it is possible to construct a line through M that is tangent to C. By Book III Proposition 18, if M is on the circle, then the line drawn from O through M is perpendicular to a line tangent to C at M. If M is inside the circle, let L be any line through M. Since there is a point on L of any distance away from M, there exists a point on L outside of C. Then L is not tangent to C by Book III, Definition 2. Consequently, B consists of all points on or outside of C. △

Example 2.4.4. It is a common exercise in linear algebra to prove that if U and W are subspaces of a vector space V, then $U \cap W$ is also a subspace of V. In this example, we prove a more general result: If

$\{W_i\}_{i \in \mathcal{I}}$ is a collection of subspaces of a real vector space V, then

$$W = \bigcap_{i \in \mathcal{I}} W_i$$

is again a subspace of V.

This generalized intersection W is a subset of V. First, notice that for all $i \in \mathcal{I}$ we have $\mathbf{0} \in W_i$. Hence, $\mathbf{0} \in W$. In particular, W is nonempty. Let $f, g \in W$. Then $f, g \in W_i$ for all $i \in \mathcal{I}$. Since each W_i is a subspace of V, then $f + g \in W_i$ for each $i \in \mathcal{I}$. Hence, $f + g \in W$. Finally, let $f \in W$ and $c \in \mathbb{R}$. Again, $f \in W_i$ for all $i \in \mathcal{I}$. Since each W_i is a subspace of V, then $cf \in W_i$ for each $i \in \mathcal{I}$. Thus $cf \in W$. We have shown that W is a nonempty subset of V that is closed under addition and scalar multiplication, so the generalized intersection W is a subspace of V. $\triangle$

EXERCISES FOR SECTION 2.4

1. Let $A_i = [i, i^2]$ for all $i \in \mathbb{N}$. Determine (a) $\bigcup_{i=1}^{8} A_i$; (b) $\bigcap_{i=4}^{10} A_i$.

2. Let $\mathcal{I} = \{2, 3, 5, 7\}$ and let A_i be the closed intervals $A_i = [2/i, 2 - 2/i]$. Determine (a) $\bigcup_{i \in \mathcal{I}} A_i$; (b) $\bigcap_{i \in \mathcal{I}} A_i$.

3. Let $\{A_n\}_{n=2}^{\infty}$ be the collection of subsets of $\mathbb{R}$ with $A_n = \left[\frac{1}{n}, 1 - \frac{1}{n}\right]$. Prove that

$$\bigcup_{n=2}^{\infty} A_n = (0, 1).$$

4. Determine $\bigcap_{i=1}^{\infty} B_i$ and $\bigcup_{i=1}^{\infty} B_i$ for each of the following collection of subsets:

   (a) $B_i = [i, 2i]$;
   (b) $B_i = \{1, 2, \ldots, i\}$;
   (c) $B_i = [2 - 1/i, 5 - 1/i]$.

5. For all $\varepsilon \in \mathbb{R}^*$, define $A_\varepsilon = \{f \in C^0(\mathbb{R}) : |f(x)| < \varepsilon \text{ for all } x \in \mathbb{R}\}$. Determine

   (a) $\bigcap_{\varepsilon \in \mathbb{R}^*} A_\varepsilon$;   (b) $\bigcup_{\varepsilon \in \mathbb{R}^*} A_\varepsilon$.

6. Let $\{A_i\}_{i \in \mathcal{I}}$ be a collection of subsets of a set S. Show that

$$\left(\bigcup_{i \in \mathcal{I}} A_i\right)^c = \bigcap_{i \in \mathcal{I}} A_i{}^c \quad \text{and} \quad \left(\bigcap_{i \in \mathcal{I}} A_i\right)^c = \bigcup_{i \in \mathcal{I}} A_i{}^c.$$

7. Let $\{A_i\}_{i \in \mathcal{I}}$ be a collection of subsets of a set S and let $B \subseteq S$. Show that

$$B \cap \left(\bigcup_{i \in \mathcal{I}} A_i\right) = \bigcup_{i \in \mathcal{I}} B \cap A_i \quad \text{and} \quad B \cup \left(\bigcap_{i \in \mathcal{I}} A_i\right) = \bigcap_{i \in \mathcal{I}} B \cup A_i.$$

8. We revisit Example 2.4.3 and prove the result analytically, i.e., using calculus. For simplicity, let C be the circle of radius R with center at the origin. Recall that C has the equation $x^2 + y^2 = R^2$.

   (a) Let $p = (a, b)$ be any point on the circle C. Use implicit differentiation (or some other strategy from calculus) to show that $ax + by = R^2$ is an equation for the tangent line L_p to C at p.

   (b) Let $M = (x_0, y_0)$ be an arbitrary point in the plane. Show that if M is on a tangent line L_p, then a satisfies the quadratic equation

   $$(x_0^2 + y_0^2)a^2 - 2R^2 x_0 a + R^2(R^2 - y_0^2) = 0.$$

   [Hint: Using the line equation $ax_0 + by_0 = R^2$ from M being on L_p, eliminate b from the identity $a^2 + b^2 = R^2$.]

   (c) Using this quadratic equation, show that there exists $p \in C$ such that M is on L_p if and only if $x_0^2 + y_0^2 \geq R^2$.

   (d) Deduce that $\bigcup_{p \in C} L_p$ consists of all the points on or outside C.

9. Let P be the parabola $P = \{(x, y) \in \mathbb{R}^2 \mid y = x^2\}$. For all points $q = (a, b) \in P$, let L_q be the tangent line to P at q.

   (a) Find an equation for L_q. [Hint: it will have $a \in \mathbb{R}$ as a parameter.]

   (b) For an arbitrary point $(x_0, y_0) \in \mathbb{R}^2$, find a quadratic equation in a that is required so that (x_0, y_0) is on L_q.

   (c) Deduce that the union of all tangent lines to the parabola P consists of the region below the parabola; more precisely,

   $$\bigcup_{q \in P} L_p = \{(x, y) \in \mathbb{R}^2 \mid y \leq x^2\}.$$

10. For all $t \in [0, 1]$, let L_t be the line segment from the point $(0, 1 - t)$ to $(t, 0)$. Show that

   $$\bigcup_{t \in [0,1]} L_t = \{(x, y) \in [0, 1]^2 \mid 0 \leq y \leq (\sqrt{x} - 1)^2\}.$$

   (The diagram below exhibits three line segments in the collection.)

11. For every $\alpha \in \mathbb{R}$, we call $L_\alpha = \mathrm{Span}(\langle \cos \alpha, \sin \alpha, 1 \rangle)$. Recall that the span of one vector $\mathbf{v} = \langle a, b, c \rangle$ consists of all multiples of that vector $t\mathbf{v} = \langle ta, tb, tc \rangle$ for any $t \in \mathbb{R}$. Prove that

   $$\bigcup_{\alpha \in [0, 2\pi]} L_\alpha = \{\langle x, y, z \rangle \in \mathbb{R}^3 \mid x^2 + y^2 - z^2 = 0\},$$

   which is the unit double cone through the origin. (See Figure 2.2.)

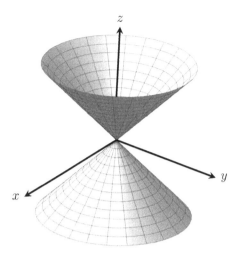

Figure 2.2: Double cone.

12. For every $\theta \in [0, 2\pi]$, consider the line in space L_θ parametrized by

$$\langle cos\theta, \sin\theta, 0\rangle + t\langle -\sin\theta, \cos\theta, 1\rangle \qquad \text{for } t \in \mathbb{R}.$$

Show that $\bigcup\limits_{\theta \in [0,2\pi]} L_\theta$ is the hyperbolid of one sheet with equation $x^2 + y^2 - z^2 = 1$.

13. (*) Let $\mathcal{E}_r = \{(x, y) \in \mathbb{R}^2 \,|\, x^2/r^2 + y^2 < 1\}$ be a collection of filled ellipses. Show that

$$\bigcap\limits_{r \in (0,1]} \mathcal{E}_r = \{(0, y) \in \mathbb{R}^2 \,|\, -1 \leq y \leq 1\}.$$

14. (*) Let $\mathcal{E}_r = \{(x, y) \in \mathbb{R}^2 \,|\, x^2/r^2 + y^2 \leq 1\}$ be a collection of filled ellipses. Show that

$$\bigcup\limits_{r \in [1,\infty)} \mathcal{E}_r = \{(0, 1), (0, -1)\} \cup \{(x, y) \in \mathbb{R}^2 \,|\, -1 < y < 1\}.$$

CHAPTER 3

Functions

German mathematician Carl Jacobi (1804-1851) pithily stated, "One should always generalize"[46]. This sentiment reflects a drive in mathematics of discovering not just properties of a specific object, but also the most general context in which such properties hold. The reader will experience this generalization movement as we study functions.

Students first encounter functions in algebra and study properties of (differentiable) functions at length in calculus. In linear algebra, students explore linear transformations, which are special types of functions from one vector space to another. The most general context for functions is set theory.

Like sets, functions play a fundamental role throughout mathematics. They offer a precise mental framework for associating one object to another. After defining functions and common related concepts, this chapter deliberately introduces a broad swath of specific examples and classes of functions that occur often in mathematics.

3.1 Functions

3.1.1 Definition and examples

Compared to the definition and uses of functions in high school algebra, the following definition may seem rather unfamiliar. However, as we will soon see, it provides a precise set theory framework for modeling unique associations.

Definition 3.1.1

Let A and B be two sets. A *function* from A to B is a subset f of $A \times B$ such that for all $a \in A$, there exists a unique $b \in B$ such that $(a, b) \in f$. The set A is called the *domain* of f. The set B is called the *codomain* of f. Instead of writing $(a, b) \in f$, we write $b = f(a)$.

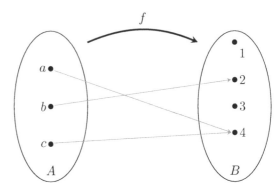

Figure 3.1: A graphical representation of a function.

If $f : A \to B$ is a function between sets, for any $a \in A$, we call $f(a)$ the *image* of a. By definition, the image of any element in the domain is unique. For any $b \in B$, if there exists $a \in A$ such that $f(a) = b$, then a is called a *pre-image* of b. Any given $b \in B$ may have one, many, or no pre-images. For example, with the function $f : \mathbb{R} \to \mathbb{R}$ with $f(x) = x^2$, the number 3 has two pre-images, namely $-\sqrt{3}$ and $\sqrt{3}$, while -3 has no pre-images.

Occasionally, we may describe what a function does by the $\mapsto$ symbol, e.g., $x \mapsto x^2$. Some authors use this symbol frequently; our habit is to use it only when describing a function in passing, without giving it a letter name like f, g, or so on.

Definition 3.1.2

> For any sets A and B we denote by $\text{Fun}(A, B)$ or B^A the set of all functions $A \to B$. Functions are also called *transformations* or *mappings* or even shorter, just *maps*.

Figure 3.1 gives a visual representation of a specific function. The key essential of this representation is that for each $x \in A$, there exists a unique arrow from the bubble for A to elements in B. On the other hand, elements in B do not need to have any arrows arriving at them but can also have more than one.

Example 3.1.3. The concept of a person's height can be described by a function in the following way. Let P be the set of all living people. We define $h : P \to \mathbb{N}$ as $h(p)$ is the height, rounded down to the closest centimeter, of the person p. Lining up to the definition, the height h is the subset of $P \times \mathbb{N}$ such that $h = \{(p, x) \in P \times \mathbb{N} \mid x$ is the height of $p\}$. $\triangle$

Example 3.1.4. The concept of a person's biological mother can be described by a function as well. Let P be the set of all people living or dead. We define $m : P \to P$ as $m(p)$ as p's biological mother. Again, lining up to the definition, the concept of mother m is the subset of $P \times P$ such that $m = \{(p, q) \in P \times P \mid q \text{ is } p\text{'s mother}\}$. △

Example 3.1.5. The floor defined in Definition 2.2.3 is a function $\lfloor\,\rfloor : \mathbb{R} \to \mathbb{Z}$ such that $\lfloor x \rfloor = n$ if and only if $n \leq x < n+1$. Similarly, we define the *ceiling* function $\lceil\,\rceil : \mathbb{R} \to \mathbb{Z}$ such that $\lceil x \rceil = n$ if and only if $n - 1 < x \leq n$. Thus, the ceiling of a real number x is the smallest integer greater than or equal to x. △

Example 3.1.6. The minimum between two real numbers is a function $\min : \mathbb{R}^2 \to \mathbb{R}$ defined by

$$\min(x, y) = \begin{cases} x & \text{if } x \leq y \\ y & \text{if } y \leq x. \end{cases}$$

The same is true for the maximum $\max(x, y)$ of two real numbers x and y. △

Before we proceed with more examples that illustrate the breadth of the function concept, we connect Definition 3.1.1 to the terminology introduced earlier courses.

In pre-calculus, we encountered exercises like this: "Determine the domain of the function $f(x) = \ln(x+1)$." This way of posing the question is imprecise compared to Definition 3.1.1. With a function $f : A \to B$, for *every* element of A, there must exist a unique element b with $f(a) = b$ so the domain and the codomain must be part of the definition of the function.

Example 3.1.7. Recall Example 1.2.2 in which we discussed the exercise of finding the domain of $f(x) = 1 + \sqrt{x + 2}$. The correct way of phrasing this pre-calculus question is: "Find the largest subset A of $\mathbb{R}$ such that $f : A \to \mathbb{R}$ with $f(x) = 1 + \sqrt{x + 2}$ is a function." To answer this question, we know we can only take square roots of non-negative numbers. Hence we need $x + 2 \geq 0$, so the domain is $[-2, \infty)$. In light of the precise definition for functions, the proper way to talk about the function is $f : [-2, \infty) \to \mathbb{R}$ with $f(x) = 1 + \sqrt{x + 2}$. △

To emphasize the point of the above example, it is not correct to say, "Consider the function $f : \mathbb{R} \to \mathbb{R}$ defined by $f(x) = \sqrt{4 - x^2}$," because it is not true that for every real number $x \in \mathbb{R}$, the quantity $\sqrt{4 - x^2} \in \mathbb{R}$.

The definition for functions in set theory makes it clear that two functions f and g are equal if and only if they have the same domain, the same codomain, and that $f(x) = g(x)$ for all x in the domain. So for example, the functions $f : \mathbb{R} \to \mathbb{R}$ with $f(x) = x^2$ and $g : \mathbb{R}^{\geq 0} \to \mathbb{R}$

with $g(x) = x^2$ are *not* equal as functions, because, lining up with the definition, they would not be equal as sets. Furthermore, this understanding for function equality gives a standard method to prove when two functions f and g with the same domain are equal: for an arbitrary element x in the domain, prove that $f(x) = g(x)$.

In multivariable calculus, we encounter functions of the form $f : \mathbb{R}^m \to \mathbb{R}$, or more precisely functions of the form $f : D \to \mathbb{R}$, where $D \subseteq \mathbb{R}^m$. These are called multivariable real-valued functions. We also study functions of the form $f : D \to \mathbb{R}^n$, where $D \subseteq \mathbb{R}^m$. As a branch of mathematics, complex analysis studies functions of the form $f : \mathbb{C} \to \mathbb{C}$.

Example 3.1.8. An important theorem at the beginning of linear algebra is that to every matrix $A \in M_{m \times n}(\mathbb{R})$ there exists one matrix in reduced row echelon form that is row equivalent to A. This theorem allowed us to define *the* reduced row echelon form of A as this unique matrix. This unique association is a function that we write as rref : $M_{m \times n}(\mathbb{R}) \to M_{m \times n}(\mathbb{R})$.   △

Example 3.1.9. The process of differentiation can be viewed as a function in the following sense. Let I be an interval of $\mathbb{R}$. Consider the function $D : C^1(I) \to C^0(I)$ defined by $D(f) = f'$.

Recall that $C^1(I)$ is the set of all real-valued functions that are differentiable on the interval I with derivatives that are continuous on I. The set $C^0(I)$ is the set of all real-valued functions that are continuous on I. To every $f \subset C^1(I)$, the function D returns the derivative. In this way, the derivative is itself a function.

Because of the possible confusion when we discuss functions on function sets, it is not uncommon to call *operators* a function whose domain is of the form $C^n(I)$ or otherwise a set of functions.   △

Example 3.1.10 (Evaluation map). We can view the process of evaluating a function as a function itself in the following sense. Let A and B be sets and let $\mathrm{Fun}(A, B)$ be the set of all functions from A to B. Fix a particular $a \in A$. The *evaluation map* is $\mathrm{ev}_a : \mathrm{Fun}(A, B) \to B$ defined by $\mathrm{ev}_a(f) = f(a)$.   △

3.1.2   Special types of functions

Occasionally, because of their uses in particular areas of mathematics, certain classes of functions may carry specific names and may even involve their unique notation. We give two examples from familiar concepts.

> **Definition 3.1.11**
>
> Let A be any set. A *sequence* in A is a function $f : \mathbb{N} \to A$. With sequences, we write f_n instead of $f(n)$.

More generally, a sequence is a function of the form $f : D \to A$, where D is any contiguous subset of $\mathbb{Z}$ with a least element but no greatest element. So a sequence could start at 1 or 10 or even -1 and have values for all successive integers.

Example 3.1.12. The *harmonic sequence* is the function $H : \mathbb{N}^* \to \mathbb{Q}$ such that $H_n = 1/n$. △

Example 3.1.13. For all integers $n \geq 0$, let L_n be the line in the plane that satisfies the equation $y = 2nx - n^2$. We can think of these lines as a sequence of lines, which means a function from $\mathbb{N}$ to the set of lines in the plane. △

With sequences, instead of describing the sequence object as f or $f : \mathbb{N} \to A$ following typical function notation, it is customary to write $(f_n)_{n \in \mathbb{N}}$, or $(f_n)_{n=0}^{\infty}$, or even more briefly (f_n) if the domain of the sequence is clear from context.

Another type of function that is ubiquitous in mathematics is a binary operation.

> **Definition 3.1.14**
>
> Let S be any set. A *binary operation* on S is any function of the form $f : S \times S \to S$.

Interestingly enough, though we express some binary operations with the expected notation $f(a, b)$ (e.g., the minimum of two real numbers a and b, written $\min(a, b)$, is a binary operation on $\mathbb{R}$), it is more common to use a symbol $\star$ and write $a \star b$ instead of $\star(a, b)$ for the image of the pair (a, b).

Example 3.1.15. Addition is a binary operation on $\mathbb{N}$. We use the symbol $+$ for addition and we write $a + b$ instead of $+(a, b)$ for the image of the pair (a, b). △

We emphasize that since a binary operation on S is a function whose domain is $S \times S$, every pair $(a, b) \in S$ must have an image under the binary operation. For example, subtraction $-$ is not a binary operation on $\mathbb{N}$ because for example $2 - 3 = -1 \notin \mathbb{N}$. On the other hand, $-$ is a binary operation on $\mathbb{Z}$ since the difference of any two integers is again an integer.

Example 3.1.16. Consider the set $V = \mathbb{R}^3$ of vectors. The cross product $\times$ is a binary operation because for any two vectors $\mathbf{a}$ and $\mathbf{b}$, the cross product $\mathbf{a} \times \mathbf{b}$ exists and is another vector in $\mathbb{R}^3$. On the other hand, the dot product is not a binary operation on V because for any two vectors $\mathbf{a}$ and $\mathbf{b}$, the dot product $\mathbf{a} \cdot \mathbf{b}$ is in $\mathbb{R}$ and not $\mathbb{R}^3$. To wit, the dot product is a function $\mathbb{R}^3 \times \mathbb{R}^3 \to \mathbb{R}$, but it is not a binary operation. $\triangle$

Most mathematics students have seen the following terms pertaining to binary operations. We repeat them here for completeness.

Definition 3.1.17

Let S be a set equipped with a binary operation $\star$. We say that the binary operation is

1) called *associative* if $\forall a, b, c \in S$, $(a \star b) \star c = a \star (b \star c)$;

2) called *commutative* if $\forall a, b \in S$, $a \star b = b \star a$;

3) said to have an *identity* if $\exists e \in S \, \forall a \in S$, $a \star e = e \star a = a$ (e is called an *identity element*);

4) called *idempotent* if $\forall a \in S$, $a \star a = a$.

It is important to note the order of the quantifiers in definition (3) for the identity element. For example, let $\star$ be the operation of geometric average on $\mathbb{R}^{>0}$, i.e., $a \star b = \sqrt{ab}$. Note that this operation is commutative and idempotent. The operation of geometric average does not have an identity because if we attempted to solve for b in $a \star b = a$, we would obtain $b = a$. However, in order for the geometric average to have an identity, this element b could not depend on a.

Proposition 3.1.18

Let S be a set equipped with a binary operation $\star$. If $\star$ has an identity element, then $\star$ has a unique identity element.

Proof. Suppose[1] that $e_1, e_2 \in S$ are identity elements for $\star$. Since e_1 is an identity element, then $e_1 \star e_2 = e_2$. Since e_2 is an identity element, then $e_1 \star e_2 = e_1$. Thus, $e_1 = e_2$ and so $\star$ has a unique identity element. $\square$

Because of this proposition, we no longer say, "an identity element" but "the identity element."

Definition 3.1.19

Let S be a set equipped with a binary operation $\star$ that has an identity e. The operation is said to *have inverses* if $\forall a \in S \, \exists b \in S$, $a \star b = b \star a = e$.

---

[1] Note the use of the Section 2.3.3 uniqueness strategy for this proof.

For example, in $\mathbb{R}$ the operation $+$ has inverses because for all $a \in \mathbb{R}$, we have $a + (-a) = (-a) + a = 0$. So $(-a)$ is the (additive) inverse of a. In $\mathbb{R}^*$, the operation $\times$ also has inverses: For all $a \in \mathbb{R}^*$, we have $a \times \frac{1}{a} = \frac{1}{a} \times a = 1$. So $\frac{1}{a}$ is the (multiplicative) inverse of a.

Definition 3.1.20

Let S be a set equipped with two binary operations $\star$ and $*$. We say that

1) $\star$ is *left-distributive over* $*$ if $\forall a, b, c \in S$, $a \star (b * c) = (a \star b) * (a \star c)$;

2) $\star$ is *right-distributive over* $*$ if $\forall a, b, c \in S$, $(b * c) \star a = (b \star a) * (c \star a)$;

3) $\star$ is *distributive over* $*$ if $\star$ is both left-distributive and right-distributive over $*$.

The quintessential example for distributivity is that, as binary operations on $\mathbb{R}$, $\times$ is distributive over $+$. However, many other pairs of operations share this property.

Example 3.1.21. Let S be any set and consider its power set $\mathcal{P}(S)$. The intersection $\cap$ and the union $\cup$ are binary operations on $\mathcal{P}(S)$. From formulas on set operations (see Table 1.2), we see that $\cap$ is associative, commutative, idempotent, and has S as its identity. The union $\cup$ is also associative, commutative, idempotent, and has $\emptyset$ as its identity. Furthermore, $\cap$ distributes over $\cup$, and $\cup$ distributes over $\cap$. $\triangle$

We wish to point out that in Definition 3.1.17(3), the requirement $a \star e = e \star a = a$ is not redundant when the operation is not commutative. A similar comment holds for the definition of the inverse element in Definition 3.1.19. The following example illustrates the necessity of the seemingly redundant statements.

Example 3.1.22. Define the operation $\star$ on $\mathbb{R}^{\geq 0}$ by

$$x \star y = \sqrt{|x^2 + xy - y^2|}.$$

We leave it to the reader to check that $\star$ is not commutative and that 0 is the identity element for this binary operation. Now suppose we are given a and wish to solve for x in $a \star x = 0$. We have

$$a \star x = 0 \implies a^2 + ax - x^2 = 0 \implies x = \frac{a \pm a\sqrt{5}}{2},$$

where we choose the $+$ in order for x to remain a nonnegative number. On the other hand, solving for x in $x \star a = 0$, gives

$$x \star a = 0 \implies x^2 + ax - a^2 = 0 \implies x = \frac{-a \pm a\sqrt{5}}{2},$$

where we choose the $+$ in order for x to remain a nonnegative number. Consequently, the binary operation $\star$ does not have an inverse.

We may sometimes call the *left-inverse* of a an element x such that $x \star a$ is the identity and call the *right-inverse* of a an element x such that $a \star x$ is the identity. In this example, the operation has a left- and a right-inverse for all elements in $\mathbb{R}^{\geq 0}$. However, since the left-inverse and the right-inverse are not equal, then $\star$ does not have inverses for any element.   △

3.1.3   Ways to describe a function

At this point, it is likely that the reader is most accustomed to describing a function by an algebraic formula. After all, this is the method used in pre-calculus and calculus. For example, describing a function $f : \mathbb{R} \to \mathbb{R}$ by $f(x) = x \cos x$ feels familiar. Even if the domain is not a set of numbers, this method may still serve us well. For example, we could define a sequence $\{A_n\}_{n \in \mathbb{N}^*}$ in $\mathcal{P}(\mathbb{R})$ as $A_n = [n, n^2]$. This is still a formula description. There are a few other common ways to describe a function.

We can also define a function *by cases*. This generalizes what we called piecewise functions in calculus or pre-calculus. For example, if $f : A \to B$ and $A = A_1 \cup A_2 \cup A_3$, with the subsets $A_i \subseteq A$ mutually disjoint, then we can specify f on each subset.

Example 3.1.23. The sign function $\text{sign} : \mathbb{R} \to \mathbb{Z}$ is defined by

$$
\text{sign}(x) = \begin{cases} -1 & \text{if } x < 0 \\ 0 & \text{if } x = 0 \\ 1 & \text{if } x > 0. \end{cases}
$$

Note that $\text{sign}(x)x = |x|$ for all $x \in \mathbb{R}$.   △

Example 3.1.24. Consider the function $f : \mathbb{R}^{\geq 0} \to \mathbb{R}$ defined by

$$
f(x) = \begin{cases} 0 & \text{if } x \in \mathbb{R} \setminus \mathbb{Q} \\ \frac{1}{b} & \text{if } x \in \mathbb{Q} \text{ and } x = a/b \text{ expressed in reduced form.} \end{cases}
$$

It is hard to graph this function, but it has a number of interesting properties. It is periodic of period 1, continuous at every irrational, but discontinuous at every rational.   △

Consider the concept of factorial of a nonnegative integer. We usually define it by $0! = 1$ and for all positive n, we set $n! = n(n-1)(n-2)\cdots 2 \cdot 1$. Alternatively, we could define the factorial function $F : \mathbb{N} \to \mathbb{N}$ by $F(0) = 1$ and $F(n) = nF(n-1)$ for all $n \geq 1$. This second method of describing the factorial function is called a *recursive definition*.

A function $f : A \to B$ is said to be defined recursively if there is a sequence of subsets of A

$$A_0 \subseteq A_1 \subseteq A_2 \subseteq \cdots \subseteq A$$

such that we explicitly specify the value of f on A_0 and for each $i \geq 1$ and each $x \in A_i$ the value of $f(x)$ is specified by x and the value of f on elements in A_{i-1}. The definition of factorial in the previous paragraph uses the sequence $A_i = \{0, 1, 2, \ldots, i\}$, as do many recursively defined sequences.

Example 3.1.25. Consider the Fibonacci sequence $\{F_n\}_{n \in \mathbb{N}}$ defined by $F_0 = 0$, $F_1 = 1$, and $F_n = F_{n-1} + F_{n-2}$ for all $n \geq 2$. The sequence begins with 0 and 1 and from then on, each term of the sequence is the sum of the previous two. We can start a table of values as follows

n	0	1	2	3	4	5	6	7	8	9	10
F_n	0	1	1	2	3	5	8	13	21	34	55

The Fibonacci sequence will offer us many interesting examples.[2] $\triangle$

The equation $F_n = F_{n-1} + F_{n-2}$ in the above example is called a *recurrence relation*. A recurrence relation is any equation that involves the terms of a sequence. The values $F_0 = 0$ and $F_1 = 1$ are called *initial conditions*. Given the recurrence relation, the initial conditions are enough to define, one at a time, all the subsequent terms of the sequence.

Example 3.1.26. Consider the following function $g : \mathbb{R}^{\geq 0} \to \mathbb{R}$ defined by

$$g(x) = \begin{cases} 3x - 2 & \text{if } 0 \leq x \leq 2 \\ g\left(\frac{x}{2}\right) + 3 & \text{if } x > 2. \end{cases}$$

Though it is not immediately clear what the sequence of subsets A_i in $\mathbb{R}$ are in this example, this is a recursively defined function. For example,

$$g(10) = g(5) + 3 = (g(5/2) + 3) + 3 = g(5/2) + 6$$
$$= (g(5/4) + 3) + 6 = 3 \times \frac{5}{4} - 2 + 3 + 6 = \frac{43}{4}.$$

We leave it to the reader to check that $A_i = [0, 2^{i+1}]$. $\triangle$

---

[2] Section 8.1.5 discusses Fibonacci's place in mathematical history.

Finally, we may define a function as the result of an algorithm or the result of an existence and uniqueness theorem. For example, consider again the reduced row echelon form of an $m \times n$ matrix, discussed above in Example 3.1.8. A linear algebra theorem tells us that to each matrix A there exists a unique row-equivalent matrix in reduced row echelon form. Before we know how to find it, we could label this unique matrix as $\text{rref}(A)$ and call rref a function. The Gauss-Jordan elimination gives us an algorithm, a list of instructions that will involve a finite number of definite steps, the result of which is $\text{rref}(A)$.

At times, it may happen that an algebraic expression might not properly define a function. Suppose for example that we tried to define the function $f : \mathbb{Q} \to \mathbb{Z}$ by $f(p/q) = p+q$. On the surface, this definition seems well formed, but it is not. For example,

$$f\left(\frac{1}{3}\right) = 1 + 3 = 4 \quad \text{but} \quad f\left(\frac{1}{3}\right) = f\left(\frac{2}{6}\right) = 2 + 6 = 8.$$

This is not possible; functions must only have one image for every element in the domain. This algebraic expression does not give a *well defined* function. We could fix the definition to perhaps capture the apparent intent by saying, let $f : \mathbb{Q} \to \mathbb{Z}$ be the function with $f(r) = p + q$, where $r = p/q$ is expressed in reduced form.

Subtleties about whether a function is well defined typically arise when there are multiple ways of describing the same object, as in $1/3 = 2/6$. We will revisit this issue in Section 6.4.

EXERCISES FOR SECTION 3.1

1. Determine whether f defines a function $f : \mathbb{R} \to \mathbb{R}$ and state your reasons.

   (a) $f(x) = (-5 \pm \sqrt{5^2 - 20})/2$ (c) $f(x) = 1/(x^2 - 3x + 2)$
   (b) $f(x) = \ln(x^2 + 1)$ (d) $f(x) = x - \lfloor x \rfloor$

2. Let P be the set of all people, living or dead. Decide whether each of the following concepts defines a function $f : P \to P$.

   (a) $f(p) =$ the eldest son of p.
   (b) $f(p) =$ the paternal grandfather of p.
   (c) $f(p) =$ the person who spent the most time with p besides themselves.

3. Let W be the set of finite strings of letters in the English alphabet. Determine whether f defines a function $f : W \to \mathbb{Z}$ and state your reasons.

   (a) $f(w) =$ the difference between the number of consonants and the number of vowels. (Vowels are A, E, I, O, and U.)
   (b) $f(w) =$ the string of letters listed backwards.

(c) $f(w)$ = the average of the number values (A is 1, B is 2, and so on).

4. Determine whether f defines a function $f : \mathbb{Q} \to \mathbb{Q}$ and state your reasons.

   (a) $f(p/q) = (p^2 + q^2)/q^2$ (c) $f(p/q) = \sqrt{|p/q|}$
   (b) $f(p/q) = q/p$ (d) $f(p/q) = 2p + q^2$

5. For the function $A \to B$, state the image of $a \in A$ and all pre-images of $b \in B$.

   (a) Function: $f : \mathbb{R} \to \mathbb{R}$ with $f(x) = \cos x + \sin x$; $a = \pi/6$; and $b = 0$.
   (b) Function: $f : \mathcal{P}(\{1,2,3,4,5\}) \to \mathbb{N}$ with $f(C)$ is the sum of elements in C; $a = \{1,2,4\}$; and $b = 7$.
   (c) Function: $D : C^1(\mathbb{R}) \to C^0(\mathbb{R})$ with $D(f) = f'$; $a(x) = x \cos x$; and $b(x) = \sin(2x)$.

6. For the function $A \to B$, state the image of $a \in A$ and all pre-images of $b \in B$.

   (a) Function: $f : \mathbb{Z} \times \mathbb{Z} \to \mathbb{Z}$ with $f(m,n) = m^2 + n^2$; $a = (5,12)$; and $b = 35$.
   (b) Function: $f : \mathcal{P}(\mathbb{Z}) \to \mathcal{P}(\mathbb{Z})$ with $f(X) = X \cup \{1,2,3\}$; $a = \{-1,1\}$ and $b = \{1,2,3,4\}$.
   (c) Function: $f : \mathbb{C} \to \mathbb{C}$ with $f(z) = z^2 + 2z$; $a = 4 + 7i$; and $b = -10$.

7. For the function $A \to B$, state the image of $a \in A$ and all pre-images of $b \in B$.

   (a) Function: $T : \mathbb{R}^2 \to \mathbb{R}^2$ with $T(\mathbf{x}) = \begin{pmatrix} 4 & -6 \\ 6 & -9 \end{pmatrix} \mathbf{x}$; $\mathbf{a} = \langle 5, 2 \rangle$; and $\mathbf{b} = \langle 2, 3 \rangle$.
   (b) Function: $S : M_{2\times3}(\mathbb{R}) \to M_{2\times2}(\mathbb{R})$ with $S(M) = MM^\top$; $a = \begin{pmatrix} 1 & 2 & 0 \\ -3 & 2 & 4 \end{pmatrix}$; and $b = \begin{pmatrix} 4 & 1 \\ 0 & 4 \end{pmatrix}$.
   (c) Function: rref : $M_{2\times3}(\mathbb{R}) \to M_{2\times3}(\mathbb{R})$; $a = \begin{pmatrix} 4 & 6 & 1 \\ 10 & 15 & 2 \end{pmatrix}$; and $b = \begin{pmatrix} 1 & 5 & 0 \\ 0 & 0 & 1 \end{pmatrix}$.

8. Calculate the following values with the floor or ceiling function.

   (a) $\lfloor 7/3 \rfloor$ (c) $\lfloor \pi + 5 \rfloor$ (e) $\lceil e \rceil$
   (b) $\lfloor -3.2 \rfloor$ (d) $\lceil 7/3 \rceil$ (f) $\lceil -10/3 \rceil$

9. Sketch the graph of the real-valued functions over the stated domain:

   (a) $f(x) = \lfloor x \rfloor$ over $[-1,6]$ (c) $f(x) = \lfloor \sin x \rfloor$
   (b) $f(x) = \lfloor 2x \rfloor$ over $[-1,4]$ (d) $f(x) = \lfloor x^2 \rfloor$ over $[-3,3]$

10. Prove that for all $x \in \mathbb{R}$ and all $m \in \mathbb{Z}$, we have $\lfloor x + m \rfloor = \lfloor x \rfloor + m$.

11. Show that $f : \mathbb{R} \to \mathbb{Z}$ defined by $f(x) = 1 - \lceil x \rceil + \lfloor x \rfloor$ is equal to 1 if $x \in \mathbb{Z}$ and 0 if $x \in \mathbb{R} \setminus \mathbb{Z}$.

12. Let $f : \mathbb{N}^* \to \mathbb{N}^*$ where $f(n)$ is the number of digits in the decimal expansion of n. Prove that $f(n) = \lfloor \log_{10} n \rfloor + 1$.

13. Use a proof by cases to show that min $: \mathbb{R} \times \mathbb{R} \to \mathbb{R}$ is an associative binary operation. (See Example 3.1.6.)

For Exercises 3.1.14 through 3.1.20, determine (with proof or counterexample) if the binary operation is associative, is commutative, has an identity, has inverses, and/or is idempotent.

14. The operation $*$ on vectors of $\mathbb{R}^n$ defined by $\mathbf{u} * \mathbf{v} = \text{proj}_{\mathbf{u}} \mathbf{v}$, i.e., projection of $\mathbf{v}$ onto $\mathbf{u}$.

15. The operation $\star$ on the open interval $[0, 1)$ described by $a \star b = a + b - \lfloor a + b \rfloor$ where $\lfloor x \rfloor$ is the greatest integer less than or equal to x.

16. The operation $\triangle$ on nonnegative integers $\mathbb{N}$ defined by $n \triangle m = |m - n|$.

17. The operation $\circledast$ on the plane $\mathbb{R}^2$ where $A \circledast B$ is the midpoint of A and B.

18. The operation $\triangle$ of symmetric difference on $\mathcal{P}(S)$, where S is any set.

19. The commutator operation $[\,,\,]$ on matrices $M_{n \times n}(\mathbb{R})$ defined by $[A, B] = AB - BA$.

20. The power operator $a^{\wedge} b = a^b$ on the set $\mathbb{N}^*$ of positive integers.

21. Show that the power operation $a^{\wedge} b = a^b$ on $\mathbb{R}^{>0}$ right-distributes over $\times$ but does not left-distribute.

22. Find all the terms of the sequence (a_n) up to $n = 10$ defined by $a_1 = 1$ and $a_n - 2a_{n-1} + n$ for all $n \geq 2$.

23. Explain how Newton's method involves a recursively defined sequence of real numbers.

24. Find all terms up to F_{10} in the sequence $\{F_n\}_{n \in \mathbb{N}}$ defined by $F_0 = 1$, $F_1 = 3$, and $F_n = F_{n-1} + F_{n-2}$ for all $n \geq 2$.

25. Find all terms up to F_{10} in the sequence $\{F_n\}_{n \in \mathbb{N}}$ defined by $F_0 = 0$, $F_1 = 1$, $F_2 = 1$ and $F_n = F_{n-1} + F_{n-2} + F_{n-3}$ for all $n \geq 3$.

26. Find all terms up to A_5 in the sequence of 2×2 matrices $\{A_n\}_{n \in \mathbb{N}}$ defined by $A_0 = I_2$, $A_1 = \begin{pmatrix} 1 & 2 \\ 3 & 4 \end{pmatrix}$, and $A_n = 3A_{n-1} - 2A_{n-2}$ for all $n \geq 2$.

27. Define $f : \mathbb{R}^{\geq 0} \to \mathbb{R}$ by $f(x) = 3x$ if $0 \leq x \leq 2$ and $f(x) = f(\sqrt{x})$ if $x > 2$.
    (a) Calculate $f(3)$, $f(5)$ and $f(100)$.
    (b) Find all the pre-images of $1/2$ and of 6.

28. (*) Prove that the function $f : \mathbb{R}^{\geq 0} \to \mathbb{R}^{\geq 0}$ defined recursively by

$$f(x) = \begin{cases} x & \text{if } 0 \leq x < 2 \\ 2f(x - 2) + 2 & \text{if } x \geq 2. \end{cases}$$

is continuous.

In calculus or earlier, students encounter the summation symbol. There also exists a similar product symbol. If $(a_k)_{k=1}^n$ is a finite sequence of numbers (or functions or matrices), then

$$\sum_{k=m}^{n} a_k \stackrel{\text{def}}{=} a_m + a_{m+1} + \cdots + a_n \quad and \quad \prod_{k=m}^{n} a_k \stackrel{\text{def}}{=} a_m a_{m+1} \cdots a_n.$$

If S is a subset of a set of numbers (functions or matrices), then the notations

$$\sum_{j \in S} a_j \quad and \quad \prod_{i \in S} a_j$$

mean the sum and product, respectively, of all the terms a_j satisfying $j \in S$. Exercises 3.1.29 through 3.1.33 involve practice with these useful symbols.

29. Calculate: (a) $\displaystyle\sum_{k=1}^{5} k^2$ (b) $\displaystyle\sum_{k=1}^{10} \lfloor \sqrt{k} \rfloor$ (c) $\displaystyle\sum_{k=-1}^{3} 3^k$.

30. Calculate: (a) $\displaystyle\prod_{k=1}^{5} 1 + 2^k$ (b) $\displaystyle\prod_{k=1}^{10} \lfloor \sqrt{k} \rfloor$ (c) $\displaystyle\prod_{k=1}^{7} 2 + (-1)^k$.

31. For $n = 1, 2, 3, 4, 5$, calculate the values of the following.

    (a) $a_n = n^2 + n + 1$;

    (b) a_n, where $a_1 = 1$ and $a_n = a_{n-1}^2 + a_{n-1} + 1$ for all $n \geq 2$;

    (c) $\displaystyle\sum_{k=1}^{n} k^2 + k + 1$;

    (d) $\displaystyle\prod_{k=1}^{n} k^2 + k + 1$.

32. (*) Find a formula for $\sum_{k=1}^{n} \lfloor \sqrt{k} \rfloor$.

33. Let $S = \{2, 3, 5, 7\}$. Calculate: (a) $\displaystyle\sum_{k \in S} k^2$; (b) $\displaystyle\sum_{k \in S} \frac{1}{k}$; (c) $\displaystyle\prod_{k \in S} k!$;

    (d) $\displaystyle\prod_{k \in S} \frac{1}{k}$.

3.2 Properties of Functions

The reader should have encountered the topics in this section in precalculus, calculus and linear algebra. However, these properties of functions play a fundamental role in set theory, and therefore throughout all mathematics.

3.2.1 Injective, surjective, bijective

Definition 3.2.1

We say that a function $f : A \to B$ is

1) injective (one-to-one) when $f(a_1) = f(a_2) \implies a_1 = a_2$;

2) surjective (onto) when for all $b \in B$, there exists $a \in A$ such that $f(a) = b$;

3) bijective (one-to-one and onto) when f is both injective and surjective.

We also say that a function f is a *injection*, a *surjection*, or a *bijection*.

The contrapositive of the definition for injective offers an alternative definition, namely $a_1 \neq a_2 \implies f(a_1) \neq f(a_2)$. This equivalent statement lines up with a colloquial way of thinking about injectivity: "different inputs give different images."

As we will see in the examples below, to prove a function is injective, solve $f(x_1) = f(x_2)$ for x_1 in terms of x_2. If the only solution is $x_1 = x_2$, then f is injective. To prove surjectivity, we must show that for any y in the domain, the equation $f(x) = y$ has a solution in the domain of f.

Definition 3.2.2

The *range* of a function $f : A \to B$ is the subset in the codomain $\{b \in B \mid \exists a \in A\, f(a) = b\}$.

The range is the set of all the images under the function. With the concept of range at our disposal, we can say that a function is surjective if and only if its range is equal to the codomain.

As the following example illustrates, the domain, the codomain, and the association itself all come into play when determining if a function is injective or surjective.

Example 3.2.3. Consider the function $f : \mathbb{R} \to \mathbb{R}$ with $f(x) = x^2$. The function is not injective, because for example $f(-2) = f(2)$. If we restrict the domain to the new function $f_1 : \mathbb{R}^{\geq 0} \to \mathbb{R}$ with $f_1(x) = x^2$, then f_1 is injective. We can see this as follows. If $f(x_1) = f(x_2)$, then $x_1^2 = x_2^2$. Hence

$$x_1^2 - x_2^2 = 0 \implies (x_1 - x_2)(x_1 + x_2) = 0 \implies x_1 = x_2 \text{ or } x_1 = -x_2.$$

Since x_1 and x_2 are the same sign, we deduce that $x_1 = x_2$.

The function f_1 is still not surjective because $f_1(x) = -1$ has no solutions. This is a matter of the codomain. If we define $f_2 : \mathbb{R}^{\geq 0} \to$

$\mathbb{R}^{\geq 0}$ by $f_2(x) = x^2$, it is still injective but it is now surjective with $f(x) = y$ implies $x = \sqrt{y}$.   △

Example 3.2.4. Let $[a, b]$ be a closed interval of $\mathbb{R}$. Consider the operator of differentiation $D : C^1([a, b]) \rightarrow C^0([a, b])$ by $D(f) = f'$. This is not injective because for example $D(x^2) = 2x = D(x^2 + 7)$. This operator is surjective, but this result follows from the Fundamental Theorem of Calculus, which implies that if g is continuous over $[a, b]$, then it has a continuous antiderivative $f : [a, b] \rightarrow \mathbb{R}$ such that $D(f) = f' = g$. In fact,

$$f(x) = \int_a^x g(u)\, du.$$

Hence, D is surjective.   △

Example 3.2.5. Consider the linear transformation $T : \mathbb{R}^4 \rightarrow \mathbb{R}^3$ defined by $T(\mathbf{x}) = A\mathbf{x}$, where

$$A = \begin{pmatrix} -2 & -1 & 0 & 1 \\ 2 & 3 & 4 & 5 \\ 6 & 7 & 8 & 9 \end{pmatrix}.$$

To decide injectivity, we first suppose that $T(\mathbf{x}_1) = T(\mathbf{x}_2)$. This means that $A\mathbf{x}_1 = A\mathbf{x}_2$ and using matrix algebra, we see this is equivalent to $A(\mathbf{x}_1 - \mathbf{x}_2) = \mathbf{0}$. Since

$$\text{rref } A = \begin{pmatrix} 1 & 0 & -1 & -2 \\ 0 & 1 & 2 & 3 \\ 0 & 0 & 0 & 0 \end{pmatrix},$$

using standard methods from linear algebra, we deduce that the possible values of $\mathbf{x}_1 - \mathbf{x}_2$ can be

$$\mathbf{x}_1 - \mathbf{x}_2 = s \begin{pmatrix} 1 \\ -2 \\ 1 \\ 0 \end{pmatrix} + t \begin{pmatrix} 2 \\ -3 \\ 0 \\ 1 \end{pmatrix} \qquad \text{for any } s, t \in \mathbb{R}.$$

In particular, $T(\mathbf{x}_1) = T(\mathbf{x}_2)$ does not imply $\mathbf{x}_1 = \mathbf{x}_2$, so T is not injective.

To decide surjectivity, the usual strategy involves considering an arbitrary element $\mathbf{y}$ in the codomain $\mathbb{R}^3$ and checking whether $T(\mathbf{x}) = \mathbf{y}$ has a solution in $\mathbf{x}$. However, theorems from linear algebra give us a different approach. The range of T consists of the span of the columns of A. Furthermore, a basis of the span of the columns of A consists of the columns that contain a pivot. From $\mathrm{rref}(A)$, we see that a basis of the range of T, also called the *image* of T, is

$$\left\{ \begin{pmatrix} -2 \\ 2 \\ 6 \end{pmatrix}, \begin{pmatrix} -1 \\ 3 \\ 7 \end{pmatrix} \right\}.$$

The range of T is a 2-dimensional subspace of $\mathbb{R}^3$, so it cannot be all of $\mathbb{R}^3$. Thus T is not surjective.   △

3.2.2   Function composition

The reader should have encountered the concept of composition of functions in high school algebra. A typical observation at this stage is that function composition is not commutative: For example, if $f(x) = x^2 + 3x$ and $g(x) = x + 1$, then

$$(f \circ g)(x) = f(g(x)) = x^2 + 5x + 4$$
$$\text{but} \quad (g \circ f)(x) = g(f(x)) = x^2 + 3x + 1.$$

Following the previous chapter, the reader should note that we did not properly introduce the functions in the previous sentence, but only because we were imitating language from earlier courses. For both these functions we should have specified that $f : \mathbb{R} \to \mathbb{R}$ and $g : \mathbb{R} \to \mathbb{R}$.

In calculus, function composition rears its head with the chain rule. We now generalize the concept to arbitrary sets and functions.

Definition 3.2.6

Let A, B, and C be sets, and let $f : A \to B$ and $g : B \to C$ be two functions. The *composition* of g with f is the function $g \circ f : A \to C$ defined by for all $x \in A$,

$$(g \circ f)(x) = g(f(x)).$$

Example 3.2.7. Let P be the set of people, let $m : P \to P$ be the function of a person's biological mother, and let $h : P \to \mathbb{N}$ be the function of a person's height rounded down to the closest centimeter. Then the composition $h \circ m : P \to \mathbb{N}$ is such that $(h \circ m)(p)$ is the height of person p's mother.   △

Figure 3.2 gives a visual representation of function composition.

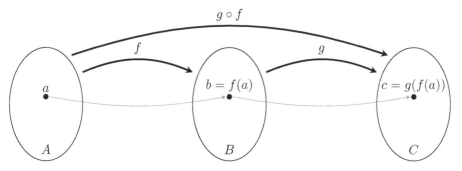

Figure 3.2: Function composition.

Proposition 3.2.8

Let $f : A \to B$, $g : B \to C$, and $h : C \to D$ be functions. Then
$$h \circ (g \circ f) = (h \circ g) \circ f.$$

Proof. Let x be an arbitrary element in A.[3] Then

$$(h \circ (g \circ f))(x) = h((g \circ f)(x)) = h(g(f(x)))$$
$$= (h \circ g)(f(x)) = ((h \circ g) \circ f)(x).$$

Since the functions are equal on all elements of A, they are equal. $\square$

Because of their order of presentation, some courses in linear algebra do not fully show why we define matrix multiplication as we do. An important theorem in linear algebra establishes that any linear transformation $S : \mathbb{R}^m \to \mathbb{R}^n$ has the form $S(\mathbf{v}) = A\mathbf{v}$, where A is an $n \times m$ matrix. We say that A represents S with respect to the standard basis. Suppose that $T : \mathbb{R}^n \to \mathbb{R}^p$ is a linear transformation such that $T(\mathbf{v}) = B\mathbf{v}$ for some $p \times n$ matrix B. Matrix multiplication is defined precisely so that the product BA represents the composition linear transformation $T \circ S$. (See Exercise 3.2.10.) This fact together with Proposition 3.2.8 prove associativity of matrix multiplication.

Definition 3.2.9

Let A be any set. The *identity* function on A is $\iota_A : A \to A$ where $\iota_A(x) = x$ for all $x \in A$.

The terminology of identity makes sense in two ways. First, we are accustomed to calling an "identity" a function that does not change its inputs. Second, identity functions serve as an identity to the operation of composition in the following way.

---

[3]Notice the strategy of taking an arbitrary element x in the domain and verifying that the functions are equal on x.

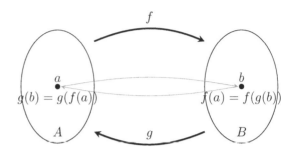

Figure 3.3: Inverse function.

Proof. The domain of $\iota_B \circ f$ is A and the codomain is B. Furthermore, for any $x \in A$ we have $(\iota_B \circ f)(x) = \iota_B(f(x)) = f(x)$. Thus $\iota_B \circ f = f$.

Similarly, the domain of $f \circ \iota_a$ is A and the codomain is B. Furthermore, for any $x \in A$ we have $(f \circ \iota_A)(x) = f(\iota_A(x)) = f(x)$. Thus $f \circ \iota_A = f$. □

3.2.3 Inverse function

Definition 3.2.11

Let $f : A \to B$ be a function. An *inverse function* to f is a function $g : B \to A$ such that $f \circ g = \iota_B$ and $g \circ f = \iota_A$.

Figure 3.3 depicts the relationship of inverse functions.

Proposition 3.2.12

Let $f : A \to B$ be a function. If an inverse function exists, then it is unique.

Proof. Suppose that g_1 and g_2 are two inverse functions to f. By definition, they both have domain B and codomain A. Then

$$
\begin{aligned}
g_2 &= \iota_A \circ g_2 && \text{by definition of identity function} \\
&= (g_1 \circ f) \circ g_2 && \text{since } g_1 \text{ is an inverse to } f \\
&= g_1 \circ (f \circ g_2) && \text{Proposition 3.2.8} \\
&= g_1 \circ \iota_B && \text{since } g_2 \text{ is an inverse to } f \\
&= g_1.
\end{aligned}
$$

Hence, if an inverse exists, it is unique. □

Because inverses to functions are unique, we can talk about "the" inverse function. We denote the inverse function of f by f^{-1}.

Until now, we could only talk about "if f has an inverse." The following proposition gives a characterization of when a function has an inverse function.

Proposition 3.2.13

A function $f : A \to B$ has an inverse function if and only if f is bijective.

Proof. First, suppose that $f : A \to B$ has an inverse function f^{-1}.   ($\Longrightarrow$)
Suppose that $f(a_1) = f(a_2)$. Then $f^{-1}(f(a_1)) = f^{-1}(f(a_2))$ so $a_1 = a_2$. Therefore, f is injective. To prove surjectivity, let $b \in B$ be arbitrary. Then $f^{-1}(b)$ is an element in A and $f(f^{-1}(b)) = b$. Hence, f is surjective.

Conversely, suppose that f is bijective. Since f is surjective, for   ($\Longleftarrow$)
all $b \in B$, there exists $a \in A$ with $f(a) = b$. Furthermore, since f is injective, there exists a unique $a \in A$ such that $g(a) = b$. Hence, we can define the function $g : B \to A$ by $g(b) = a$ such that $f(a) = b$. Then, for all $b \in B$ we have $f(g(b)) = b$ and for all $a \in A$ we have $g(f(a)) = a$. Thus $f \circ g = \iota_B$ and $g \circ f = \iota_A$. Hence, g is an inverse function to f.   □

It is common in advanced mathematics to define a property in a way that is natural for how we use it, but to then prove a theorem that characterizes all objects with that property. Definition 3.2.11 followed by Proposition 3.2.13 exemplifies this habit.

Example 3.2.14. Consider the function $f : \mathbb{R} \to \mathbb{R}^{>0}$ defined by $f(x) = e^x$. This function is a bijection and its inverse function is $f^{-1} : \mathbb{R}^{>0} \to \mathbb{R}$ with $f^{-1}(x) = \ln x$.   △

Example 3.2.15 (Arcsine function). Consider the function $f : \mathbb{R} \to \mathbb{R}$ defined by $f(x) = \sin x$. It is not bijective. If we restrict the codomain to $[-1, 1]$, then the resulting function is surjective. In order to consider an inverse function, we also need to restrict the domain to a subset over which the resulting function is injective. We have choices here. For example, $f_1 : [-\pi/2, \pi/2] \to [-1, 1]$ with $f_1(x) = \sin x$ does happen to be bijective. So would be $f_2 : [\pi/2, 3\pi/2] \to [-1, 1]$ with $f_2(x) = \sin x$ or even $f_3 : [3\pi/2, 5\pi/2] \to [-1, 1]$ with $f_3(x) = \sin x$. The arcsine function, denoted $\arcsin x$ or $\sin^{-1} x$ specifically makes the choice of the domain $[-\pi/2, \pi/2]$. Thus, by convention, $\sin^{-1}$ refers to the inverse function to f_1.   △

We mention another property of composition and taking inverses, but we leave the proof to the reader.

Proposition 3.2.16

Suppose that $f : A \to B$ and $g : B \to C$ are bijective functions. Then $g \circ f : A \to C$ is also a bijection and $(g \circ f)^{-1} = f^{-1} \circ g^{-1}$.

3.2.4   Restriction, images, fibers

It is not uncommon to consider how a function behaves on a subset of elements in the domain. If $f : A \to B$ is a function and $S \subseteq A$, we regularly use the following shorthand notation:

$$f(S) \stackrel{\text{def}}{=} \{f(s) \mid s \in S\} = \{b \in B \mid \exists s \in S, f(s) = b\}. \tag{3.1}$$

This is the image set of S under f.

We have emphasized how the domain and codomain are part of the function's definition. At times, we would like to consider the function but on a smaller domain. If $S \subseteq A$, we define the *restriction* of f to S, and denote it by $f|_S$, the function $f|_S : S \to B$ such that $f|_S(x) = f(x)$ for all $x \in S$.

If $T \subseteq B$, we also use the shorthand notation

$$f^{-1}(T) \stackrel{\text{def}}{=} \{a \in A \mid f(a) \in T\}. \tag{3.2}$$

This set is called the *pre-image* of T by f. It is essential to note that using this latter notation does not presume that f is bijective; the definition in (3.2) is a matter of notation. If $T = \{b\}$, consisting of a single element $b \in B$, then $f^{-1}(\{b\})$ is called the *fiber* of b.

For example, if $f : \mathbb{R} \to \mathbb{R}$ is defined by $f(x) = \sin x$, then the fiber of 2 is $f^{-1}(\{2\}) = \emptyset$, and the fiber of $\frac{1}{2}$ is

$$f^{-1}(\{1/2\}) = \left\{\frac{\pi}{6} + 2\pi k \mid k \in \mathbb{Z}\right\} \cup \left\{\frac{5\pi}{6} + 2\pi k \mid k \in \mathbb{Z}\right\}.$$

EXERCISES FOR SECTION 3.2

1. For each of the following real-valued functions, determine the largest possible domain D as a subset of $\mathbb{R}$ and then prove whether $f : D \to \mathbb{R}$ is an injection, surjection, both, or neither.

   (a) $f(x) = -3x + 4$ (c) $f(x) = (x + 1)/(x + 2)$

   (b) $f(x) = -3x^2 + 7$ (d) $f(x) = x^5 + 1$

2. Give an explicit example of a function $f : \mathbb{Z} \to \mathbb{Z}$ that is

   (a) surjective but not injective;   (c) neither injective nor surjec-

   (b) injective but not surjective;   tive;

                                        (d) bijective.

3. Given an explicit example of a function $f : \mathbb{N} \to \mathbb{N}$ that is
   (a) bijective;
   (b) surjective but not injective;
   (c) injective but not surjective;
   (d) neither injective nor surjective.

4. Determine if the function $f : \mathbb{N} \times \mathbb{N} \to \mathbb{N}$ is onto when f is
   (a) $f(a, b) = a + 2b$;
   (b) $f(a, b) = |a - b|$;
   (c) $f(a, b) = a^2 + b^2$;
   (d) $f(a, b) = |3^a - 2^b|$.

5. Show that the function $f : M_{3 \times 2}(\mathbb{R}) \to M_{2 \times 2}(\mathbb{R})$ defined by $f(A) = A^\top A$ is not surjective.

6. Let m and n be positive integers. Is rref : $M_{m \times n}(\mathbb{R}) \to M_{m \times n}(\mathbb{R})$ injective or surjective?

7. Prove that if $f : A \to B$ and $g : B \to C$ are both injective, that $g \circ f : A \to C$ is also injective.

8. Prove that if both $f : A \to B$ and $g : B \to C$ are surjective, that $g \circ f : A \to C$ is also surjective.

9. Prove that if $g \circ f$ is injective, then f is injective.

10. Let $S : \mathbb{R}^m \to \mathbb{R}^n$ and $T : \mathbb{R}^n \to \mathbb{R}^p$ be linear transformations represented respectively by an $n \times m$ matrix A and a $p \times n$ matrix B. Prove that the matrix product BA represents the composition $T \circ S$.

11. Prove Proposition 3.2.16.

12. For the following functions f, find the image $f(S)$ of the given set S in the domain.
   (a) $f : \mathbb{R} \to \mathbb{R}$ with $f(x) = \lfloor x^2 \rfloor$ and $S = [2, 5]$.
   (b) $f : \mathbb{R} \to \mathbb{R}$ with $f(x) = \cos x$ and $S = \emptyset$.
   (c) $f : \mathbb{R} \to \mathbb{R}$ with $f(x) = x/(x^2 + 1)$ and $S = [4, \infty)$.

13. For the following functions f, find the pre-image $f^{-1}(T)$ of the given set T in the codomain.
   (a) $f : \mathbb{R} \to \mathbb{R}$ with $f(x) = \sin x$ and $T = \{\sqrt{3}/2\}$.
   (b) $f : \mathbb{R} \to \mathbb{R}$ with $f(x) = x^2 - 2$ and $T = [1, 2]$.
   (c) $f : \mathbb{R} \to \mathbb{R}$ with $f(x) = x^3 + 2x$ and $T = [-1, 0]$.

14. Let $f : A \to B$ be a function. Let S and T be subsets of the domain A. Recall the notation in (3.1).
   (a) Show that $f(S \cup T) = f(S) \cup f(T)$.
   (b) Show that $f(S \cap T) \subseteq f(S) \cap f(T)$.
   (c) Show that if f is injective, then $f(S \cap T) = f(S) \cap f(T)$.

15. Let $f : A \to B$ be a function. Let V and W be subsets of the codomain B. Recall the notation in (3.2). Show the following.
   (a) $f^{-1}(V \cup W) = f^{-1}(V) \cup f^{-1}(W)$.
   (b) $f^{-1}(V \cap W) = f^{-1}(V) \cap f^{-1}(W)$.

16. Let $f : A \to B$ be a function and let $T \subseteq B$. Prove that $f^{-1}(T^c) = f^{-1}(T)^c$.

17. Let $D : C^1(\mathbb{R}) \to C^0(\mathbb{R})$ be the derivative operator.

    (a) Explain why P, the set of polynomial functions, is a subset of $C^1(\mathbb{R})$.

    (b) Show that $D(P) \subseteq P$.

    (c) Show that $D|_P : P \to P$ is a surjection but not an injection.

18. Let $M_n(\mathbb{R})$ be the set of $n \times n$ matrices with coefficients in $\mathbb{R}$. Recall that the subset of symmetric matrices is $\mathcal{S} = \{B \in M_n(\mathbb{R}) \mid B^\top = B\}$ and that the subset of antisymmetric matrices is $\mathcal{A} = \{B \in M_n(\mathbb{R}) \mid B^\top = -B\}$. Prove that the function

$$\psi : M_n(\mathbb{R}) \to \mathcal{S} \times \mathcal{A} \text{ where } \psi(B) = \left(\frac{1}{2}(B + B^\top), \frac{1}{2}(B - B^\top)\right)$$

is a bijection and find the inverse function ψ^{-1}.

19. Let I be an interval of $\mathbb{R}$. Prove that if f is either strictly increasing or decreasing on I, then f is injective.

20. Show that the sum of two increasing functions is increasing. Also show that the sum of two decreasing functions is decreasing.

21. Show that the product of two positive increasing functions is increasing.

22. Prove that the composition of two strictly increasing functions is again strictly increasing.

23. Prove that the composition of two strictly decreasing functions is again strictly decreasing.

24. Let I be an interval in $\mathbb{R}$ and suppose that $f : I \to \mathbb{R}$ is increasing and convex.

    (a) Suppose that $g : f(I) \to \mathbb{R}$ is increasing and convex over the range of f. Show that $g \circ f$ is again convex.

    (b) Show that if $g : f(I) \to \mathbb{R}$ is convex but not necessarily increasing, then $g \circ f$ is not necessarily convex.

25. Prove that $f(x) = x^2$ is convex over $\mathbb{R}$ *without* using the theorem that $f''(x) > 0$ over an interval implies that f is convex over that interval.

26. (*) Prove that $f(x) = e^x$ is convex over $\mathbb{R}$ *without* using the theorem that $f''(x) > 0$ over an interval implies that f is convex over that interval. [Hint: Fix $x_1 < x_2$ in $\mathbb{R}$; consider $g(t) = (1-t)e^{x_1} + te^{x_2} - e^{(1-t)x_1 + tx_2}$; use Rolle's Theorem.]

27. Let $f_1, f_2 : \mathbb{R} \to \mathbb{R}$ be functions with $f_1(x) = x/3$ and $f_2(x) = 2/3 + (1/3)x$. Define the function $F : \mathcal{P}(\mathbb{R}) \to \mathcal{P}(\mathbb{R})$ by $F(A) = f_1(A) \cup f_2(A)$. Determine (a) $F([0,1])$; (b) $F(F([0,1]))$; (c) $F([-1,2])$.

3.3 Choice Functions; The Axiom of Choice

We finish this chapter with a brief discussion about *choice functions*. When we introduced set theory in Section 1.2, we mentioned that a thorough treatment of the **ZFC** axioms of set theory would detract from our presentation. However, it is valuable to discuss the Axiom of Choice, because it and its many equivalent variants play a key role in the proofs of various important theorems in advanced mathematics.

Definition 3.3.1

Let $\mathcal{C}$ be a nonempty collection of nonempty sets. A *choice function* on $\mathcal{C}$ is a function $f : \mathcal{C} \to \bigcup_{S \in \mathcal{C}} S$ such that $f(S) \in S$ for all $S \in \mathcal{C}$.

Example 3.3.2. Let $\mathcal{C} = \{\{1,2,3,4\}, \{5,6,7\}, \{4,7,8,42\}\}$. The association f with

$$f(\{1,2,3,4\}) = 3, \quad f(\{5,6,7\}) = 7, \quad \text{and } f(\{4,7,8,42\}) = 42$$

is a choice function on the collection $\mathcal{C}$. $\triangle$

Example 3.3.3. Let $\mathcal{C} = \{\mathbb{Z}, \mathbb{Q} \setminus \mathbb{Z}, \mathbb{R} \setminus \mathbb{Q}\}$. The function $f : \mathcal{C} \to \mathbb{R}$ with $f(\mathbb{Z}) = 3$, $f(\mathbb{Q}\setminus\mathbb{Z}) = \frac{22}{7}$, and $f(\mathbb{R}\setminus\mathbb{Q}) = \pi$ is a choice function. $\triangle$

From these first two examples, it seems obvious that choice functions should exist for any collection of sets X. In fact, mathematicians implicitly referred to choice functions in proofs well before Ernst Zermelo formulated the Axiom of Choice in 1904, while working on the development of axiomatic set theory. The axioms is as follows.

Axiom 3.3.4 (Axiom of Choice)

For every collection $\mathcal{C}$ of nonempty sets, there exist choice functions.

Inclusion of this axiom into Zermelo-Frankel set theory stirred up considerable controversy in the early 20th century, not because of simple examples such as those given above, but because of some implications with infinite collections of sets that felt counterintuitive. Consider the following two examples.

Example 3.3.5. Let $\mathcal{L}$ be the set of all lines in the Euclidean plane. The union of all the lines in the Euclidean plane is the whole Euclidean plane, $\mathbb{E}^2$. A choice function of $\mathcal{L}$ is a function $f : \mathcal{L} \to \mathbb{E}^2$ (the Euclidean plane) such that $f(L)$ is a point on L. Here is one example of a choice function on $\mathcal{L}$: If O is any point in $\mathbb{E}^2$, define $f_O(L)$ to be the point on L that is closest to O. Note that if $O \in L$, then

$f_O(L) = O$, and if $O \notin L$, then $f_O(L) = L \cap M$, where M is the perpendicular line to L through O. △

Example 3.3.6. Example 3.3.5 offers a constructive example of a choice function. However, consider the collection $\mathcal{S}$ of all nonempty subsets of $\mathbb{R}$. It is hard (impossible?) to imagine an explicit choice function on $\mathcal{S}$. For a given $A \in \mathcal{S}$, we cannot talk about, the least element in A, or the largest element in A, or the center of A since these concepts might not be defined. Even in this example, the Axiom of Choice allows us to posit the existence of such a choice function.△

The Axiom of Choice has many variants. It does not just imply, but is equivalent to, the proposition that every vector space has a basis. That finite dimensional vector spaces have bases is obvious. However, that the vector space of all continuous real-valued functions over an interval has a basis should feel quite strange; indeed it is difficult to image such a basis. Some equivalents to the Axiom of Choice feel natural, intuitive, or almost obvious such as Zorn's Lemma, the Trichotomy Law of Cardinalities, and the Well-Ordering Theorem. Other consequences like the Banach-Tarski Paradox strain the imagination.

Because of the controversy, some mathematicians took care to be clear as to whether they worked with **ZF**, which means Zermelo-Frankel set theory without the Axiom of Choice, or **ZFC**, which means Zermelo-Frankel set theory with the Axiom of Choice.

Work by Kurt Gödel and Paul Cohen established that the Axiom of Choice is independent of **ZF**. More specifically, Gödel [37] proved that if **ZF** is consistent, then **ZF** with the Axiom of Choice is consistent, and later Cohen [19] proved that if **ZF** is consistent, then **ZF** with the negation of the Axiom of Choice is also a consistent axiomatic system. So, in some ways, we can opt to do mathematics with the Axiom of Choice or without it. Today, however, the vast majority of mathematicians use **ZFC**.

We can approach the Axiom of Choice from a different direction: through the concept of Cartesian products. We discussed the Cartesian product of a finite collection of sets in Section 1.2.3. The reader should wonder what we mean by a Cartesian product of an arbitrary collection of sets. We clarify this presently.

Suppose that $X = \{A, B, C\}$ is a collection of three sets. By definition, the Cartesian product $A \times B \times C$ consists of all triples (a, b, c) such that $a \in A$, $b \in B$, and $c \in C$. Any triple $t = (a, b, c)$ defines a choice function f_t via $f_t(A) = a$, $f_t(B) = b$ and $f_t(C) = c$. Conversely, any choice function $f : X \to A \cup B \cup C$ defines a triple $t_f = (f(A), f(B), f(C))$. This thought process extends easily to any positive integer n. For a collection X of n sets, a choice function on X

is tantamount to an n-tuple in the Cartesian product. This inspires a definition for the Cartesian product of an arbitrary collection of sets.

Definition 3.3.7

Let $\mathcal{C}$ be a collection of sets. The Cartesian product of $\mathcal{C}$ is the set of all choice functions on $\mathcal{C}$, namely

$$\prod_{A \in \mathcal{C}} A \overset{\text{def}}{=} \left\{ f : \mathcal{C} \to \bigcup_{A \in \mathcal{C}} A \, \middle| \, \forall A \in \mathcal{C}, f(A) \in A \right\}.$$

Sometimes, we use an indexing set for the collection of sets $\mathcal{C}$ and write $\mathcal{C} = \{A_i\}_{i \in \mathcal{I}}$. With this notation, we write the Cartesian product of $\mathcal{C}$ as

$$\prod_{i \in \mathcal{I}} A_i \overset{\text{def}}{=} \left\{ f : \mathcal{I} \to \bigcup_{i \in \mathcal{I}} A_i \, \middle| \, \forall i \in \mathcal{I}, f(i) \in A_i \right\}.$$

The concept of a generalized Cartesian product leads to an alternate form of the Axiom of Choice.

Axiom 3.3.8 (Axiom of Choice, alternate form)

The Cartesian product of any collection $\mathcal{C}$ of nonempty sets is nonempty.

As the student might see in various advanced mathematics topics, the Axiom of Choice sometimes provides the key step for a variety of deep theorems. The following proposition illustrates a simple application.

Proposition 3.3.9

Let $f : A \to B$ be a surjective function. Then there exists a function $h : B \to A$ such that $f \circ h = \iota_B$.

Proof. Consider the collection of fibers $\mathcal{C}$ of elements in B. In other words,

$$\mathcal{C} = \left\{ f^{-1}(\{b\}) \, \middle| \, b \in B \right\}.$$

Since f is surjective, then every set in the collection $\mathcal{C}$ is nonempty. By the Axiom of Choice, there exists a function $g : \mathcal{C} \to A$ with $h(x) \in x$ for all $x = f^{-1}(\{b\})$. We define the function $h : B \to A$ by $h(b) = g(f^{-1}(\{b\}))$. Then for all $b \in B$, since $h(b) \in f^{-1}(\{b\})$, then by definition $f(h(b)) = b$. Thus, $f \circ h = \iota_B$. $\square$

We point out that this does not mean that f is invertible and hence a bijection. The function h is not an inverse function to f but only a right-inverse to f.

1. We revisit Example 3.3.5. Let $O = (0,0)$. Using the general equation of a line, every line L in the plane can be described by an equation $ax + by = c$, where $(a,b) \neq (0,0)$.

   (a) Express $f_O(L)$ in coordinates in terms of a, b, c, where L is given by $ax + by = c$.

   (b) Show that this expression creates a well-defined choice function. [Hint: If L is the solution to $ax + by - c$, then it is also the solution to $(\lambda a)x + (\lambda b)y = \lambda c$, where $\lambda \neq 0$.]

2. Consider the set of all choice functions f_O, defined in Example 3.3.5, each corresponding to a point $O \in \mathbb{E}^2$. Show that not all choice functions $\mathcal{L} \to \mathbb{E}^2$ are equal to f_O for some point $O \in \mathbb{E}^2$.

3. Let $\mathcal{C}$ be the collection of circles (of positive radius) in the Euclidean plane $\mathbb{E}^2$. Describe a choice function $f : \mathcal{C} \to \mathbb{E}^2$.

4. Let $\mathcal{C}$ be a collection of sets that are mutually disjoint, i.e., for all $A, B \in \mathcal{C}$ if $A \neq B$, then $A \cap B = \emptyset$. Show that there exists a set Z such that $Z \cap A$ consists of a single element, for all $A \in \mathcal{C}$.

5. (*) Prove that the Axiom of Choice is equivalent to the following statement: For any nonempty set S, there exists a function $f : \mathcal{P}(S) \setminus \{\emptyset\} \to S$ such that $f(B) \in B$ for all $B \in \mathcal{P}(S) \setminus \{\emptyset\}$.

CHAPTER 4

Properties of the Integers

Leopold Kronecker, the 19th century mathematician, famously said, "God gave us the integers; all else is the work of man." Some philosophers of mathematics balk at this statement's modernist hubris and underscore that mathematicians discover and label, but do not create the patterns in mathematics. Nevertheless, the integers form a cornerstone of mathematics. Studying their properties should feel safe, tame. After all, we learn about integers from the earliest grades in primary school. However, the integers never end. Consequently, to prove properties about all of them, we need a structural definition.

Section 4.1 presents a list of axioms of the natural numbers, $\mathbb{N}$, using the most common modern approach, by Peano. Sections 4.2 through 4.4 study divisibility and primality. It is an unfortunate necessity of education that when students first learn arithmetic properties of integers, they do not see the proofs; we will emphasize proofs. Section 4.5 presents a powerful proof technique, called induction. Finally, Section 4.6 discusses congruence of integers.

4.1 A Definition of the Integers

4.1.1 Peano's axioms of the natural numbers, $\mathbb{N}$

Definition 4.1.1 (Peano's Axioms)

The set $\mathbb{N}$ satisfies the following properties:

1) $\mathbb{N}$ contains an element denoted by 0.

2) There exists an injective function $S : \mathbb{N} \to \mathbb{N}$ called the *successor* function.

3) For all $n \in \mathbb{N}$, it is not true that $S(n) = 0$.

4) If K is any set such that $0 \in K$, and $n \in K$ implies $S(n) \in K$, then K contains $\mathbb{N}$.

Peano [62] wrote his original formulation purely in the context of logic without set theory. Using set theory, the above list of axioms for $\mathbb{N}$ can be more concise than Peano's original definition.

The first three axioms in Peano's definition of $\mathbb{N}$ establish a set that contains an element called 0 (zero), and where every element besides 0 has a unique successor. This definition lines up with our intuitive notion for integers in the following sense. We designate the element $S(0)$ as 1, the element $S(S(0))$ as 2, and so on. Consequently, the positive integers correspond to how often we apply the successor function S to the element 0.

The last axiom in the definition is called the *axiom of induction*. The thrust of this axiom is to make $\mathbb{N}$ the smallest set that satisfies the first three axioms. The axiom of induction has multiple reformulations, each with profound consequences.

Proposition 4.1.2

Assuming that a set satisfies the first three of Peano's axioms, then the fourth axiom is equivalent to the following restatement: If P is any predicate on $\mathbb{N}$ such that $P(0)$ is true and $\forall n \in \mathbb{N}$, $P(n) \to P(S(n))$, then $\forall n \in \mathbb{N}$, $P(n)$.

Proof. Let us first suppose that $\mathbb{N}$ satisfies all four of Peano's original axioms. Let $P(n)$ be a predicate on $\mathbb{N}$ such that $P(0)$ is true and such that $P(n) \implies P(S(n))$. Define the set $K = \{n \in \mathbb{N} \mid P(n) \text{ is true}\}$. From the hypothesis on P, we know that $0 \in K$ and also, if $n \in K$ then $S(n) \in K$. Thus, by (4) in Peano's axioms, $\mathbb{N} \subseteq K$. Since $K \subseteq \mathbb{N}$ by definition, then $K = \mathbb{N}$. Thus $P(n)$ is true for all $n \in \mathbb{N}$.

Conversely, suppose instead that we defined $\mathbb{N}$ as satisfying the first three axioms in Definition 4.1.1 along with the fourth axiom that if P is any predicate on $\mathbb{N}$ such that $P(0)$ is true and $\forall n \in \mathbb{N}$, $P(n) \to P(S(n))$, then $\forall n \in \mathbb{N}$, $P(n)$. Let K be any set such that $0 \in K$, and $n \in K$ implies $S(n) \in K$. Then let $P'(n)$ be the specific predicate "$n \in K$." Clearly, $P'(0)$ is true and for all $n \in \mathbb{N}$, if $P'(n)$, then $P'(S(n))$. From our assumption, this implies that $P'(n)$ is true for all $n \in \mathbb{N}$, which means that $\mathbb{N} \subseteq K$. $\square$

Because of the equivalence in Proposition 4.1.2, the alternate axiom is also called the axiom of induction. In fact, this equivalent formulation leads to an very effective proof technique called an induction proof, which we discuss in Section 4.5.

Despite the abstraction of Definition 4.1.1, it is not particularly difficult to recast all common notions about the nonnegative integers $\mathbb{N}$ using Peano's axioms. We consider addition, multiplication and inequalities.

Definition 4.1.3

Let $m, n \in \mathbb{N}$. If $m \neq 0$, then $m = (S \circ S \circ \cdots \circ S)(0)$ for some sequence of compositions of the successor functions. Call this composition T. Then we define the binary operation of addition $+$ on $\mathbb{N}$ by

$$n + m \stackrel{\text{def}}{=} \begin{cases} n & \text{if } m = 0 \\ T(n) & \text{otherwise.} \end{cases}$$

From this definition, it is easy to prove the well-known results that $+$ is associative, commutative, and has 0 as an identity. Note in particular that according to this notation, $S(n) = n + 1$.

In this formulation of $\mathbb{N}$, the concept of less than or equal to has the following definition.

Definition 4.1.4

Let $m, n \in \mathbb{N}$. We say that m is less than or equal to n, and we write $m \leq n$, if there exists $k \in \mathbb{N}$ such that $m + k = n$. We also say that m is less than n and write $m < n$, if $m \leq n$ and $m \neq n$.

In other words, $m < n$ if there exists a sequence of successor functions such that $n = (S \circ S \circ \cdots \circ S)(m)$. With the concept of inequality at our disposal, we say that an element n of $\mathbb{N}$ is *positive* if $0 < n$.

We can use Peano's axioms to define multiplication as follows.

Definition 4.1.5

Define the binary operation of multiplication $\cdot$ on $\mathbb{N}$ recursively by:

- $\forall a \in \mathbb{N} \, a \cdot 0 = 0$;
- $\forall a, b \in \mathbb{N} \, a \cdot S(b) = a + a \cdot b$.

All of the familiar properties about $+$, $\cdot$, and $\leq$ on $\mathbb{N}$ can be proven from this perspective. Though we explore a few of these in the exercises, we do not prove them all.

As we mentioned in this book's preface, our Introduction to Proofs enters in the middle of a story. The reader was already familiar with $\mathbb{N}$, $\mathbb{Z}$, $\mathbb{Q}$, $\mathbb{R}$ and $\mathbb{C}$, and we used these sets for many examples in the previous chapters. Peano's axioms provide a structural definition of $\mathbb{N}$, setting our understanding of the natural numbers in a solid set theory foundation.

With the concept of equivalence relations, Section 6.4.2 provides precise set theoretic definitions for $\mathbb{Z}$ based on $\mathbb{N}$, for $\mathbb{Q}$ based on $\mathbb{Z}$, for $\mathbb{R}$ based on $\mathbb{Q}$, and finally $\mathbb{C}$ based on $\mathbb{R}$. Until then, we continue to remain in the middle of a story that establishes precise definitions for standard number sets. In the meantime, we can think of the set of integers $\mathbb{Z}$ as pairs $\{-, +\} \times \mathbb{N}$, but where we identify $(-, 0)$ with $(+, 0)$ and we use the familiar arithmetic on the integers, simply bolstered now by Peano's definition of $\mathbb{N}$.

4.1.2   Well-ordering property

The axiom of induction has another equivalent formulation.

> **Theorem 4.1.6 (Well-Ordering Property on $\mathbb{N}$)**
>
> Let A be any nonempty subset of nonnegative integers. There exists $m \in A$ such that $m \le a$ for all $a \in A$. In other words, A contains a *minimal* element.

Proof. Assume[1] that A does not contain a minimal element. Let $P(n)$ be the predicate on $\mathbb{N}$ defined by $P(n) = \forall k \le n, k \in A^c$. Since $0 \le a$ for all $a \in A$, then 0 is not in A so $P(0)$ is true. Now suppose that $P(n)$ is true. This means that $k \in A^c$ for all integers k with $0 \le k \le n$. Since A does not have a minimal element, we cannot have $n + 1 \in A$. Hence $n + 1 \in A^c$, which implies that $P(n+1)$ is true. Hence, applying the axiom of induction, Proposition 4.1.2, to $P(n)$, we deduce that $P(n)$ is true for all $n \in \mathbb{N}$. This means that $A^c = \mathbb{N}$ and so $A = \emptyset$. This is a contradiction since, by hypothesis, A is a nonempty subset of $\mathbb{N}$. Hence, A must contain a minimal element. $\square$

This property is called the *well-ordering property* of $\mathbb{N}$ or simply the *well-ordering principle*. We leave it as an exercise to prove that it not only follows from the axiom of induction but is equivalent to it. (See Exercise 4.1.8.)

There is a good intuitive reason for why some authors state the well-ordering property instead of Axiom (4) in Definition 4.1.1. One structural particularity about $\mathbb{N}$ is that there is no element in $\mathbb{N}$ where other elements can get arbitrarily close to it. The well-ordering property expresses this intuitive concept with precision.

Example 4.1.7. Let $A = \{n \in \mathbb{N} \mid n^2 > 167\}$. A is nonempty because for example $20^2 = 400 > 167$ so $20 \in A$. Theorem 4.1.6 allows us to conclude that A has a minimal element. By trial and error or using some basic algebra, we can find that the minimal element of A is 13. $\triangle$

---

[1]Notice the proof by contradiction.

We temporarily reach beyond $\mathbb{N}$ and point out that neither $\mathbb{Z}$ nor $\mathbb{Q}^{\geq 0}$ satisfy the well-ordering principle. For example, in $\mathbb{Z}$, the set $A = \mathbb{Z}$ does not have a minimal element. In $\mathbb{Q}^{\geq 0}$, though 0 is a minimal element of $\mathbb{Q}^{\geq 0}$, not every subset has a minimal element. For example, let $A = \{x \in \mathbb{Q}^{\geq 0} \mid x > 1\}$. Assume that A has a minimal element x_0. Since $x_0 + 1 > 2$, we have $(x_0 + 1)/2 > 1$. Furthermore, since x_0 is rational, then so is $(x_0 + 1)/2$, so $(x_0 + 1)/2 \in A$. However,

$$x_0 > 1 \implies 2x_0 > x_0 + 1 \implies x_0 > (x_0 + 1)/2.$$

This contradicts the minimality of x_0 in A. Therefore, A does not contain any minimal element.

Example 4.1.7 gives a very rudimentary application of the well-ordering principle. The following example provides a more substantive one.

Example 4.1.8. We prove that $1 \cdot 2 + 2 \cdot 3 + \cdots + n(n+1) = n(n+1)(n+2)/3$ for all positive integers n. We call the formula $P(n)$. We observe that the formula holds for $n = 1$, i.e., $P(1)$ is true because the left-hand side is $1 \cdot 2 = 2$, while the right-hand side is $(1 \cdot 2 \cdot 3)/3 = 2$.

Assume that the formula does not always hold, namely assume that there exists $n \in \mathbb{N}^*$ with $P(n)$ is not true. Define the set

$$C = \left\{ n \in \mathbb{N}^* \;\middle|\; \sum_{k=1}^{n} k(k+1) \neq \frac{n(n+1)(n+2)}{3} \right\}.$$

The assumption means that C is not empty. By the well-ordering principle, C has a least element c. By the observation that $P(1)$ is true, we know that $c \geq 2$. Then

$$1 \cdot 2 + 2 \cdot 3 + \cdots + c(c+1) \neq \frac{c(c+1)(c+2)}{3}$$

and subtracting $c(c+1)$ on both sides implies

$$1 \cdot 2 + 2 \cdot 3 + \cdots + (c-1)c \neq \frac{c(c+1)(c+2)}{3} - c(c+1).$$

Combining fractions, we see that $c(c+1)(c+2)/3 - c(c+1) = (c-1)c(c+1)/3$. Hence, $1 \cdot 2 + 2 \cdot 3 + \cdots + (c-1)c \neq (c-1)c(c+1)/3$. But

$$\frac{(c-1)c(c+1)}{3} = \frac{(c-1)((c-1)+1)((c-1)+2)}{3},$$

which implies that c is not the least element of C. By contradiction, we conclude that C must be empty, which means that $\forall n \in \mathbb{N}^* \; P(n)$. $\triangle$

This example illustrates the common strategy to combine a proof by contradiction with the well-ordering principle. More explicitly, to prove a statement p, we assume its opposite $\neg p$ is true; show how $\neg p$ leads to a nonempty subset $A \subseteq \mathbb{N}$; invoke the well-ordering principle to affirm that A contains a minimal element m; then do something to show that A contains elements less than m. This contradicts the minimality of m as given by the well-ordering principle. This contradiction implies that $\neg p$ is false, and hence that p is true.

We will see this strategy at work a few times in key theorems. However, we would typically prove the formula in this example using induction (Section 4.5), but we used the well-ordering principle for pedagogical purposes.

4.1.3   Finite and infinite

We finish this first section about properties of integers by discussing a fundamental dichotomy in set theory.

Definition 4.1.9

A set A is called *finite* if either $A = \emptyset$ or there exists a bijection from A to $\{1, 2, \ldots, n\}$, where n is a positive integer. In this case, we write in shorthand $|A| = n$, and we also write $|\emptyset| = 0$.

If a set A is not finite, it is called *infinite*.

Having given a definition for an infinite set, we should show that such sets exist.

Proposition 4.1.10

The set $\mathbb{N}$ is infinite.

Proof. We prove this by contradiction. Clearly $\mathbb{N} \neq \emptyset$. Assume that there exists a positive integer n and a bijection $f : \{1, 2, \ldots, n\} \to \mathbb{N}$. We claim that

$$1 + \sum_{i=1}^{n} f(i)$$

cannot be in the range of f. Since $f(i) \geq 0$ for all $i \in \{1, 2, \ldots, n\}$, then $f(j) \leq \sum_{i=1}^{n} f(i)$ for all $j \in \{1, 2, \ldots, n\}$. Hence, $f(j) < 1 + \sum_{i=1}^{n} f(i)$. So the integer $1 + \sum_{i=1}^{n} f(i)$ is not in the range of f. Thus, f is not surjective and therefore not a bijection. By contradiction, the proposition follows.   □

We leave a number of properties concerning finite and infinite sets to the exercises, but we prove one first result here.

Proposition 4.1.11

The union of two finite sets is finite.

Proof. Let A and B be two finite sets. Since $A \cup B = A \cup (B \setminus A)$ and $A \cap (B \setminus A) = \emptyset$, we can suppose without loss of generality that A and B are disjoint finite sets.

We consider a first case in which, $A = \emptyset$ or $B = \emptyset$. In this case, $A \cup B$ is equal to A and B. Therefore, $A \cup B$ is finite.

Now suppose that neither A nor B is empty. Then by definition there exist positive integers m and n and bijections

$$f : \{1, 2, \ldots, m\} \to A \text{ and } g : \{1, 2, \ldots, n\} \to B.$$

We define the function $h : \{1, 2, \ldots, m + n\} \to A \cup B$ by

$$h(k) = \begin{cases} f(k) & \text{if } k \leq m \\ g(k - m) & \text{if } k \geq m + 1. \end{cases}$$

We need to prove that this is a bijection. To prove that h is a surjection, consider $c \in A \cup B$. If $c \in A$, then $c = f(k)$ for some $1 \leq k \leq m$ and then $c = h(k)$; if $c \in B$, then $c = g(\ell)$ for some $1 \leq \ell \leq n$ and then $c = h(m + \ell)$. This shows that h is surjective. To prove that h is injective, suppose that $h(i) = h(j)$. If $h(i) \in A$, then $1 \leq i \leq m$ and so $f(i) = h(i) = h(j) = f(j)$ and thus $i = j$ since f is a bijection; if $h(i) \in B$, then $m + 1 \leq i \leq m + n$ and so $g(i - m) = h(i) = h(j) = g(j - m)$ and thus $i - m = j - m$ so $i = j$. Therefore, h is injective. Since $h : \{1, 2, \ldots, m + n\} \to A \cup B$ is bijective and $m + n$ is a positive integer, we conclude that $A \cup B$ is finite. $\qquad \square$

EXERCISES FOR SECTION 4.1

1. From Peano's axioms, prove that $+$ is commutative and associative on $\mathbb{N}$.

2. From Peano's axioms, prove that if $m, n, p \in \mathbb{N}$ with $m \leq n$ and $n \leq p$, then $m \leq p$.

3. From Peano's axioms, prove that if $m, n, p \in \mathbb{N}$ with $m \leq n$, then $m + p \leq n + p$.

4. From Peano's axioms, prove that $1 = S(0)$ is the multiplicative identity.

5. From Peano's axioms, prove that $\cdot$ distributes over $+$, i.e., $\forall a, b, c \in \mathbb{N} \, (a \cdot (b + c) = a \cdot b + a \cdot c)$.

6. From Peano's axioms and any of the above exercises, prove that $a \cdot b = 0$ implies that $a = 0$ or $b = 0$.

7. Using the well-ordering principle, prove that $\sum_{j=0}^{n} 2^j = 2^{n+1} - 1$ for all $n \in \mathbb{N}$.

8. Prove that the well-ordering principle on $\mathbb{N}$ is equivalent to the axiom of induction. Theorem 4.1.6 establishes one direction. To prove the converse, suppose that $\mathbb{N}$ is defined as satisfying the first three of Peano's axioms, equip it with $\leq$ as defined by Definition 4.1.4, suppose that $\mathbb{N}$ also satisfies the statement of the well-ordering principle, and then prove from these hypotheses that the axiom of induction as given in Proposition 4.1.2 holds.

9. Prove that the intersection of two finite sets is finite.

10. Prove that the union of an infinite set with a finite set is infinite.

11. Prove that the union of two infinite sets is infinite.

4.2 Divisibility

4.2.1 Definition

The notion of divisibility of integers is often introduced in elementary school. However, in order to prove theorems about divisibility, we need a rigorous definition.

Definition 4.2.1

If $a, b \in \mathbb{Z}$ with $a \neq 0$, we say that a divides b if $\exists k \in \mathbb{Z}$ such that $b = ak$. We write $a \mid b$ if a divides b and $a \nmid b$ otherwise. We also say that a is a *divisor* (or a *factor*) of b and that b is a *multiple* of a.

We write $a\mathbb{Z}$ for the set of all multiples of the integer a.

We can prove many basic properties about divisibility directly from the definition. We mention only a few in the following proposition.

Proposition 4.2.2

Let a, b, and c be integers with $a \neq 0$.

1) Any nonzero integer divides 0.

2) If $a|b$ and $a|c$, then $a|(b+c)$.

3) Suppose that $b \neq 0$. If $a|b$ and $b|c$, then $a|c$.

4) Suppose that $b \neq 0$. If $a|b$ and $b|a$, then $a = b$ or $a = -b$.

Proof. For (1), setting $k = 0$, then for any nonzero integer a has $ak = 0$. Thus, a divides 0.

For (2), by $a|b$ and $a|c$, there exist $k, \ell \in \mathbb{Z}$ such that $ak = b$ and $a\ell = c$. Then
$$b + c = ak + a\ell = a(k + \ell).$$
Since $k + \ell$ is an integer, $a|(b + c)$.

For (3), by $a|b$ and $b|c$, there exist $k, \ell \in \mathbb{Z}$ such that $ak = b$ and $b\ell = c$. Then $c = b\ell = a(k\ell)$. Since $k\ell \in \mathbb{Z}$, then $a|c$.

For (4), $a|b$ and $b|a$ means that there exist $k, \ell \in \mathbb{Z}$ such that $ak = b$ and $b\ell = a$. Thus, $a = ak\ell$. We can rewrite this as $ak\ell - a = a(k\ell - 1) = 0$. Since $a \neq 0$, we deduce $k\ell = 1$. (See Exercise 4.1.6.) There are only two ways to multiply two integers to get 1: Either $k = \ell = 1$ or $k = \ell = -1$. The result follows. □

Example 4.2.3. It is easy to check that 3 divides 51 because $3 \cdot 17 = 51$. So we can see that $3 \mid 51$. From the product, we also see that $17 \mid 51$. △

Note that the notation $b \mid a$ is not a binary operation, but rather states a relationship between a and b.

To discuss how divisibility interacts with the sign of integers, note that since $dk = a$ implies $d(-k) = -a$, then d is a divisor of a if and only if d is a divisor of $-a$.

Let d be a nonzero integer. To get a positive multiple dk of d, we need $k \neq 0$ of the same sign as d. Then

$$dk = |d||k| \geq |d|$$

because $|k| \geq 1$. Thus, any nonzero multiple a of d satisfies $|a| \geq |d|$. By the same token, any divisor d of a satisfies $|d| \leq |a|$.

4.2.2 Integer division

The following theorem is one of those propositions about number theory that elementary school children learn early on. In fact, education ingrains it so well that most students believe it with some sort of faith, not realizing that there is a proof for it, one that depends on the well-ordering principle. Even though this is a theorem and not an algorithm, this is often called the Division Algorithm.

Theorem 4.2.4

For all $a, b \in \mathbb{Z}$ with $a \neq 0$, there exist unique integers q and r such that

$$b = aq + r \qquad \text{where } 0 \leq r < |a|.$$

The integer q is called the *quotient* and r is called the *remainder*.

Proof. Given $a, b \in \mathbb{Z}$ with $a \neq 0$. Define the set

$$S = \{b - ak \mid k \in \mathbb{Z}\} \cap \mathbb{N}.$$

Let $b - ak_1$ and $b - ak_2$ be two elements of S. We have

$$(b - ak_1) - (b - ak_2) = a(k_2 - k_1),$$

so the difference of any two elements in S is a multiple of a. Furthermore, to each element $b - ak$ in S there corresponds only one integer k.

By the well-ordering principle, S has a least element $r = b - aq$. Now if $r \geq |a|$, then

$$b - a(q + \operatorname{sign}(a)) = b - aq - |a| = r - |a| \geq 0$$

and so $r - |a|$ contradicts the minimality of r as a least element of S. Thus $r < |a|$. Since any two elements of S differ by a multiple of $|a|$, then r is the unique element n of S with $0 \leq n < |a|$. Hence, the integers q and r satisfy the conclusion of the theorem. $\square$

Techniques for integer division are taught in elementary school. It is obvious that $d|n$ if and only if the remainder of the integer division of n by d is 0.

Borrowing notation from some programming languages, it is common to write $b \bmod a$ or $b \operatorname{\mathbf{mod}} a$ for the remainder of the division of b by a.

4.2.3   Base representation of integers

Most of the modern world represents integers in base 10. Some ancient civilizations used different number bases to represent integers. For example, the Babylonians used base 60 to represent their integers [49]. This is likely the reason why we still have 60 minutes in an hour, 60 seconds in a minute, and 360° in a circle. (Babylonian geometers considered the equilateral triangle a fundamental geometrical figure, so they put 60 angle units in one angle of an equilateral triangle. This led to 360° around a point.)

Definition 4.2.5

Let b be an integer $b \geq 2$. We say that a nonnegative integer n is *represented in base* b if we write

$$n = d_k \cdot b^k + d_{k-1} \cdot b^{k-1} + \cdots + d_1 \cdot b + d_0$$

with $0 \leq d_i \leq b - 1$. We say that d_i is the ith digit of n in the base b representation. For short, we write this as $n = (d_k d_{k-1} \cdots d_1 d_0)_b$.

From this definition, it is straightforward to convert the base b representation of an integer into the usual base 10 representation.

Example 4.2.6. The integer expressed in base 5 as $(23401)_5$ is in base 10:
$$2 \cdot 5^4 + 3 \cdot 5^3 + 4 \cdot 5^2 + 0 \cdot 5 + 1 = 1726. \qquad \triangle$$

Example 4.2.7 (Binary). The *binary representation* of an integer means representing in base 2. For example, the number $(1101011)_2$ in base 10 is
$$1 \cdot 2^6 + 1 \cdot 2^5 + 0 \cdot 2^4 + 1 \cdot 2^3 + 0 \cdot 2^2 + 1 \cdot 2^1 + 1 \cdot 2^0 = 107.$$

The digits in binary are called bits, from the contraction of "binary digits."

The binary representation of integers is foundational for computer science. Originally using transistors but now involving more technical devices, computers store information using a sequence of switches, each turned either on or off. In this perspective, we usually think of the bits in a binary as 0 representing "off" and 1 representing "on." $\triangle$

Example 4.2.8 (Hexadecimal). In a representation in base b, we need to have b distinct symbols, one for each of the digits in the base b representation corresponding to $\{0, 1, \ldots, b-1\}$. In base 10, we employ our customary 10 digits 0, 1, 2, 3, 4, 5, 6, 7, 8, and 9. In base 5 or base 2, we can still use digits from this set. However, another common base used in computer science is *hexadecimal*, where $b = 16$. Hexadecimal requires 16 different symbols for digits. By convention, we use the digits 0 through 9 for their usual value and then use a for 10, b for 11, and so on up to f for 15.

For example, the integer expressed in hexadecimal as $(2\mathsf{bad})_{16}$ in base 10 is

$$2 \cdot 16^3 + 11 \cdot 16^2 + 10 \cdot 16 + 13 = 11,181.$$

Hexadecimal is particularly useful in computer science because, besides the bit, one of the fundamental units of information is the *byte*, which corresponds to 8 bits. One byte of storage has $2^8 = 16^2$ possible fillings. Hence, every byte can be represented by a two-digit hexadecimal number. Instead of writing $(2\mathsf{bad})_{16}$, most programming languages use the prefix 0x in front of the hexadecimal to signal the beginning of a hexadecimal number. Hence, 11,181 would be written as 0x2bad. $\triangle$

Converting a number from decimal (base 10) representation to its representation in any other base begins with an observation. If

$n = (d_k d_{k-1} \cdots d_1 d_0)_b$ then we can easily find the integer division of n by b via

$$n = \sum_{j=0}^{k} d_j b^j = d_0 + \sum_{j=1}^{k} d_j b^j = b \left(\sum_{j=1}^{k} d_j b^{j-1} \right) + d_0.$$

So the remainder is d_0 and the quotient is $(d_k d_{k-1} \cdots d_1)_b$. Repeating this, d_1 is the remainder of the integer division of $(d_k d_{k-1} \cdots d_1)_b$ by b, and so on. We create two finite sequences (a_n) and (d_n) defined recursively as follows.

- Set $a_0 = n$.
- For all j, take the integer division $a_j = b a_{j+1} + d_j$, so a_{j+1} is the quotient and d_j is the remainder.
- Stop when $a_{j+1} = 0$. So k is the smallest integer $j = k$ with $a_{j+1} = 0$.

Example 4.2.9. As an example, we find the base 8 (octal) representation of 4,567. We need to perform a sequence of integer divisions:

$$4,567 = 8 \times 570 + 7$$
$$570 = 8 \times 71 + 2$$
$$71 = 8 \times 8 + 7$$
$$8 = 8 \times 1 + 0$$
$$1 = 8 \times 0 + 1.$$

We now stop since we see a quotient that is 0. (If we continued the process, all the subsequent quotients and remainders would be 0 too.) Hence, the base 8 representation of 4,567 is $(10727)_8$. △

EXERCISES FOR SECTION 4.2

1. Show that if $a, b, c \in \mathbb{Z}$ with $a \neq 0$ and $a \mid b$, then $a \mid bc$.
2. Show that if $a, c \in \mathbb{Z}^*$ and $b, d \in \mathbb{Z}$ with $a \mid b$ and $c \mid d$, then $ac \mid bd$.
3. Determine the quotient and remainder for b divided by a for each pair.
   - (a) $(a, b) = (6, 73)$
   - (b) $(a, b) = (31, 120)$
   - (c) $(a, b) = (13, -100)$
   - (d) $(a, b) = (23, 3)$
4. Determine the quotient and remainder for b divided by a for each pair.
   - (a) $(a, b) = (8, 75)$
   - (b) $(a, b) = (13, -314)$
   - (c) $(a, b) = (12345, 54321)$
   - (d) $(a, b) = (101, 10001)$
5. Define the function $f : \mathbb{N} \to \mathbb{N}$ by

$$f((a_k a_{k-1} \cdots a_1 a_0)_2) = ((2a_k)(2a_{k-1}) \cdots (2a_1)(2a_0))_3.$$

(a) Determine the values of: (i) $f(22)$; (ii) $f(131)$.

(b) Decide if the function is (i) injective; (ii) surjective.

6. For all integers $n \geq 2$, determine the integer division of n^3 by $n^2 - 1$.

7. For all integers $n \geq 2$, determine the integer division of n^3 by $n - 1$.

8. Find the decimal representation of the following numbers.

   (a) $(120120)_3$ (c) $(12335)_8$
   (b) $(555)_7$ (d) $(111010101)_2$

9. Find the decimal representation of the following numbers.

   (a) $(23124)_5$ (c) $(d1ed)_{16}$
   (b) $(1000)_7$ (d) $(a1a2a)_{11}$

10. Represent 1000 in base 7 and in base 8.

11. Represent the following integers in base 9: (a) 700; (b) 99; (c) 1234.

12. Express the following integers in binary: (a) 78; (b) 100; (c) 1013; (d) 255.

13. Express the following integers in hexadecimal: (a) 95; (b) 4215; (c) 2730; (d) 3,124,461.

14. Describe a method to quickly convert a number in binary to hexadecimal. Apply your method to convert the following to hexadecimal: $(1000101)_2$; $(11000010010)_2$; $(111110101101)_2$.

15. Prove that for all integers n, the number $n(n + 1)$ is divisible by 2.

16. Prove that for all integers n, the number $n(n + 1)(n + 2)$ is divisible by 6.

17. Prove that if n is an odd integer, then $8 \mid (n^2 - 1)$.

18. Prove that $3 \mid (n^3 - n)$ for all positive integers n.

19. Show that the difference of two consecutive cubes (an integer of the form n^3) is never divisible by 3.

20. Show that if a and b are distinct integers and $k \in \mathbb{N}^*$, then $a - b$ divides $a^k - b^k$.

21. Show that if $a, b \in \mathbb{Z}$ with $a \geq 1$, then the quotient and remainder in the integer division of b by a are $\lfloor b/a \rfloor$ and $b - a\lfloor b/a \rfloor$, respectively.

4.3 Greatest Common Divisor; Least Common Multiple

The concepts of greatest common divisor and the least common multiple arise early in math education. It may come as a surprise that we introduce them without the concept of inequality and calculate them without referring to prime numbers (the next section's topic). This approach may appear novel but is equivalent with the elementary school formulation. However, this approach has the benefit of allowing us to later generalize the notion of divisibility to other algebraic contexts besides integers.

4.3.1   Greatest common divisor

Definition 4.3.1

If $a, b \in \mathbb{Z}$ with $(a, b) \neq (0, 0)$, a *greatest common divisor* of a and b is an element $d \in \mathbb{Z}$ such that:

- $d|a$ and $d|b$ (d is a common divisor);

- if $d'|a$ and $d'|b$ (d' is another common divisor), then $d'|d$.

Because of condition (2) in this definition, it is not obvious that two integers not both 0 have a greatest common divisor. (If we had said $d' \leq d$ in the definition, the proof that two integers not both 0 have a greatest common divisor is a simple application of the well-ordering principle on $\mathbb{N}$.) By Proposition 4.2.2(4), we deduce from this condition (2) the following result.

Proposition 4.3.2

If d_1 and d_2 are greatest common divisors of integers a and b, then $d_2 = d_1$ or $d_2 = -d_1$.

Though Definition 4.3.1 uses the grammar "a" greatest common divisor, this simple proposition shows that there is a unique positive greatest common divisor of two integers, not both 0.

Definition 4.3.3

Let $a, b \in \mathbb{Z}$ with $(a, b) \neq (0, 0)$. We denote by $\gcd(a, b)$ the positive greatest common divisor of a and b.

The key to showing that integers possess a greatest common divisor relies on the *Euclidean Algorithm*, which we describe here below.

Let a and b be two positive integers with $a \geq b$. The Euclidean Algorithm starts by setting $r_0 = a$ and $r_1 = b$ and then repeatedly performs the following integer divisions

$$r_0 = r_1 q_1 + r_2 \qquad \text{(where } 0 \leq r_2 < b)$$
$$r_1 = r_2 q_2 + r_3 \qquad \text{(where } 0 \leq r_3 < r_2)$$
$$\vdots$$
$$r_{n-2} = r_{n-1} q_{n-1} + r_n \qquad \text{(where } 0 \leq r_3 < r_2)$$
$$r_{n-1} = r_n q_n + 0 \qquad \text{(and } r_n > 0).$$

We stop the Euclidean Algorithm when we obtain a remainder of 0; the output of the algorithm is r_n, the last nonzero remainder.

Example 4.3.4. We perform the Euclidean Algorithm on the pair of positive integers $(a, b) = (516, 354)$.

$$516 = 354 \times 1 + 162$$
$$354 = 162 \times 2 + 30$$
$$162 = 30 \times 5 + 12$$
$$30 = 12 \times 2 + 6$$
$$12 = 6 \times 2 + 0$$

The output of the Euclidean Algorithm is 6. △

Though we do not discuss algorithms at length in this text, an *algorithm* is a finite sequence of well-defined instructions to accomplish a particular goal. We also describe this essential feature of involving a finite number of steps by saying that the "algorithm terminates." When we implement an algorithm with a computer program, it must stop in a finite amount of time.

Theorem 4.3.5

For any pair of positive integers a and b, the Euclidean Algorithm terminates. Furthermore, the output is equal to $\gcd(a, b)$.

The first part of the theorem simply justifies the terminology of "algorithm."

Proof. We observe first that $a = r_0 \geq r_1 = b$. By integer division, we see that the remainders satisfy $r_1 > r_2 > r_3 > \ldots$, so they form a strictly decreasing sequence of positive integers. There must be b or fewer positive remainders in this sequence. This means that the algorithm terminates.

Now consider the output r_n of the Euclidean Algorithm. By the last line of the Euclidean Algorithm we see that $r_n | r_{n-1}$. From the second to last row, of the Euclidean Algorithm, $r_n | r_{n-1}q_{n-1}$ and by Proposition 4.2.2(2), $r_n | r_{n-2}$. Repeatedly applying this process ($n - 1$ times, and hence a finite number), we see that $r_n | r_1$ and $r_n | r_0$, so r_n is a common divisor of a and b.

Suppose that d' is any common divisor of a and b. Then $d'k_0 = a = r_0$ and $d'k_1 = b = r_1$. We have

$$r_2 = r_0 - r_1q_1 = d'k_0 - d'k_1q_1 = d'(k_0 - k_1q_1).$$

Hence, d' divides r_2. Repeating this process ($n - 1$ times), we deduce that $d'|r_n$. Thus, r_n is a positive greatest common divisor of a and b. □

This proof by algorithm is a new form of an existence proof. Though it does not provide a formula or exhibit an object explicitly, it provides an algorithm to obtain the desired object. We alluded to this when we discussed constructive existence proofs, but this is a first example.

This proof affords us the opportunity to emphasize another key point. Computational methods are common in modern mathematics and play an increasing role both in applied mathematics and theoretical fields. However, showing that a sequence of instructions terminates and returns a certain desired object (and hence provides an algorithm) always requires a proof.

Note that if $b \mid a$, then $n = 1$ and the Euclidean Algorithm has one line. We could also generalize the algorithm slightly by removing the assumption that $a \geq b$. If we run the algorithm with $a < b$, the first integer division would be $a = b \cdot 0 + a$, so the second division would involve dividing b, the larger integer, by a.

Proposition 4.3.6

There exists a unique positive greatest common divisor for all pairs of integers $(a, b) \in \mathbb{Z} \times \mathbb{Z} \setminus \{(0, 0)\}$.

Proof. First suppose that either a or b is 0. Without loss of generality, assume that $b = 0$ and $a \neq 0$. Since any integer divides 0, common divisors of a and b consist of divisors of a. Then greatest common divisors of a and 0 consist of a and $-a$ and $|a|$ is the unique positive greatest common divisor of a and 0.

Now suppose neither a nor b is 0. By Theorem 4.3.5, the Euclidean Algorithm applied to $|a|$ and $|b|$ gives $\gcd(|a|, |b|)$. Since the set of divisors of an integer c is the same set as the divisors of $-c$, this is the same as $\gcd(a, b)$. $\qquad\square$

Definition 4.3.7

If a and b are integers with $(a, b) \neq (0, 0)$, we say that a and b are *relatively prime* if $\gcd(a, b) = 1$.

Lemma 4.3.8

Let a and b be positive integers. Then if k and l are integers such that $a = k \gcd(a, b)$ and $b = l \gcd(a, b)$, then k and l are relatively prime.

Proof. Consider $c = \gcd(k, l)$, and write $k = ck'$ and $l = cl'$ for some integers k' and l'. Then

$$a = k'c \gcd(a, b) \quad \text{and} \quad b = l'c \gcd(a, b).$$

Therefore, $c \gcd(a, b)$ would be a divisor of $\gcd(a, b)$ so for some integer h, we have $c \gcd(a, b) h = \gcd(a, b)$. Hence, $ch = 1$. Since c is a positive integer, this is only possible if $c = 1$. □

There is an alternative characterization of the positive greatest common divisor. Let $a, b \in \mathbb{Z}^*$ and define $S_{a,b}$ as the set of integer linear combinations in integers of a and b, i.e.,

$$S_{a,b} = \{sa + tb \,|\, s, t \in \mathbb{Z}\}.$$

Proposition 4.3.9

The set $S_{a,b}$ is the set of all integer multiples of $\gcd(a, b)$. Consequently, $\gcd(a, b)$ is the least positive linear combination in integers of a and b.

Proof. By Proposition 4.2.2, any common divisors to a and b divide sa, tb, and $sa + tb$. This shows that $S_{a,b} \subseteq \gcd(a, b)\mathbb{Z}$. We need to show the reverse inclusion.

By the well-ordering principle, the set $S_{a,b}$ has a least positive element. Call this element d_0 and write $d_0 = s_0 a + t_0 b$ for some $s_0, t_0 \in \mathbb{Z}$. We show by contradiction that d_0 is a common divisor of a and b. Suppose that d_0 does not divide a. Then by integer division,

$$a = q d_0 + r \qquad \text{where } 0 < r < d_0.$$

Then $a - r = q d_0 = q s_0 a + q t_0 b$. Then after rearranging, we get

$$r = (1 - q s_0)a + (-q t_0)b.$$

This writes r, which is positive and less than d_0, as a linear combination of a and b. This contradicts the assumption that d_0 is the least positive element in $S(a, b)$. Hence, the assumption that d_0 does not divide a is false so d_0 divides a. By a symmetric argument, d_0 divides b as well. Since $k d_0 = (k s_0)a + (k t_0)b$, then every multiple of d_0 is in $S_{a,b}$ in particular every multiple of $\gcd(a, b)$ is too. Hence, $\gcd(a, b)\mathbb{Z} \subseteq S_{a,b}$. We conclude that $S_{a,b} = \gcd(a, b)\mathbb{Z}$ and the proposition follows. □

The characterization of the greatest common divisor as given in Proposition 4.3.9 is often useful to prove results about the greatest common divisor of two integers. The following proposition gives one such example.

Proposition 4.3.10

Let a and b be nonzero integers that are relatively prime. For any integer c, if $a | bc$, then $a | c$.

Proof. Since a and b are relatively prime, then $\gcd(a,b) = 1$. By Proposition 4.3.9, there exist integers $s,t \in \mathbb{Z}$ such that $sa + tb = 1$. Since $a \mid bc$, there exists $k \in \mathbb{Z}$ such that $ak = bc$. Then

$$atk = tbc = c(1 - as) = c - cas,$$

which implies that

$$c = atk + acs = a(tk + cs).$$

From this, we conclude that $a \mid c$. $\qquad\Box$

4.3.2   Extended Euclidean algorithm

A corollary to Proposition 4.3.9 is that there exists a pair $(s,t) \in \mathbb{Z}^2$ such that $sa + tb = \gcd(a,b)$. However, the proof is not constructive. If a and b are small then one can find s and t by inspection. For example, by inspecting the divisors of 22, it is easy to see that $\gcd(22,14) = 2$. A linear combination that illustrates Proposition 4.3.9 for 22 and 14 is

$$2 \times 22 - 3 \times 14 = 44 - 42 = 2.$$

However, it is possible to backtrack the steps of the Euclidean Algorithm and find s and t such that $sa + tb = \gcd(a,b)$.

The Extended Euclidean Algorithm of two positive integers a and b keeps track of more information than the Euclidean Algorithm so that it can express each remainder r_i as a linear combination of a and b. We define four sequences $(q_i)_{i \geq 1}$, $(r_i)_{i \geq 0}$, $(s_i)_{i \geq 0}$, and $(t_i)_{i \geq 0}$ recursively as follows, with $i \geq 2$.

$$
\begin{array}{llll}
r_0 = a & s_0 = 1 & t_0 = 0 & \\
r_1 = b & s_1 = 0 & t_1 = 1 & q_1 = \lfloor r_0/r_1 \rfloor \\
r_i = r_{i-2} - r_{i-1}q_{i-1} & s_i = s_{i-2} - s_{i-1}q_{i-1} & t_i = t_{i-2} - t_{i-1}q_{i-1} & q_i = \lfloor r_{i-1}/r_i \rfloor
\end{array}
$$

Like the Euclidean Algorithm, we terminate these sequences at the smallest positive integer n such that $r_{n+1} = 0$. Notice that for all $i \geq 1$, this algorithm, like the Euclidean Algorithm, involves a sequence of integer divisions $r_{i-1} = r_i q_i + r_{i+1}$, with $0 \leq r_{i+1} < r_i$.

Proposition 4.3.11

In the Extended Euclidean Algorithm, for all i with $0 \leq i \leq n$, we have $r_i = s_i a + t_i b$.

Proof. The claim is clearly true for $i = 0$ and $i = 1$. Assume that the proposition is false. Then there exists a least element $j \in \{0, 1, \ldots, n\}$

such that $r_j \neq s_j a + t_j b$. For the beginning remark, we must have $j \geq 2$. So $r_i = s_i a + t_i b$ is true for $i = j - 1$ and $i = j - 2$. Hence,

$$r_j = r_{j-2} - r_{j-1} q_{j-1} = (s_{j-2} a + t_{j-2} b) - (s_{j-1} a + t_{j-1} b) q_{j-1}$$
$$= (s_{j-2} - s_{j-1} q_{j-2}) a + (t_{j-2} - t_{j-1} q_{j-1}) b$$
$$= s_j a + t_j b.$$

This is a contradiction so the proposition follows. $\square$

Consequently, when we perform the Extended Euclidean Algorithm, we obtain $r_n = \gcd(a, b)$ and s_n and t_n are integers such that $\gcd(a, b) = s_n a + t_n b$.

Example 4.3.12. In Example 4.3.4, the Euclidean Algorithm gave the value of $\gcd(516, 354) = 6$. We perform the Extended Euclidean Algorithm on this situation. It is convenient to use a table to keep track of the data.

i	r_i	s_i	t_i	q_i
0	516	1	0	
1	354	0	1	1
2	162	1	-1	2
3	30	-2	3	5
4	12	11	-16	2
5	6	-24	35	2
6	0			

We conclude that $\gcd(522, 408) = 6$, but also that $-24 \times 516 + 35 \times 354 = 6$. $\triangle$

4.3.3 Least common multiple

Definition 4.3.13

If $a, b \in \mathbb{Z}^*$, a *least common multiple* is an element $m \in \mathbb{Z}$ such that:

- $a|m$ and $b|m$ (m is a common multiple);

- if $a|m'$ and $b|m'$, then $m|m'$.

Similar to our presentation of the greatest common divisor, we should note that from this definition, it is not obvious that a least common multiple always exists. Again, we must show that it exists.

Proposition 4.3.14

There exists a unique positive least common multiple m for all pairs of integers $(a, b) \in \mathbb{Z} \times \mathbb{Z} \setminus \{(0, 0)\}$.

Proof. If m_1 and m_2 are least common multiples of a and b, then $m_1|m_2$ and $m_2|m_1$. Therefore, by Proposition 4.2.2(4), if a and b have a least common multiple m, the integer $-m$ is the only other least common multiple.

Without loss of generality, assume that a and b are positive.

Since $\gcd(a,b)$ divides a and divides b, then $\gcd(a,b)|ab$. Also, we can write $a = k\gcd(a,b)$ and $b = l\gcd(a,b)$. Let M be the positive integer such that $M\gcd(a,b) = ab$. From $M\gcd(a,b) = \gcd(a,b)kb$, we get $M = bk$ and similarly $M = al$ and hence M is a common multiple of a and b.

Let m' be another common multiple of a and b with $m' = pa$ and $m' = qb$. Since $pa = qb$ then

$$pk\gcd(a,b) = ql\gcd(a,b)$$

and hence $pk = ql$. Since $\gcd(k,l) = 1$, by Proposition 4.3.10 we conclude that $k|q$ with $q = kc$ for some integer c. Then $m' = (kc)b = c(bk) = cM$ and thus $M \mid m'$.

This shows that $M = ab/\gcd(a,b)$ satisfies the criteria of a least common multiple and the proposition follows. $\square$

We regularly call this unique positive least common multiple of a and b "the" least common multiple. We denote this positive integer as $\mathrm{lcm}(a,b)$. The proof of Proposition 4.3.14 establishes the following proposition.

Proposition 4.3.15

For all $a, b \in \mathbb{N}^*$, $\gcd(a,b)\,\mathrm{lcm}(a,b) = ab$.

This proposition also provides a method to calculate the least common multiple of two positive integers a and b: Use the Euclidean Algorithm to calculate $\gcd(a,b)$, and then $\mathrm{lcm}(a,b) = ab/\gcd(a,b)$.

EXERCISES FOR SECTION 4.3

1. Perform the Euclidean Algorithm on the following pairs of integers:
   (a) $a = 321$ and $b = 123$ (c) $a = 635$ and $b = 117$
   (b) $a = 3000$ and $b = 66$ (d) $a = 5291$ and $b = 858$

2. Determine the least common multiples of the pairs in the previous exercise.

3. Perform the Euclidean Algorithm on the following pairs of integers:
   (a) $a = 482$ and $b = 751$ (c) $a = 10101$ and $b = 1011$
   (b) $a = 861$ and $b = 1617$ (d) $a = 54321$ and $b = 1728$

4. Determine the least common multiples of the pairs in the previous exercise.

5. Recall the Fibonacci sequence $\{f_n\}_{n\geq 0}$ by $f_0 = 0$, $f_1 = 1$ and $f_n = f_{n-1} + f_{n-2}$ for all $n \geq 2$. Let f_n and f_{n+1} be two consecutive terms in the Fibonacci sequence. Prove that $\gcd(f_{n+1}, f_n) = 1$ and show that for all $n \geq 2$, the Euclidean algorithm requires exactly $n - 1$ integer divisions.

6. Apply the Extended Euclidean Algorithm to:
   (a) $(a, b) = (863, 517)$;   (b) $(a, b) = (1010, 315)$.

7. Apply the Extended Euclidean Algorithm to:
   (a) $(a, b) = (4321, 1234)$;   (c) $(a, b) = (10901, 495)$.

8. Suppose that we only consider positive integers. Show that Definition 4.3.1 is equivalent to "$d \mid a$ and $d \mid b$; and $d' < d$ for all other integers with $d' \mid a$ and $d' \mid b$. (This is the elementary school definition.) Show that this elementary school definition is not equivalent to Definition 4.3.1 if we consider negative integers.

9. Let a, b, c be positive integers. Prove that $\gcd(ab, ac) = a\gcd(b, c)$.

10. Prove or disprove that $\gcd(a, \gcd(b, c)) = \gcd(\gcd(a, b), c)$ for all $a, b, c \in \mathbb{N}^*$.

11. Show that $\{105s + 26t \mid s, t \in \mathbb{Z}\} = \mathbb{Z}$.

12. Show that if a and b are positive integers, then $a\mathbb{Z} \cap b\mathbb{Z} = \mathrm{lcm}(a, b)\mathbb{Z}$.

13. For any finite subset $\{a_1, a_2, \ldots, a_n\} \subseteq \mathbb{Z}$ with not all $a_i = 0$, we define a greatest common divisor of $\{a_1, a_2, \ldots, a_n\}$ as an integer d such that $d \mid a_i$ for all $i \in \{1, \ldots, n\}$; and such that if $d' \mid a_i$ for all i, then $d' \mid d$. Prove that the least positive integer s in

$$S = \{t_1 a_1 + t_2 a_2 + \cdots + t_n a_n \mid t_i \in \mathbb{Z}\}$$

is a greatest common divisor of $\{a_1, a_2, \ldots, a_n\}$.

4.4   Prime Numbers

Prime numbers are like building blocks for integers when it comes to multiplication. With addition on the positive integers, we can begin with 1 and then add it to itself as many times as necessary to produce any other positive integer. Or in another direction, any positive integer greater than 1 can be written as the sum of two positive integers. Not so with the operation of multiplication. There exist plenty of positive integers that cannot be expressed as a product of smaller positive integers.

Definition 4.4.1

An element $p \in \mathbb{Z}$ is called a *prime number* if $p > 1$ and the only divisors of p are 1 and itself. If an integer $n > 1$ is not prime, then n is called *composite*.

For short, we often say "p is prime" instead of "p is a prime number." Properties about prime numbers have intrigued mathematicians since Euclid and before. The distribution of prime numbers in $\mathbb{N}$ or in sequences of integers, additive properties of prime numbers, fast primality-testing algorithms, and many other questions offer active areas of research in number theory.

It should be obvious that every integer $n \geq 2$ is divisible by a prime number. We can use well-ordering to see this clearly. Note that every positive integer is either 1, a prime number, or a composite number. If $n \geq 2$ is prime, then it is trivially divisible by a prime: itself. If n is composite, consider the set $S = \{d \geq 2 \mid d \mid n\}$. By the well-ordering principle, S has a least element d_0. Assume that d_0 is composite. Then $d_0 = ab$ with $a, b > 1$. By Proposition 4.2.2(3) $a \mid n$ with $1 < a < d_0$, so $a \in S$ and this contradicts the minimality of d_0. Hence, d_0 is a prime divisor of n.

One of the earliest results about prime numbers dates back to Euclid, who cleverly applied an argument by contradiction to prove that there exists an infinite number of prime numbers.

Theorem 4.4.2 (Euclid's Prime Number Theorem)
The set of prime numbers is infinite.

Proof. Assume the contrary, that the set of prime numbers is finite. Write the set of prime numbers as $\{p_1, p_2, \ldots, p_n\}$. Consider the integer

$$Q = (p_1 p_2 \cdots p_n) + 1.$$

The integer Q is obviously larger than 1, so it is divisible by a prime number, say p_k. Then since

$$1 = Q - (p_1 p_2 \cdots p_n)$$

we conclude by Proposition 4.2.2(2) that 1 is divisible by p_k. This is a contradiction since 1 is only divisible by 1 and -1. The theorem follows.   □

The problem of finding a fast algorithm for determining whether a given integer n is prime is difficult. This problem is not just one of simple curiosity but has applications to information security. It is possible to simply take in sequence all integers $1 < d \leq n$ and perform an integer division of n by d. The least integer $d \geq 2$ that divides n is a prime number. This d satisfies $d = n$ if and only if n is prime. This method offers an exhaustive algorithm to determine whether n is prime, but the following proposition shortens it.

Proposition 4.4.3

If n is composite, then it has a divisor d such that $1 < d \le \sqrt{n}$.

Proof. Suppose that all the divisors of n are greater than $\sqrt{n}$. Since n is composite, there exist positive integers a and b greater than 1 with $n = ab$. The supposition that $a > \sqrt{n}$ and $b > \sqrt{n}$ implies that $ab = n > n$, a contradiction. The proposition follows. □

Consequently, when looking for nontrivial divisors of an integer n, it suffices to only look up to $\lfloor \sqrt{n} \rfloor$.

The following proposition gives an alternative characterization for primality.

Proposition 4.4.4 (Euclid's Lemma)

If $p > 1$, then p is prime if and only if $p|ab \implies p|a$ or $p|b$.

Proof. Suppose that p is a prime number. Now suppose that $p|ab$. If $p|a$ then the implication in the conclusion is true and we are done. Suppose instead that $p \nmid a$. Then, since the only divisors of p are 1 and itself, $\gcd(p, a) = 1$. By Proposition 4.3.10, $p|b$ and the implication is still true.   ($\implies$)

Conversely, suppose that for all integers $a, b \in \mathbb{Z}$ the integer $p > 1$ has the property that $p \mid ab$ implies $p \mid a$ or $p \mid b$. Assume that p is composite with $p = cd$ and $1 < c, d < p$. Then $p \mid cd$ and $p \nmid c$ and $p \nmid d$. This contradicts the property we supposed about p. Hence, we conclude that p is prime.   ($\impliedby$)   □

We mention the following result about prime numbers not only because it is useful, but also because the proof uses Euclid's Lemma.

Proposition 4.4.5

If p is a prime number, then $\sqrt{p}$ is irrational.

Proof. Assume $\sqrt{p}$ is rational. Then $\sqrt{p} = a/b$, where a and b are relatively prime nonzero integers. Then $pb^2 = a^2$ and so $p \mid a^2$. By Euclid's Lemma, $p \mid a$. Hence, $a = pc$ for some $c \in \mathbb{N}$. Then $pb^2 = (pc)^2$, which is equivalent to $b^2 = pc^2$. Thus $p \mid b^2$ and again $p \mid b$ by Euclid's Lemma.[2] Hence $p \mid \gcd(a, b)$, so a and b are not relatively prime. This is a contradiction. Thus, $\sqrt{p}$ is not rational.□

---

[2]It is proper to cite any theorem that is not extremely common.

Example 4.4.6. As a follow-up example to Proposition 4.4.5, we prove the following more complicated result: Let p and q be distinct primes; then $\sqrt{p}$ is not equal to $r + s\sqrt{q}$ for any $r, s \in \mathbb{Q}$.

We prove this by contradiction. Assume that $\sqrt{p} = (a/b) + (c/d)\sqrt{q}$ for some $a, b, c, d \in \mathbb{Z}$, with a/b and c/d reduced fractions. By Proposition 4.4.5, we cannot have $c = 0$. We consider two cases, for whether $a = 0$ or $a \neq 0$.

First, suppose that $a = 0$. Then $pd^2 = qc^2$. Since p and q are distinct primes, $\gcd(p, q) = 1$. Since $p \mid qc^2$, by Proposition 4.3.10 we have $p|c^2$. By Euclid's Lemma, $p|c$, so $c = pk$ for some $k \in \mathbb{Z}$. Thus $pd^2 = qp^2k^2$ and hence $d^2 = pqk^2$. Thus $p \mid d^2$ and again by Euclid's Lemma, $p \mid d$. This contradicts c and d being relatively prime.

Now, suppose that $a \neq 0$. Squaring $\sqrt{p} = (a/b) + (c/d)\sqrt{q}$, we get

$$p = \left(\frac{ad + bc\sqrt{q}}{bd} \right)^2 = \frac{(ad)^2 + abcd\sqrt{q} + (bc)^2}{(bd)^2}.$$

Since $abcd \neq 0$, we have

$$\sqrt{q} = \frac{pb^2d^2 - a^2d^2 - b^2c^2}{abcd}.$$

This contradicts Proposition 4.4.5 because the numerator and denominator are integers. We conclude that $\sqrt{p}$ is not equal to $r + s\sqrt{q}$ for any $r, s \in \mathbb{Q}$. △

The study of prime numbers has a very rich history but also practical applications in information security. The ability to quickly produce very large primes is part of the RSA public key algorithm that many browsers use for secure communication. We know that there are an infinite number of primes but, not unlike a sport where athletes continue to push the boundaries, there is an ongoing computational effort to find larger and larger primes. Because of the Lucas-Lehmer Test, there are fast algorithms to verify if a Mersenne number is prime.

Definition 4.4.7

A *Mersenne number* is a number of the form $2^n - 1$. If such a number is prime, it is called a *Mersenne prime*.

The GIMPS (Great Internet Mersenne Prime Search) is a loose community of people who allow their computers to run specific large computational tests when the computer is not otherwise in use. Occasionally, these find a new prime number.

Proposition 4.4.8

If $2^n - 1$ is a prime number, then n is a prime number.

Proof. Suppose that n is composite with $n = ab$ for integers $a, b \geq 2$. Then

$$2^n - 1 = 2^{ab} - 1 = (2^a)^b - 1$$
$$= (2^a - 1)\left((2^a)^{b-1} + (2^a)^{b-2} + \cdots + 2^a + 1\right). \quad (4.1)$$

Since $a \geq 2$, then $2^a \geq 4$ so $2^a - 1 \geq 3$. Since $b \geq 2$ and $2^a \geq 4$, then

$$(2^a)^{b-1} + (2^a)^{b-2} + \cdots + 2^a + 1 \geq 2^a + 1 \geq 5.$$

Hence, both products on the right-hand side of (4.1) are greater than 1. Hence $2^n - 1$ is composite. We have proven that n is composite implies that $2^n - 1$ is composite. Taking the contrapositive, the proposition follows. $\square$

4.4.1 The fundamental theorem of arithmetic

Every high school student learns the Fundamental Theorem of Arithmetic, though not so many remember its formal name. Not unlike the Division Algorithm, this is a theorem that educators teach students long before students are mathematically mature enough to understand a proof.

Theorem 4.4.9 (Fundamental Theorem of Arithmetic)

If $n \in \mathbb{Z}$ and $n \geq 2$, then there is a unique factorization (up to rearrangement) of n into a product of prime numbers. More precisely, if n can be written as the product of primes in two different ways as

$$n = p_1 p_2 \cdots p_r = q_1 q_2 \cdots q_s \quad (4.2)$$

with p_i and q_j primes, then $r = s$ and there is a bijective function $f : \{1, 2, \ldots, n\} \to \{1, 2, \ldots, n\}$ such that $p_i = q_{f(i)}$.

Proof. Let n be a positive integer greater than 1.

Define a sequence (p_k) recursively as follows: $p_0 = 1$ and for $i \geq 1$ set p_i as the smallest prime dividing $n/(p_0 p_1 \cdots p_{i-1})$. We terminate the sequence at $i = r$ when $n/(p_1 p_2 \cdots p_r) = 1$. By construction, n is the product of primes $p_1 p_2 \cdots p_r$.

To prove the uniqueness of the factorization, suppose that $n = p_1 p_2 \cdots p_r$ and $n = q_1 q_2 \cdots q_s$ where each p_i is a prime number for all $1 \leq i \leq r$ and q_j is a prime number for all $1 \leq j \leq s$. Since multiplication is associative and commutative, we can suppose without loss of generality that $p_1 \leq p_2 \leq \cdots \leq p_r$ and $q_1 \leq q_2 \leq \cdots \leq q_s$.

We prove the uniqueness by contradiction. Assume that the finite sequences $(p_1, p_2, \ldots, p_r)$ and $(q_1, q_2, \ldots, q_s)$ differ for the first time at index k. Then

$$p_k p_{k+1} \cdots p_r = q_k q_{k+1} \cdots q_s$$

with $p_k \neq q_k$. Without loss of generality, suppose that $p_k < q_k$. (If $p_k > q_k$, we would reverse the following argument.) Then $p_k \mid q_k q_{k+1} \cdots q_s$. By a repeated application of Euclid's Lemma, we deduce that $p_k \mid q_j$ for some $k \leq j \leq s$. However, since $p_k \neq 1$ and q_j is prime, then $p_k = q_j$. This is a contradiction since $p_k < q_k \leq q_{k+1} \leq \cdots \leq q_s$. Consequently, when listed in nondecreasing order, the sequences of prime factors of n are identical and the theorem follows. $\qquad\square$

In the factorizations in (4.2), we do not assume that the prime numbers p_i are unique. It is common to write the generic factorization of integers as

$$n = p_1^{\alpha_1} p_2^{\alpha_2} \cdots p_n^{\alpha_n} \qquad (4.3)$$

where the primes p_i are distinct, listed in increasing order, and where α_i are positive integers. Using these habits, we call the expression in (4.3) *the prime factorization* of n. There exist a variety of methods to find the prime factorization of an integer. Since they are taught early on, we will not dwell on them here.

Example 4.4.10. The prime factorization of 360 is $2^3 \cdot 3^2 \cdot 5$. $\qquad\triangle$

Example 4.4.11. The prime factorization of 777,777 is $3 \cdot 7^2 \cdot 11 \cdot 13 \cdot 37$. $\qquad\triangle$

Example 4.4.12. As another application of The Fundamental Theorem of Arithmetic, we provide another proof of Proposition 4.4.5. (By contradiction) Assume that $\sqrt{p}$ is rational. Then $\sqrt{p} = a/b$ for positive rational numbers. Then $pb^2 = a^2$. We consider the prime decomposition of a and b. Then in the equality $pb^2 = a^2$, the left-hand side involves an odd number of prime factors, while the right-hand side involves an even number of prime factors. By the Fundamental Theorem of Arithmetic, this is a contradiction. Hence, $\sqrt{p}$ is irrational. $\qquad\triangle$

The prime factorization inspires the so-called *prime order* function $\mathrm{ord}_p : \mathbb{N}* \to \mathbb{N}$ defined by

$$\mathrm{ord}_p(n) = k \quad \Longleftrightarrow \quad p^k \mid n \text{ and } p^{k+1} \nmid n. \qquad (4.4)$$

Example 4.4.13. Let $n = 2016$. By dividing about by appropriate primes, we find that the prime factorization is $2016 = 2^5 \times 3^2 \times 7$. Thus,

$$\mathrm{ord}_2(2016) = 5, \quad \mathrm{ord}_3(2016) = 2, \quad \mathrm{ord}_7(2016) = 1,$$

and the $\mathrm{ord}_p(2016) = 0$ for all $p \notin \{2, 3, 7\}$.   △

The order function extends to a function $\mathrm{ord}_p : \mathbb{Q}^{>0} \to \mathbb{Z}$ in the following way. Let $\frac{m}{n}$ be a fraction written in reduced form. Then

$$\mathrm{ord}_p \left(\frac{m}{n} \right) = \mathrm{ord}_p(m) - \mathrm{ord}_p(n).$$

Example 4.4.14. Let $n/m = 48/55$. Supposing p is prime, we have

$$\mathrm{ord}_2 \left(\frac{48}{55} \right) = 4, \quad \mathrm{ord}_3 \left(\frac{48}{55} \right) = 1, \quad \mathrm{ord}_5 \left(\frac{48}{55} \right) = -1,$$

$$\mathrm{ord}_{11} \left(\frac{48}{55} \right) = -1, \text{ and } \mathrm{ord}_p (48/55) = 0 \text{ for all } p \notin \{2, 3, 5, 11\}. \triangle$$

EXERCISES FOR SECTION 4.4

1. Find the prime factorization of the following integers: (a) 56; (b) 97; (c) 126; (d) 399; (e) 255; (f) 1728.

2. Find the prime factorization of the following integers: (a) 111; (b) 470; (c) 289; (d) 743; (e) 2345; (f) 101010.

3. Prove that any integer greater than 3 that is 1 less than a square cannot be prime.

4. Find a counterexample to the converse of Proposition 4.4.8. In other words, find a prime number p such that $2^p - 1$ is not prime.

5. Prove or disprove that $p_1 p_2 \cdots p_n + 1$ is a prime number where $p_1, p_2, \ldots, p_n$ are the n smallest consecutive prime numbers.

6. Suppose that the prime factorizations of a and b are

$$a = p_1^{\alpha_1} p_2^{\alpha_2} \cdots p_n^{\alpha_n} \quad \text{and} \quad b = p_1^{\beta_1} p_2^{\beta_2} \cdots p_n^{\beta_n},$$

with p_i distinct primes and $\alpha_i, \beta_i \geq 0$.

   (a) Prove that $\gcd(a, b) = p_1^{\min(\alpha_1, \beta_1)} p_2^{\min(\alpha_2, \beta_2)} \cdots p_n^{\min(\alpha_n, \beta_n)}$.
   (b) Prove that $\mathrm{lcm}(a, b) = p_1^{\max(\alpha_1, \beta_1)} p_2^{\max(\alpha_2, \beta_2)} \cdots p_n^{\max(\alpha_n, \beta_n)}$.

7. Use the result of Exercise 4.4.6 to prove Proposition 4.3.15 for positive integers a and b.

8. A prime number of the form $2^n + 1$ is called a Fermat prime. Prove that if $2^n + 1$ is a prime number, then $n = 2^\ell$ for some nonnegative integer ℓ.

9. Prove that for all primes p and all integers $k \geq 2$, the number $\sqrt[k]{p}$ is irrational.

10. Determine all the nonzero $\mathrm{ord}_p(n)$, defined in (4.4), for all primes p, where n is one of the following

   (a) 450;   (b) 392;   (c) 2310;   (d) 121212.

11. Find $\text{ord}_5(200!)$. Use this to determine the number of 0s to the right in the decimal (the usual base 10) expansion of 200!. (In other words, determine the highest power k of 10, such that 10^k divides 200!.)

12. Prove that $\text{ord}_p(n!) = \sum_{k \geq 1} \left\lfloor \dfrac{n}{p^k} \right\rfloor$. (Note that this sum consists of only a finite number of nonzero terms since $\lfloor n/p^k \rfloor = 0$ for $k > \log_p(n)$.)

13. Let p be a prime number. Prove that the function $\text{ord}_p : \mathbb{Q} \to \mathbb{Z}$ defined in (4.4) satisfies the following logarithmic-type properties.

    (a) $\text{ord}_p(mn) = \text{ord}_p(m) + \text{ord}_p(n)$ for all $m, n \in \mathbb{Z}$;
    (b) $\text{ord}_p(m^k) = k \, \text{ord}_p(m)$ for all $m \in \mathbb{Z}$ and $k \in \mathbb{N}^*$.

4.5   Induction

The principle of induction is a proof technique that arises from the structural definition of the integers. As we will see from the many examples in this section, this method of proof is surprisingly effective. Proofs by induction are ubiquitous throughout mathematics, sometimes providing the key technique in surprisingly useful ways.

4.5.1   Weak induction

Theorem 4.5.1 (Principle of Induction)

Let $P(n)$ be a predicate about integers. Suppose that $P(n_0)$ is true for some integer n_0 and that for all $n \geq n_0$, if $P(n)$ then $P(n+1)$. Then the statement $P(n)$ is true for all $n \geq n_0$.

Proof. Consider the predicate $Q(n)$ on $\mathbb{N}$ defined by $Q(n) = P(n + n_0)$. Since $P(n_0)$ is true, then $Q(0)$ is true. We also know that for all $n \geq n_0$, $P(n)$ implies $P(n+1)$. Equivalently, for $n \geq 0$, $Q(n) = P(n + n_0)$ implies $P(n + n_0 + 1) = Q(n+1)$. Hence, the axiom of induction, Proposition 4.1.2 gives that $Q(n)$ is true for all $n \in \mathbb{N}$. Hence, $P(n)$ is true for all $n \geq n_0$.   □

The Principle of Induction as stated above is also called *weak induction* in contrast to *strong induction*, discussed below. A proof that uses the principle of induction, whether weak or strong, is called an induction proof. In an induction proof, we call the step of proving $P(n_0)$ the *basis step*. This is usually easy, especially when it requires just a calculation check. The part of an induction proof that involves proving $P(n) \to P(n+1)$ for all $n \geq n_0$ is called the *induction step*. While proving the induction step, we refer to $P(n)$ as the *induction hypothesis*.

Example 4.5.2. When studying the definition of the Riemann integral, we often come across the formula

$$\sum_{i=1}^{n} i^2 = \frac{n(n+1)(2n+1)}{6} \qquad \text{for all positive integers } n. \qquad (4.5)$$

We prove this result by induction.

The predicate in this case is $P(n)$ is $\displaystyle\sum_{i=1}^{n} i^2 = \frac{n(n+1)(2n+1)}{6}$.

Its universe of discourse is positive integers. We wish to prove the universally quantified proposition $\forall n \in \mathbb{N}^* \, P(n)$.

Basis Step: Here, we wish to prove $P(1)$. In many cases, this is a matter of simply checking. Suppose $n = 1$. The left-hand side of (4.5) is $1^2 = 1$, while the right-hand side is $(1 \cdot 2 \cdot 3)/6 = 1$. The formula holds for $n = 1$, i.e., $P(1)$ is true.

Induction Step: Now we wish to show that $\forall n \in \mathbb{N}^* \, (P(n) \rightarrow P(n+1))$. In this case, we will prove this with a direct proof. This means that we suppose that $P(n)$ is true for some $n \geq 1$ and show that this implies $P(n+1)$ is true.

Suppose that (4.5) is true for some $n \geq 1$. Then

$$\sum_{i=1}^{n+1} i^2 = \left(\sum_{i=1}^{n} i^2\right) + (n+1)^2 = \frac{n(n+1)(2n+1)}{6} + (n+1)^2,$$

where the last equality holds by the induction hypothesis. Then,

$$\sum_{i=1}^{n+1} i^2 = \frac{n+1}{6}\left(n(2n+1) + 6(n+1)\right) = \frac{n+1}{6}(2n^2 + 7n + 6)$$

$$= \frac{(n+1)(n+2)(2n+3)}{6}$$

$$= \frac{(n+1)\left((n+1)+1\right)\left(2(n+1)+1\right)}{6}.$$

This proves the induction step. Hence, by induction (4.5) is true for all integers $n \geq 1$. $\triangle$

As the reader might expect, when using a proof by induction, we do not usually draw out the explanations that we used when starting the basis step or when starting the induction step. Someone familiar with this proof technique understands these steps.

Example 4.5.3. Recall the result from linear algebra that if A is an invertible matrix, then for all positive integers n the matrix A^n is invertible and $(A^n)^{-1} = (A^{-1})^n$. (Recall that a square matrix A is invertible if there exists a square matrix B such that $AB = BA = I$.

The matrix B is called the inverse of A.) We prove this result by induction.

Basis Step: with $n = 1$, the result is obvious: $A^{-1} = A^{-1}$.

Induction Step: Suppose that for some $n \in \mathbb{N}^*$, we have $(A^n)^{-1} = (A^{-1})^n$. Consider $n + 1$. Then

$$A^{n+1}(A^{-1})^{n+1} = A^n A A^{-1}(A^{-1})^n$$
$$= A^n I(A^{-1})^n = A^n(A^{-1})^n$$
$$= I.$$

The first $=$ holds by associativity and the last $=$ follows from the induction hypothesis. By an identical reasoning, we also show that $(A^{-1})^{n+1} A^{n+1} = I$. Hence, for this n, the matrix A^{n+1} is invertible with inverse $(A^{n+1})^{-1} = (A^{-1})^{n+1}$.

By induction, we deduce that for all $n \in \mathbb{N}^*$, the matrix A^n is invertible with inverse $(A^n)^{-1} = (A^{-1})^n$. $\triangle$

This example follows the standard pattern of writing a proof by induction. When a student learns the technique of proofs by induction for the first time, it is common to state "We prove this result by induction" and then to write explicitly "Basis step" and "Induction step." However, since induction is such a common proof method, after a while, we tend not to write "Basis step:" and "Induction step:" and we expect the reader to recognize these.

Example 4.5.4. Recall the Fibonacci sequence in Example 3.1.25. The sequence satisfies a recurrence relation $f_{n+2} = f_{n+1} + f_n$ for all $n \geq 0$, along with initial values $f_0 = 0$ and $f_1 - 1$.

In this example, we prove that if $5 \mid n$, then $5 \mid f_n$.

We need to rephrase the problem in a manner that is suitable for an induction proof. We propose to prove that 5 divides f_{5k} for all nonnegative integers k.

Basis step: Suppose that $k = 0$. Since 5 certainly divides $f_0 = 0$, the basis step holds.

Induction step: Now, suppose that $5 \mid f_{5k}$ for some nonnegative integer k. This means that $f_{5k} = 5m$ for some integer m. Furthermore,

$$f_{5(k+1)} = f_{5k+5} = f_{5k+4} + f_{5k+3}$$
$$= (f_{5k+3} + f_{5k+2}) + f_{5k+3} = 2f_{5k+3} + f_{5k+2}$$
$$= 2(f_{5k+2} + f_{5k+1}) + f_{5k+2} = 3f_{5k+2} + 2f_{5k+1}$$
$$= 3(f_{5k+1} + f_{5k}) + 2f_{5k+1} = 5f_{5k+1} + f_{5k}$$
$$= 5f_{5k+1} + 5m = 5(f_{5k+1} + m).$$

Hence, 5 divides $f_{5(k+1)}$. By induction on k, the Fibonacci number f_{5k} is divisible by 5 for all k, which we can restate as $5|n \implies 5|f_n$. $\triangle$

Example 4.5.5. We prove that $2^n \geq n^2$ for all $n \geq 4$. For the basis step, we notice that $2^4 = 16$, and $4^2 = 16$ so equality holds.

For the induction step, we assume that for some $n \geq 4$, the inequality $2^n \geq n^2$ holds. Then $2^{n+1} = 2 \cdot 2^n \geq 2n^2$ by the induction hypothesis. We would be done if we knew that $2n^2 \geq (n+1)^2$. We consider the following logical equivalences (Backward-and-forward proof.)

$$2n^2 \geq (n+1)^2 \iff 2n^2 - (n^2 + 2n + 1) \geq 0 \iff n^2 - n - 1 \geq 0$$
$$\iff (n-1)^2 - 2 \geq 0$$
$$\iff n - 1 \leq -\sqrt{2} \text{ or } n - 1 \geq \sqrt{2}$$
$$\iff n \leq 1 - \sqrt{2} \text{ or } n \geq \sqrt{2} + 1. \tag{4.6}$$

Since we are under the assumption that $n \geq 4$, then $n \geq 1 + \sqrt{2}$ and hence $2n^2 \geq (n+1)^2$. We conclude that

$$2^{n+1} \geq (n+1)^2$$

and, by induction, we conclude that $2^n \geq n^2$ for all integers $n \geq 4$.

By checking the validity of the inequality for $n = 0, 1, 2, 3$, we find that the inequality also holds for all integers $0 \leq n \leq 3$. So we can actually conclude that $2^n \geq n^2$ for all $n \in \mathbb{N}$. If we had attempted to establish the result for $n \geq 0$ with a proof by induction, then we would have found that the necessary inequality in (4.6) to show $P(n) \to P(n+1)$ fails when $n = 1$ or 2.   △

Example 4.5.6. Consider the function $C : \mathbb{N} \times \mathbb{N} \to \mathbb{N}$ defined recursively by

$$C(n,k) = \begin{cases} 1 & \text{if } k = 0 \\ 0 & \text{if } n = 0 \text{ and } k \geq 0 \\ C(n-1, k-1) + C(n-1, k) & \text{if both } n, k \geq 1. \end{cases}$$

We plot the values for this recursive function with $0 \leq n, k \leq 5$.

n/k	0	1	2	3	4	5
0	1	0	0	0	0	0
1	1	1	0	0	0	0
2	1	2	1	0	0	0
3	1	3	3	1	0	0
4	1	4	6	4	1	0
5	1	5	10	10	5	1

The resulting numbers, with the nonzero numbers organized in the shape of a triangle, form Pascal's Triangle. We discuss these numbers more at length in Section 5.2. However, in this example, we illustrate another proof by induction that establishes the formula

$$C(n,k) = \begin{cases} \dfrac{n!}{k!(n-k)!} & \text{if } n \geq k \\ 0 & \text{otherwise} \end{cases} \tag{4.7}$$

for all $n, k \in \mathbb{N}^2$. This formula involves two indices of nonnegative integers. To describe the strategy, let $P(n,k)$ be the predicate that (4.7) is true for some pair $(n,k) \in \mathbb{N}^2$ and let $Q(n)$ be the predicate $\forall k \in \mathbb{N}, P(n,k)$. By proving that $Q(n)$ is true for all $n \geq 0$, we establish the desired result for all $(n,k) \in \mathbb{N}^2$.

(Basis step)   Let $n = 0$. By definition, $C(0,0) = 1$ and $0!/(0!(0-0)!) = 1$ so (4.7) holds for $(n,k) = (0,0)$. If $k \geq 1$, then by definition $C(0,k) = 0$ and this is also true in (4.7). Hence, (4.7) holds when $n = 0$ and for all $k \in \mathbb{N}$.

(Induction step)   Now suppose that for some $n \geq 0$, (4.7) holds for the specific n and for all $k \in \mathbb{N}$. We have four cases when we try to prove (4.7) for $n + 1$.

Case 1, $k = 0$. By definition, $C(n+1,0) = 1$ and $(n+1)!/(0!(n+1-0)!) = 1$, so (4.7) holds for the pair $(n+1,0)$.

Case 2, $1 \leq k \leq n$. By definition, $C(n+1,k) = C(n,k-1)+C(n,k)$ and by the induction hypothesis

$$\begin{aligned} C(n+1,k) &= \frac{n!}{(k-1)!(n-k+1)!} + \frac{n!}{k!(n-k)!} \\ &= \frac{k(n!)}{k!(n-k+1)!} + \frac{(n!)(n-k+1)}{k!(n-k+1)!} \\ &= \frac{n!(k+n-k+1)}{k!(n-k+1)!} = \frac{(n+1)!}{k!((n+1)-k)!}. \end{aligned}$$

Hence, (4.7) holds for $1 \leq k \leq n$.

Case 3, $k = n+1$. By definition, $C(n+1,n+1) = C(n,n)+C(n,n+1)$ and by the induction hypothesis $C(n,n) = 1$, while $C(n,n+1) = 0$. Hence, $C(n+1,n+1) = 1 = (n+1)!/((n+1)!(n+1-(n+1))!)$. Hence, (4.7) holds for $k = n+1$.

Case 4, $k > n+1$. In this case, $k - 1 > n$. By definition, $C(n+1,k) = C(n,k-1) + C(n,k)$ and by the induction hypothesis both $C(n,k)$ and $C(n,k-1)$ are 0. Hence, $C(n+1,k) = 0$, which conforms to (4.7) for $k > n+1$.

We have proven that (4.7) holds for $n + 1$ and for all $k \in \mathbb{N}$. By induction on n, this establishes that (4.7) is true for all $(n,k) \in \mathbb{N}^2$. $\triangle$

In the above example, we used a proof by cases to establish that $Q(n+1) = \forall k \in \mathbb{N}, P(n+1, k)$ in the induction step. However, in another similar situation, it might be useful or necessary to use induction on k to show that $Q(n)$ implies $Q(n+1)$. We might refer to such a strategy as a double induction.

4.5.2 Strong induction

Theorem 4.5.7 (Principle of Strong Induction)

Let $Q(n)$ be a predicate about integers. Suppose that $Q(n_0)$ is true for some integer n_0 and that $Q(k)$ for all k with $n_0 \leq k \leq n$ implies $Q(n+1)$. Then the statement $Q(n)$ is true for all $n \geq n_0$.

The principle of strong induction might appear more powerful than the first principle of induction. The conjunctive statement

$$Q(n_0) \text{ and } Q(n_0 + 1) \text{ and } \cdots \text{ and } Q(n)$$

is false not only when $Q(n)$ is false, but when any of the other instantiated predicates are false. A conditional statement $p \to q$, meaning "if p then q" or "p implies q," is false when p is true but q is false and is true otherwise. Hence, the conditional statement

$$\Big(Q(n_0) \text{ and } Q(n_0 + 1) \text{ and } \cdots \text{ and } Q(n)\Big) \text{ implies } Q(n+1)$$

will be false if all of the $Q(k)$ with $n_0 \leq k \leq n$ are true and $Q(n+1)$ is false, whereas

$$Q(n) \text{ implies } Q(n+1)$$

is false only when $Q(n)$ is true and $Q(n+1)$ is false. Thus, the induction step of strong induction is less likely to occur than the induction step of weak induction.

However, we can see that the strong induction and weak induction are in fact the same. If the induction hypothesis holds in weak induction, then the strong induction hypothesis holds. On the other hand, by setting the predicate $P(n)$ to be "$Q(k)$ is true for all integers k with $n_0 \leq k \leq n$," we see that the principle of strong induction immediately follows from weak induction.

Example 4.5.8. Consider the sequence $\{a_n\}_{n \geq 0}$ defined by $a_0 = 0$, $a_1 = 1$, and $a_{n+2} = a_{n+1} + 2a_n$ for all $n \geq 0$. We prove that for all $n \geq 0$,

$$a_n = \frac{1}{3}(2^n - (-1)^n). \tag{4.8}$$

Notice that (4.8) holds for $n = 0$ and $n = 1$. (These observations serve as the basis step.) For the induction step, assume that $n \geq 1$

and that (4.8) is true for all indices k with $0 \le k \le n$. According to the induction hypothesis,

$$a_n = \frac{1}{3}\left(2^n - (-1)^n\right) \quad \text{and} \quad a_{n-1} = \frac{1}{3}\left(2^{n-1} - (-1)^{n-1}\right).$$

Thus,

$$\begin{aligned} a_{n+1} = a_n + 2a_{n-1} &= \frac{1}{3}\left(2^n - (-1)^n\right) + \frac{2}{3}\left(2^{n-1} - (-1)^{n-1}\right) \\ &= \frac{1}{3}\left(2^n + 2 \cdot 2^{n-1} - (-1)^n - 2(-1)^{n-1}\right) \\ &= \frac{1}{3}\left(2^n + 2^n - (-1)^n + 2(-1)^n\right) \\ &= \frac{1}{3}\left(2^{n+1} + (-1)^n\right) = \frac{1}{3}\left(2^{n+1} - (-1)^{n+1}\right). \end{aligned}$$

This proves the induction hypothesis and hence, by strong induction, (4.8) is true for all $n \ge 0$. $\triangle$

Example 4.5.8 exemplifies one of the key reasons why some situations may require strong induction and not just weak induction. The proof of the induction hypothesis required using the proposed formula (4.8) for $Q(n)$ and $Q(n-1)$ to establish $Q(n+1)$. However, since the proof used more than just the one previous step $Q(n)$ to establish $Q(n+1)$, it is not a (weak) induction proof. The following example, taken from a Putnam Competition, shows an even more general use of strong induction.

Example 4.5.9 (Putnam 2009, B1). The problem states: Show that every positive rational number can be written as a quotient of products of factorials of (not necessarily distinct) primes. For example

$$\frac{10}{9} = \frac{2! \cdot 5!}{3! \cdot 3! \cdot 3!}.$$

By the Fundamental Theorem of Arithmetic, every positive rational number r can be expressed as

$$\frac{p_1 p_2 \cdots p_m}{p'_1 p'_2 \cdots p'_n}, \tag{4.9}$$

where p_i and p'_j are primes, not necessarily distinct. Consequently, in order to prove the desired result, it suffices to prove it where r is a prime number; expressing p_i and p'_j as quotients of products of factorials of primes and putting these expressions into (4.9) produces a quotient of products of factorials of primes. We use induction.

(Basis step)   Suppose that $r = 2$. Then the identity $2 = 2!$ gives 2 as a quotient of products of factorials of primes.

Now suppose that r is a prime and that every prime number less   (Induction
than r can be expressed as a quotient of products of factorials of   step)
primes. Then

$$r = \frac{r!}{(r-1)!}.$$

Every prime number in the product $(r-1)!$ is less than r so by the
induction hypothesis, it can be written as a quotient of products of
factorials of primes. Putting all of these into the expression $r!/(r-1)!$ again gives a quotient of products of factorials of primes. By
induction, every prime number can be expressed as a quotient of
products of factorials of primes. This establishes the result.   △

The following proposition is important in its own right. We provide it here because its proof relies on strong induction.

Proposition 4.5.10

Let S be a set and let $\star$ be a binary operation on S that is
associative. A finite operation expression

$$a_1 \star a_2 \star \cdots \star a_n \qquad (4.10)$$

results in the same elements of S regardless of how valid
pairs of parentheses are inserted.

Proof. We define a temporary notation. Given elements $a_1, a_2, \ldots, a_k$
in S, by analogy with the $\sum$ notation, we define

$$\bigstar_{i=1}^{k} a_i \overset{\text{def}}{=} (\cdots((a_1 \star a_2) \star a_3) \cdots a_{k-1}) \star a_k.$$

In this notation, we perform the operations in (4.10) from left to right.
Note that if $k = 1$, the expression is equal to the element a_1.

We prove by (strong) induction on n, that every operation expression in (4.10) is equal to $\bigstar_{i=1}^{n} a_i$.

If $n = 3$, the proposition simply restates the assumption that $\star$ is   (Basis step)
associative.

We now assume that the proposition is true for all integers k with   (Induction
$3 \le k \le n$. Consider an operation expression (4.10) involving $n+1$ step)
terms. Suppose without loss of generality that the last operation
performed occurs between the jth and $(j+1)$th term, i.e.,

$$q = \overbrace{(\text{operation expression}_1)}^{j \text{ terms}} \star \overbrace{(\text{operation expression}_2)}^{n-j \text{ terms}}.$$

Since both operation expressions involve n terms or less, by the induction hypothesis

$$q = \left(\bigstar_{i=1}^{j} a_i \right) \star \left(\bigstar_{i=j+1}^{n} a_i \right).$$

Furthermore,

$$
\begin{aligned}
q &= \left(\bigstar_{i=1}^{j} a_i \right) \star \left(a_{j+1} \star \left(\bigstar_{i=j+2}^{n} a_i \right) \right) && \text{by induction hypothesis} \\
&= \left(\left(\bigstar_{i=1}^{j} a_i \right) \star a_{j+1} \right) \star \left(\bigstar_{i=j+2}^{n} a_i \right) && \text{by associativity} \\
&= \left(\bigstar_{i=1}^{j+1} a_i \right) \star \left(\bigstar_{i=j+2}^{n} a_i \right).
\end{aligned}
$$

Repeating this $n - j - 2$ more times, we conclude that $q = \bigstar_{i=1}^{n+1} a_i$. The proposition follows. $\qquad\square$

EXERCISES FOR SECTION 4.5

1. Prove that $\displaystyle\sum_{i=1}^{n} i^3 = \frac{n^2(n+1)^2}{4}$ for all integers $n \geq 1$.

2. Use mathematical induction to prove the geometric summation formula:
   $$\sum_{i=0}^{n} Ar^i = \frac{A(r^{n+1} - 1)}{r - 1} \qquad \text{where } r \neq 1$$
   for all nonnegative integers n.

3. Prove that for every positive integer n,
   $$1 \cdot 2 + 2 \cdot 3 + \cdots + n(n+1) = \frac{n(n+1)(n+2)}{3}.$$

4. Prove that $1 + nh \leq (1 + h)^n$ for all $h \geq -1$ and for nonnegative integers $n \geq 0$.

5. Prove that $5 \mid (n^5 - n)$ for all nonnegative integers n.

6. Prove by induction that $\displaystyle\sum_{i=0}^{n} (2i + 1) = (n + 1)^2$.

In Exercises 4.5.7 through 4.5.12, let $\{f_n\}_{n \geq 0}$ be the sequence of Fibonacci numbers.

7. Prove that $f_{n-1}f_{n+1} - f_n^2 = (-1)^n$ for all $n \geq 1$.

8. Prove that $\displaystyle\sum_{i=0}^{n} f_i = f_{n+2} - 1$.

9. Prove $2 \mid f_n$ if and only if $3 \mid n$.

10. Prove that $\displaystyle\sum_{i=0}^{n} f_i^2 = f_n f_{n+1}$.

11. (*) Generalize Example 4.5.4 to prove that if $k \mid n$, then $f_k \mid f_n$.

12. (*) Prove that $f_n^2 + f_{n-1}^2 = f_{2n-1}$ for all $n \geq 1$.

13. Prove that for all real numbers $r \neq 1$,

$$\sum_{k=0}^{n} kr^k = \frac{((r-1)n - 1)\, r^{n+1} + r}{(r-1)^2}.$$

14. Let $f : \mathbb{R} \to \mathbb{R}$ with $f(x) = x^2 e^x$. Prove by induction that $f^{(n)}(x) = (x^2 + 2nx + n(n-1))e^x$ for all $n \geq 0$.

15. Let $f : \mathbb{R} \to \mathbb{R}$ with $f(x) = x \sin x$. Prove by induction that

$$f^{(n)}(x) = \begin{cases} (-1)^{n/2}(x \sin x - n \cos x) & \text{if } n \text{ is even} \\ (-1)^{(n-1)/2}(n \sin x + x \cos x) & \text{if } n \text{ is odd.} \end{cases}$$

[Hint: Use strong induction.]

16. Let $f : \mathbb{R} \to \mathbb{R}$ be any infinitely differentiable function and let $g : \mathbb{R} \to \mathbb{R}$ with $g(x) = f(x^2)$. Prove that

$$g^{(n)}(0) = \begin{cases} 0 & \text{if } n \text{ is odd,} \\ \frac{n!}{(n/2)!} f^{(n/2)}(0) & \text{if } n \text{ is even.} \end{cases}$$

17. (*) Consider the function $B : \mathbb{N} \times \mathbb{N} \to \mathbb{N}$ defined recursively by

$$B(n,k) = \begin{cases} 1 & \text{if } k = 0 \text{ or if } n = 0 \\ B(n-1,k) + B(n,k-1) & \text{if both } n, k \geq 1. \end{cases}$$

   (a) As in Example 4.5.6, give a chart of the values of $B(n,k)$ for $0 \leq k, n \leq 5$.

   (b) Prove that $B(n,k) = \frac{(n+k)!}{k!n!}$ for all $n, k \in \mathbb{N}$.

18. Prove that 13 divides $3^{n+1} + 4^{2n-1}$ for all $n \geq 1$.

19. Let $A \in M_n(\mathbb{R})$ be a square matrix with eigenvalue λ. Prove that for all $k \geq 1$, the number λ^k is an eigenvalue of A^k.

20. Let A_n be the $n \times n$ matrix that consists of 2s on the main diagonal, 1s on the diagonals immediately above and immediately below the main diagonal, and 0s in all other entries. Prove that $\det(A_n) = n+1$ for all $n \geq 1$.

21. Prove the following matrix power identity for all $n \geq 1$. If

$$A = \begin{pmatrix} \lambda & 1 & 0 \\ 0 & \lambda & 1 \\ 0 & 0 & \lambda \end{pmatrix} \text{ then } A^n = \begin{pmatrix} \lambda^n & n\lambda^{n-1} & n(n-1)\lambda^{n-2}/2 \\ 0 & \lambda^n & n\lambda^{n-1} \\ 0 & 0 & \lambda^n \end{pmatrix}.$$

22. A well-know property of eigenvalues is that for any square matrix A, if $\mathbf{v}_1, \mathbf{v}_2, \ldots, \mathbf{v}_k$ are eigenvectors of A corresponding to distinct eigenvalues, then $\{\mathbf{v}_1, \mathbf{v}_2, \ldots, \mathbf{v}_k\}$ is linearly independent. Prove this theorem by induction on k. [Hint: Use the predicate $P(k)$ to be "For all matrices, any set of k eigenvectors $\{\mathbf{v}_1, \mathbf{v}_2, \ldots, \mathbf{v}_k\}$ each corresponding to a distinct eigenvalues is linearly independent."]

23. A set of lines in the plane is said to be in general position if no two lines are parallel and no three lines intersect at a single point. Prove that for any set $\{L_1, L_2, \ldots, L_n\}$ of lines in $\mathbb{R}^2$ in general position, the complement $\mathbb{R}^2 - (L_1 \cup L_2 \cup \cdots \cup L_n)$ consists of $(n^2 + n + 2)/2$ disjoint regions in the plane.

24. Let $H_n = 1 + \dfrac{1}{2} + \dfrac{1}{3} + \cdots + \dfrac{1}{n}$ be the nth harmonic number. Prove that $H_{2^n} \leq 1 + n$.

25. Prove that $n! < n^{n-1}$ for all positive integers n.

26. Show that $\displaystyle\sum_{i=1}^{n} i(i!) = (n+1)! - 1$.

27. Show that any amount of postage of value 48 cents or higher can be formed using just 5-cent and 12-cent stamps.

28. Show that the Fibonacci sequence satisfies

$$f_n = \frac{1}{\sqrt{5}}\left(\frac{1+\sqrt{5}}{2}\right)^n - \frac{1}{\sqrt{5}}\left(\frac{1-\sqrt{5}}{2}\right)^n$$

for all $n \geq 0$.

29. Suppose that we have at our disposal an unlimited supply of 7-cent stamps and of 13-cent stamps. Show that using just these stamps we can get the exact postage for n cents, for any integer $n \geq 72$.

30. Let α be any real number such that $\alpha + \frac{1}{\alpha} \in \mathbb{Z}$. Prove that for all nonnegative integers n,

$$\alpha^n + \frac{1}{\alpha^n} \in \mathbb{Z}.$$

31. Prove that every integer $n \geq 2$ can be written as the sum of terms of the form $2^a \cdot 3^b$ with $(a,b) \in \mathbb{N}^2 \setminus \{(0,0)\}$ and in which none of these terms divide each other. For example, $61 = 2^4 + 2 \cdot 3^2 + 3^3$. [Hint: For a given integer n, consider the highest power of 2 such that $n - 2^k$ is divisible by 3.]

32. (*) Prove that the n-th derivative of $1/(x^k - 1)$ has the form

$$P_n(x)/(x^k - 1)^{n+1},$$

where $P_n(x)$ is a polynomial. Also prove that $P_n(1) = (-1)^n n! k^n$ for all $n \in \mathbb{N}$. [Note: This is a modified version of problem A1 on the 2002 Putnam Competition.]

4.6   Modular Arithmetic

In this section, n represents an integer greater than or equal to 2.

One of the theorems often presented as a highlight in an introduction to modular arithmetic, Fermat's Little Theorem, dates as far back as to 1640. However, many properties of congruences, a fundamental notion in modular arithmetic, appear in Leonhard Euler's early work on number theory, circa 1736 (see [13, p.131] and [69, p.45]). The modern formulation of congruences first appeared in Gauss' *Disquisitiones Arithmeticae* in 1801 [35]. He applied the theory of congruences to the study of Diophantine equations, algebraic equations in which we look for only integer solutions.

4.6.1 Congruence

Definition 4.6.1

Let a and b be integers. We say that a *is congruent to* b *modulo* n if $n \mid (b - a)$ and we write

$$a \equiv b \pmod{n}.$$

If n is understood from context, we simply write $a \equiv b$. The integer n is called the *modulus*.

(Notice that the symbol $\equiv$ in the context of congruences has nothing to do with propositional equivalences. This is one of the many instances in mathematics of the same symbol being reused for different purposes.)

Proposition 4.6.2

Congruence modulo n satisfies the following:

1) $\forall a \in \mathbb{Z} \, (a \equiv a)$.
2) $\forall a, b \in \mathbb{Z} \, (a \equiv b \longrightarrow b \equiv a)$.
3) $\forall a, b, c \in \mathbb{Z} \, (a \equiv b \wedge b \equiv c \longrightarrow a \equiv c)$.

Proof. (1) For any integer $a \in \mathbb{Z}$, we have $n \mid (a - a) = 0$.

(2) Suppose that $a \equiv b$. Then $n|(b - a)$, so there exists $k \in \mathbb{Z}$ with $nk = b - a$. Then $n(-k) = a - b$ and so $n|(a - b)$ and hence $b \equiv a$.

(3) Suppose that $a \equiv b$ and $b \equiv c$. Then $n \mid (b - a)$ and $n \mid (c - b)$. By Proposition 4.2.2(2),

$$n \mid ((b - a) + (c - b)) \quad \text{so} \quad n \mid (c - a).$$

Hence, $a \equiv c$. $\square$

Example 4.6.3. Consider 16 and 37 modulo 7. Since $37 - 16 = 21$ and $7 \mid 21$, then $16 \equiv 37 \pmod 7$. On the other hand, $100 \not\equiv 56$ because $7 \nmid -44 = (56 - 100)$. $\triangle$

Proposition 4.6.4

For all $a, b \in \mathbb{Z}$, $a \equiv b \pmod{n}$ if and only if a and b have the same remainder when divided by n.

Proof. (Left as an exercise for the reader. See Exercise 4.6.6.) □

This proposition means that when working with congruences modulo n, we can remember that any integer is always uniquely congruent to its remainder when divided by n.

It is important for applications of congruences to note that

$$a \equiv 0 \pmod{n} \iff n \mid a.$$

One of the useful applications of congruences stems from this property and how congruence interacts with addition and multiplication on $\mathbb{Z}$.

Proposition 4.6.5

Fix a modulus n. Let $a, b, c, d \in \mathbb{Z}$ such that $a \equiv c$ and $b \equiv d$. Then

$$a + b \equiv c + d \qquad \text{and} \qquad ab \equiv cd.$$

Proof. By definition, $n \mid (c - a)$ and $n \mid (d - b)$. Thus, there exist k, ℓ such that

$$c - a = nk \tag{4.11}$$
$$d - b = n\ell. \tag{4.12}$$

Adding these two expressions, we get

$$(d + c) - (b + a) = nk + n\ell = n(k + \ell).$$

This shows that $n \mid (d + c) - (b + a)$ so $a + b \equiv c + d$.

To show the multiplication, multiply Equation (4.12) by c and subtract from it Equation (4.11) multiplied by b. We obtain

$$c(d - b) - b(c - a) = cn\ell - bnk \iff cd - ab = n(c\ell - bk).$$

This shows that $n \mid (cd - ab)$, which means that $ab \equiv cd \pmod{n}$. □

A little induction proof extends Proposition 4.6.5 to show that if $a \equiv b \pmod{n}$, then $a^k \equiv b^k \pmod{n}$ for all $k \in \mathbb{N}$.

The term *modular arithmetic* (or *congruence arithmetic*) refers to arithmetic on integers involving congruence identities. Instead of caring about the specific value integer, we care more about an integer's *congruence class*, what it is congruent to. By Proposition 4.6.4, there are exactly n congruence classes modulo n. Modular arithmetic plays an important role in number theory and algebra because it offers an approach to study patterns within integers. We illustrate this claim already with a simple example about divisibility by 3.

Example 4.6.6. As an application, we consider divisibility of integers by 3. Hence, we work modulo 3.

Let $m = b_k b_{k-1} \cdots b_1 b_0$ be a positive integer expressed in base 10. This means that

$$m = b_k 10^k + b_{k-1} 10^{k-1} + \cdots b_1 10 + b_0,$$

where $0 \le b_i \le 9$ for all b_i. We note that $10 \equiv 1 \pmod 3$. Then

$$10^2 \equiv 1^2 = 1, \quad 10^3 \equiv 1^3, \quad \ldots \quad 10^k \equiv 1 \text{ for all } k \in \mathbb{N}.$$

Then

$$m \equiv b_k 10^k + b_{k-1} 10^{k-1} + \cdots b_1 10 + b_0 \pmod 3$$
$$\equiv b_k + b_{k-1} + \cdots + b_1 + b_0 \pmod 3.$$

Hence, an integer n has the same remainder when divided by 3 as the remainder of the sum of its digits when divided by 3. In particular, an integer n is divisible by 3 if and only if the sum of its digits is divisible by 3. The reader might notice that since $10 \equiv 1 \pmod 9$, that this result also holds for modular arithmetic modulo 9, namely that an integer m has the same remainder when divided by 9 as does the sum of the digits of m. $\triangle$

4.6.2 Inverses and units

Consider the congruence equation $x + 17 \equiv 3 \pmod{25}$. Congruence only pertains to integers, so we are looking for integers x as solutions. Adding -17 to both sides gives $x \equiv 3 - 17 \equiv -14 \equiv 11$. So the solution set of this congruence equation consists of all integers with a remainder of 11 when divided by 25.

On the other hand, if we try to solve the equation $3x + 17 \equiv 3 \pmod{25}$, we come to $3x \equiv 11$. We are still considering integer solutions, so we arrive at a curious situation: We cannot simply divide both side by 3. This motivates the next definition.

Definition 4.6.7

An integer b is an *inverse* to a modulo n whenever $ab \equiv 1 \pmod n$. If a has an inverse, it is said to be a *unit*.

Following the discussion immediately before the definition, we remark that 17 is an inverse to 3 modulo 25 because $3 \cdot 17 = 51 \equiv 1$

(mod 25). Consequently, we solve the congruence equation $3x \equiv 11$ by multiplying both sides by 17, an inverse to 3 modulo 25. We get $51x \equiv 187$, which is equivalent to $x \equiv 12$ (mod 25). This is the solution to the congruence equation.

It is not too hard to notice that not every integer a has an inverse modulo any n. For example, 4 does not have an inverse modulo 10. Assume otherwise; then there exist b with $2b \equiv 1$ (mod 10). Thus, there exists $k \in \mathbb{Z}$ with $2b - 1 = 10k$. Thus $2b - 10k = 2(b - 5k) = 1$. This is a contradiction because the left-hand side is even, while the right-hand side is odd.

Proposition 4.6.8

Let $n \geq 2$ and let $a \in \mathbb{Z}$. The integer a is a unit modulo n if and only if $\gcd(a, n) = 1$.

($\Longrightarrow$)   *Proof.* First, suppose that there exists $b \in \mathbb{Z}$ with $ab \equiv 1$ (mod n). Then there exists $k \in \mathbb{Z}$ such that $ab = 1 + kn$. Thus, $ab - kn = 1$. Hence, there is a linear combination of a and n that is 1. The number 1 is the least positive integer, so by Proposition 4.3.9, $\gcd(a, n) = 1$.

($\Longleftarrow$)   Conversely, suppose that $\gcd(a, n) = 1$. Then again by Proposition 4.3.9, there exists $s, t \in \mathbb{Z}$ such that $sa + tn = 1$. Then $sa = 1 - tn$ and so $sa \equiv 1$ (mod n), making s an inverse to a modulo n.   □

Example 4.6.9. We consider the question of inverses modulo $n = 21$. We first point out that an integer a is a unit modulo 21 if and only if its remainder when divided by n is a unit. Hence, as we look for inverses modulo 21, by Proposition 4.6.8, we only need to consider $1, 2, 4, 5, 8, 10, 11, 13, 16, 17, 19, 20$. The unit and inverse pairs are:

$$1 \cdot 1 \equiv 1, \qquad 2 \cdot 11 \equiv 1, \qquad 4 \cdot 16 \equiv 1, \qquad 5 \cdot 17 \equiv 1$$
$$8 \cdot 8 \equiv 1, \qquad 10 \cdot 19 \equiv 1, \qquad 13 \cdot 13 \equiv 1. \qquad \triangle$$

If n is small, it is often possible to guess and check inverses modulo n, but as n gets larger that method becomes inefficient. The proof of this proposition gives us a fast method to find an inverse to a modulo n. Supposing that $\gcd(a, n) = 1$, we know there exist $s, t \in \mathbb{Z}$ such that $sa + tn = 1$. Considering this expression modulo n, since $n \equiv 0$ (mod n), we have $sa \equiv 1$. Thus s is an inverse to a. In particular, the Extended Euclidean Algorithm helps in two ways: (1) checks whether $\gcd(a, n) = 1$ (whether a has an inverse); and (2) finds a multiplicative inverse s to a modulo n.

Example 4.6.10. We look for the inverse of 79 modulo 123. We use the Extended Euclidean Algorithm (EEA) with $a = 123$ and $b = 79$.

At the outset, we might not know if 79 has an inverse, but we will find $\gcd(79, 123)$ as a part of the algorithm.

i	r_i	s_i	t_i	q_i
0	123	1	0	
1	79	0	1	1
2	44	1	-1	1
3	35	-1	2	1
4	9	2	-3	3
5	8	-7	11	1
6	1	9	-14	8
7	0			

From $r_6 = 1$, we see that $\gcd(123, 79) = 1$ and from row $i = 6$, we also see that
$$1 = 123 \times 9 - 79 \times 14.$$
Since $123 \equiv 0 \pmod{123}$, this gives $1 \equiv -14 \times 79 \pmod{123}$. Hence, any integer congruent to -14 modulo 123 is an inverse to 79 modulo 123. If we prefer to give answers to congruence problems with integers between 0 and $n-1$, we point out that 109 is an inverse to 79 modulo 123. △

Example 4.6.11. Suppose we work in modulo $n = 15$ and we propose to solve the equation $7x + 10 \equiv y$ for x in terms of y. We point out that modular arithmetic is in integers, so all our answers should be in integers.

Note first that $2 \cdot 7 \equiv 14 \equiv -1$. So $-2 \equiv 13$ are multiplicative inverses of 7 modulo 15. Now we have

$$7x + 10 \equiv y \implies 7x \equiv y - 10 \implies -2 \cdot 7x = -2(y - 10)$$
$$\implies x \equiv -2y + 20 \equiv 13y + 5.$$

A key in this reasoning is that at the stage $7x \equiv y - 10$, it would **not** be appropriate to divide by 7. Dividing by 7 means we work in the context of rational numbers, whereas modular arithmetic only applies to integers. △

4.6.3 Patterns in powers

We finish the section with a brief discussion about powers in modular arithmetic.

In regular arithmetic in $\mathbb{Z}$, if an integer a is greater than 1 in absolute value, then the powers a^k increase without bound. In modular arithmetic, there are only a finite number of possible congruence classes, which leads to patterns in the powers of elements in modular arithmetic.

Example 4.6.12. We calculate the powers of 2 and 3 modulo 7. The following table lists the remainder of 2^k and 3^k when divided by 7, for $k = 0, 1, \ldots, 8$.

k	0	1	2	3	4	5	6	7	8
2^k	1	2	4	1	2	4	1	2	4
3^k	1	3	2	6	4	5	1	3	2

We notice that the powers follow a repeating pattern. This is because if $a^k \equiv a^{k+l}$, then

$$a^{k+2l} \equiv a^{k+l}a^l \equiv a^k a^l \equiv a^{k+l} \equiv a^k$$

and, by induction, we can prove that $a^{k+ml} \equiv a^k$ for all $m \in \mathbb{N}$. In particular, if $a^k \equiv 1$ for some k, then $a^{mk} \equiv 1$ for all $m \in \mathbb{N}$. Observing the pattern for 3, we see for example that

$$3^{3201} \equiv 3^{6 \times 533 + 3} \equiv (3^6)^{533} \cdot 3^3 \equiv 1^5 33 \cdot 3^3 \equiv 6 \pmod 7.$$

Therefore, using congruences, we have easily calculated the remainder of 3^{3201} when divided by 7, without ever calculating 3^{3201}. This might come as a surprise since 3^{3201} has $\lfloor \log_{10}(3^{3201}) \rfloor + 1 = \lfloor 3201 \log_{10} 3 \rfloor + 1 = 1,528$ digits.   △

Some of the patterns in powers of a number in a given modulus are not always easy to detect. Though we could prove the following profound theorem now, we leave its proof as a guided exercise in a future section. (See Exercise 6.4.6.)

Theorem 4.6.13 (Fermat's Little Theorem)

Let p be a prime number and a an integer with $p \nmid a$. Then

$$a^{p-1} \equiv 1 \pmod p.$$

Example 4.6.14. We find a simplified congruence for $3^{10,000}$ modulo 19 using Fermat's Little Theorem.

By Fermat's Little Theorem, $3^{18} \equiv 1 \pmod{19}$. The integer division of 10,000 by 18 is $10,000 = 18 \times 555 + 10$. We have a chain of calculations, all of which we can do without technology:

$$3^{10,000} \equiv 3^{18 \times 555 + 10} \equiv 3^{18 \times 555} \cdot 3^{10}$$

$$\equiv (3^{18})^{555} \cdot 3^{10} \equiv 1^{555} \cdot 3^{10} \equiv 3^{10}$$

$$\equiv 3 \cdot (3^3)^3 \equiv 3 \cdot 27^3 \equiv 3 \cdot 8^3 \equiv (3 \cdot 8) \cdot (8^2)$$

$$\equiv 24 \cdot 64 \equiv 5 \cdot 7 \equiv 35 \equiv 16.$$

Since $0 \leq 16 < 19$ we deduce that 16 is the remainder of $3^{10,000}$ when divided by 19.   △

EXERCISES FOR SECTION 4.6

1. Decide whether each statement is true or false.
   (a) $18 \equiv 43 \pmod 5$ (c) $34 \equiv -16 \pmod{10}$
   (b) $100 \equiv 43 \pmod 7$ (d) $-62 \equiv -34 \pmod{11}$

2. Decide whether each statement is true or false.
   (a) $275 \equiv 132 \pmod 8$ (c) $-1 \equiv 70 \pmod{23}$
   (b) $20 \equiv 215 \pmod{13}$ (d) $1 \equiv 70 \pmod{23}$

3. List ten elements congruent to 3 modulo 7.

4. Prove that if $n \mid m$ and $a \equiv b \pmod m$, then $a \equiv b \pmod n$.

5. Prove that if a, b, c, and m are integers with $m \geq 2$ and $c > 0$, then $a \equiv b \pmod m$ implies that $ac \equiv bc \pmod{mc}$.

6. Prove Proposition 4.6.4.

7. Find all the units and their inverses modulo (a) 13; (b) 24.

8. Perform the Extended Euclidean Algorithm to find an inverse of 52 modulo 101.

9. Perform the Extended Euclidean Algorithm to find an inverse of 72 modulo 125.

10. Solve for x in the congruence equation $7x + 13 \equiv 10 \pmod{23}$.

11. Solve for x in the congruence equation $13x + 25 \equiv 32 \pmod{100}$.

12. Solve for x in terms of y in the congruence equation $y \equiv 2x + 3 \pmod{17}$.

13. Solve for x in terms of y in the congruence equation $y \equiv 17x + 20 \pmod{29}$.

14. Show that for all integers a, we have $a^2 \equiv 0$ or $1 \pmod 4$. Show how this implies that for all integers $a, b \in \mathbb{Z}$, the sum of squares $a^2 + b^2$ never has a remainder of 3 when divided by 4. [Hint: By cases.]

15. Find the smallest positive integer n such that $2^n \equiv 1 \pmod{17}$.

16. Find the smallest positive integer n such that $3^n \equiv 1 \pmod{19}$.

17. Show that a number is divisible by 11 if and only if the alternating sum of its digits is divisible by 11. (An alternating sum means that we alternate the signs in the sum $+ - + - \ldots$) [Hint: $10 \equiv -1 \pmod{11}$.]

18. Prove that if n is odd then $n^2 \equiv 1 \pmod 8$.

19. Use Fermat's Little Theorem to determine the remainder of 73^{4171} modulo 13.

20. Use Fermat's Little Theorem to determine the remainder of 5^{2020} modulo 19.

21. Use Fermat's Little Theorem to determine the remainder of 721^{2378} modulo 17.

22. Without a computer, find the unit's digit of 73^{357}. [Hint: Use modulo 10.]

23. Without a computer, determine the last two digits of 7^{12345}.

24. Prove that if $ac \equiv bc \pmod{m}$, then $a \equiv b \pmod{(m/d)}$ where $d = \gcd(m, c)$.

25. Let $P(x)$ be a polynomial with integer coefficients. Prove that if $a \equiv b \pmod{n}$, then $P(a) \equiv P(b) \pmod{n}$.

26. (*) Let $\{b_n\}_{n \geq 1}$ be the sequence of integers defined by $b_1 = 1$, $b_2 = 11$, $b_3 = 111$, and in general

$$b_n = \overbrace{111 \cdots 1}^{n \text{ digits}}.$$

Prove that for all prime numbers p different from 2 or 5, there exists a positive n such that $p \mid b_n$.

27. (*) Consider the sequence of integers $\{c_n\}_{n \geq 0}$ defined by

$$c_0 = 1, \quad c_1 = 101, \quad c_2 = 10101, \quad c_3 = 1010101, \quad \ldots$$

Prove that for all integers $n \geq 2$, the number c_n is composite.

CHAPTER 5

Counting and Combinatorial Arguments

In an interview, Italian-American mathematician Gian-Carlo Rota [66] said, "Combinatorics is an honest subject. No adèles, no sigma-algebras. You count balls in a box, and you either have the right number or you haven't."

Combinatorics is the name of the field involving counting problems. In many branches of mathematics, e.g., algebra, number theory, probability, representation theory, we wish to count the number of objects with a certain property. Rota's statement reflects that, unlike some areas of mathematics, counting feels tangible. And yet, he goes on to say, "Counting finite sets can be a highbrow undertaking, with sophisticated techniques."

Section 5.1 applies set theory operations to principles of counting. Section 5.2 extends counting concepts related to the binomial coefficients and introduces the concept of a combinatorial proof. Section 5.3 connects counting techniques to the proof strategy called the pigeonhole principle. Finally, Section 5.4 steps beyond finite sets to introduce set cardinality and the property of countability.

5.1 Counting Techniques

The expression "counting techniques" may feel shallow: Do I count in my head or on my fingers? That is not what we mean by counting techniques. In fact, we use techniques to determine the cardinality of large finite sets without listing out all the elements. Some of the terminology that we use, e.g., options, decisions, ways, permutations, combinations, arrangement, and so on, may feel unfamiliar. They are simply an artifact of the field of combinatorics.

5.1.1 Addition principle

What we will soon label the "addition principle" comes from a simple observation from set theory. Embedded in the proof of Proposition 4.1.11, we find the property that if two finite sets A and B are disjoint, then

$$|A \cup B| = |A| + |B|.$$

In combinatorics, this is sometimes restated in the following less theoretical way.

Proposition 5.1.1 (Addition Principle)

If a decision involves two disjoint types of options with one type having n_1 options and another type having n_2 options, then there are $n_1 + n_2$ ways of making the decision.

In the above statement, the term "decision" comes from the idea of selecting an element from a set.

Example 5.1.2. Suppose there are 12 men in the course and 17 women taking a course. The total number of people in the course is $12 + 17 = 29$. △

The addition principle extends to a finite number of disjoint sets. If $A_1, A_2, \ldots, A_n$ are sets such that $A_i \cap A_j = \emptyset$ whenever $i \neq j$, then $|A_1 \cup A_2 \cup \cdots \cup A_n| = |A_1| + |A_2| + \cdots + |A_n|$. The addition principle also leads to the following simple result.

Proposition 5.1.3

Let S be a finite set and let $A \subseteq S$. Then $|A^c| = |S| - |A|$.

Proof. Since $A \cap A^c = \emptyset$ and $A \cup A^c = S$, the addition principle gives $|S| = |A| + |A^c|$. The result follows after solving for $|A^c|$. □

The addition principle carries the label of "principle" because it is so intuitive and connects so closely to how we use addition in applications. If the possible sets of options do intersect, i.e., if $A \cap B \neq \emptyset$, then saying $|A \cup B| = |A| + |B|$ would be incorrect because the sum $|A| + |B|$ double-counts all elements in $A \cap B$. This leads to the inclusion-exclusion principle.

Proposition 5.1.4 (Inclusion-Exclusion Principle)

If A and B are finite sets, then

$$|A \cup B| = |A| + |B| - |A \cap B|.$$

Example 5.1.5. A customer wants to buy a car that is red or has leather seats. Suppose a dealership has 26 cars that are red 45 that have leather seats and just 8 that are both red and have leather seats. Then the total number of cars that are red or have leather seats is

$$26 + 45 - 8 = 63. \qquad \triangle$$

The inclusion-exclusion principle generalizes to more than two sets, but the formula gets more complicated. Suppose we have three finite sets A, B, and C and wished to find $|A \cup B \cup C|$. If we hypothesized the generalization to Proposition 5.1.4 of

$$|A| + |B| + |C| - |A \cap B| - |A \cap C| - |B \cap C|,$$

we can easily check cases depending on whether this properly counts elements in each region of the triple Venn diagram. We find that this is correct, except that it does not count any elements in the triple intersection $A \cap B \cap C$. This leads to the correct generalization of Proposition 5.1.4.

Proposition 5.1.6 (Inclusion-Exclusion for Three Sets)

If A, B, and C are finite sets, then

$$|A \cup B \cup C| = |A| + |B| + |C| - |A \cap B|$$
$$- |A \cap C| - |B \cap C| + |A \cap B \cap C|.$$

5.1.2 Multiplication principle

The multiplication principle has its theoretical roots in the following property of sets.

Proposition 5.1.7

Let A and B be finite sets. Then $|A \times B| = |A|\,|B|$.

Proof. First suppose that A or B is the empty set. Without loss of generality, suppose that $A = \emptyset$. Then $\emptyset \times B$ contains no elements so $\emptyset \times B = \emptyset$. Hence, $|A \times B| = 0$ and $|A|\,|B| = 0|B| = 0$, so this product formula holds.

Suppose now that neither A nor B is empty but that $|A| = m$ and $|B| = n$ for some positive integers m and n. For each $a \in A$, call $S_a = \{(a,b) \in A \times B \mid b \in B\}$. It is clear that the function $f : B \to S_a$ defined by $f(b) = (a,b)$ is a bijection. Hence, $|S_a| = |B| = n$ for all $a \in A$. Furthermore, if $a \neq a'$, then $S_a \cap S_{a'} = \emptyset$. Since

$$A \times B = \bigcup_{a \in A} S_a, \quad \text{we have} \quad |A \times B| = \sum_{a \in A} |S_a|$$

by the addition principle. Thus $|A \times B| = mn = |A||B|$.   □

As with the addition principle, we often restate this proposition in a more intuitive manner, using the terminology of "decisions."

Proposition 5.1.8 (Multiplication Principle)

If a decision of two independent parts has n_1 options the first part of the decision and n_2 options the second part of the decision, then there are $n_1 n_2$ ways to make the decision.

The connection of this terminology to set theory is that we model a two-part decision by selecting an element from the Cartesian product $A \times B$, where A represents the options for the first part and B represents the options for the second part. Again, the multiplication principle carries the label of "principle," because it connects so closely to how we use multiplication in applications.

Example 5.1.9. Suppose that a kitchen cabinet maker offers 30 styles of wood for the cabinet door and 8 styles of cut on the panels. How many different types of looks could the cabinet maker offer customers? The answer is $30 \times 8 = 240$. In this case, unlike the addition principle, the multiplication principle involves a two-part decision, and the options for the style of wood are independent from the style of cut.

Suppose that in addition to cut and wood type, we took into account the 25 different types of hardware (knobs and handles) that are offered. Taking this extra information into account, how many different types of looks could the cabinet maker offer customers? There would now be $30 \times 8 \times 25 = 6,000$.   △

This example illustrates that though we phrased the principle in terms of two parts to the decision, we can extend the principle to a decision with any finite number of parts.

Example 5.1.10. To contrast with the addition principle, consider Examples 5.1.2. Instead of asking for the total number of students in the class, we could have asked how many ways could we select a representative pair of students involving one man and one woman. Then the decision would involve a two-part process, and there would be $12 \cdot 17 = 204$ possibilities.   △

Example 5.1.11. Suppose that you toss a coin 10 times and record the H (Heads) or T (Tails) result of the toss. The number of possible outcomes is $2^{10} = 1024$. From a set theory perspective, we are calculating the cardinality of $\{H, T\}^{10}$.   △

Despite the simplicity of the multiplication principle, it leads to a number of interesting theoretical results.

Proposition 5.1.12

Let S be a finite set. Then $|\mathcal{P}(S)| = 2^{|S|}$.

Proof. First, suppose that $S = \emptyset$. Then $\mathcal{P}(S) = \{\emptyset\}$ so we have $|S| = |\emptyset| = 0$ and $|\mathcal{P}(\emptyset)| = 1 = 2^0$.

Now, suppose that $S \neq \emptyset$ and $|S| = n$, a positive integer. Then we can write $S = \{s_1, s_2, \ldots, s_n\}$. Counting the number of subsets A of S entails counting the number of decisions involved in whether each s_i is in A or not in A. This becomes an n-part decision with 2 options for each part. Hence, there are $2^n = 2^{|S|}$ subsets of S. □

The detailed set theory behind Proposition 5.1.12 involves a bijection $f : \mathcal{P}(S) \to \{0,1\}^n$. It is not hard to show that defining f by $f(A) = (b_1, b_2, \ldots, b_n)$ with

$$b_i = \begin{cases} 1 & s_i \in A \\ 0 & s_i \notin A \end{cases}$$

gives this bijection.

Proposition 5.1.13

If A and B are finite sets, then there exist $|B|^{|A|}$ functions $f : A \to B$.

Proof. Let A and B be finite sets with $A - \{a_1, a_2, \ldots, a_m\}$. Counting the number of functions involves a decision with m parts, with exactly n options for the image of each element a_i. Hence, the number of functions is $n^m = |B|^{|A|}$. □

This proposition inspires the notation B^A to represent the set of function $A \to B$. For example, if A is any set, we denote by $A^{\mathbb{N}}$ the set of sequences in A. Sometimes, this notation is not convenient, e.g., the set of functions from $M_n(\mathbb{R})$ to $\mathbb{R}$ would be $\mathbb{R}^{M_n(\mathbb{R})}$, so some authors write $\mathrm{Fun}(A, B)$ for the set of functions from A to B.

The following example shows how certain counting situations require us to combine the multiplication principle with the addition principle, sometimes in different ways.

Example 5.1.14. Let $\mathcal{A}$ be the set of letters in the English alphabet and consider strings of elements in $\mathcal{A}$.

1) How many strings of length 7 exist? Answer: 26^7. This is a product rule situation where the "decision" has 7 independent parts and each part has 26 possibilities. So the answer is

$$\overbrace{26 \times 26 \times \cdots \times 26}^{7 \text{ times}} = 26^7 = 8,031,810,176.$$

2) How many strings of consonants of length 7 can be created? Answer: 21^7. We first observe that there are 21 consonants. Notice that if we call C the set of consonants and $V = \{a, e, i, o, u\}$ the set of vowels, then $C \cap V = \emptyset$.

3) How many strings of length 7 exist, where (just for this question) we distinguish between lower case and upper case? Answer: 52^7. Here, we first use an addition rule to determine that we have $26 + 26 = 52$ letter symbols. Then, using the product rule in the same way as above, we get 52^7.

4) How many strings of length 7 can we make that have 4 consonants followed by 3 vowels? Answer: $21^4 \cdot 5^3$. We have 7 parts (letter selections) to our decision. For each of the first 4, we have 21 options, and for the last 3, we have 5 choices each. Then we use the product rule. In terms of set theory, we are calculating the cardinality of $C^4 \times V^3$.

5) How many strings of length 7 can be made that begin or end with a vowel? We take this question in parts. There are $5 \cdot 26^6$ options for strings beginning with a vowel; this is the set $B = V \times \mathcal{A}^6$. Similarly, there are $5 \cdot 26^6$ options for strings that end with a vowel; this is the set $E = \mathcal{A}^6 \times V$. However, $V \times \mathcal{A}^6$ and $\mathcal{A}^6 \times V$ are not disjoint. They overlap in strings of letters that begin and end with a vowel, namely $V \times \mathcal{A}^5 \times V$. The answer to our question is

$$|B \cup E| = |B| + |E| - |B \cap E|$$
$$= 5 \cdot 26^6 + 5 \cdot 26^6 - 5^2 \cdot 26^5 = 2,792,123,360.$$

6) How many strings of length 7 can we make in which there are no repetitions of letters? Answer: $26 \cdot 25 \cdot 24 \cdot 23 \cdot 22 \cdot 21 \cdot 20 = 3,315,312,000$. This is because our decision has 7 parts with 26 options for the first part; once the first part is decided, there are 25 options for the second part of the decision; once the first two parts are decided, there are 24 ways for the third part, and so on.   △

The last list item in the previous example illustrated something new. It does not count a Cartesian product of any collection of sets.

The decision for the second part and the third part and so on are affected by the choices for the previous parts of the compound decision. However, the number of possibilities for each part is not affected. This last example is a counting situation known as a permutation.

5.1.3 Permutations

Recall that the Cartesian product A^k of a set A with itself k times is the set of functions $\{1, \ldots, k\} \to A$. A k-tuple $(a_1, a_2, \ldots, a_k)$ corresponds to a function $f : \{1, \ldots, k\} \to A$ with $f(1) = a_1$, $f(2) = a_2$, and so on. Instead of counting just functions, we could ask to count injective functions. In the last list itme of Example 5.1.14, the counting problem asks for just that.

Definition 5.1.15

> A *permutation* is an arrangement of a certain number of objects taken from a lager set of objects. We say there are $P(n,r)$ permutations of r objects taken out of a set of n objects.

In counting problems, the term "arrangement" signals a list of objects where order matters, but repetitions are not allowed. An arrangement of r objects in A is an injection $\{1, 2, \ldots, r\} \to A$. So $P(n,r)$ is the number of injective functions $\{1, \ldots, r\} \to \{1, \ldots, n\}$.

Proposition 5.1.16

> $$P(n,r) = n \times (n-1) \times (n-2) \times \cdots \times (n-r+1) = \frac{n!}{(n-r)!}.$$

Proof. That $P(n,r) = n \times (n-1) \times (n-2) \times \cdots \times (n-r+1)$ follows from the multiplication rule. Furthermore,

$$\frac{n!}{(n-r)!} = \frac{n(n-1)\cdots(n-r+1)(n-r)(n-r-1)\cdots 2 \cdot 1}{(n-r)(n-r-1)\cdots 2 \cdot 1}.$$

After canceling every term in the denominator with a corresponding term on the numerator, the proposition follows.   □

Example 5.1.17. You decide to visit 3 out of 7 cities and the order in which you visit matters. How many itineraries could you make this way? Answer: This is an arrangmnet or a permutation, so the answer is $P(7,3) = 7 \cdot 6 \cdot 5 = 210$.   △

Example 5.1.18. We pick from the letters M, A, T, H to make strings of letters. In how many ways can we arrange the four letters? The answer is $P(4,4) = 4!/0! = 24$.

Recall $0! = 1$.

In contrast to Definition 5.1.15, where we refer to a permutation *from* a set, this last example illustrates a permutation *on* a set or *of* a set. We can view this case as a bijection on a set. As per the example, a permutation on the set $\{\mathsf{M}, \mathsf{A}, \mathsf{T}, \mathsf{H}\}$ is any bijection on this set.

Corollary 5.1.19

There exist $n!$ permutations on (bijections of) a set of size n.

5.1.4 Combinations

As a motivating example, consider the 400-yard dash in the Olympics. During the final heat, we care about the top 3 out of 8 runners and the order they place. The number of ways the podium could be filled is $8 \times 7 \times 6 = 336$. However, in qualifying heats, the officials only care about the top 3, and the order they come in is irrelevant. This is a situation for which we would want to count the options, but it is not a permutation.

Definition 5.1.20

A *combination* is a subset of a set of objects. We denote by $C(n, r)$ or also $\binom{n}{r}$ the number of r-combinations from n objects (or subsets of size r taken from a set of n elements).

The notation $C(n, r)$ tends to figure in finite mathematics or elementary statistics textbooks. Most combinatorics and algebra textbooks use the notation $\binom{n}{r}$ and pronounce this symbol as "n choose r." This is called the *binomial coefficient*. Because it is more standard in advanced mathematics, we will continue to emphasize the latter notation.

By definition, the *binomial coefficient* $\binom{n}{r}$ is a positive integer when $0 \leq r \leq n$. We can also easily see that $\binom{n}{0} = 1$ because in any set, the only subset containing 0 elements is the empty set. It is also clear from the definition the $\binom{n}{n} = 1$ for all $n \geq 0$. However, the binomial coefficients extend naturally to all pairs of integers by defining

$$\binom{n}{r} = 0 \qquad \text{if it is not true that } 0 \leq r \leq n.$$

In every other case, we have a formula for them.

Proposition 5.1.21

If $0 \leq r \leq n$, then $\binom{n}{r} = C(n,r) = \dfrac{P(n,r)}{r!} = \dfrac{n!}{r!(n-r)!}$.

Proof. Using the multiplication principle, we can break down the count for permutations $P(n,r)$ into a two-part decision: the first part being the combination and the second part being the permutations on this given combination. By definition, $C(n,r)$ counts the combinations of r objects taken from n. There are $r!$ permutations on each arrangement. Hence, $P(n,r) = C(n,r)r!$. $\square$

To reiterate the difference between permutations and combinations, the number of combinations $C(n,r)$ is the number of ways to choose r objects out of a set of size n (where we do not care about order), whereas $P(n,r)$ is the number of ways to choose r objects out of a set of size n *and* order this selection.

Corollary 5.1.22

For all pairs $(n,r) \in \mathbb{N}^2$, we have $\binom{n}{r} = \binom{n}{n-r}$.

Proof. (Left as an exercise for the reader.) $\square$

Example 5.1.23. Let us return to the motivating example of our subsection, involving qualifying heats of 3 runners taken out of 8. There are

$$C(8,3) = \frac{8 \cdot 7 \cdot 6}{3 \cdot 2 \cdot 1} = 56$$

different groups that could emerge from the heat. $\triangle$

To understand the set theory behind this problem, for any finite set S we consider

$$\mathcal{P}_r(S) = \{A \subseteq S \,\big|\, |A| = r\}.$$

This is a way of describing all subsets of S of size r. Then $|\mathcal{P}_r(S)| = C(|S|,r)$.

Example 5.1.24. A painter prepared 10 pieces for a gallery show but must select 4 of them. How many different collections of 4 paintings can the artist form? The answer is $\binom{10}{4} = 210$. $\triangle$

Counting may involve mixtures of permutation and combination problems.

Example 5.1.25. Suppose you have 12 people to choose from as you form a committee of 4.

1) In how many ways could you select a president, secretary, treasurer, and VP? The selection process is a permutation since all the roles are distinct. Hence, the number of options is $P(12, 4) = 11,880$.

2) If we do not select officers, in how many ways could we form the committee? In this case, we count combinations, because we are essentially counting the numbers of possible subsets of size 4 from 12 people. Hence, the number of such committees is $C(12, 4) = 495$.

3) In how many ways could we select a president, VP, and 2 members at-large? In this counting problem, we only care about order for the first two selections, and once those are chosen, we do not care about order. Hence, with a multiplication principle connecting them, the number of possible committees with this structure is $P(12, 2) \cdot C(10, 2) = 132 \cdot 45 = 5,940$. We only use 10 in $C(10, 2)$ because once 2 have been selected for president and VP, there are only 10 people remaining to pick from.

The astute reader might wonder whether we would obtain a different number if we selected the at-large members first. If we determine the number of at-large members first, our multiplication principle gives $C(12, 2)P(10, 2) = 66 \cdot 90 = 5,940$. Though the factors in the product are different, the final answer remains the same.   △

EXERCISES FOR SECTION 5.1

1. Suppose that Allison's library consists of 30 novels and 15 non fiction books.

   (a) In how many ways can Allison pick one book from among her novels or non fiction books?

   (b) In how many ways can Allison pick a pair of books consisting of one novel and one non fiction book?

2. How many different ways are there to write 12 as a sum of 1 or 2? (We assume $1 + 2$ is the same as the sum $2 + 1$.)

3. An apartment building has 20 floors of 12 standard apartments apiece and then 5 floors of 8 luxury apartments apiece. How many apartments are in the building?

4. How many integers less than 1000 have an odd highest digit?

5. How many strings of 10 bits exist?

6. How many strings of 3 letter initials can be created?

7. How many license plates can be made using two letters followed by 5 digits?

8. How many passwords of length 12 can we make from 52 upper- or lower-case letters, 10 digits, and 8 special characters?

9. How many integers exist between 1 and 1000 that are
   (a) divisible by 5;   (d) divisible by 5 or by 11;
   (b) divisible by 11;   (e) divisible by 5 but not 11;
   (c) divisible by 5 and by 11;   (f) divisible by neither 5 nor 11.

10. Suppose there is a football conference with 8 teams in it. If each team plays each other once, how many games will occur?

11. There are 10 viable candidates for a job and the hiring agent will bring in 4 for interviews. In how many ways can she arrange 4 out of the 10 candidates?

12. A photographer arranges the students of first-grade class for a picture. There are 7 girls and 5 boys. Count the possible arrangements if
    (a) they all stand in one line;
    (b) the boys stand behind the girls;
    (c) they stand in one line but two girls stand at either end;
    (d) they stand in two rows of 6, with the tallest in the back.

13. Maddie prepares her playlist for a 5 km run. She has 13 songs to choose from but will only need 7 of them. How many playlists can she make if
    (a) (no other conditions)?
    (b) she already knows which song she will end with?

14. Suppose that Allison's library consists of 20 novels and 8 non fiction books. She takes groups of books with her on a trip.
    (a) How many groups of 5 books can she take?
    (b) How many groups of 3 novels and 3 non fiction can she take?

15. A pizza restaurant allows customers to build their own pizza using: 2 crust styles, 3 categories of cheese (none, light, heavy), 4 sizes, 5 types of sauces, 8 meat toppings, and 17 non meat toppings. How many types of pizzas can be created if we
    (a) pick one of each of these features, including exactly 1 meat topping and 1 non meat topping.
    (b) choose the hand-tossed crust style, the light cheese, the extra large size, and the Robust marinara sauce, but then we just want a 3-topping pizza (meat or non meat).
    (c) choose the hand-tossed crust style, the light cheese, medium size, but still need to choose the sauce, 1 meat option, and 2 non meat toppings.

16. Suppose that a math class has 42 students with 18 men and 24 women.

(a) How many subsets of size 5 exist?

(b) For each integer k with $0 \leq k \leq 5$, decide how many subsets of size 5 have k women? [Hint: Break this into a two-part decision to select a combination of women of the right size and combination of men of the right size.]

(c) In the subset, we also elect a president. We want a subset of size 5 with 2 men and 3 women in which the president is a woman. How many scenarios can we make this way?

17. Jasper's department at work includes 12 people: 7 men and 5 women.

(a) In how many ways can the company rank a top 4 of these coworkers?

(b) In how many ways can a committee of 4 be selected?

(c) If Jasper's coworkers competed in 4 events of office-Olympics and only recorded winners for each event, how many outcomes could there be to such office-Olympics?

(d) In how many ways could a committee of 4 be selected if exactly 2 are women?

(e) In how many ways could a manager select a single gender committee of 4 of these coworkers?

(f) In how many ways could a manager select a committee of 4 with at least 1 woman?

18. Suppose that you play with a deck of cards that includes 52 cards, involving 4 suits, each with 13 number/face cards. How many hands of 5 cards can you be dealt

(a) (no other conditions)?

(b) that are a full house (3 cards of the same number/face with 2 cards of the same number/face)?

(c) that are a set of three (but not a full house)?

(d) that are a flush (all the same suit)?

(e) that are a straight (5 cards in numerical order, irregardless of suits)?

19. Let S be a finite set. Determine the number of binary operations on S.

20. Referring to Proposition 5.1.13, let A and B be finite sets with $|A| = m$. Construct an explicit bijection between the set of functions $\mathrm{Fun}(A, B) = B^A$ and B^m. Prove that it is a bijection.

Exercises 5.1.21 through 5.1.23 introduce multinomial coefficients, *defined by*

$$\binom{n}{k_1, k_2, \ldots, k_\ell} = \frac{n!}{k_1! \cdot k_2! \cdots k_\ell!} \qquad where\ k_1 + k_2 + \cdots + k_\ell = n. \quad (5.1)$$

This multinomial coefficient counts the number of distinct permutations of n objects in which there are ℓ types of objects with k_1 objects viewed as identical, k_2 objects viewed as identical, and so on.

21. Express with a multinomial coefficient and calculate the number of distinct words that can be created by rearranging the letters of MIS-SISSIPPI.

22. Express with a multinomial coefficient and calculate the number of distinct words that can be created by rearranging the letters of AD-DIS ABABA.

23. Show that the multinomial coefficient in (5.1) counts the number of ways of putting n objects into ℓ boxes, with k_i objects in the ith box, for $1 \leq i \leq \ell$.

5.2   Concept of a Combinatorial Proof

In this section, we explore more applications of counting techniques. In so doing, we will encounter many interesting formulas. However, the more important aspect of this section is that we introduce a new style of proof.

5.2.1   Combinatorial proofs

A *combinatorial proof* is a proof that arises from rephrasing the given problem so that its solution emerges from a counting problem, sometimes viewed in more than one way. The proof of the following propositions gives us a flavor of this style.

Proposition 5.2.1

Let n be a positive integer. Then $\displaystyle\sum_{k=0}^{n} \binom{n}{k} = 2^n$.

Proof. Let A be a set with $|A| = n$. We know that A has 2^n subsets from Proposition 5.1.12. We count up this number of subsets in another way: We observe that each subset can have size k for $0 \leq k \leq n$. There are $\binom{n}{k}$ subsets of size k. Since the collection of subsets of a given cardinality is disjoint from the collection of subsets of a different cardinality, the addition principle implies that

$$\sum_{k=0}^{n} \binom{n}{k} = 2^n.$$

□

The lesson to learn from this proof is that we did not use the formula $\binom{n}{k} = n!/k!(n-k)!$ and challenging work with summation. Instead we counted up the number of subsets in two different ways. The following two examples illustrate this further.

Example 5.2.2. As an easier example, consider these two proofs for Corollary 5.1.22, which affirms

$$\binom{n}{k} = \binom{n}{n-k} \quad \text{for all } 0 \le k \le n.$$

First proof method: We use Proposition 5.1.21 that

$$\binom{n}{k} = \frac{n!}{k!(n-k)!}.$$

Then,

$$\binom{n}{n-k} = \frac{n!}{(n-k)!(n-(n-k))!} = \frac{n!}{(n-k)!k!} = \binom{n}{k}.$$

Second proof method: Let S be a set with $|S| = n$. The function $f : \mathcal{P}_k(S) \to \mathcal{P}_{n-k}(S)$ with $f(A) = A^c$ is a bijection with $B \mapsto B^c$ as its inverse. Hence $|\mathcal{P}_k(S)| = |\mathcal{P}_{n-k}(S)|$, which means $\binom{n}{k} = \binom{n}{n-k}$.

The second proof method avoids using a formula for the binomial coefficient but deduces the formula directly from its meaning. $\triangle$

Theorem 5.2.3 (Pascal's Identity)

If n and k are nonnegative integers, then

$$\binom{n+1}{k} = \binom{n}{k-1} + \binom{n}{k}.$$

We previously proved this result by induction in Example 4.5.6. The reader may recall that the proof given there was not particularly simple. However, the following combinatorial proof establishes this formula much faster.

Proof (of Theorem 5.2.3). First, consider the case $k = 0$. By definition with $n \ge 0$, we have $\binom{n+1}{0} = 1 = \binom{n}{0} = \binom{n}{0} + \binom{n}{-1}$, so the formula holds for all $n \ge 0$ and $k = 0$. Now suppose that $k \ge 1$. Let S be any set with $|S| = n$ and let $a \notin S$. We count $|\mathcal{P}_k(S \cup \{a\})|$, the subsets of size k in $S \cup \{a\}$. Note that if $B \subseteq S \cup \{a\}$, then either $a \notin B$, in which case $B \subseteq S$ and $|B| = k$, or $a \in B$, in which case $B = C \cup \{a\}$ for some $C \in S$ with $|C| = k - 1$. The first of these two cases has size $|\mathcal{P}_k(S)|$ and the second has $|\mathcal{P}_{k-1}(S)|$ elements. Since $|S \cup \{a\}| = n + 1$,

$$\binom{n+1}{k} = |\mathcal{P}_k(S \cup \{a\})|$$

$$= |\mathcal{P}_k(S)| + |\mathcal{P}_{k-1}(S)| = \binom{n}{k} + \binom{n}{k-1}. \qquad \square$$

Example 5.2.4. We show that $\sum_{k=0}^{n} k \binom{n}{k} = n2^{n-1}$ for all $n \in \mathbb{N}^*$.

To prove this formula with a combinatorial proof, we must devise a counting problem answered by $n2^{n-1}$. Suppose that, for this example alone, we call a "committee" a subset with a distinguished element (that is the chairperson).[1] From a set of n people, we can count the number of possible committees in two ways. First, select the chairperson (n options) and then select the rest of the committee, consisting of any subset of the remaining people (2^{n-1} options). Hence, there are $n2^{n-1}$ such committees. We count the same number of committees by selecting the whole committee first. For each k there are $\binom{n}{k}$ options, but once we have chosen this overall committee then we need to select a chairperson (k options since the committee only has k people). So there are $k\binom{n}{k}$ "committees" of size k. Summing through all possible values of k, we get

$$\sum_{k=0}^{n} k \binom{n}{k} = n2^{n-1}.$$

$\triangle$

There are a variety of ways we could prove the following theorem, but a combinatorial proof is the shortest.

Theorem 5.2.5 (Vandermonde Theorem)

For any positive integers a, b and n, we have

$$\binom{a+b}{n} = \sum_{k=0}^{n} \binom{a}{k}\binom{b}{n-k} = \sum_{i+j=k} \binom{a}{i}\binom{b}{j}.$$

Proof. Let A and B be finite disjoint sets of cardinality a and b, respectively. Consider the number of subsets of $|A \cup B|$ of size n. Since $|A \cup B| = |A| + |B| = a + b$, this number is $\binom{a+b}{n}$.

We can approach the same counting problem in a different way. Let k be any integer with $0 \leq k \leq n$. If, in a subset N of size n taken from $A \cup B$ we have k elements in A, then there are $\binom{a}{k}$ options for $N \cap A$ and $\binom{b}{n-k}$ options for $N \cap B$. Hence, the multiplication principle gives us $\binom{a}{k}\binom{b}{n-k}$ possibilities for when $|N \cap A| = k$. Since k can range between 0 and n, we deduce the formula

$$\binom{a+b}{n} = \sum_{k=0}^{n} \binom{a}{k}\binom{b}{n-k}$$

once we apply the addition principle on these different cases for k. $\square$

---

[1] A pair (S, a), where S is a set and $a \in S$, is called a *pointed set*. So in this example, a committee is a pointed set of people.

5.2.2   Binomial theorem

Before we present the Binomial Theorem, we present a counting problem that serves as a motivating example.

Example 5.2.6. We create passwords of length 8 using letters (26) and digits (10).

1) How many passwords exist that use exactly 3 digits? We need to break this problem down into a multiplication principle in a 9-part decision. The first part of the decision consists of choosing which of the 8 spots in the password we will use the digits. There are $\binom{8}{3}$ ways to choose 3 spots from 8 password characters. Then for any of the given configurations, we have to choose 3 digits, for a total of 10^3 options; and we also have to choose the letters, for a total of 26^{8-3}. According to the multiplication principle, we have $\binom{8}{3}10^3 \cdot 26^5 = 665,357,056,000$ such passwords.

2) How many passwords exist that have 3 or fewer digits? The problem is similar to the previous one, but instead of finding the number for $k = 3$ digits, we need to solve it for $k = 0, 1, 2, 3$ digits and add these numbers. Hence, the answer is

$$\sum_{k=0}^{3} \binom{8}{k} 10^k \cdot 26^{8-k} = 2,381,693,107,456.$$

$\triangle$

Theorem 5.2.7 (Binomial Theorem)

For any reals a and b, and for any positive integer n,

$$(a+b)^n = \sum_{k=0}^{n} \binom{n}{k} a^{n-k} b^k. \qquad (5.2)$$

Proof. Consider the n-fold distribution for

$$(a+b)^n = \overbrace{(a+b)(a+b)\cdots(a+b)}^{n \text{ times}}.$$

Every product will involve a total of n of the symbols a and b. We count how many terms of the form $a^{n-k}b^k$ appear in the expansion by counting how many ways we could select one of the $(a+b)$ parentheses from which to select the symbol b (and from the remaining $n - k$ parentheses groups we select the symbol a): there are $\binom{n}{k}$ such ways to make the selection. Hence, for each k with $0 \le k \le n$, there are $\binom{n}{k}$ terms $a^{n-k}b^k$ in the distributed expansion. The theorem follows when we sum over the possible values of k. $\quad\square$

We can see this formula at work in Example 5.2.6. In that example, $n = 8$, $a = 26$ for the number of letters, and $b = 10$ for the number of digits. The answer to part (1) is precisely $\binom{8}{3}a^5b^3$, which is one of the terms in the right-hand side of (5.2). Part (2) in the example also involves taking the portion of (5.2) involving $k = 0, 1, 2, 3$.

Example 4.5.6 introduced Pascal's triangle. The Binomial Theorem establishes that the entries of Pascal's triangle provide the coefficients of a binomial expansion. For example, the first few rows of Pascal's triangle are:

$$
\begin{array}{ccccccccccc}
 & & & & & 1 & & & & & \\
 & & & & 1 & & 1 & & & & \\
 & & & 1 & & 2 & & 1 & & & \\
 & & 1 & & 3 & & 3 & & 1 & & \\
 & 1 & & 4 & & 6 & & 4 & & 1 & \\
1 & & 5 & & 10 & & 10 & & 5 & & 1
\end{array}
$$

The reader should notice that 1s descend at approximately $45°$ angles from the apex, and that underneath them every entry is the sum of the two above it. The last row corresponds to $n = 5$ so using the coefficients, the Binomial Theorem gives

$$(a + b)^5 = a^5 + 5a^4b + 10a^3b^2 + 10a^2b^3 + 5ab^4 + b^5.$$

EXERCISES FOR SECTION 5.2

1. What is the coefficient of $x^{11}y^{19}$ in $(x + y)^{30}$?

2. What is the coefficient of x^7 in $(3 + x)^{10}$?

3. What is the coefficient of x^3y^9 in $(4x + 5y)^{12}$?

4. Use the 4th row of Pascal's triangle to determine 11^4 without performing any multiplications.

5. Prove Proposition 5.2.1 by induction.

6. Show that $\sum_{k=2}^{n} \binom{k}{2}\binom{n}{k} = \binom{n}{2}2^{n-2}$ for all integers $n \geq 2$ in two ways:
   (a) by induction on n;
   (b) with a combinatorial proof.

7. Use a combinatorial proof to show that $k\binom{n}{k} = n\binom{n-1}{k-1}$.

8. Prove the formula $\binom{n}{r}\binom{r}{k} = \binom{n}{k}\binom{n-k}{r-k}$ in two ways:
   (a) using a combinatorial argument;
   (b) using Proposition 5.1.21.

9. Prove the Binomial Theorem (Theorem 5.2.7) by induction on n.

10. In the case when a and b are positive integers, give a combinatorial proof for the Binomial Theorem (Theorem 5.2.7) by counting functions $N \to A \cup B$, where $|N| = n$, $|A| = a$ and $|B| = b$.

11. Suppose that we have n symbols 1 and $k - 1$ symbols $+$.

   (a) Prove that there are $\binom{n+k-1}{n}$ distinct ways to put the symbols we have into $n + k - 1$ boxes.

   (b) Interpret this counting problem to conclude that $\binom{n+k-1}{n}$ is the number of solutions to the equation $a_1 + a_2 + \cdots + a_k = n$ in nonnegative integers a_i.

   (c) Deduce that there are $\binom{n+k-1}{n}$ monomials $x_1^{a_1} x_2^{a_2} \cdots x_k^{a_k}$ of total degree n with k variables.

12. Give a combinatorial proof that $\sum_{k=0}^{n} k \binom{n}{k}^2 = n \binom{2n}{n-1}$.

13. Let p be a prime.

   (a) Prove that p divides the binomial coefficient $\binom{p}{k}$ for all k with $1 \le k \le p - 1$.

   (b) Use the Binomial Theorem to conclude that $(a + b)^p \equiv a^p + b^p \pmod{p}$ for all integers $a, b \in \mathbb{Z}$. [This formula is called *Frobenius' Formula*.]

14. If a, b, c are commuting variables, prove that

$$(a + b + c)^n = \sum_{k_1 + k_2 + k_3 = n} \binom{n}{k_1, k_2, k_3} a^{k_1} b^{k_2} c^{k_3}.$$

   [See (5.1).]

15. Consider the multinomial coefficient in (5.1). Prove that for any $n \ge 1$ and for any nonnegative integers $k_1, k_2, \ldots, k_\ell$ such that $k_1 + k_2 + \cdots + k_\ell = n$, we have

$$\binom{n}{k_1, k_2, \ldots, k_\ell} = \binom{n-1}{k_1 - 1, k_2, \ldots, k_n} + \cdots + \binom{n-1}{k_1, k_2, \ldots, k_n - 1}.$$

   [Hint: Use Exercises 5.1.23.]

16. Use the previous exercise to describe a "Pascal's pyramid" that generalizes Pascal's triangle but to obtain the trinomial coefficients (multinomial but with 3 variables).

17. Prove that for all positive integers $a, b, c \ge 1$,

$$\binom{a + b + c}{n} = \sum_{i+j+k=n} \binom{a}{i} \binom{b}{j} \binom{c}{k}.$$

18. Prove that for all nonnegative integers $n \ge 0$,

$$\sum_{k=0}^{n} \binom{k}{i} \binom{n-k}{j} = \binom{n+1}{i+j+1} \qquad \text{for all } i, j \ge 0.$$

   [Hint: By induction on n.]

19. Use Exercise 5.2.18 to prove the following formulas. Fix a nonnegative integer n.

(a) Let S be the set of pairs $(i, j) \in \mathbb{N}^2$ such that $i + j = n$. Prove that

$$\sum_{(i,j) \in S} ij = \binom{n+1}{3}.$$

(b) Let T be the set of triples $(i, j, k) \in \mathbb{N}^3$ such that $i + j + k = n$. Use part (a) to prove that

$$\sum_{(i,j,k) \in T} ijk = \binom{n+2}{5}.$$

5.3 Pigeonhole Principle

Suppose that a young man has a drawer full of socks and that each sock is either black or white. If he dresses in a rush, how many socks must he pull from the sock drawer to be sure he has a matching pair? The answer to the simple riddle is three socks. If he only pulls two, he might draw a pair, but he might also pull one white and one black. In the latter case, if he draws one more, then that third sock must be either white or black, and he will have a pair. The thought process to this solution generalizes to a proof technique called the pigeonhole principle.

The evocative name "pigeonhole principle' stems from the thought experiment that if a pigeon fancier[2] has a fixed number of enclosures for each pigeon and he or she happens to have on hand more pigeons than enclosures, then two pigeons will need to squeeze into one enclosure.

5.3.1 The Principle

Theorem 5.3.1 (Pigeonhole Principle)

If $n + 1$ or more objects are placed in n boxes, then there is at least one box with more than one object.

The set theoretic framing is that if A and B are finite sets with $|A| > |B|$, there exists no injection $f : A \to B$. It is not so much the complexity of the concept that warrants its own name but rather the surprising number of proof applications.

Example 5.3.2. A company has 700 employees. There must be two employees with the same first and last name initial pair. We can see this first by counting the number of possible first and last name initial pairs. Assuming we use the letters of the English alphabet $\mathcal{A}$, then

---

[2]Someone who keeps pigeons is called a "pigeon fancier."

$|\mathcal{A} \times \mathcal{A}| = 26^2 = 676$. However, $700 > 676$, so by the pigeonhole principle, at least two people must have the same pair of initials. $\triangle$

Example 5.3.3. Suppose that there are 10 points placed at random in a square of side length 1. At least two of the points must be at most $\sqrt{2}/3$ apart.

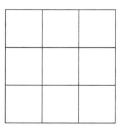

Cut the square up into 9 squares as depicted above. (The squares are the pigeonholes and the points are the pigeons.) By the pigeonhole principle, since we have 10 points and 9 squares, at least two points must lie in the same small square. The diagonal of each small square is $\sqrt{2}/3$, so those two points must be at most $\sqrt{2}/3$ apart. $\triangle$

Example 5.3.4. Recall the Fibonacci sequence $(F_n)_{n \geq 0}$ defined in Example 3.1.25. We show that the last two digits of the Fibonacci sequence follow a periodic pattern.

Since we are looking at the last two digits of the Fibonacci sequence, we consider F_n modulo 100. Furthermore, since $F_n = F_{n-1} + F_{n-2}$ for all $n \geq 2$, the terms of the sequence are uniquely determined by the previous two terms. This is still true modulo 100, since $F_n \equiv F_{n-1} + F_{n-2} \pmod{100}$. We consider as pigeonholes pairs of remainders of integers when divided by 100, and the pigeons are successive pairs of the Fibonacci sequence. There are $100^2 = 10,000$ possible holes but the Fibonacci sequence involves an infinite number of terms. Hence, there must exist $i \neq j$ with $(F_i \bmod 100, F_{i+1} \bmod 100) = (F_j \bmod 100, F_{j+1} \bmod 100)$. Once we see a repetition in successive pairs, then we know that the sequence modulo 100 will be periodic.

So far this only shows that the sequence is eventually periodic of some period p. Assume that $k > 0$ is the first nonnegative integer for which $F_{i+p} \equiv F_i \pmod{100}$ for all $i \geq k$. But

$$F_{k-1} \equiv F_{k+1} - F_k \equiv F_{k+1+p} - F_{k+p} \equiv F_{k-1+p} \pmod{100},$$

which contradicts a minimality of $k > 0$. Hence $k = 0$, so we deduce that $(F_n \bmod 100)_{n \geq 0}$ is periodic. $\triangle$

The Pigeonhole Principle extends naturally in the following way.

> **Theorem 5.3.5 (Extended Pigeonhole Principle)**
> If n or more objects are placed in m boxes, then there is a box containing at least $1 + \lfloor (n-1)/m \rfloor = \lceil n/m \rceil$ objects.

A simple proof by contradiction establishes this theorem. We stated the theorem as such because $1 + \lfloor (n-1)/m \rfloor = \lceil n/m \rceil$ for all positive integers m and n.

Example 5.3.6. Suppose that we have a deck of 52 playing cards (consisting of 4 suits and 13 cards per suit). How many cards do we need to draw to guarantee a hand that contains at least 5 cards in a suit?

We apply the extended pigeonhole principle using the cards we draw as pigeons and the suits representing the pigeonholes. We have $m = 4$ boxes and n must satisfy $5 = \lceil n/4 \rceil$ in order to have a hand with at least 5 cards of one suit. Hence, $\lceil n/4 \rceil = 5$ and the smallest integer that satisfies this has $n = 17$. We must draw $n = 17$ cards.△

Example 5.3.7. Prove that in any set of 100 integers, there is a subset of size 8 such that the difference of any two integers in the subset is divisible by 13.

To say that the difference of two integers a and b is divisible by 13 is the same as $a \equiv b \pmod{13}$. In other words, applying the Pigeonhole Principle to this problem we take remainders when divided by 13 as the pigeonholes, there are 13 of them, and the pigeons are the 100 integers. Since $\lceil 100/13 \rceil = 8$, then there is a subset of remainders (pigeonhole) with at least 8 integers in it. Since these are the same congruence class, the difference of any two of them is divisible by 13.   △

Example 5.3.8. Suppose that we have 15 positive integers, not necessarily distinct, whose sum is 400. We show that there exists a subset of 5 of them with a sum of at least 134.

There are $\binom{15}{5} = 3{,}003$ subsets of size 5 of this S set of integers. From the information given, we cannot know any of the sums of elements in the subsets of size 5 but we can determine the average of all the sums from each of these subsets. Every one of the 15 integers appears in $\binom{14}{4} = 1{,}001$ subsets of size 5. This is because once we know a given integer is in a subset, then we need to choose 4 out of the remaining 14 to make up the rest of the subset of size 5. Hence, in the sum of all sums of subsets of size 5, each integer from S appears exactly 1,001 times, but since the sum of elements in S is 400, then the sum of all sums of subsets of size 5 is $400 \cdot 1{,}001 = 400{,}400$. We consider 400,400 as that same number of 1s and think of distributing them among the 3,003 subsets of size 5. By the (extended) pigeonhole principle, one of the subsets must contain

$$\left\lceil \frac{400,400}{3,003} \right\rceil = 134$$

of these 1s, which means that one of the subsets of size 5 has a sum of at least 134.   △

One of the challenges of applying the pigeonhole principle is identifying what might work as pigeons and what might serve as pigeonholes. In this last example, the application ignored the fact that our pigeonholes (subsets of size 5) overlap. Some of our examples may feel contrived to work well with this counting principle but, like many other proof techniques, exposure to examples helps with an intuition for when the pigeonhole principle might serve us well.

5.3.2   Further tricks

The following examples illustrate a few elegant strategies that use the principle.

First we point out that the pigeonholes do not necessarily have to be disjoint.

Example 5.3.9. Show that given any 7 points on a sphere, there is a closed hemisphere that contains 5 of them. (By closed hemisphere we mean the hemisphere and its bounding equator.)

Take any two points A and B of the 7. There is a great circle Γ (circle on the sphere whose center is also the center of the sphere) that goes through A and B. This great circle Γ defines two closed hemispheres S_1 and S_2, which we consider as pigeonholes. We have five remaining points. By the pigeonhole principle, one of the closed hemispheres contains at least $1 + \lfloor (5-1)/2 \rfloor = 3$ of these remaining 5 points. Including A and B, we conclude that this hemisphere contains a total of 5 points.   △

In the previous example the "pigeonholes" are not disjoint but have $S_1 \cap S_2 = \Gamma$. However, we could still use the principle because the contradiction argument behind the extended pigeonhole principle still applies as follows. Assume the contrary of the conclusion of Theorem 5.3.5. Then, if we think of the m boxes as sets A_i with $i = 1, \ldots, m$, then $|A_i| \leq \lfloor (n-1)/m \rfloor$ for all i. Then we have

$$\left| \bigcup_{i=1}^{m} A_i \right| \leq \sum_{i=1}^{m} |A_i| \leq m \left\lfloor \frac{n-1}{m} \right\rfloor < n.$$

This contradicts having n different objects distributed into the sets A_i.

Example 5.3.10. We revisit Example 5.3.3 and show a slightly different result: Show that if we have 5 points in a square of side length 1, then there are 2 points that are at most $\sqrt{2}/2$ apart.

Cover the square of unit side length with 4 disks as shown below.

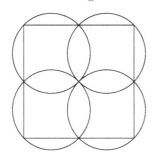

The 4 disks completely cover the square. Since we have 5 points, one of the disks must contain 2 points. However, the diameter of each disk is $\sqrt{2}/2$, so any two points in one of the disks must be at most $\sqrt{2}/2$ apart. $\triangle$

The following example illustrates a well-known trick with pigeonhole principle.

Example 5.3.11. Suppose that in a 100 game season, a baseball player scores at least 1 point per game and a total of 150 points. Show that there was a sequence of consecutive games during which he scored exactly 40 points.

Define the sequence of integers $a_1, a_2, \ldots, a_{100}$ as a_i is the number of points he scored after i games. Notice that the finite sequence is increasing and has

$$1 \leq a_1 < a_2 < \cdots < a_{100} = 150.$$

Let us also consider the sequence $a_i + 40$,

$$41 \leq a_1 + 40 < a_2 + 40 < \cdots < a_{100} + 40 = 190.$$

Take as the pigeons then 200 integers that occur in the two sequences (a_i) and $(a_i + 40)$. We see that the values can fall anywhere between 1 and 190. Since there are more integers than possible values, 2 of these integers must be equal. However, since (a_i) and $(a_i + 40)$ are both strictly increasing sequences, the two equal integers must come from separate sequences. Hence, there exist i and j such that $a_i = a_j + 40$. So $a_i - a_j = 40$, but this is precisely the baseball players points earn during the consecutive games $\{j+1, j+2, \ldots, i\}$.

Note that in this proof strategy, we don't even know how long was the sequence of games with at least 1 point. This argument avoids considering all the possibilities of streak length. $\triangle$

The following example appeared on a Putnam Mathematical Competition and illustrates a particularly clever application of the pigeonhole principle.

Example 5.3.12. Let n be a positive number. Show that given $n+1$ positive integers not exceeding $2n$, there are at least two integers such that one divides the other.

Let $a_1, a_2, \ldots, a_{n+1}$ be the positive integers. Write each one as $a_i = 2^{k_i} q_i$, where q_i is odd. In other words, we factor out the highest power of 2 and q_i is the odd part of the integer. Consider the odd integers q_i. They satisfy $1 \le q_i \le 2n$. There are precisely n odd integers between 1 and $2n$, but since we have $n + 1$ integers, two of these must be equal. Suppose that $q_i = q_j$. Then, we have $a_i = 2^{k_i} q_i$ and $a_j = 2^{k_j} q_i$. Without loss of generality, suppose that $k_j \ge k_i$. Then, $a_j = 2^{k_j - k_i} a_i$ so $a_i \mid a_j$. The result follows.   △

EXERCISES FOR SECTION 5.3

1. Suppose that a set of 300 athletes arrive to register at the Olympics at the same time. Show that at least two of them come from the same National Olympic Committee.

2. Prove that among 14 integers, two of them have a difference that is divisible by 13.

3. Let n be a positive integer. Show that in any subset of $n+1$ integers taken from $\{1, 2, \ldots, 2n\}$, there are two integers that add to $2n + 1$.

4. Show that among 5 points taken from the vertices of a regular dodecagon, two must form an edge to an equilateral triangle.

5. Let (x_i, y_i, z_i) with $i = 1, 2, \ldots, 9$ be 9 points with integer coefficients. Show that the midpoint of at least one pair of these also has integer coefficients.

6. Let a, b, c, d be any integers. Prove that 12 divides $(b-a)(c-a)(d-a)(c-b)(d-b)(d-c)$.

7. Frank is a customer service representative at a bank. He managed to sell at least one mortgage every week and sold 90 mortgages over the course of a 52-week year. Prove that there were consecutive weeks during which he sold exactly 10 mortgages.

8. Suppose that 37 students take a particular class. Show that at least 10 of them must be the same class (freshman, sophomore, junior, or senior).

9. Show that if 100 campers spend the week at a camp with 8 cabins, then at least one cabin will need to house 13 or more people.

10. At a college with an incoming class of 1,001 students from the United States, show that at least 21 come from the same state.

11. Prove that on a circle of diameter 10, if we have 9 points, then at least 3 must be at most 8 units away from each other.

12. Prove that in any set $\{x_1, x_2, \ldots, x_n\}$ of n integers, there is a subset whose sum is a multiple of n.

13. Show that for any group of n people, at least two people have the same number of friends within the group.

14. Show that for any irrational number x and for any positive integer n, there exists a rational number p/q with $1 \leq q \leq n$ and

$$\left| x - \frac{p}{q} \right| < \frac{1}{nq}.$$

[Hint: Consider real numbers of the form $ax - b$ with $a, b \in \mathbb{Z}$.]

15. Show that there exist integers a, b with $-1000 \leq a, b \leq 1000$ such that $a\sqrt{2} + b\sqrt{3}$ is within 10^{-6} of the closest integer.

16. Every point in the plane is colored black, red, or yellow. Show that there is a rectangle with vertices that are all the same color.

17. Decide whether it is possible or not to number the vertices of an icosahedron with the integers 1 through 20 such that around each vertex there is the same sum.

18. (*) Prove that among 7 real numbers $y_1, y_2, \ldots, y_7$ there are two of them y_i and y_j such that

$$0 \leq \frac{y_i - y_j}{1 + y_i y_j} \leq \frac{1}{\sqrt{3}}.$$

[Hint: Consider $y_i = \tan \theta_i$ with $-\pi/2 < \theta_i < \pi/2$.]

5.4 Countability and Cardinality

We end this chapter with a section that moves away from combinatorics to notions of size and countability with infinite sets.

In Section 7.3.2, we mention how mathematics often borrows words from everyday language and attaches to them new and precise definitions. We employ this habit not to confuse but help with intuition. However, the adjective "infinite" as applied to sets is one of those borrowed terms that carries unhelpful connotations in its wake. The English definition of "infinite" is: limitless or boundless in space, extent or size; impossible to measure. Consequently, it might come as as surprise, that there exists structure within the concept of infinite sets.

5.4.1 Countable sets

Section 4.1.3 defines a set as finite if it is empty or if there exists a positive integer n and a bijection between that set and $\{1, 2, \ldots, n\}$. This set theory definition reflects the mental process of enumerating all the elements in the set. The notion of countable extends this further.

Definition 5.4.1

A set A is called *countable* if it is finite, or if it is infinite and there exists a bijection $f : A \to \mathbb{N}^*$. In the latter case, we say A is *countably infinite*. If a set is not countable, it is called *uncountable*.

This definition models the mental process of counting out all the elements in the set A, labeling them as "first" (1), "second" (2), "third" (3), and so on. If a set is finite, this process stops at some point; if the set is infinite, the process proceeds indefinitely. The function $f : \mathbb{N} \to \mathbb{N}^*$ defined by $f(n) = n + 1$ sets up a bijection between $\mathbb{N}$ and $\mathbb{N}^*$ so $\mathbb{N}$ is countable. Often, to show that some other sets are countable requires a little more creativity.

We remark that $f : A \to \mathbb{N}^*$ is a bijection if and only if $f^{-1} : \mathbb{N}^* \to A$ is a bijection. Hence, finding a bijection $\mathbb{N}^* \to A$ suffices to prove that A is countable.

Proposition 5.4.2

The set of integers $\mathbb{Z}$ is countable.

Proof. Consider the function $f : \mathbb{N}^* \to \mathbb{Z}$ defined by

$$f(n) = (-1)^{n-1} \left\lfloor \frac{n}{2} \right\rfloor$$

Note that $f(n) = (n-1)/2$ is n is odd and $f(n) = -(n/2)$ is n is even. We can see that f is surjective as follows. Let $m \in \mathbb{Z}$. If $m \geq 0$, then $f(2m + 1) = m$ but if $m < 0$, then $f(-2m) = m$. Furthermore, f is injective: If $f(n_1) = f(n_2) \geq 0$, then $(n_1 - 1)/2 = (n_2 - 1)/2$, which implies that $n_1 = n_2$. If $f(n_1) = f(n_2) < 0$, then $-n_1/2 = -n_2/2$ and again we solve this and find that $n_1 = n_2$. Since f is both injective and surjective, it is bijective. Thus $\mathbb{Z}$ is countable. $\square$

Countably infinite sets possess a number of counter intuitive properties. However, even some of the intuitive properties require careful proofs.

Proposition 5.4.3

Any subset of a countable set is countable.

Proof. Let S be a countable set and let $A \subseteq S$. As a first case, if A is finite, then it is countable by definition. Suppose that A is infinite. Clearly, this means that S is infinite. Since S is countable and infinite, there exists a bijection $f : \mathbb{N}^* \to S$. By the well-ordering principle, every subset $C \subseteq \mathbb{N}$ has a least element, which we denote here by $\min(C)$. We define a function $h : \mathbb{N}^* \to A$ recursively as follows

$$h(i) = \begin{cases} f(\min(f^{-1}(A))) & \text{if } i = 1 \\ f(\min(f^{-1}(A \setminus \{h(1), h(2), \ldots, h(i-1)\}))) & \text{if } i > 1. \end{cases}$$

Since A is infinite, $A \setminus \{h(1), h(2), \ldots, h(n-1)\} \neq \emptyset$ for all positive integers n, so h is defined on $\mathbb{N}^*$. Since f is a bijection, $f(m) \in A$ for any $m \in f^{-1}(C)$, where $C \subseteq A$, so the $h(\mathbb{N}^*) = A$, which shows that h is a function from $\mathbb{N}^*$ to A.

We claim that $h : \mathbb{N}^* \to A$ is a bijection. By construction, the sequence of subsets of $\mathbb{N}^*$

$$f^{-1}(A) \supsetneq f^{-1}(A \setminus \{h(1)\}) \supsetneq f^{-1}(A \setminus \{h(1), h(2)\}) \supsetneq \cdots$$

is strictly decreasing with the minimum of each set strictly larger than the previous one. Since $f^{-1} \circ h(i) = \min(f^{-1}(A \setminus \{h(1), h(2), \ldots, h(i-1)\}))$, we deduce that $f^{-1} \circ h : \mathbb{N}^* \to \mathbb{N}^*$ is strictly increasing. In particular, if $i \neq j$, without loss of generality $i < j$, so $f^{-1}(h(i)) < f^{-1}(h(j))$ and thus $f^{-1}(h(i)) \neq f^{-1}(h(j))$. Since f^{-1} is a bijection, we deduce that $h(i) \neq h(j)$. This shows that h is injective.

To see that h is surjective, let $a \in A$ be arbitrary. Since f is a bijection, then $f^{-1}(a) = m$ is a positive integer in $f^{-1}(A)$. Since the minima of each subset in the above chain is increasing, we will need to remove at most $m - 1$ elements from A before m is the minimum element in some $f^{-1}(A \setminus \{h(1), h(2), \ldots, h(i-1)\})$. Thus, there exists some i such that $h(i) = a$. In fact, if

$$i = |\{1, 2, \ldots, m\} \cap f^{-1}(A)|,$$

then $h(i) = a$. This shows that h is a bijection. Thus, by definition A is countably infinite and the proposition follows. □

Proposition 5.4.4

The union of two countable sets is countable.

Proof. By Proposition 4.1.11, the union of two finite sets is finite so it suffices to prove the situation when A or B is countably infinite.

For any two sets A and B, we have $A \cup B = A \cup (B \setminus A)$ and $A \cap (B \setminus A) =$. Since $B \setminus A \subseteq B$, by Proposition 5.4.3, we know that $B \setminus A$ is countable. So without loss of generality, we can suppose that A and B are disjoint. The proof breaks down into two cases.

Case 1: Either A or B is finite. Without loss of generality, suppose that A is finite and B is countably infinite. Then there exist a positive integer m and bijections $f : \{1, 2, \ldots, m\} \to A$ and $g : \mathbb{N}^* \to B$. We define the function $h : \mathbb{N}^* \to A \cup B$ by

$$h(i) = \begin{cases} f(i) & \text{if } i \leq m \\ g(i - m) & \text{if } i \geq m + 1. \end{cases}$$

To see that h is surjective, let $c \in A \cup B$. If $c \in A$, then $c = f(k)$ for some $k \in \{1, \ldots, m\}$ and then $c = h(k)$; if $c \in B$, then $c = g(k)$ for some $k \in \mathbb{N}^*$ and then $c = h(m+k)$. Hence, h is surjective. To see that h in injective, suppose that $h(i) = h(j)$. If $h(i) \in A$, then $1 \leq i, j \leq m$ and $f(i) = h(i) = h(j) = f(j)$, but since f is an injection, $i = j$. If $h(i) \in B$, then $i, j \geq m + 1$ and $g(i) = h(i + m) = h(j + m) = g(j)$, but since g is an injection, $i = j$. Thus, h is an injection and therefore a bijection, so $A \cup B$ is countable.

Case 2: Suppose that A and B are disjoint countably infinite sets. By definition, there exist bijections $f : \mathbb{N}^* \to A$ and $g : \mathbb{N}^* \to B$. Consider the function $h : \mathbb{N}^* \to A \cup B$ defined by

$$h(n) = \begin{cases} f((n + 1)/2) & \text{if } n \text{ is odd} \\ g(n/2) & \text{if } n \text{ is even.} \end{cases}$$

The function h is surjective: First, for all $a \in A$, there exists $m \in \mathbb{N}^*$ such that $f(m) = a$. Then $h(2m - 1) = f(m) = a$. *Mutatis mutandis* for all $b \in B$. The function h is also injective: Suppose that $h(i) = h(j)$. Since $A \cap B = \emptyset$, this is only possible if either i and j are both even or both odd. Case 1: If i and j are both even, we deduce that $g(i/2) = g(j/2)$. Since g is bijective, then $i/2 = j/2$, and therefore $i = j$. Case 2: If i and j are both odd, then $h(i) = h(j)$ means that $f((i + 1)/2) = f((j + 1)/2)$. Since f is a bijection, we deduce that $(i + 1)/2 = (i + 1)/2$, from which it follows that $i = j$. In either case, $h(i) = h(j)$ implies that $i = j$, so h is injective. Since $h : \mathbb{N}^* \to A \cup B$ is a bijection, by definition $A \cup B$ is countable. □

Proposition 5.4.4 helps us prove a counterintuitive property about countably infinite sets.

Theorem 5.4.5

The set of rationals $\mathbb{Q}$ is countable.

Proof. We first prove that $\mathbb{Q}^{>0}$ is countable.

Depict $\mathbb{N}^* \times \mathbb{N}^*$ as an infinite two-dimensional grid. We put into the (p, q) box of this grid the fraction p/q. Define subsets $A_n \subseteq \mathbb{N}^* \times \mathbb{N}^*$ for $n \geq 2$ by $A_n = \{(p, q) \in \mathbb{N}^* \times \mathbb{N}^* \mid p + q = n\}$.

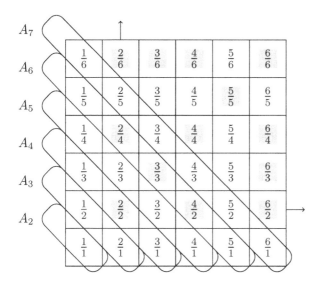

We enumerate the elements of $\mathbb{Q}^{>0}$ as follows:

- Proceed in order through the subsets A_n;
- In each A_n, proceed with increasing values of the numerator;
- If in the box (p, q) the fraction p/q is not expressed in reduced form, then skip the corresponding rational number.

This defines a bijection $h : \mathbb{N}^* \to \mathbb{Q}^{>0}$. Referring to the above diagram, we see that the first few values of this function are as follows.

n	1	2	3	4	5	6	7	8	9	10	11
$h(n)$	$\frac{1}{1}$	$\frac{1}{2}$	$\frac{2}{1}$	$\frac{1}{3}$	$\frac{3}{1}$	$\frac{1}{4}$	$\frac{2}{3}$	$\frac{3}{2}$	$\frac{4}{1}$	$\frac{1}{5}$	$\frac{5}{1}$

Our construction shows that $h : \mathbb{N}^* \to \mathbb{Q}^{>0}$ is bijective, which establishes that $\mathbb{Q}^{>0}$ is countable.

The function $f : \mathbb{Q}^{>0} \to \mathbb{Q}^{<0}$ defined by $f(x) = -x$ is a bijection so $\mathbb{Q}^{<0}$ is also countable. By Proposition 5.4.4, $\mathbb{Q}^* = \mathbb{Q}^{>0} \cup \mathbb{Q}^{<0}$ is countable. By the same proposition, $\mathbb{Q} = \mathbb{Q}^* \cup \{0\}$ is also countable.□

This theorem should surprise us. After all, the countably infinite set of reciprocals of the integers $\{1/n \mid n \in \mathbb{N}^*\}$ is a subset of $(0, 1]$ and appears to consist of so many fewer elements than the full set of rational numbers. Properties with infinite sets get stranger though.

Theorem 5.4.6

The set of reals $\mathbb{R}$ is uncountable.

Proof. We first prove by contradiction that the subset

$B = \{x \in (0, 1) \mid \text{the decimal expansion of } x \text{ involves only 1 and 2}\}$

is uncountable. Notice that because we are not using 9s, no $x \in B$ has a decimal expansion with a tail of 9s. Hence, the decimal expansions of elements in B are unique.

Assume that B is countable. Then we can write

$$B = \{x_1, x_2, x_3, \ldots\}.$$

We consider the decimal expansion of each of the x_i:

$$x_1 = 0.d_{11}d_{12}d_{13}d_{14}\cdots$$
$$x_2 = 0.d_{21}d_{22}d_{23}d_{24}\cdots$$
$$x_3 = 0.d_{31}d_{32}d_{33}d_{34}\cdots$$

and so on. Define the number α by its digits $\alpha = 0.b_1b_2b_3b_4\cdots$ where

$$b_i = \begin{cases} 2 & \text{if } d_{ii} = 1 \\ 1 & \text{if } d_{ii} = 2. \end{cases}$$

Clearly, $\alpha \in B$ since its decimal expansion only has 1s and 2s. By construction, $b_i \neq d_{ii}$ for all $i \in \mathbb{N}^*$. Hence, for all i, the ith digit of α is different from the ith digit of x_i. Thus $\alpha \neq x_i$ for all $i \in \mathbb{N}^*$. This is a contradiction.

We conclude that B is uncountable. So by modus tollens on Proposition 5.4.3, we deduce that $\mathbb{R}$ is not countable. □

The strategy in this proof is called *Cantor's diagonalization argument*.

Taken together, the previous two theorems illustrate something particularly counterintuitive. It is not hard to show that between any two irrational numbers there is a rational number and that between any two rational numbers, there is an irrational number. From intuition inspired by working with finite sets, this seemingly symmetric relationship between $\mathbb{Q}$ and $\mathbb{R} \setminus \mathbb{Q}$ might lead us to suspect that in some sense they have the same number of elements. However, by Propositions 5.4.3 and 5.4.4, whereas $\mathbb{Q}$ is countable, $\mathbb{R} \setminus \mathbb{Q}$ is not.

5.4.2   Cardinality

Until now, we have considered two dichotomies concerning sets: finite versus infinite; and countable versus uncountable. The concept of cardinality, inspired by the definitions for finite, and infinitely countable, allows us to take these comparisons further still.

Definition 5.4.7

We say that two sets A and B have the same *cardinality* if there exists a bijection $f : A \to B$. We write $|A| = |B|$. If there does not exist a bijection between A and B, then we write $|A| \neq |B|$. If there exists an injection $f : A \to B$, then we write $|A| \leq |B|$. If $|A| \leq |B|$ and $|A| \neq |B|$, then we write $|A| < |B|$.

The notion of cardinality, generalizes the concept of counting the number of elements in a set beyond the context of finite sets. From the set-theoretic definitions, it is easy to prove the following properties, which we call transitivity of $=$ and $\leq$ on cardinality.

Proposition 5.4.8

Let A, B and C be any sets. Then
1) If $|A| = |B|$ and $|B| = |C|$, then $|A| = |C|$.
2) If $|A| \leq |B|$ and $|B| \leq |C|$, then $|A| \leq |C|$.

Proof. (1) Suppose that $|A| = |B|$ and $|B| = |C|$. Then there exist bijections $f : A \to B$ and $g : B \to C$. By Proposition 3.2.16, $g \circ f : A \to C$ is a bijection so $|A| = |C|$.

(2) Suppose that $|A| \leq |B|$ and $|B| \leq |C|$. Then there exist injections $f : A \to B$ and $g : B \to C$. By Exercise 3.2.7, $g \circ f : A \to C$ is an injection so $|A| \leq |C|$. $\qquad\qquad\square$

Since any subset of a countable set is countable, the cardinality of $\mathbb{N}^*$ is the smallest infinite cardinality. It enjoys a special symbol, namely $|\mathbb{N}^*| = \aleph_0$ and read it as "aleph naught."

The following theorem, proved at the end of the 19th century leads to yet another unexpected consequence about infinite sets.

Theorem 5.4.9 (Cantor's Theorem)

Let A be any set. Then $|A| < |\mathcal{P}(A)|$.

Proof. Consider the function $g : A \to \mathcal{P}(A)$ defined by $g(a) = \{a\}$. Suppose that $g(a) = g(b)$. Then $\{a\} = \{b\}$ and hence $a = b$. Thus g is injective. Since there exists an injection $g : A \to \mathcal{P}(A)$, we may write $|A| \leq |\mathcal{P}(A)|$.

To prove the strict inequality between cardinalities, we must prove that there does not exist a bijection between the two sets. In particular, we prove that if $f : A \to \mathcal{P}(A)$, then f is not surjective. Assume the contrary: assume there exists a surjective function $f : A \to \mathcal{P}(A)$. Consider the set $B = \{x \in A \mid x \notin f(x)\}$. Since f is surjective, there exists $a \in A$ such that $f(a) = B$. We consider two cases. Case 1: if $a \in B$, then by construction $a \notin f(a) = B$, which is a contradiction. Case 2: if $a \notin B$, then $a \notin f(a)$ and hence $a \in B$, which is again a contradiction. This contradicts the assumption that there exists a surjective function $A \rightarrow \mathcal{P}(A)$.

Since there cannot exist a bijection between A and $\mathcal{P}(A)$, then $|A| \neq |\mathcal{P}(A)|$. Since $|A| \leq |\mathcal{P}(A)|$, then we conclude that $|A| < |\mathcal{P}(A)|$. $\square$

Suppose that we define a sequence of sets as follows: $A_0 = \mathbb{N}$, and $A_n = \mathcal{P}(A_{n-1})$. This sequence of infinite subsets satisfies

$$|\mathbb{N}| < |\mathcal{P}(\mathbb{N})| < |\mathcal{P}(\mathcal{P}(\mathbb{N}))| < |\mathcal{P}(\mathcal{P}(\mathcal{P}(\mathbb{N})))| \cdots$$

This means not only that there are different cardinalities of infinite sets, but, that there is an infinite number of different infinite cardinalities.

We conclude this section by stating two deep results about cardinality. Though the statements involve the symbols of $=$, $\leq$, and $<$ that are reminiscent of inequalities over the integers, the reader should not be lulled into thinking that these theorems are easy to prove. We only provide citations for the proofs.

> **Theorem 5.4.10 (Schröder-Bernstein Theorem)**
> Let A and B be sets. If $|A| \leq |B|$ and $|B| \leq |A|$, then $|A| = |B|$

Proof. (See [31, 13.10].) $\square$

> **Theorem 5.4.11 (Trichotomy Law)**
> For any two sets A and B, exactly one of the following is true:
> $$|A| < |B|, \quad |A| = |B|, \quad |A| > |B|.$$

Proof. (The Trichotomy Law is equivalent to the Axiom of Choice. See [67, p. 9].) $\square$

EXERCISES FOR SECTION 5.4

1. Let A be a countably infinite, and let B be a finite subset. Show that $A \setminus B$ is countably infinite.

2. Find a recursive definition for the function $h : \mathbb{N}^* \rightarrow \mathbb{Q}$ described in the proof of Proposition 5.4.5.

3. Show that A is countable if and only if there exists a surjection $f : \mathbb{N}^* \to A$.

4. Prove that if A is uncountable but that B is countable, then $A \setminus B$ is uncountable.

5. Prove that if A and B are countable, then the Cartesian product $A \times B$ is countable.

6. Prove that for a finite collection of countable sets $\{A_1, A_2, \ldots, A_n\}$, the Cartesian product $A_1 \times A_2 \times \cdots \times A_n$ is countable. [Hint: Cite the previous exercise and use induction.]

7. Prove that if $A \subseteq B$, then $|A| \leq |B|$.

8. Prove that if $|A| = |B|$ for two sets, then $|\mathcal{P}(A)| = |\mathcal{P}(B)|$.

9. Prove that $|\mathbb{R}| = |\mathbb{R}^{>0}|$.

10. Suppose that A is infinite and that B contains more than one element. Prove that the set of functions B^A, from A to B, is uncountable.

11. Let S be any set. For every $A \in \mathcal{P}(S)$, define the function $\chi_A : S \to \{0, 1\}$ by

$$\chi_A(s) = \begin{cases} 1 & \text{if } s \in A \\ 0 & \text{if } s \notin A. \end{cases}$$

Prove the following properties of the function.

(a) For all $A, B \in \mathcal{P}(S)$ and for all $s \in S$, $\chi_{A \cap B}(s) = \chi_A(s)\chi_B(s)$.

(b) For all $A \in \mathcal{P}(S)$ and for all $s \in S$, $\chi_{A^c}(s) = 1 - \chi_A(s)$.

(c) For all $A, B \in \mathcal{P}(S)$ and for all $s \in S$, $\chi_{A \cup B}(s) = \chi_A(s) + \chi_B(s) - \chi_{A \cap B}(s)$.

(d) Show that the function $\Psi : \mathcal{P}(S) \to \{0, 1\}^S$ defined by $\Psi(A) = \chi_A$ is a bijection.

12. Let $\mathbb{Q}[x]$ be the set of polynomials with coefficients in $\mathbb{Q}$. Prove that $\mathbb{Q}[x]$ is countable.

13. Prove that the set $\mathcal{P}_{\text{fin}}(\mathbb{N}^*)$ of all finite subsets of $\mathbb{N}^*$ is countable. [This contrasts with Cantor's Theorem, which affirms that $\mathcal{P}(\mathbb{N}^*)$ is uncountable.]

14. Let $c \in \mathbb{R}$ and consider the function $f : \mathbb{R} \to \mathbb{R}$ with $f(x) = c + e^x$. Define the subset $A \subseteq \mathbb{R}$ as $A = \{c, f(c), f(f(c)), f(f(f(c))), \ldots\}$, and define the function $f : \mathbb{R} \to \mathbb{R} \setminus \{c\}$ by

$$h(x) = \begin{cases} f(x) & \text{if } x \in A \\ x & \text{otherwise.} \end{cases}$$

(a) Prove that $h : \mathbb{R} \to \mathbb{R} \setminus \{c\}$ is a bijection. Deduce that $|\mathbb{R}| = |\mathbb{R} \setminus \{c\}|$.

(b) Use part (a) to show that $|\mathbb{R}| = |\mathbb{R} \times \mathbb{R}|$.

(c) Prove that for all positive integers n, the set $\mathbb{R}$ has the same cardinality as $\mathbb{R}^n$.

15. (*) Prove that every infinite set has a countably infinite subset. [Hint: Axiom of Choice.]

16. (*) Prove that a countable union of countable sets is again countable. [Hint: If $(A_n)_{n \in \mathbb{N}^*}$ is a countable collection of countable sets, then for each $n \in \mathbb{N}^*$ there is a bijection $f_n : \mathbb{N}^* \to A_n$. Consider the function $\phi : \mathbb{N}^* \times \mathbb{N}^* \to \bigcup_{n \in \mathbb{N}^*} A_n$ defined by $\phi(m, n) = f_n(m)$.]

CHAPTER 6

Relations

Widely regarded as the most influential mathematician at the turn of the 20th century, David Hilbert (1862-1943) said about set theory, "No one shall expel us from the paradise that Cantor has created for us." In this last chapter of Part I, we explore relations, another central topic in set theory. The term relation draws from the intuitive concept of when some things are related to others. Set theory gives this concept a precise meaning. As general as the concept will feel at the beginning, this chapter underscores how important relations are for precise definitions of many common objects and patterns in mathematics.

Section 6.1 discusses the general concept of a relation, while Sections 6.2 and 6.3 look in more detail at two specific types of relations that each play particularly important roles in mathematics. We conclude the chapter with the concept of a quotient set and end with formal constructions of all the standard number sets.

6.1 Relations

6.1.1 Definition and examples

A relation is not unlike a function, except that any given object can be related to more than one thing. It is surprising that from such a simple definition there is such rich theory.

Definition 6.1.1

A *relation from* a set A to a set B is a subset R of $A \times B$. A *relation on* a set A is a subset of A^2. If $(a, b) \in R$, we often write $a \, R \, b$ and say that a is related to b via R. The set A is called the *domain* of R and B is the *codomain*.

At first sight, this definition may appear strange. We typically think of a relation as some statement about pairs of objects that is true or false: are they related in the desired fashion or are they not? By gathering all the true statements about a relation into a subset of the Cartesian product, this definition gives the notion of a relation (in mathematics) the same rigor as sets and as Boolean logic.

The empty relation occurs when $R = \emptyset$. Intuitively, this means that nothing is in relation to anything.

Example 6.1.2. Let U be the set of Excellent State University students registered now and C the set of classes offered now. Let T be the relation of "taking classes." T is a relation from U to C and we write $w\,T\,c$ if student w is taking class c. A major function of the registrar's office involves keeping track of T at any given point in time. $\triangle$

Example 6.1.3. In contrast to the previous example, we should not consider the concept of friends as a relation on the set of people. For any pair of people, it might not be clear whether they are friends. We could, on the other hand, define the relation R on the set P of all people by $a\,R\,b$ if right now b is a friend on any of a's social media accounts. This has the logical precision for R to be a relation.

Most social media sites use the label of "friend" for a relation that is symmetric, meaning $a\,R\,b \implies b\,R\,a$. In contrast, the social media concept of "following" is not symmetric because it is possible for a to follow b, but for b not to follow a. $\triangle$

Example 6.1.4. Consider the relation $\leq$ on $S = \{1, 2, 3, 4, 5\}$. According to Definition 6.1.1, the relation $\leq$ is the subset of $S \times S$, given by

$$\{(1,1), (1,2), (1,3), (1,4), (1,5), (2,2), (2,3),$$
$$(2,4), (2,5), (3,3), (3,4), (3,5), (4,4), (4,5), (5,5)\}. \quad \triangle$$

When we consider relations on reasonably small sets, we may depict them in a variety of ways. We illustrate the following four descriptions with the same relation R from $A = \{1, 2, 3, 4, 5\}$ to $B = \{a, b, c\}$.

List. Sticking close to the definition, we can depict the R by writing out all its elements as a subset of $A \times B$. For example,

$$R = \{(1,b), (1,c), (2,a), (2,b), (4,a), (4,b), (4,c), (5,c)\}.$$

Chart. In a chart with the columns labeled for the elements of A and the rows labeled with the elements of B, we can mark a check in the box of column x and row y if $x\,R\,y$. The chart for our running example is the following.

	1	2	3	4	5
a		x		x	
b	x	x		x	
c	x			x	x

Matrix. For the matrix representation, label the elements of A and B by $A = \{a_1, a_2, \ldots, a_m\}$ and $B = \{b_1, b_2, \ldots, b_n\}$. Define the $m \times n$ matrix M_R associated to R by the matrix with entries m_{ij} defined by

$$m_{ij} = \begin{cases} 1 & \text{if } a_j \, R \, b_i \\ 0 & \text{otherwise.} \end{cases}$$

For our example, we have

$$M_R = \begin{pmatrix} 0 & 1 & 0 & 1 & 0 \\ 1 & 1 & 0 & 1 & 0 \\ 1 & 0 & 0 & 1 & 1 \end{pmatrix}.$$

Note that the chart and matrix descriptions are very similar.

Arrow Diagram. Use a Venn diagram with a bubble for A and a bubble for B, illustrating the elements in each as points. Draw an arrow from a point $a \in A$ to a point $b \in B$ if $a \, R \, b$. The arrow diagram for our running example is given below.

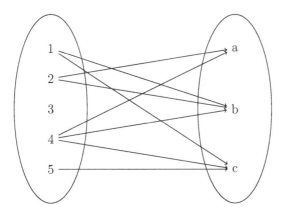

When we consider relations from a set A to itself, the first three methods of depicting a relation remain unchanged but for the arrow diagram we use a directed graph. In the directed graph of a relation on A, we have one point in the plane for each element of A and we draw an arrow from a_1 to a_2 if and only if $(a_1, a_2) \in R$.

Example 6.1.5. Consider the relation R on $\{1, 2, 3, 4, 5\}$ described in list form as

$$R = \{(1,2), (1,5), (2,3), (2,5), (3,1), (4,4), (4,5), (5,2)\}.$$

The directed graph for this relation is

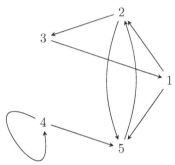

How the graph is laid out on the page does not technically matter, though some placings may be more visually helpful than others. For example, in this relation, if we put 5 above 2, we could lay out this graph so that none of the arrows cross.   •   △

Example 6.1.6. Consider the relation ⋔ on $\mathbb{R}$ defined by $x \pitchfork y$ whenever

$$\begin{cases} y = 10 - x^2 \\ x^2 + y^2 = 4. \end{cases}$$

Setting $x^2 = 10 - y$ and plugging into the second equation, we have $y^2 - y - 6 = 0$, which has two roots: -2 and 3. Referring to the first equation, we find that

$$\pitchfork = \{(\sqrt{7}, 3), (-\sqrt{7}, 3), (\sqrt{12}, -2), (-\sqrt{12}, -2)\}.$$

In this relation, no other real numbers are in relation to each other.△

In this book, we discussed functions before introducing relations. A relation generalizes the concept of a function by removing the requirement $\forall a \in A \exists! b \in B \, (a, b) \in f$. (See Definition 3.1.1.) Many presentations of set theory reverse the order to go from the general to the specific, but our presentation moves from the more familiar to the less familiar.

Example 6.1.7. Let P be the set of living people and let E be the set of working email accounts. Let R be the relation from P to E so that $p \, R \, e$ stands for person p owns the email e. Some people own multiple email accounts, so R could not be a function. R also fails to be a function because some people do not own any email accounts. Conversely, some email accounts are used by more than one person so it would not be possible to create a function from E to P either.△

Consider the many symbols that we use in mathematics. Some, like $\lfloor \rfloor$ represent functions but many either represent a relation or a binary operation. Some common relation symbols on real numbers are $=$, $\leq$, $\geq$, $<$, $>$, and $\neq$. In fact, on any set A the symbol $=$ is the relation given as the diagonal of the $A \times A$, namely $\{(a, a) \mid a \in A\}$. If S is any set, $\subseteq$, $\subsetneq$, $\not\subseteq$ are relations on $\mathcal{P}(S)$. The set theory symbols $\in$ and $\notin$ represent relations from S to $\mathcal{P}(S)$.

When defining relations, it is common to create a symbol to signify a relation. We already did this in Example 6.1.6. Another standard of notation is that if a symbol represents a relation R, then the symbol with an angle slash through it represents the complement relation $\overline{R}$. For example, $\neq$ is the symbol not equals and $\not\trianglelefteq$ is the complement of whatever the relation $\trianglelefteq$ represents.

6.1.2 Operations on relations

Suppose that R_1 and R_2 are two relations from A to B. Since R_1 and R_2 are subsets of $A \times B$, then all the usual operations on subsets are at our disposal: the union $R_1 \cup R_2$, the intersection $R_1 \cap R_2$, the set difference $R_1 \setminus R_2$, and so on. As we mentioned above, the complement $R_1{}^c$ carries the notation $\not{R}_1$.

Example 6.1.8. Consider a given university and let U consists of all registered students and let C be the set of classes that are offered. Consider the relation T standing for $s\,T\,c$ means that student s has taken class c, but now also consider the relation N where $s\,N\,c$ means that s needs class c for his or her major.

The relation $T \cap N$ pairs each student with all the classes they have taken and need for the major. Also $N \setminus T$ is the relation with $s(N \setminus T)c$ when s still needs to take class c for the major. $\triangle$

Interestingly, we can also define the notion of a composition of relations.

Definition 6.1.9

Let A, B, and C be sets. Let R_1 be a relation from A to B and let R_2 be a relation from B to C. The *composite* relation of R_2 with R_1 is the relation $R = R_2 \circ R_1$ from A to C such that

$$aRc \iff \exists b \in B, a\,R_1\,b \text{ and } b\,R_2\,c.$$

Definition 6.1.10

If R is a relation from A to B, then the *inverse relation* R^{-1} is the relation from B to A such that $b\,R^{-1}\,a \iff a\,R\,b$.

Example 6.1.11. Let P be the set of all people, living or deceased, and let H be the set of all deeded properties (houses, lots, commercial buildings, and so on). Define R as the relation from P to H of $p\,R\,h$ if p owned/owns property h. (To precisely define ownership, we could say that p's name was recorded on the deed to the property.) Then R^{-1} is the relation of having been owned by.

It is interesting to think about the relation $R^{-1} \circ R$ from P to P. It is not the equals relation. Rather $p_1(R^{-1} \circ R)p_2$ if p_1 owns/owned a property that was owned (is owned) by p_2. △

6.1.3   Properties of relations

In Example 6.1.3, we introduced in passing the concept of a symmetric relation. The word made intuitive sense in that example. More generally, certain classes of relations from a set A to itself play important roles due to a combination of specific properties. In the next sections, we introduce equivalence relations and partial orders, both of which are essential in many areas of mathematics and applications. However, we list here below some of the properties for relations on a set A that are often of particular interest.

Definition 6.1.12

Let R be a relation on a set A. The relation R is called

1) *reflexive* if $\forall a \in A$, $a\,R\,a$ (Reflexivity);

2) *symmetric* if $a\,R\,b \Longrightarrow b\,R\,a$ (Symmetry);

3) *antisymmetric* if $a\,R\,b$ and $b\,R\,a \Longrightarrow a = b$ (Antisymmetry);

4) *transitive* if $a\,R\,b$ and $b\,R\,c \Longrightarrow a\,R\,c$ (Transitivity).

Example 6.1.13. Let $\mathcal{L}$ be the set of lines in the Euclidean plane $\mathbb{E}^2$. The notion of perpendicularity $\perp$ is a relation on $\mathcal{L}$. It is a symmetric relation, but it does not satisfy any of the other properties described in Definition 6.1.12. △

Example 6.1.14. Let B be the set of blood types. Encode the blood types by $B = \{\mathsf{o}, \mathsf{a}, \mathsf{b}, \mathsf{ab}\}$ and consider the donor relation $\to$, such that $t_1 \to t_2$ means (disregarding all other factors) someone with blood type t_1 can donate to someone with blood type t_2. As a subset of B^2, the donor relation is

$$\{(\mathsf{o},\mathsf{o}),(\mathsf{o},\mathsf{a}),(\mathsf{o},\mathsf{b}),(\mathsf{o},\mathsf{ab}),(\mathsf{a},\mathsf{a}),(\mathsf{a},\mathsf{ab}),(\mathsf{b},\mathsf{b}),(\mathsf{b},\mathsf{ab}),(\mathsf{ab},\mathsf{ab})\}.$$

It is not hard to check exhaustively or logically that $\to$ is reflexive, antisymmetric, and transitive but not symmetric. △

Example 6.1.15. Let P be the set of people. As we meet people, depending on context, we will sometimes ask, "Are you related to so and so?". We can make the concept of relation precise by setting R as the relation $p\,R\,q$ if p and q are in the same nuclear family, or in other words, p is to q self, sibling, parent, child, or spouse. We could consider R to be biological relatives or legal relatives. (When we ask the question, we usually mean legal relatives.)

This R does not cover everyone we think of as related to us. We can get to other relatives in this way by composing the relation R with itself. The composition $R \circ R$ joins to R, grandparents, grandchildren, aunts and uncles, nieces and nephews. The relation of cousin does not come in until we consider $R \circ R \circ R$. $\triangle$

EXERCISES FOR SECTION 6.1

1. Describe four relations from the set of living people to the set of cities (townships).

2. Let $A = \{1, 2, 3, 4, 5\}$ and $B = \{a, b, c\}$. Define the relation R from A to B by the set $R_1 = \{(1, a), (1, b), (2, a), (3, b), (3, c), (5, c)\}$. Draw the chart, write the matrix, and draw the arrow diagram for this relation.

3. Let $B = \{a, b, c\}$ and $C = \{6, 7, 8, 9\}$. Define the relation R from B to C by the set $R_2 = \{(a, 7), (a, 8), (b, 6), (c, 7), (c, 9)\}$. Draw the chart, write the matrix, and draw the arrow diagram for this relation.

4. Using the relations in the previous two exercises, consider the composite relation $R_2 \circ R_1$ from A to C. Draw the chart, write the matrix, and draw the arrow diagram for $R_2 \circ R_1$.

5. Let $A = \{[-1, 2], [1, \pi], [3, 4], [6, 9]\}$ and $B = \{[0, \sqrt{2}/2], [1, 3], [3, 5]\}$ be two sets of intervals of real numbers. Define the relation R from A and B by $a\,R\,b$ if the interval $a \cap b \neq \emptyset$. Draw the chart, write the matrix, and draw the arrow diagram for this relation.

6. Let A and B be finite sets. Determine the number of distinct relations from A to B.

7. Consider the relation R on $A = \{1, 2, 3, 4, 5\}$ defined by

$$R = \{(1, 3), (1, 4), (2, 1), (2, 3), (2, 5), (3, 3), (4, 2), (4, 3), (5, 1)\}$$

   (a) Write the matrix representing R.
   (b) Sketch the arrow diagram for R.

8. Consider the relation R on $A = \{1, 2, 3, 4, 5, 6\}$ given by $x R y$ when $|x - y| = 1$.
   (a) Write the matrix representing R.
   (b) Sketch the arrow diagram for R.

9. Let R be a relation from A and B and let S be a relation from B to C. Prove that the matrix of $S \circ R$ is

$$M_{S \circ R} = M_S \odot M_R,$$

where the $P \odot Q$ product is the usual matrix product followed by the function $x \mapsto \min(1, x)$ applied to each entry.

For Exercises 6.1.10 through 6.1.16, determine (with proof) which of the properties reflexivity, symmetry, antisymmetry, and transitivity hold for each of the following relations.

10. For any set S, consider the relation $\emptyset$ on $\mathcal{P}(S)$ defined by $A \mathbin{\emptyset} B$ to mean that $A \cap B \neq \emptyset$.

11. The relation $\succsim$ on S the set of people defined by $p_1 \succsim p_2$ if p_1 is taller than or the same height as p_2.

12. The relation R on $\mathbb{Z}$ defined by nRm if $n \geq m^2$.

13. The relation Γ on $S = \mathbb{R}^2$ defined by $(x_1, y_1) \mathrel{\Gamma} (x_2, y_2)$ to mean $x_1^2 + y_1^2 \leq x_2^2 + y_2^2$.

14. The relation $\overset{\circ}{=}$ on $\mathbb{R}$ defined by $a \overset{\circ}{=} b$ to mean $ab = 0$.

15. For any set S, consider the relation $\rightsquigarrow$ on $\mathcal{P}(S)$ defined by $A \rightsquigarrow B$ to mean that $A \cup B = S$.

16. The relation $\leftrightharpoons$ on the set of pairs of points in the plane $S = \mathbb{R}^2 \times \mathbb{R}^2$ defined by $(P_1, Q_1) \leftrightharpoons (P_2, Q_2)$ if the segment $[P_1, P_2]$ intersects $[Q_1, Q_2]$.

17. Let S be a set and let R be a relation on S. Prove that if a relation is reflexive, symmetric, and antisymmetric, then it is the $=$ relation on S.

18. Let A be a finite set with n elements.

    (a) Prove that the number of reflexive relations on A is $2^{n^2 - n}$.

    (b) Prove that the number of symmetric relations on A is $2^{n(n+1)/2}$.

19. We can define the graph of a relation R from $\mathbb{R}$ to itself as the subset of $\mathbb{R}^2$

$$\{(x, y) \in \mathbb{R}^2 \mid x \mathrel{R} y\}.$$

    (a) Sketch the graph of the relation $\leq$.

    (b) Sketch the graph of the relation $\overset{\frown}{=}$ defined by $x \overset{\frown}{=} y$ if $|x - y| = 1$.

    (c) Provide defining geometric characteristics of a subset of $\mathbb{R}^2$ for a relations on $\mathbb{R}$ that are (i) reflexive; (ii) symmetric; (iii) transitive; (iv) antisymmetric.

20. Let $S = \{a, b, c, d, e\}$ and consider the relation R on S described by

$$R = \{(a, a), (a, c), (a, d), (b, c), (b, e), (c, b), (c, d), (e, a), (e, b)\}.$$

    Determine as a list in $S \times S$, the composite relation $R \circ R$.

21. Let P be the set of people who are living now. Let R be the relation on P defined by aRb if a and b are in the same nuclear family as defined in Example 6.1.15.

    (a) Decide whether R is reflexive, symmetric, antisymmetric, or transitive.

    (b) List all the family relations included in $R^{(2)} = R \circ R$.

    (c) Give four commonly used family terms for relations in $R^{(3)} = R \circ R \circ R$ though not in $R^{(2)}$.

22. Let R be a relation on a set A. Denote by $R^{(n)}$ the n-composite relation of R with itself:

$$R^{(n)} \overset{\text{def}}{=} \overbrace{R \circ R \circ \cdots \circ R}^{n \text{ times}}.$$

Prove that the relation R is transitive if and only if $R^{(n)} \subseteq R$ for all $n = 1, 2, 3, \ldots$.

23. In Example 6.1.13, we observed that $\perp$ is not transitive on the set $\mathcal{L}$ of lines in the Euclidean plane $\mathbb{E}^2$. Show that that the relation $\perp^{(3)} = \perp$. Show that this fact is no longer true for lines in Euclidean space.

6.2 Partial Orders

6.2.1 Definition and first examples

Section 6.1 presented relations in general. In that section, we remarked how functions form a subclass of relations. Another particularly useful subclass of relations is that of partial orders, which generalizes the inequality $\leq$ on $\mathbb{R}$ to a mental model of ordering objects in a set.

Definition 6.2.1

A *partial order* on a set S is a relation $\preccurlyeq$ that is reflexive, antisymmetric, and transitive.

Motivated by the notations for inequalities over $\mathbb{R}$, we use the symbol $\prec$ to mean

$$x \prec y \iff x \preccurlyeq y \text{ and } x \neq y$$

and the symbol $\npreccurlyeq$ to mean that it is not true that $x \preccurlyeq y$.

Example 6.2.2. Consider the relation $\leq$ on $\mathbb{R}$. For all $x \in \mathbb{R}$, $x \leq x$ so $\leq$ is reflexive. For all $x, y \in \mathbb{R}$, if $x \leq y$ and $y \leq x$, then $x = y$ and hence $\leq$ is antisymmetric. It is also true that $x \leq y$ and $y \leq z$ implies that $x \leq z$ and hence $\leq$ is transitive. Thus, the inequality $\leq$ on $\mathbb{R}$ is a partial order. $\triangle$

Note that $\geq$ is also a partial order on $\mathbb{R}$ but that $<$ and $>$ are not. The strict inequality $<$ is not reflexive though it is both antisymmetric and transitive. ($<$ is antisymmetric vacuously: because there do not exist any $x, y \in \mathbb{R}$ such that $x < y$ and $y < x$, the conditional statement "$x < y$ and $y < x$ implies $x = y$" is trivially satisfied.)

Though modeled after the relation of $\leq$, we need to consider a few other examples to begin to see the numerous possibilities.

Example 6.2.3. Let S be any set. The subset relation $\subseteq$ on $\mathcal{P}(S)$ is a partial order. Reflexivity: For all $A \in \mathcal{P}(S)$, if $x \in A$, then $x \in A$, so $A \subseteq A$. Antisymmetry: For all $A, B \in \mathcal{P}(S)$, if $A \subseteq B$ and $B \subseteq A$, then $A = B$. (This is a standard way of showing sets are equal.) Transitivity: For all $A, B \in \mathcal{P}(S)$, suppose that $A \subseteq B$, and $B \subseteq C$. Then for all $x \in A$, we know that $x \in B$ since $A \subseteq B$ and also $x \in C$ because $B \subseteq C$. Therefore $A \subseteq C$. $\triangle$

Example 6.2.4. Consider the set $\mathbb{N}^*$ of positive integers. The divisibility relation $\mid$ is a partial order. Reflexivity: For all $a \in \mathbb{N}^*$, since $a \cdot 1 = a$, then $a \mid a$. Antisymmetry: Let $a, b \in \mathbb{N}^*$ with $a \mid b$ and $b \mid a$. By Proposition 4.2.2(4), then $a = b$ or $a = -b$. However, since we do not have $a = -b$ in positive integers, by resolution, we deduce that $a = b$. Transitivity: This is Proposition 4.2.2(3). $\triangle$

Both of these familiar examples exhibit a situation that does not occur with $\leq$. With a partial order $\preccurlyeq$ on a set S, given two arbitrary elements $a, b \in S$, it is possible that neither $a \preccurlyeq b$ nor $b \preccurlyeq a$. For example, neither $2 \mid 3$ nor $3 \mid 2$ is true.

Definition 6.2.5

Let $\preccurlyeq$ be a partial order on a set S. If for some pair $\{a, b\}$ of distinct elements, either $a \preccurlyeq b$ or $b \preccurlyeq a$, then we say that a and b are *comparable*; otherwise, a and b are called *incomparable*. A partial order in which every pair of elements is comparable is called a *total order*.

In the partial order of $\subseteq$ on $\mathcal{P}(S)$, many pairs of subsets in S are incomparable. In fact, two subsets A and B are incomparable if and only if $A \setminus B$ and $B \setminus A$ are both nonempty.

Example 6.2.6. Consider the donor relation $\to$ defined on the set of blood types $B = \{\mathsf{o}, \mathsf{a}, \mathsf{b}, \mathsf{ab}\}$ as discussed in Example 6.1.14. We saw that $\to$ is reflexive, antisymmetric, and transitive and so it is a partial order on the set of blood types. Note that a and b are incomparable, meaning that neither can donate to the other. $\triangle$

Example 6.2.7. Consider the relation $\preccurlyeq$ on $\mathbb{R}^2$ defined by

$$(x_1, y_1) \preccurlyeq (x_2, y_2) \iff 2x_1 - y_1 < 2x_2 - y_2 \text{ or } (x_1, y_1) = (x_2, y_2).$$

That $(x_1, y_1) \preccurlyeq (x_1, y_1)$ is built into the definition so $\preccurlyeq$ is reflexive. It is impossible for $2x_1 - y_1 < 2x_2 - y_2$ and $2x_2 - y_2 < 2x_1 - y_1$, so the only way $(x_1, y_1) \preccurlyeq (x_2, y_2)$ and $(x_2, y_2) \preccurlyeq (x_1, y_1)$ can occur is if $(x_1, y_1) = (x_2, y_2)$. Finally, the relation is also transitive so that $\preccurlyeq$ is a partial order on $\mathbb{R}^2$.

In this partial order, two elements (x_1, y_1) and (x_2, y_2) are incomparable if and only if $2x_2 - y_2 = 2x_1 - y_1$ and $(x_1, y_1) \neq (x_2, y_2)$, namely they are distinct points on the same line of slope 2. $\triangle$

Example 6.2.8 (Another Order on $\mathbb{Q}$). Consider the bijection $h : \mathbb{N}^* \to \mathbb{Q}^{>0}$ defined in Example 5.4.5 where we discussed the countability of the rational numbers. We can define a new partial order $\preccurlyeq$ on $\mathbb{Q}^{>0}$ by $r \preccurlyeq s$ if and only if $h^{-1}(r) \leq h^{-1}(s)$. This is a total order on $\mathbb{Q}^{>0}$ since $\leq$ is a total order on $\mathbb{N}^*$ and h is a bijection. As an example $6 \preccurlyeq 2/5$. $\triangle$

Besides the distinction between total and not total orders, another dichotomy arises when comparing properties of the partial order $\leq$ on $\mathbb{N}$ versus $\mathbb{R}$ or even $\mathbb{Q}$. With real numbers, given any $x \leq y$ with $x \neq y$, there always exists an element z such that $x < z < y$; in contrast, with integers for example $2 \leq 3$ but for all $z \in \mathbb{N}$, if $2 \leq z \leq 3$, then $z = 2$ or $z = 3$.

Definition 6.2.9

Let $\preccurlyeq$ be a partial order on S and let $x \in S$. We call $y \in S$ an *immediate successor* (resp. *immediate predecessor*) to x if $y \neq x$ with $x \preccurlyeq y$ (resp. $y \preccurlyeq x$) and for all $z \in S$ such that $x \preccurlyeq z \preccurlyeq y$ (resp. $y \preccurlyeq z \preccurlyeq x$), either $z = x$ or $z = y$.

For $\leq$ on $\mathbb{N}$, all elements have both immediate successors and immediate predecessors, except for 0 that does not have an immediate predecessor. For $\leq$ on $\mathbb{Z}$, all elements have both immediate successors and immediate predecessors. In contrast, as commented above, in $\mathbb{R}$ no element has either an immediate successor or an immediate predecessor.

A partial order does not have to be a total order to have immediate successors or predecessors. In the blood donor relation $(B, \to)$ in Example 6.2.6, o has two immediate successors, namely a and b.

6.2.2   Posets

Definition 6.2.10

A pair $(S, \preccurlyeq)$, where S is a set and $\preccurlyeq$ is a partial order on S is often succinctly called a *poset*.

The name *poset* abbreviates "partially ordered **set**." This idea
of pairing a set with an associated object or objects is a common
construction in mathematics. In linear algebra, we define a vector
space as a set equipped with a binary operation called addition and
an operation called scalar multiplication that satisfy a specific list of
axioms. The concept of a poset is similar; it is a set equipped with a
relation satisfying the axioms of a partial order.

If $(S, \preccurlyeq)$ is a poset and T any subset of S, then when we restrict
$\preccurlyeq$ only to elements of T, the quantifiers in the definition of a partial
order still hold when restricted to T. Hence $(T, \preccurlyeq)$ is also a poset. We
sometimes call $(T \preccurlyeq)$ a *subposet* of $(S, \preccurlyeq)$. (This contrast with vector
spaces, where the concept of a subspace required that the subset be
closed under addition and under scalar multiplication.)

Though a generic poset $(S, \preccurlyeq)$ need not be a total order, many
of the familiar terms associated to inequalities $\mathbb{R}$ have corresponding
definitions in any poset.

Definition 6.2.11

Let $(S, \preccurlyeq)$ be a poset, and let A be a subset of S.
1) A *maximal* element of A is an $M \in A$ such that if
   $t \in A$ with $M \preccurlyeq t$, then $t = M$.
2) A *minimal* element of A is an $m \in A$ such that if $t \in A$
   with $t \preccurlyeq m$, then $t = m$.

As an example, consider the blood donor relation $(B, \rightarrow)$ described
in Example 6.2.6 and consider the subset $A = \{o, a, b\}$. Then A has
one minimal element o and two maximal elements a and b.

Definition 6.2.12

Let $(S, \preccurlyeq)$ be a poset, and let A be a subset of S.
1) An *upper bound* of A is an element $u \in S$ such that
   $\forall t \in A, t \preccurlyeq u$.
2) A *lower bound* of A is an element $\ell \in S$ such that
   $\forall t \in A, \ell \preccurlyeq t$.
3) A *least upper bound* of A is an upper bound u of A
   such that for all upper bounds u' of A, we have $u \preccurlyeq u'$.
4) A *greatest lower bound* of A is a lower bound ℓ of A
   such that for all lower bounds ℓ' of A, we have $\ell' \preccurlyeq \ell$.

We say that a subset $A \subseteq S$ is *bounded above* if A has an upper
bound, is *bounded below* if A has a lower bound, and is *bounded* if A
is bounded above and bounded below.

If u_1 and u_2 are two least upper bounds to A, then by definition $u_1 \preccurlyeq u_2$ and $u_2 \preccurlyeq u_1$. Thus, $u_1 = u_2$ and we conclude that least upper bounds are unique. Similarly, greatest lower bounds for a set are unique. Therefore, if a subset A has a least upper bound, we talk about *the* least upper bound of A and denote this element by $\mathrm{lub}(A)$. Similarly, if a subset A has a greatest lower bound, we talk about *the* greatest lower bound of A and denote this element by $\mathrm{glb}(A)$.

From the perspective of analysis, one of the most important differences between the posets $(\mathbb{R}, \leq)$ and its subposet $(\mathbb{Q}, \leq)$ is that any bounded subset of $\mathbb{R}$ has a least upper bound whereas this does not hold in $\mathbb{Q}$. Consider for example, the subset

$$A = \left\{ \frac{p}{q} \in \mathbb{Q} \mid p^2 < 2q^2 \right\}.$$

In $\mathbb{R}$, $\mathrm{lub}(A) = \sqrt{2}$ whereas in $\mathbb{Q}$ for any upper bound $u = r/s$ of A we have

$$\sqrt{2} < \frac{1}{2}\left(\frac{r}{s} + \frac{2s}{r} \right) < \frac{r}{s}.$$

(We leave the proof to the reader.) Hence, A has no least upper bound in $(\mathbb{Q}, \leq)$.

Definition 6.2.13

In a poset $(S, \preccurlyeq)$ any subposet $(T, \preccurlyeq)$ that is a total order is called a *chain*.

The concept of a chain allows us to introduce a theorem that is essential in a variety of contexts in algebra.

Theorem 6.2.14 (Zorn's Lemma)

Let $(S, \preccurlyeq)$ be a poset. Suppose that every chain in S has an upper bound. Then S contains a maximal element.

In the context of **ZF**-set theory, Zorn's Lemma is equivalent to the Axiom of Choice. (See [79, Theorem 5.13.1] for a proof.)

Definition 6.2.15

A poset $(S, \preccurlyeq)$ is called a *lattice* if for all pairs $(a, b) \in S \times S$, both $\mathrm{lub}(a, b)$ and $\mathrm{glb}(a, b)$ exist.

Lattices are a particularly nice class of partially order sets. They occur frequently in various areas of mathematics. For example, given any set S, the power set $(\mathcal{P}(S), \subseteq)$ is a lattice with $\mathrm{lub}(A, B) = A \cup B$ and $\mathrm{glb}(A, B) = A \cap B$.

Example 6.2.16. The poset $(\mathbb{N}^*, |)$ is a lattice. The least common multiple is precisely the least upper bound in the poset; the greatest common divisor serves the role of a greatest lower bound. It is partly because of how well gcd and lcm fit into this general theory with partial orders that advanced mathematics used Definition 4.3.1, even though it differs from the elementary school definition of greatest common divisor.   △

Example 6.2.17. Early in Part I, in Example 1.2.6 we talked about the smallest subset of $\mathbb{Z}$ satisfying a certain property. In all honesty, the term "smallest" at that stage was imprecise and premature. Only now can we explain it precisely.

Let $\mathcal{C}$ be a collection of subsets of a set S, i.e., $\mathcal{C} \subseteq \mathcal{P}(S)$. By "the smallest set in $\mathcal{C}$," we mean a set $A \in \mathcal{C}$ such that $A \subseteq B$ for all $B \in \mathcal{C}$. Because of this, writers will often say "the smallest set by inclusion" in order to explicitly mention the partial order. This is a stronger condition than a minimal element and, since it is in $\mathcal{C}$, we do not mean a lower bound either.

The concept of "the smallest set" in a collection occurs in many fields, but it has another equivalent formulation that is often useful in proofs.

> **Proposition 6.2.18**
>
> Let $\mathcal{C}$ be a collection of subsets of a set S. Then $\mathcal{C}$ has a smallest set if and only if
>
> $$\bigcap_{A \in \mathcal{C}} A \in \mathcal{C}.$$
>
> Furthermore, if this is true, then this generalized intersection is the smallest set in $\mathcal{C}$.

($\Longleftarrow$)   *Proof.* We first prove the converse. Suppose that $T = \bigcup_{A \in \mathcal{C}} A \in \mathcal{C}$. Then, $T \subseteq A$ for all $A \in \mathcal{C}$, and hence T is the smallest element.

($\Longrightarrow$)   Suppose now that $\mathcal{C}$ has a smallest element V. Then, $V \subseteq A$ for all $A \in \mathcal{C}$ and so $V \subseteq \bigcap_{A \in \mathcal{C}} A$. However, since $V \in \mathcal{C}$, then $\bigcap_{A \in \mathcal{C}} A \subseteq V$, and so V is equal to this generalized intersection. In particular, $\bigcap_{A \in \mathcal{C}} A \subseteq \mathcal{C}$.   □

6.2.3   Hasse diagrams

For posets with a small number of elements, it is possible to easily visualize the relation via a *Hasse diagram*.

Let $(S, \preccurlyeq)$ be a poset in which S is finite. In a Hasse diagram, each element of S corresponds to a point in the plane, with the points placed on the page so that if $a \preccurlyeq b$, then b appears higher on the page. The points of the Hasse diagram are also called *nodes* or *vertices*.

Finally, we draw an edge between two points (corresponding to) a and b with b above a if b is an immediate successor of a.

As a first example, Figure 6.1 gives the Hasse diagram for the donor relation described in Example 6.2.6.

Figure 6.1: Hasse diagram for the donor relation on blood types.

By reflexivity, we know that $a \preccurlyeq a$ for all $a \in S$; nevertheless, we do not draw a loop at each vertex. Because of transitivity, $p \preccurlyeq q$ if and only if there is a (rising) path through the Hasse diagram from p to q. For example, in Figure 6.1, the diagram does not show an explicit edge between o and ab, but we can see that o $\to$ ab because there is a rising path in the diagram from o to ab.

Example 6.2.19. Let $S = \{a, b, c, d, e, f, g, h, i\}$ and consider the partial order on S described by the Hasse diagram shown in Figure 6.2. The diagram makes it clear what relations hold between elements. For example, notice that all elements in $\{a, b, c, d, e, f, g\}$ are incomparable with the elements in $\{h, i\}$. The maximal elements in S are d and i. The minimal elements are a, e, g, and h. As a least upper bound calculation, $\mathrm{lub}(a, f) = c$ because c is the first element in a chain above a that is also in a chain above f. We also see that $\mathrm{lub}(e, h)$ does not exist because there is no chain above e that intersects with a chain above h.   △

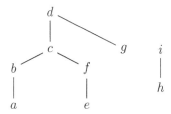

Figure 6.2: An example of a poset defined by a Hasse diagram

Hasse diagrams allow for easy visualization of properties of the poset. For example, a poset will be a lattice if and only if taking any two points p_1 and p_2 in the diagram, there exists a chain rising from p_1 that intersects with a chain rising from p_2 (existence of the least upper bound) and a chain descending from p_1 that intersects with a chain descending from p_2 (existence of the greatest upper bound). Figure 6.3 illustrates three different lattices.

Figure 6.3: Lattice types.

6.2.4 Lexicographic orders

Suppose that $(A_1, \preccurlyeq_1)$ and $(A_2, \preccurlyeq_2)$ are two posets, the *lexicographic order* on the Cartesian product $A_1 \times A_2$ is the partial order $\preccurlyeq$ defined by $(a_1, a_2) \preccurlyeq (b_1, b_2)$ if $a_1 \neq b_1$ and $a_1 \preccurlyeq b_1$ or $a_1 = b_1$ and $a_2 \preccurlyeq b_2$.

More generally, if $(A_i, \preccurlyeq_i)$ are posets for $i = 1, 2, \ldots, n$, then the associated lexicographic order on $A_1 \times A_2 \times \cdots \times A_n$ is the order $\preccurlyeq$ such that

$$(a_1, a_2, \ldots, a_n) \preccurlyeq (b_1, b_2, \ldots, b_n)$$

if and only if the first i such that $a_i \neq b_i$ we have $a_i \preccurlyeq b_i$.

The lexicographic order on the Cartesian product derives its name from how it mimics the total order used on words expressed in alphabetical languages for organization in a dictionary. For example, in English, we can think of the alphabet $\mathcal{A}$ as consisting of the 26 letters "A" through "Z" as well as the space character. We define a total order on $\mathcal{A}$ with the space being the least element and all the other letters in their usual order. Words are elements of $\mathcal{A}^n$ for some large enough n, where we can think of using spaces at the end if the word is shorter than length n. So "cantaloupe" comes before "cantilever" in the lexicographic order since at the first position in which the letters differ (the fifth letter) "a" comes before "i."

Example 6.2.20. Consider the poset $(\mathbb{Z}, \leq)$ and let $\preccurlyeq$ be the lexicographic ordering on $\mathbb{Z} \times \mathbb{Z} \times \mathbb{Z}$. Then, for example,

$(2, 170, -5) \preccurlyeq (4, -30, 2)$ at the first differing entry, $2 \leq 4$,

$(-5, 4, -10) \preccurlyeq (-5, 4, 0)$ at the first differing entry, $-10 \leq 0$. $\triangle$

Example 6.2.21. Consider the posets $(\mathbb{Z}, \leq)$ and $(\mathcal{P}(\{1, 2, 3, 4\}), \subseteq)$ and let $\preccurlyeq$ be the lexicographic ordering on $\mathbb{Z} \times \mathcal{P}(\{1, 2, 3, 4\})$. Then, for example,

$(3, \{1, 3\}) \preccurlyeq (5, \{2, 3, 4\})$ at the first differing entry, $3 \leq 5$,

$(-2, \{4\}) \preccurlyeq (-2, \{2, 3, 4\})$ at the first entry, $\{4\} \subseteq \{2, 3, 4\}$.

On the other hand $(-2, \{1,4\})$ and $(-2, \{2,3,4\})$ are incomparable. $\triangle$

EXERCISES FOR SECTION 6.2

1. Let $S = \{a, b, c, d, e\}$ (where we consider all the labels unique elements). In the following relations on S, determine with explanation whether or not the relation is a partial order. If it fails antisymmetry, then remove a least number of pairs, and if it fails transitivity, then add some pairs to make the relation a partial order.
   (a) $R = \{(a,a), (b,b), (c,c), (d,d), (e,e), (a,c)\}$
   (b) $R = \{(a,a), (b,b), (c,c), (d,d), (e,e), (a,c), (a,d)\}$
   (c) $R = \{(a,a), (b,b), (c,c), (d,d), (e,e), (a,c), (d,a)\}$
   (d) $R = \{(a,a), (b,b), (c,c), (d,d), (e,e), (b,c), (c,d), (d,e), (a,e)\}$

2. In microeconomics (the study of consumer behavior), one considers consumer's utility (preference) in regards to pairs of commodities. Let $(q_1, q_2) \in \mathbb{N}^2$ be a pair of nonnegative integers representing quantities of two commodities. Explain why, given two specific commodities and a given consumer, the relation of preferable (or equal) is a partial order.

3. Let $S = \mathbb{R}^{>0} \times \mathbb{R}^{>0}$ be the positive first quadrant in the Cartesian plane. Consider the relation R on S defined by

$$(x_1, y_1) \, R \, (x_2, y_2) \implies x_1 y_1 \geq x_2 y_2.$$

   Prove or disprove that R is a partial order.

4. Prove that for any real $x > \sqrt{2}$, the inequality $\sqrt{2} < \frac{1}{2} \left(x + \frac{2}{x} \right) < x$ holds.

5. Let $(S, \preccurlyeq)$ be a partial order in which every element has an immediate successor. Prove that it is not necessarily true that for any two elements $a \preccurlyeq b$ that any chain between a and b has finite length.

6. Draw the Hasse diagram of the partial order $\subseteq$ on $\mathcal{P}(\{1,2,3,4\})$.

7. Draw the Hasse diagram for the poset $(\{1,2,3,4,5,6\}, \leq)$.

8. Let $A = \{a, b, c, d, e, f, g\}$. Draw the Hasse diagram for the partial order $\preccurlyeq$ given as a subset of $A \times A$ as

$$\preccurlyeq = \{(a,a), (b,b), (c,c), (d,d), (e,e), (f,f), (g,g), (a,c),$$
$$(b,c), (d,g), (a,e), (b,e), (c,e), (d,h), (g,h)\}.$$

9. A person's blood type is usually listed as one of the eight elements in the set

$$B' = \{\mathsf{o+}, \mathsf{o-}, \mathsf{a+}, \mathsf{a-}, \mathsf{b+}, \mathsf{b-}, \mathsf{ab+}, \mathsf{ab-}\}.$$

   We define the donor relation $\to$ on B' as follows. The relation $t_1 \to t_2$ holds if the letter portion of the blood type donates according to Examples 6.1.14 and 6.2.6, and if someone with a $+$ designation can only give to someone else with $+$, while someone with $-$ can give to anybody.

(a) Draw the Hasse diagram for $(B', \rightarrow)$.

(b) Show that the $(B', \rightarrow)$ poset does not have the lexicographic order on $B \times \{+, -\}$.

10. Draw the Hasse diagram of the partial order of divisibility $|$ on the set $S = \{1, 2, \ldots, 12\}$.

11. Consider the two posets $(\{1, 2, 3\}, \leq)$ and $(\{1, 2, 3, 4\}, |)$. Draw the Hasse diagram of the lexicographic order on $\{1, 2, 3\} \times \{1, 2, 3, 4\}$.

12. Consider the set of triples of integers $\mathbb{Z}^3$. Define the relation $\preccurlyeq$ on $\mathbb{Z}^3$ by

$$(a_1, a_2, a_3) \preccurlyeq (b_1, b_2, b_3) \iff$$

$$\begin{cases} a_1 + a_2 + a_3 < b_1 + b_2 + b_3 & \text{if } a_1 + a_2 + a_3 \neq b_1 + b_2 + b_3; \\ a_1 + a_2 + a_3 \preccurlyeq_{\text{lex}} b_1 + b_2 + b_3 & \text{if } a_1 + a_2 + a_3 = b_1 + b_2 + b_3, \end{cases}$$

where $\preccurlyeq_{\text{lex}}$ is the lexicographic order on $\mathbb{Z}^3$ (with each copy of $\mathbb{Z}$ equipped with the partial order $\leq$). Prove that $\preccurlyeq$ is a partial order on $\mathbb{Z}^3$. Prove also that $\preccurlyeq$ is a total order.

13. Let $(A_i, \preccurlyeq_i)$ be posets for $i = 1, 2, \ldots, n$ and define $\preccurlyeq_{\text{lex}}$ as the lexicographic order on $A_1 \times A_2 \times \cdots A_n$. Prove that $\preccurlyeq_{\text{lex}}$ is a total order if and only if $\preccurlyeq_i$ is a total order on A_i for all i.

14. Let $\preccurlyeq$ be the lexicographic order on $\mathbb{R}^3$, where each $\mathbb{R}$ is equipped with the usual $\leq$. Prove or disprove the following statement: For all vectors $\mathbf{a}, \mathbf{b}, \mathbf{c}, \mathbf{d}$, if $\mathbf{a} \preccurlyeq \mathbf{b}$ and $\mathbf{c} \preccurlyeq \mathbf{d}$, then $\mathbf{a} + \mathbf{c} \preccurlyeq \mathbf{b} + \mathbf{d}$.

15. Answer the following questions pertaining to the poset described by the Hasse diagram below.

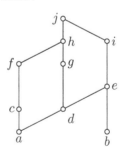

(a) List all the minimal elements.

(b) List all the maximal elements.

(c) List all the maximal elements in the subposet $\{a, b, c, d, e, f, g\}$.

(d) Determine the length of the longest chain and find all chains of that length.

(e) Find the least upper bound of $\{a, b\}$, if it exists.

(f) Find the greatest lower bound of $\{b, c\}$, if it exists.

(g) List all the upper bounds of $\{f, d\}$.

16. Consider the partial order on $\mathbb{R}^2$ given in Example 6.2.7. Let A be the unit disk

$$A = \{(x, y) \in \mathbb{R}^2 \mid x^2 + y^2 \leq 1\}.$$

   (a) Show that A has both a maximal and minimal element. Find all of them.

   (b) Find all the upper bounds and all the lower bounds of A.

17. Consider the lexicographic order on $\mathbb{R}^2$ coming from the standard $(\mathbb{R}, \leq)$. Let A be the closed disk of center $(1, 2)$ and radius 5.

   (a) Show that A has both a maximal and minimal element. Find all of them.

   (b) Find all the upper bounds and all the lower bounds of A.

   (c) Show that A has both a least upper bound and a greatest lower bound.

18. Prove that for all sets S, the poset $(\mathcal{P}(S), \subseteq)$ is a lattice. Also show that if X is any subset of $\mathcal{P}(S)$, then $\bigcup_{A \in X} A$ is $\mathrm{lub}(X)$.

19. There are 5 distinct lattices for sets of 5 elements. Find all of them and sketch the Hasse diagram for each one.

20. Prove that in a finite lattice, there exists exactly one maximal element and one minimal element.

21. Determine whether the posets corresponding to the following Hasse diagrams are lattices. If they are not, explain why.

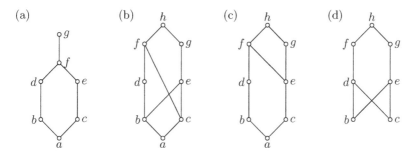

22. Let R_1 and R_2 be partial orders on a set S.

   (a) Prove that $R_1 \cap R_2$ is a partial order.

   (b) Show by a counterexample that $R_1 \cup R_2$ is not necessarily a partial order.

The remaining exercises involve the concept of monotonic functions. Let $(S, \preccurlyeq)$ and $(T, \preccurlyeq')$ be two posets. A function $f : S \to T$ is called a monotonic if $s_1 \preccurlyeq s_2 \implies f(s_1) \preccurlyeq' f(s_2)$.

23. Let S be a finite set. Consider the function $f : \mathcal{P}(S) \to \mathbb{N}$ defined by $f(A) = |A|$. Show that this is a monotonic function from $(\mathcal{P}(S), \subseteq)$ to $(\mathbb{N}, \leq)$.

24. Let $(S, \preccurlyeq)$, $(T, \preccurlyeq')$, and $(U, \preccurlyeq'')$ be three posets. Let $f : S \to T$ and $g : T \to U$ be monotonic functions. Prove that the composition $g \circ f : S \to U$ is monotonic.

25. Let $(S, \preccurlyeq_1)$ and $(T, \preccurlyeq_2)$ be two partially ordered sets and let $f : S \to T$ be a monotonic function.

(a) Prove that if A is a subset of S with an upper bound u, then $f(u)$ is an upper bound of $f(A)$.

(b) Prove with a counterexample that $f(\text{lub}(A))$ is not necessarily $\text{lub}(f(A))$.

6.3   Equivalence Relations

Just as partial orders generalize the concept of inequality, equivalence relations generalize the concept of equality. Intuitively speaking, an equivalence relation models the common mental process of viewing things as the same along some perspective. Consequently, equivalence relations are ubiquitous in mathematics.

6.3.1   Definition and examples

Definition 6.3.1

An *equivalence relation* on a set S is a relation $\sim$ that is reflexive, symmetric, and transitive.

Example 6.3.2. Let S be any set. The equal relation is reflexive, symmetric, antisymmetric, and transitive. In particular, $=$ is an equivalence relation. Two elements are in relation via $=$ if and only if they are the same object. $\triangle$

Example 6.3.3. Let $\mathcal{L}$ be the set of lines in $\mathbb{R}^2$. Consider the relation of parallelism $\parallel$ on $\mathcal{L}$. Any line is parallel to itself so $\parallel$ is reflexive. If $L_1, L_2 \in \mathcal{L}$ with $L_1 \parallel L_2$, then $L_2 \parallel L_1$. Hence, $\parallel$ is symmetric. Finally, by Proposition I.30 in Euclid's *Elements*, for any lines $L_1, L_2, L_3 \in \mathcal{L}$, if $L_1 \parallel L_2$ and $L_2 \parallel L_3$, then $L_1 \parallel L_3$. This means that $\parallel$ is transitive. Thus $\parallel$ is an equivalence relation on $\mathcal{L}$. $\triangle$

Example 6.3.4. Define C as the set of intersections in New York City and define the relation R on C to be "within walking distance." As stated, this is not precise. Let us say that two intersections in NYC are within walking distance if and only if they are two miles or fewer apart. This relation is reflexive and symmetric but not transitive. If three intersections a, b, and c lie successively in a straight line with a and b two miles apart and b and c also two miles apart, then a and c are four miles apart. Thus R is not an equivalence relation. $\triangle$

Example 6.3.5. Let $X = \{1, 2, 3, \ldots, 10\}$ and consider $S = \mathcal{P}(X)$. Define the relation $\sim_1$ on S by $A \sim_1 B$ if $|A| = |B|$. It is easy to notice that this is an equivalence relation. Another natural equivalence relation on the same set S is $A \sim_2 B$ if the sum of elements in A and B is the same. △

6.3.2 Equivalence classes

Since an equivalence relation furnishes some notion of sameness, it is natural to gather similar elements into classes. Such classes formalize the sameness property.

Definition 6.3.6

Let $\sim$ be an equivalence relation on a set S. For $a \in S$, the *equivalence class of a* is

$$[a] \stackrel{\text{def}}{=} \{s \in S \mid s \sim a\}.$$

A subset $U \subseteq S$ is called an *equivalence class* if $U = [a]$ for some $a \in S$. An element a of an equivalence class U is called a *representative* of U.

We sometimes write $[a]_\sim$ to clarify if a certain context considers more than one equivalence relation at a time.

Example 6.3.7. Let $n \geq 2$ be an integer. Proposition 4.6.2 showed that congruence modulo n is an equivalence relation on $\mathbb{Z}$. Equivalence classes for the congruence relation are called *congruence classes* modulo n. In number theory, we always write $\bar{a}$ (instead of $[a]_\equiv$) for the congruence class of the integer. For example, modulo 5, the congruence class of 2 is

$$\bar{2} = \{\ldots, -8, -3, 2, 7, 12, \ldots\} = \{2 + 5k \mid k \in \mathbb{Z}\}.$$

In this context, we see that $\bar{a} = \bar{b}$ if and only if $a \equiv b \pmod{n}$. So, congruence becomes the notion of equality when we consider the congruence classes. △

Example 6.3.8. Consider the set $\mathcal{T}$ of triangles in the Euclidean plane $\mathbb{E}^2$. Many theorems in Euclidean geometry deal with congruent triangles. Two triangles $\triangle ABC$ and $\triangle A'B'C'$ are congruent, and we write $\triangle ABC \cong \triangle A'B'C'$ if $\angle A = \angle A'$, $\angle B = \angle B'$, $\angle C = \angle C'$, $AB = A'B'$, $BC = B'C'$ and $CA = C'A'$. From the equivalence of equality, it is clear that $\cong$ is an equivalence relation on $\mathcal{T}$. Any equivalence class of $\cong$ consists of all triangles that are congruent to each other. △

Example 6.3.9. Consider the set $M_n(\mathbb{R})$ of $n \times n$ matrices with real coefficients. For two matrices $A, B \in M_n(\mathbb{R})$, we say that B is similar to A if there exists an invertible $n \times n$ matrix S such that $B = SAS^{-1}$. We leave it as an exercise for the reader to show that similarity is an equivalence relation on $M_n(A)$. The similarity class of a given matrix A consists of all matrices that are similar to A. $\triangle$

Definition 6.3.10

We call a subset T of S a *complete set of distinct representatives* of $\sim$ if any equivalence class U has $U = [a]$ for some $a \in T$ and for any two $a_1, a_2 \in T$, $[a_1] = [a_2]$ implies that $a_1 = a_2$.

The Axiom of Choice implies that a complete set of distinct representatives exists for any equivalence relation. However, with many specific equivalence relations, we do not need to cite such deep set theory to find a complete set of distinct representatives. For example, with congruence modulo n, we regularly use the remainders when we divide by n, namely $\{0, 1, \ldots, n-1\}$. With the relation of parallelism on the set of lines in the Euclidean plane (Example 6.3.3), we can think of using all the lines through the origin, or any fixed point for that matter, as a complete set of distinct representatives.

6.3.3   Partitions

Let $\sim$ be an equivalence relation on a set S. For any two elements $a, b \in S$, by definition $a \in [b]$ if and only if $a \sim b$. However, since $\sim$ is symmetric, this implies that $b \sim a$ and hence that $b \in [a]$. By transitivity, if $a \in [b]$ then $s \sim a$ implies that $s \sim b$, so $a \in [b]$ implies that $[a] \subseteq [b]$. Consequently, we have proven that the following statements are logically equivalent:

$$a \in [b] \iff [a] \subseteq [b] \iff b \in [a] \iff [b] \subseteq [a] \iff [a] = [b].$$

This observation leads to the following important proposition.

Proposition 6.3.11

Let S be a set equipped with an equivalence relation $\sim$. Then

1) distinct equivalence classes are disjoint;

2) the union of distinct equivalence classes is all of S.

Proof. Suppose that $[a] \cap [b] \neq \emptyset$. Then there exists $c \in [a] \cap [b]$, so $c \in [a]$ and $c \in [b]$. Hence, $[a] = [c] = [b]$. Hence, if two equivalence classes overlap, then they are equal.

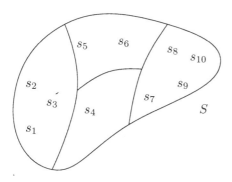

Figure 6.4: A partition of a set.

Let T be a complete set of representatives of $\sim$ in S. Obviously, since $a \in [a]$, every element of S is in some equivalence class. Thus, we have

$$S = \bigcup_{a \in S} [a] = \bigcup_{a \in T} [a]$$

and the result follows.   □

The property described in Proposition 6.3.11 has a particular name in set theory.

Definition 6.3.12

Let S be a set. A collection $\mathcal{A} = \{A_i\}_{i \in \mathcal{I}}$ of subsets of S is called a *partition* of S if

(1) $A_i \cap A_j \neq \emptyset \implies i = j$; and   (2) $\bigcup_{i \in \mathcal{I}} A_i = S$.

Partitions of sets may be visualized by a diagram akin to Figure 6.4. In this figure, S is a set with ten elements and the sets of the partition are $\{s_1, s_2, s_3\}$, $\{s_4\}$, $\{s_5, s_6\}$, and $\{s_7, s_8, s_9, s_{10}\}$.

The concept of a partition simply models the mental construction of subdividing a set into parts without losing any elements of the set and without any parts overlapping. Partitions and equivalence relations are closely connected. Proposition 6.3.11 establishes that the set of distinct equivalence classes of an equivalence relation on S forms a partition of S. The following proposition establishes the converse.

Proposition 6.3.13

Let $\mathcal{A} = \{A_i\}_{i \in \mathcal{I}}$ be a partition of a set S. Define the relation $\sim_{\mathcal{A}}$ on S by

$$a \sim_{\mathcal{A}} b \implies \exists i \in \mathcal{I} \text{ with } a \in A_i \text{ and } b \in A_i.$$

Then $\sim_{\mathcal{A}}$ is an equivalence relation. Furthermore, the sets in $\mathcal{A}$ are the distinct equivalence classes of $\sim_{\mathcal{A}}$.

Proof. Let $a \in S$ be arbitrary. Since $\mathcal{A}$ is a partition, then $a \in A_i$ for some $i \in \mathcal{I}$. Hence, $a \sim_{\mathcal{A}} a$ and $\sim_{\mathcal{A}}$ is reflexive.

Suppose that $a \sim_{\mathcal{A}} b$. Then for some $i \in \mathcal{I}$, we have $a \in A_i$ and $b \in A_i$. Obviously, this implies that $b \sim_{\mathcal{A}} a$, showing that $\sim_{\mathcal{A}}$ is symmetric.

Suppose that $a \sim_{\mathcal{A}} b$ and $b \sim_{\mathcal{A}} c$. Then for some $i \in \mathcal{I}$, we have $a \in A_i$ and $b \in A_i$ and for some $j \in \mathcal{I}$, we have $b \in A_j$ and $c \in A_j$. However, since $b \in A_i \cap A_j$, then $A_i \cap A_j \neq \emptyset$. By definition of a partition, $i = j$ and so $a \in A_i$ and $c \in A_i$ and thus $a \sim_{\mathcal{A}} c$. This shows transitivity and establishes that $\sim_{\mathcal{A}}$ is an equivalence relation.

Let A_i be a set in $\mathcal{A}$ and let s be any element in A_i. By construction, $[s] = A_i$ and so the elements of $\mathcal{A}$ are precisely the equivalence classes of $\sim_{\mathcal{A}}$. $\square$

Example 6.3.14. Consider the set $S = \{1,2,3,4,5,6,7,8\}$. The following collections of subsets of S are partitions on S:

$$\{\{1,4,5\},\{2,6,7\},\{3,8\}\} \text{ and } \{\{1,3,5,7\},\{2,4,6,8\}\}.$$

However, $\{\{1,3,5\},\{2,8\},\{4,7\}\}$ is not a partition because the union of the subsets does not contain 6. On the other hand,

$$\{\{1,2,3,5\},\{3,6,8\},\{4,6,7\}\}$$

is not a partition because some of the subsets have nonempty intersections, namely $\{1,2,3,5\} \cap \{3,6,8\} = \{3\}$ and $\{3,6,8\} \cap \{4,6,7\} = \{6\}$. $\triangle$

Example 6.3.15. Revisiting Example 6.3.7, we notice that modulo 5, the set of congruence classes is $\{\overline{0},\overline{1},\overline{2},\overline{3},\overline{4}\}$. As subsets of $\mathbb{Z}$, the intersection of congruence classes is empty and the union of all of them is $\mathbb{Z}$. So congruence classes form a partition of $\mathbb{Z}$. $\triangle$

Example 6.3.16. Consider the unit sphere in $\mathbb{R}^3$, denoted by $\mathbb{S}^2$. Consider the partition on $\mathbb{S}^2$ given by

$$\mathcal{A} = \{\{p,-p\} \mid p \in \mathbb{S}^2\}.$$

The partition $\mathcal{A}$ consists of pairs of points that are diametrically opposite each other. According to Proposition 6.3.13, there exists a unique equivalence relation $\sim_1$ on $\mathbb{S}^2$ that has $\mathcal{A}$ as a set of distinct equivalence classes. $\triangle$

EXERCISES FOR SECTION 6.3

1. Let $S = \mathbb{Z} \times \mathbb{Z}$ and let R be the relation on S defined by $(a, b) R (c, d)$ means that $a + d = b + c$. Show that R is an equivalence relation. Concisely describe the equivalence classes of R.

2. Let C be the set of people in your math class. Describe a "natural" relation satisfying each of the combination of properties listed below.

   (a) Reflexive and symmetric, but not transitive.

   (b) Reflexive and transitive, but not symmetric.

   (c) Symmetric and transitive, but not reflexive.

   (d) An equivalence relation.

For Exercises 6.3.3 through 6.3.11, prove or disprove whether the described relation is an equivalence relation. If the relation is not an equivalence relation, determine which properties it lacks.

3. Let P be the set of living people. For all $a, b \in P$, define the relation $a \, R \, b$ if a and b have met.

4. Let P be the set of living people. For all $a, b \in P$, define the relation $a \, R \, b$ if a and b live in a common town.

5. Let $\mathcal{C}$ be the set of circles in $\mathbb{R}^2$ and let R be the relation of concentric on $\mathcal{C}$.

6. Let $S = \mathbb{Z} \times \mathbb{Z}$ and define the relation R on S by $(m_1, m_2) \, R \, (n_1, n_2)$ if $m_1 m_2 = n_1 n_2$.

7. Let $S = \mathbb{Z} \times \mathbb{Z}$ and define the relation R on S by $(m_1, m_2) \, R \, (n_1, n_2)$ if $m_1 n_1 = m_2 n_2$.

8. Let $S = \mathbb{Z} \times \mathbb{Z}$ and define the relation R on S by $(m_1, m_2) \, R \, (n_1, n_2)$ if $m_1 n_2 = m_2 n_1$.

9. Consider the set $C^0(\mathbb{R})$ of continuous functions over $\mathbb{R}$. Define the relation R on $C^0(\mathbb{R})$ by $f \, R \, g$ if there exist some $a, b \in \mathbb{R}$ such that

$$g(x) = f(x + a) + b \qquad \text{for all } x \in \mathbb{R}.$$

10. Let $\ell^\infty(\mathbb{R})$ be the set of sequences of real numbers. Define the relation R on $\ell^\infty(\mathbb{R})$ by $(a_n) \, R \, (b_n)$ if

$$\lim_{n \to \infty} (b_n - a_n) = 0.$$

11. Let P_3 be the set of polynomials with real coefficients and of degree 3 or less. Define the relation R on P_3 by $p(x) \, R \, q(x)$ to mean that $q(x) - p(x)$ has 5 as a root.

12. Let R be any relation. Show that $R^{-1} \circ R$ is reflexive and symmetric but that it is not necessarily transitive. [Hint: See Example 6.1.11.]

13. Let $C^0([0,1])$ be the set of continuous real-valued functions on $[0,1]$. Define the relation $\sim$ on $C^0([0,1])$ by

$$f \sim g \quad \Longleftrightarrow \quad \int_0^1 f(x)\,dx = \int_0^1 g(x)\,dx.$$

Show that $\sim$ is an equivalence relation and describe (with a precise rule) a complete set of distinct representatives of $\sim$.

14. Let $C^1([a,b])$ be the set of continuously differentiable functions on the interval $[a,b]$. Define the relation $\sim$ on $C^1([a,b])$ as $f \sim g$ if and only if $f'(x) = g'(x)$ for all $x \in (a,b)$. Prove that $\sim$ is an equivalence relation on $C^1([a,b])$. Describe the elements in the equivalence class for a given $f \in C^1([a,b])$.

15. Let $C^\infty(\mathbb{R})$ be the set of all real-value functions on $\mathbb{R}$ such that all its derivatives exist and are continuous. Define the relation R on $C^\infty(\mathbb{R})$ by $f\,R\,g$ if $f^{(n)}(0) = g^{(n)}(0)$ for all positive, even integers n.

    (a) Prove that R is an equivalence relation.

    (b) Describe concisely all the elements in the equivalence class $[\sin x]$.

16. Let $S = \{1,2,3,4\}$ and the relation $\sim$ on $\mathcal{P}(S)$, defined by $A \sim B$ if and only if the sum of elements in A is equal to the sum of elements in B, is an equivalence relation. List the equivalence classes of $\sim$.

17. Let $\mathcal{T}$ be the set of (nondegenerate) triangles in the plane.

    (a) Prove that the relation $\sim$ of similarity on triangles in $\mathcal{T}$ is an equivalence relation.

    (b) Concisely describe a complete set of distinct representatives of $\sim$.

18. Let $M_n(\mathbb{R})$ be the set of $n \times n$ matrices with real coefficients. For two matrices $A, B \in M_n(\mathbb{R})$, we say that B is similar to A if there exists an invertible $n \times n$ matrix S such that $B = SAS^{-1}$.

    (a) Prove that similarity $\sim$ is an equivalence relation on $M_{n \times n}(\mathbb{R})$.

    (b) Prove that if A is a constant multiple of the identity matrix, i.e., $A = cI$ for some $c \in R$, then the similarity class $[A]$ is just $\{A\}$, the set consisting of only the matrix A.

19. Let V be a vector space and let W be a subspace. Define the relation $\mathbf{v}_1 \sim_W \mathbf{v}_2$ to mean $\mathbf{v}_2 - \mathbf{v}_1 \in W$. Show that $\sim_W$ is an equivalence relation. Show also that $W = [\mathbf{0}]$.

20. Let R_1 and R_2 be equivalence relations on a set S. Determine (with a proof or counterexample) which of the following relations are also equivalence relations on S. (a) $R_1 \cap R_2$; (b) $R_1 \cup R_2$; (c) $R_1 \triangle R_2$. [Note that $R_1 \cup R_2$, and similarly for the others, is a relation as a subset of $S \times S$.]

21. Let $S = \{1, 2, 3, 4, 5, 6\}$. For the partitions of S given below, write out the equivalence relation as a subset of $S \times S$.

    (a) $\{\{1, 2\}, \{3, 4\}, \{5, 6\}\}$ (c) $\{\{1, 2\}, \{3\}, \{4, 5\}, \{6\}\}$
    (b) $\{\{1\}, \{2\}, \{3, 4, 5, 6\}\}$

22. Which of the following collections of subsets of the integers form partitions? If it is not a partition, explain which properties fail.

    (a) $\{p\mathbb{Z} \mid p \text{ is prime}\}$ (c) $\{\{k \mid n^2 \leq k \leq (n + 1)^2\} \mid n \in \mathbb{N}\}$
    (b) $\{\{3n, 3n+1, 3n+2\} \mid n \in \mathbb{Z}\}$ (d) $\{\{n, -n\} \mid n \in \mathbb{N}\}$

23. Let S be a set. Prove that there is a bijection between the set of partitions of S and the set of equivalence classes on S.

24. Let S be a set and let $\mathcal{A} = \{A_i\}_{i \in \mathcal{I}}$ be a partition of S. Another partition $\mathcal{B} = \{B_j\}_{j \in \mathcal{J}}$ is called a *refinement* of $\mathcal{A}$ if

$$\forall j \in \mathcal{J}, \ \exists i \in \mathcal{I}, \quad B_j \subseteq A_i.$$

Let $\mathcal{A}$ and $\mathcal{B}$ be two partitions of a set S and let $\sim_{\mathcal{A}}$ (resp. $\sim_{\mathcal{B}}$) as the equivalence relation corresponding to $\mathcal{A}$ (resp. $\mathcal{B}$). Prove that $\mathcal{B}$ is a refinement of $\mathcal{A}$ if and only if $s_1 \sim_{\mathcal{B}} s_2 \implies s_1 \sim_{\mathcal{A}} s_2$.

25. Let S be a set. If we write $\mathcal{B} \preccurlyeq \mathcal{A}$ to mean that $\mathcal{B}$ is a refinement of $\mathcal{A}$. Show that $\preccurlyeq$ is a partial order on the set of partitions of S. [See the previous exercise.]

6.4   Quotient Sets

We conclude this chapter and the first part of this book with a section on quotient sets. As abstract as it may seem at first pass, students have secretly worked with quotient sets since they started learning properties of fractions, and even earlier. Because these arise so often in mathematics, we introduce them here. Furthermore, quotient sets precisely define the structure of all our common number sets, so we conclude the section with a precise theoretical construction of $\mathbb{Z}$, $\mathbb{Q}$, $\mathbb{R}$ and $\mathbb{C}$.

6.4.1   Definition and examples

Exercise 6.3.23 summarized two key propositions in the previous section. They established that the equivalence classes of an equivalence relation $\sim$ on a set S form a partition of S, and conversely that to any partition of S there corresponds an equivalence relation with that partition for equivalence classes. As straightforward as this seems, considering the set of equivalence classes is so common in mathematics that it has a name.

Definition 6.4.1

Let S be a set and $\sim$ an equivalence relation on S. The set of $\sim$-equivalence classes on S is called the *quotient set of S by $\sim$* and is denoted by $S/\sim$.

We remark that there is a bijection between $S/\sim$ and a complete set of representatives T of $\sim$ via the function

$$\psi : T \to S/\sim$$
$$a \mapsto [a].$$

However, we do not consider these sets as equal since their objects are different.

Example 6.4.2 (Rational Numbers). Consider the set of pairs $S = \mathbb{Z} \times \mathbb{Z}^*$ and the relation $\sim$ defined by

$$(a,b) \sim (c,d) \iff ad = bc.$$

We leave it to the reader to show that this is an equivalence relation. (See Exercise 6.4.1.) The quotient set $S/\sim$ is a rigorous definition for $\mathbb{Q}$. The equivalence relation is precisely the condition that is given when two fractions are considered equal. Hence, the fraction notation $\frac{a}{b}$ for rational numbers represents the equivalence class $[(a,b)]_\sim$. $\triangle$

Example 6.4.3 (Modular Arithmetic). Fix an integer $n \geq 2$. In Example 6.3.7, we pointed out that congruence modulo n is an equivalence relation. Instead of writing $\mathbb{Z}/\equiv$ for the associated quotient set, we write $\mathbb{Z}/n\mathbb{Z}$. This latter notation benefits from providing the modulus. In particular, $\mathbb{Z}/n\mathbb{Z} = \{\overline{0}, \overline{1}, \ldots, \overline{n-1}\}$. Proposition 4.6.5 shows that

$$\overline{a} + \overline{b} \stackrel{\text{def}}{=} \overline{a+b} \quad \text{and} \quad \overline{a} \cdot \overline{b} \stackrel{\text{def}}{=} \overline{a \cdot b}$$

give well-defined operations of $+$ (addition) and $\cdot$ (multiplication) on $\mathbb{Z}/n\mathbb{Z}$. Consequently, we can recast all of the modular arithmetic introduced in Section 4.6 as an arithmetic on elements of $\mathbb{Z}/n\mathbb{Z}$. $\triangle$

Example 6.4.4 (Projective Space). Fix a positive integer n. Let $\mathcal{L}(\mathbb{R}^{n+1})$ be the set of lines in $\mathbb{R}^{n+1}$ and consider the equivalence relation of parallelism $\parallel$ on $\mathcal{L}(\mathbb{R}^{n+1})$. If L is a line in $\mathcal{L}(\mathbb{R}^{n+1})$, then $[L]$ is the set of all lines parallel to L. The quotient set $\mathcal{L}(\mathbb{R}^{n+1})/\parallel$ is called the *real projective space* and is denoted as $\mathbb{P}^n$ or $\mathbb{P}^n_{\mathbb{R}}$. It consists of all the directions lines can possess.

There are other ways to understand the projective space. Every line in $\mathbb{R}^{n+1}$ is parallel to a unique line through the origin. Possible direction vectors for lines through the origin consist simply of nonzero vectors, so we consider the set $\mathbb{R}^{n+1} \setminus \{(0, \cdots, 0)\}$. Lines through the origin are the same if and only if their given direction vector differs by a nonzero multiple. Hence, we define the equivalence relation $\sim$ on $\mathbb{R}^{n+1} \setminus \{(0, \cdots, 0)\}$ by

$$\mathbf{a} \sim \mathbf{b} \iff \mathbf{a} = \lambda \mathbf{b}, \text{ for some } \lambda \in \mathbb{R} \setminus \{0\}.$$

Our comments on direction vectors show that as sets

$$(\mathbb{R}^{n+1} \setminus \{(0, \cdots, 0)\}) / \sim = \mathbb{P}^n.$$

Instead of using the generic set theory notation of $[(a_0, a_1, \ldots, a_n)]$ for elements of $\mathbb{P}^n$, we write $(a_0 : a_1 : \cdots : a_n)$. These are called *homogeneous coordinates*. (The notation $(a_0, a_1, \ldots, a_n)$ for an element of $\mathbb{R}^{n+1}$ is common practice when constructing projective spaces.) In many situations, homogeneous coordinates are more natural than Cartesian coordinates in $\mathbb{R}^n$. For example, consider the set of equations of degree 2 or less given by $ax^2 + bx + c = 0$. We might say it is determined by the triple of coefficients (a, b, c). However, if $\lambda \neq 0$, then $(\lambda a)x^2 + (\lambda b)x + (\lambda c) = 0$ corresponds to the same solution set. Furthermore, it would be natural to discard $0 = 0$ as a meaningful equation. The set of equations of degree 2 or less, naturally can be parametrized by $(a : b : c) \in \mathbb{P}^2$.

Consider briefly the simple example of homogeneous coordinates with $n = 1$, which give rise to the projective line $\mathbb{P}^1$. If $x \neq 0$, then $(x : y) = (1 : y/x)$, by multiplying by $\lambda = 1/x$. The origin of homogeneous coordinates should evoke the concept of rise over run as (run : rise) $= (1 : \text{rise/run})$. The points $(x : y)$ with $x \neq 0$ account for all but one point in $\mathbb{P}^1$. The one remaining point is $(0 : y)$ with $y \neq 0$. By multiplying by $\lambda = 1/y$ we have $(0 : y) = (0 : 1)$. This point in $\mathbb{P}^1$ corresponds to vertical lines, when the run is 0. The fraction $1/0$ does not exist but $(0 : 1)$ is a completely normal point in $\mathbb{P}^1$. $\triangle$

Remark 6.4.5. When working with functions whose domains are quotient sets, it is natural to wish to define a function of an equivalence class based on a representative of the class. More precisely, if S and T are sets and $\sim$ is an equivalence relation on S, we may wish to define a function

$$\begin{aligned} F : S/\sim &\longrightarrow T \\ [a] &\longmapsto f(a) \end{aligned} \tag{6.1}$$

where $f : S \to T$ is some function. This construction does not always produce a function. We say that a function defined according to (6.1) is *well-defined* if whenever $a \sim b$ then $f(a) = f(b)$. Thus, using any representative of the equivalence of $[a]$ will return the same value for $F([a])$. $\triangle$

As an example to illustrate the above remark, consider the equivalence relation $\sim$ on $\mathbb{Z} \times \mathbb{Z}^*$ described in Example 6.4.2. We use the typical notation of a/b for the equivalence classes. Suppose that we attempted to define a function $f : \mathbb{Q} \to \mathbb{Z}$ by $f(a/b) = a + b$. This is not a well-defined function (and hence not a function at all) because for example $\frac{1}{2} = \frac{3}{6}$, whereas $1 + 2 = 3$ and $3 + 6 = 9$. Depending on the representative chosen for the fraction a/b, the value of $a + b$ may differ. We introduced this issue at the end of Section 3.1.3, but the most general context in which to concern ourselves with a function being well defined comes from functions whose domains are quotient sets.

6.4.2   Definition of standard number sets

With quotient sets at our disposal, we are in a position to define all our number sets structurally.

We begin with whole numbers, namely $\mathbb{N}$. Peano's axioms of the integers provide a structural definition. We remind the reader, that Peano's axioms simply rely on set theory, the concept of a successor element, and an axiom making the integers a minimal set with the other three axioms.

Construction of $\mathbb{Z}$

If we only used numbers to count things, then why do we bother with negative numbers? From a theoretical perspective, how do we go from $\mathbb{N}$ to $\mathbb{Z}$?

Note that $+$ is a binary operation on $\mathbb{N}$. When discussing Peano's axioms, we showed how to define this binary operation on $\mathbb{N}$ by referring only to the successor function and associativity of function composition. Now consider the relation $\sim_1$ on $\mathbb{N} \times \mathbb{N}$ defined by $(a, b) \sim_1 (c, d)$ if $a + d = b + c$. This is an equivalence relation. We observe that for all $n \in \mathbb{N}$, we have $(a, 0) \sim_1 (a + n, n)$ and $(0, a) \sim_1 (n, a + n)$. Furthermore, a complete set of distinct representatives of $\sim_1$ is

$$\{(0,0)\} \cup \{(a,0) \mid a \in \mathbb{N}^*\} \cup \{(0,a) \mid a \in \mathbb{N}^*\}.$$

It is not hard to show that defining an operation of addition on the quotient set $\mathbb{N} \times \mathbb{N}/\sim_1$ by

$$[(a, b)] + [(c, d)] \overset{\text{def}}{=} [(a + c, b + d)]$$

is well defined and that furthermore $[(a, 0)] + [(0, a)] = [(a, a)] = [(0, 0)]$. So the new elements $[(0, a)]$ serve the role of negative numbers. We *define* the set of integers $\mathbb{Z}$ as the quotient set $\mathbb{N} \times \mathbb{N}/\sim_1$.

Once we have $\mathbb{Z}$, from addition and multiplication on $\mathbb{N}$ as described in Section 4.1.1 using Peano's formula, we can show that $+$ and $\cdot$ extend to the familiar operations on $\mathbb{Z}$.

Numbers were also used for measurement, which are not always integers. Though some ancient civilizations used base representations of integers, it was a long time before people imagined negative powers and what we call digits right of the period. However, people felt comfortable with the concept of a ratio, which follows from the concept of product of integers. But how do we rigorously get from $\mathbb{Z}$ to $\mathbb{Q}$? Construction of $\mathbb{Q}$

We answered this question in Example 6.4.2. We considered the equivalence relation $\sim_2$ on $\mathbb{Z} \times \mathbb{Z}^*$ defined by $(a, b) \sim_2 (c, d)$ if $ad = bc$. We *define* the set of rationals $\mathbb{Q}$ as the quotient set $(\mathbb{Z} \times \mathbb{Z}^*)/\sim_2$. For elements $[(n, d)]_{\sim_2}$ in this set we call n the numerator and d the denominator of the fraction, which we denote by n/d, instead of $[(n, d)]_{\sim_2}$. Part of the education around fractions is to properly work with the $\sim_2$ relation.

The reader might notice the similarity in definition between $\sim_1$ and $\sim_2$. We use the word fraction for the $\sim_2$ equivalence class, so it makes sense there should be a word for the $\sim_1$ equivalence class. We call $[(a, b)]_{\sim_1}$ the *difference* of a by b.

There is a long history behind the discovery of irrational numbers. Some accounts claim that Pythagoras proved that $\sqrt{2}$ (if it existed) was irrational, but he demanded that his disciples keep this as a secret of sacred importance because it might lead to philosophical confusion. So how do we step outside the rationals and rigorously define the set of real numbers? Construction of $\mathbb{R}$

Unlike the definition of $\mathbb{Z}$ from $\mathbb{N}$, there are a few ways to define the set of real numbers $\mathbb{R}$. We present one method that builds on $\mathbb{Q}$. When we think of expressing real numbers, we sometimes reference decimal expansions and simply say that for real numbers we let the decimal digits go on indefinitely. Any finite decimal expansion represents a rational numbers. For example, 1.41 represents $141/100$ and 1.41421 represents $141421/100,000$. So we can think of decimal expansions with more and more digits of accuracy as a sequence of rational numbers. This inspires our construction using an approach called *completion*.

Definition 6.4.6

A *Cauchy sequence* of rational numbers is a sequence in $\mathbb{Q}$ such that for all $\varepsilon > 0$, there exists $N \in \mathbb{N}$ such that if $m, n > N$, then $|x_m - x_n| < \varepsilon$.

If a sequence is Cauchy, it is bounded: if $\varepsilon = 1$, and if N is an integer that satisfies the criterion of the above definition, then for all $n \geq N$, the definition implies that $x_N - \varepsilon < x_n < x_N + \varepsilon$. Furthermore, Cauchy sequences do not oscillate wildly like $a_n = \sin(1/n)$, which remain bounded but continue to oscillate between -1 and 1 no matter how high n goes. Even if a sequence is Cauchy, it does not necessarily converge. Key examples of this are decimal expansions of irrational numbers, e.g.,

$$1, \ 1.4, \ 1.41, \ 1.414, \ 1.4142\ldots$$

This sequence does not converge to a rational number.

Let S be the set of Cauchy sequences of $\mathbb{Q}$ and define the relation $\sim_3$ on S by

$$(a_n) \sim_3 (b_n) \iff \lim_{n \to \infty} a_n - b_n = 0.$$

This is an equivalence relation. (See Exercise 6.3.10.) The real numbers are *defined* as the quotient set $S/\sim_3$. This process is called completion because it "fills in the holes" of $\mathbb{Q}$ with respect to the usual way of measuring distance between rational numbers. The study of real numbers, their properties, and properties of functions on $\mathbb{R}$ forms a core part of an analysis course. The study of this completion process and generalizations to it form a regular theme in analysis.

Construction of $\mathbb{C}$

Even before the formal Quadratic Formula, scholars speculated about the meaning of square roots of negative numbers. For centuries, mathematicians enjoyed the luxury of simply saying that the formula did not apply if the quantity under the square root (discriminant) was negative. However, Cardano and Ferrari's formula for the solution to a cubic equation involved square roots of negative numbers, even if all the roots turned out to be real. Intuitively speaking, their formula steps out of the set of real numbers, into some other set, and comes back down into $\mathbb{R}$. Mathematicians could no longer ignore these other numbers (called complex numbers), even if the existence of square roots of negative numbers does not line up with our intuition of numbers used for measurement. So how do we obtain the set of complex numbers from $\mathbb{R}$?

Call $\mathbb{R}[x]$ the set of polynomials with coefficients in $\mathbb{R}$. Define on $\mathbb{R}[x]$ the relation $\sim_4$ given by

$$a(x) \sim_4 b(x) \iff x^2 + 1 \text{ divides } a(x) - b(x).$$

Notice that with the relation, $(x^3 + 2x^2 - 7) \sim_4 (-x - 9)$ because

$$(x^3 + 2x^2 - 7) - (-x - 9) = x^3 + 2x^2 + x + 2 = (x^2 + 1)(x + 2).$$

This relation is an equivalence relation. The set of complex numbers is *defined* as the quotient set $\mathbb{R}[x]/\sim_4$. In this quotient set the equivalence class of x serves the algebraic role of the imaginary number i because in $\mathbb{R}[x]/\sim_4$ we have $[x^2 + 1] = [0]$. After all, referring to the above example, $i^3 + 2i^2 - 7 = -i - 2 - 7 = -9 - i$. The generalization of the arithmetic of $\mathbb{C}$ follows from $\mathbb{R}$, and the arithmetic on $\mathbb{R}[x]$ is a theme in ring theory, a central topic in abstract algebra.

EXERCISES FOR SECTION 6.4

1. Show that the relation defined in Example 6.4.2 is an equivalence relation.

2. Prove that the association $f : \mathbb{Z}/n\mathbb{Z} \to \mathbb{N}$ defined by $f(\bar{a}) = \gcd(a, n)$ is a well-defined function.

3. In high school algebra, we learn that $-1/m$ is the slope of a line that is perpendicular to a line with slope m. In light of the fact that a slope is best expressed as an element of $\mathbb{RP}^1$, prove that $f : \mathbb{RP}^1 \to \mathbb{RP}^1$ expressed by $f(a : b) = (-b : a)$ describes a well-defined function.

4. Prove that $f : \mathbb{RP}^2 \to \mathbb{RP}^2$ expressed by

$$f(a : b : c) = (a + b + c : 3a + b + 4c : 2a + 7b + c)$$

describes a well-defined function.

5. Prove that $f : \mathbb{RP}^1 \to \mathbb{RP}^3$ expressed by $f(s : t) = (s^3 : s^2 t : st^2 : t^3)$ describes a well-defined function.

6. In this exercise, we guide a proof for Fermat's Little Theorem, Theorem 4.6.13. For any prime p, call $U(p) = \{\bar{a} \in \mathbb{Z}/p\mathbb{Z} \mid p \nmid a\}$ is the set of elements $\bar{a} \in \mathbb{Z}/p\mathbb{Z}$ such that a is a unit modulo n. Note that $|U(p)| = p - 1$. Let $\bar{a} \in U(p)$.

   (a) Prove that the sequence of powers $\bar{1}, \bar{a}, \bar{a}^2, \ldots$ is periodic of some period k. [Hint: Pigeonhole Principle.]

   (b) Define the relation $\sim$ on $U(p)$ by $\bar{b} \sim \bar{c}$ when $\bar{c} = \bar{b}\bar{a}^j$ for some integer j. Prove that $\sim$ is an equivalence relation.

   (c) Prove that all equivalence classes of $\sim$ contain exactly k elements.

   (d) Deduce that $k \mid (p-1)$ and conclude that for all integers a such that $p \nmid a$ we have $a^{p-1} \equiv 1 \pmod{p}$, thereby proving Fermat's Little Theorem.

7. Define the relation $\sim$ on $C^1(\mathbb{R})$ by $f \sim g$ when $g(x) = f(x) + c$ for some $c \in \mathbb{R}$.

   (a) Show that $\sim$ is an equivalence relation on $C^1(\mathbb{R})$.

   (b) Show that the association $D : (C^1(\mathbb{R})/\sim) \to C^0(\mathbb{R})$ defined by $D([f]) = f'$ is a well-defined function.

8. Define $\sim$ as the relation of similarity on $M_n(\mathbb{R})$, the set of $n \times n$ real-valued matrices. (See Exercise 6.3.18.) Prove that $f : (M_n(\mathbb{R})/\sim) \to \mathbb{R}$ defined by $f([A]) = \det(A)$ is a well-defined function.

9. Let V be a real vector space and let W be a subspace. Consider the equivalence relation described in Exercise 6.3.19. We denote by V/W the quotient set $V/\sim_W$. Show that the operations on V/W with $c \in \mathbb{R}$

$$[\mathbf{u}] + [\mathbf{v}] \stackrel{\text{def}}{=} [\mathbf{u} + \mathbf{v}] \quad \text{and} \quad c \cdot [\mathbf{v}] \stackrel{\text{def}}{=} [c\mathbf{v}]$$

are well defined. Show also that these operations make V/W into a vector space.

10. Refer to the previous exercise. Let V be a finite dimensional vector space and let W be a subspace. Prove that $\dim V/W = \dim V - \dim W$. [Hint: Given a basis $\mathcal{B}'$ of W, show that there exists a set of vectors $\mathcal{B} \subseteq V$ with $\mathcal{B}' \subseteq \mathcal{B}$ and $\mathcal{B}$ is a basis of V. Then show that

$$\{[\mathbf{v}] \mid \mathbf{v} \in \mathcal{B} \setminus \mathcal{B}'\}$$

is a basis of V/W.]

The remaining exercises pertain to the construction of number sets described in Section 6.4.2.

11. In the paragraph on the construction of $\mathbb{Z}$, prove the claim that $\sim_1$ is an equivalence relation on $\mathbb{N} \times \mathbb{N}$.

12. Let n_0 be any integer in $\mathbb{N}$ and let $S = \{n_0, n_0 + 1, n_0 + 2, \ldots\}$. Prove that the quotient set $S \times S/\sim_1$, where $\sim_1$ is defined in Section 6.4.2, is again $\mathbb{Z}$. [Hint: Find a bijection between the quotient sets.] Explain why it is irrelevant to the construction of $\mathbb{Z}$ whether we start from $\mathbb{N} = \{0, 1, 2, 3 \ldots\}$ or $\mathbb{N}^* = \{1, 2, 3, \ldots\}$.

13. Consider the equivalence relation $\sim_1$ defined on $\mathbb{N} \times \mathbb{N}$. Prove that the operation

$$[(a,b)] \cdot [(c,d)] = [ac + bd, ad + bc]$$

is well defined on $\mathbb{N} \times \mathbb{N}/\sim_1$. Then show that this operation is associative, commutative, and has $[(1,0)]$ as an identity. [Note: This defines multiplication on $\mathbb{Z}$ as the quotient set $\mathbb{N} \times \mathbb{N}/\sim_1$.]

14. Consider the relation $\sim_4$ on $\mathbb{R}[x]$. Prove that $\sim_4$ is an equivalence relation on $R[x]$.

15. Consider the equivalence relation $\sim_4$ on $\mathbb{R}[x]$. Suppose that $a(x) \sim_4 b(x)$ and $p(x) \sim_4 q(x)$.

    (a) Prove that for every polynomial $p(x) \in \mathbb{R}[x]$, there exists a unique pair $(c,d) \in \mathbb{R}^2$ such that $p(x) \sim_4 cx + d$.

    (b) Prove that $(a(x) + p(x)) \sim_4 (b(x) + q(x))$. Deduce that

    $$[a(x)] + [p(x)] \stackrel{\text{def}}{=} [a(x) + p(x)]$$

    is a well defined binary operation on $\mathbb{R}[x]/\sim_4$.

    (c) Prove that $(a(x)p(x)) \sim_4 (b(x)q(x))$. Deduce that

    $$[a(x)] \cdot [p(x)] \stackrel{\text{def}}{=} [a(x)p(x)]$$

    is a well defined binary operation on $\mathbb{R}[x]/\sim_4$.

    (d) Using the "product" defined in the previous part, prove that in $\mathbb{R}[x]/\sim_4$ we have $[cx + d] \cdot [sx + t] = [(ct + ds)x + (dt - cs)]$.

Part II:

Culture, History, Reading, and Writing

CHAPTER 7

Mathematical Culture, Vocation, and Careers

In a 2019 interview, in response to a question about the changing nature of the mathematics profession, Jill Pipher, then president of the American Mathematical Society, began her response with, "Over time, mathematics has become more collaborative. There's an increase in social and team approaches. I think that, on the whole, this is very healthy for the profession." Every academic field develops a unique culture that determines how and where people engage with it. Mathematics is no exception.

The men and women who study, discover, invent, and apply mathematics develop certain ways of viewing, approaching, and interacting with their field. The mathematical culture is alive, under constant evolution, influenced by its historical roots and shaped by its connections with the other fields of human knowledge.

This chapter focuses on how people do mathematics today. As students transition into advanced mathematics, they step into the vibrant culture of contemporary mathematics, a world they may be unfamiliar with if they have only been exposed to the elementary areas of mathematics developed centuries ago. The 21st century is indeed a Golden Age for mathematics; the field's rate of growing is at an all-time high, and mathematics continues to permeate every aspect of our daily lives in unprecedented ways.

Section 7.1 provides a snapshot of the status of modern mathematics. Section 7.2 focuses on the intensely collaborative aspect of contemporary mathematics and describes how mathematicians who belong to local, national, or international associations gather regularly at conferences to share their ideas and learn from others. Section 7.3 offers insight and advice to students on how to study upper-level mathematics and successfully navigate the difficulties of upper-level coursework. Finally, Section 7.4 presents the vocational possibilities granted by pursuing a degree in mathematics and suggests strategies to prepare for vocational pathways while in college.

7.1   21st Century Mathematics

The field of study we call "mathematics" stands as the result of a complex evolution over 4000 years of history across many cultures. Mathematics is a unique area of human inquiry in that scholars consider it as simultaneously a part of the sciences, humanities, and arts. Historical shifts in human thought, philosophy, science, and art have all left their mark on mathematics, but mathematics has also shaped many aspects of the world as it stands today. To properly grasp the nature of mathematics in the 21st century, it is necessary to understand its history, which we outline in Chapter 8, spanning from its beginnings around 5000 BC to the 20th century.

In this section, we focus on the end result of this long history and describe the field as it currently stands. The world of contemporary mathematics is often unfamiliar to undergraduate students, since the curriculum through the calculus sequence and linear algebra consists of topics developed prior to the 19th century. Although developments of the past 100 years may be uncommon topics in undergraduate curricula, scholars who study the field estimate that these developments account for over 95% of the current body of mathematical knowledge [25]. Many mathematics popularizers labeled the 20th century as the beginning of a New Golden Age for mathematics [25], noting how it permeated virtually every field of natural and social science, and indeed every aspect of human life. Mathematical activity and research stands at an all-time high with thousands of scholarly articles containing new mathematics published every year.

Subsection 7.1.1 provides a contemporary snapshot of the field and describes some of the distinctive aspects of how mathematicians approach the discipline today as compared to the past. Subsection 7.1.2 describes what constitutes research in mathematics and outlines the process by which new results enter the body of mathematical knowledge. Finally, Subsection 7.1.3 provides an overview of nine fields of intense mathematical growth and activity in the 21st century. We briefly present each one with its background, origins, motivation, major advances, open problems, applications, and examples.

7.1.1   The landscape of mathematics in the 21st century

Specialization and fragmentation

At the turn of the 20th century, it was believed that all the mathematics discovered throughout human history could fit in a set of approximately 80 volumes and that a person with a sharp mind and extensive memory could claim to know all of it [24]. In fact, certain people before 1900 called *polymaths*, such as *Isaac Newton* (1642-1726), *Leonhard Euler* (1707-1783), and *Carl Gauss* (1777-1855) were commonly held to possess such a breadth of knowledge that they were not only proficient in all areas of mathematics, but equally knowledgeable in science, literature, and philosophy.

The early 20th century saw such a rapid growth of mathematics that it soon became impossible for a single human mind to contain all of it. The incredible explosion of recent mathematical activity has forced mathematicians to evolve from being generalists to specialists in a chosen subfield. Furthermore, the efforts made in the early 20th century to unify all areas of mathematics into a unique, consistent field were unsuccessful and largely abandoned (in no small part caused by Kurt Gödel's publication of his *Incompleteness Theorems* in 1931). Mathematicians accept their discipline as a fragmented collection of distinct theories, leading many 20th-century mathematicians to "choose" a field of specialization.[1]

The 20th century also saw mathematics applied as a universal tool in all natural and social sciences, a continuation of the trend established during the Scientific Revolution called *mathematization of science*. As mathematics gradually infiltrated all fields of science, so did the amount of mathematics developed to assist in scientific progress. This type of mathematical research, driven by its applications, is now referred to as *applied mathematics*, to differentiate it from research in *theoretical* or *pure mathematics*.[2] This pure/applied duality extends to the methods by which new mathematical knowledge develops. In many situations, the *scientific method*, the universal framework for progress in science, draws on tools developed in applied mathematics. In contrast, progress in pure mathematics does not appeal to the empirical world of experiments and data collection but grows within the confines of the *axiomatic method* and deductive reasoning.

Applied and pure mathematics

Applied mathematics refers to the type of mathematics used in science as a tool to solve "real-world" problems in the three-dimensional, physical universe. Historically, applied mathematics engaged problems almost exclusively in physics, but since the late 20th century,

---

[1] For this reason, mathematics is understood as a plural noun, and declined this way in British English.

[2] However, this boundary is not impermeable: most theoretical mathematics can trace its roots back to applied problems, and conversely many results developed as pure mathematical theories with no application in mind later offer the key steps to solve practical problems.

all natural and social sciences have become increasingly *mathematized*, notably since the advent of computing. Even the newer sciences such as ecology, epidemiology, pharmacology, psychology, and management science, now make heavy use of mathematical and statistical tools. All fields of science advance by the scientific method, and mathematics has become an essential tool to assist in their progress by this method. Moreover, a subject like linear algebra, which is developed as pure mathematics, is indispensible to applied areas such as machine learning, computer graphics, and coding theory.

The process of creating a *mathematical model* lies at the heart of applied mathematics. A model in this sense translates an observed real-world phenomenon into a set of equations (see Subsection 8.2.2). Statistics serves as the primary tool for analyzing and interpreting experimental data, while differential equations is the primary method to describe the underlying mechanisms that govern an observed behavior in time and space.

Although mathematics finds its roots in humanity's attempts to solve "real-world" problems, it reaches far beyond the role of a "toolbox for science." Unlike other sciences, mathematics is not confined to the three-dimensional, visible, physical world. Mathematician Keith Devlin defines mathematics as the field that "makes the invisible visible," allowing us to appreciate patterns, derive truth, and appreciate beauty beyond the reach of empirical science [24]. *Theoretical* or *Pure Mathematics* refers to the study of abstract patterns that have no apparent application to the physical world, although some results derived as pure mathematics are later used to unlock problems in applied mathematics, thus crossing the pure-applied boundary.

Mathematics generalizes intuitive concepts, providing a framework for the human mind to progress beyond the world of empirical science by using logic and deductive reasoning. Mathematicians organize their ideas into systems of axioms, definitions, theorems, and proofs, not unlike how philosophy organizes and analyzes ideas. The study of these axiomatic structures cultivates the mind's ability to organize and synthesize thoughts and to solve problems sequentially and creatively.

Computation

The invention of computers in the 20th century profoundly revolutionized mathematics. Whereas mathematicians formerly solved problems by analytical methods and symbolic manipulation, the advent of computing promoted the development of *numerical methods*. These methods, in turn, elevated the fields of discrete mathematics, linear algebra, and numerical analysis to central importance because of their role in providing efficient algorithms to solve computational problems.

Computers offer unparalleled visualization tools, allowing us to picture datasets and represent mathematical objects in new ways. For example, the study of fractals, discovered mathematically in the early 20th century, only really took off once computer graphics could reveal their visual beauty.

Computers also deeply impact pure mathematics. Computational simulation allows mathematicians to test conjectures that guide their research. For example, when studying whether numbers of the form $n^2 + 1$ are prime, we can use a computer to test the primality of such numbers for several million values of n. Although this approach does not prove anything, it can offer valuable clues on which way to proceed with a formal proof.

Finally, computer applications called *computer algebra systems* (CAS) are designed to solve mathematical problems by symbolic manipulations, much in the same way that a human would do by hand with paper and pencil. For example, a CAS can evaluate derivatives and integrals analytically or find exact solutions to differential equations.

Interdisciplinarity, collaboration, and diversity

The increased specialization of mathematicians, whether applied or pure, promotes a great degree of collaboration between mathematicians, and between applied mathematicians and experimental scientists. This trend is common to all areas of modern business, science, and technology. Such interdisciplinary collaborations blur the traditional boundary lines between fields, often requiring the expertise of several specialists in different fields working together. This trend has caused mathematicians to evolve from engaging mostly in solitary research to joining interdisciplinary research teams. Local, national, and international mathematical associations (such as the American Mathematical Society and the Mathematical Association of America, among others) provide networks and platforms for information sharing and collaboration. Section 7.2 digs deeper into the collaborative dimension of mathematics.

Another encouraging recent trend is the increased ethnic and gender diversity in mathematics. Female and minority mathematicians experienced heavy discrimination in the West until well into the 20th century. The history of mathematics presents myriad stories of women and minorities barred from attending mathematics classes at universities or unable to hold academic positions. Although there remains a long road ahead, the second half of the 20th century witnessed great strides toward equality (see Section 8.4.4).

7.1.2   Research in mathematics

Understanding the processes leading to the creation of new mathematics (called *original research*) is essential for understanding upper-level mathematics. As mentioned above, research in applied mathematics often takes place within the framework of experimental research by the scientific method in an area of application, whereas interesting questions internal to mathematics and efforts to generalize and extend known results drive research in pure mathematics. In either case, just as in all other academic fields, new results in both pure and applied mathematics are formally accepted once published as an article of original research in a peer-reviewed academic journal. These publications, written by the researchers themselves, are *primary sources*; their purpose is to disseminate fresh results that have not been published before. The same researchers often present their results orally in person at academic conferences.

Most of the original research published in academic journals and presented at conferences does not immediately make headlines or generate wide interest. Particularly interesting or insightful results may, however, pique the interest of other mathematicians, who may decide to extend a result or generalize a theorem. Little by little, over the course of several years, new areas of mathematics may develop into a subbranch. To attract further interest, leading researchers in the field might summarize the key ideas and results in a review article, expository paper, mathematical encyclopedia, website, book, or other *secondary sources*.

Mathematicians who wish to remain on the cutting edge of a field must follow ongoing developments in "real time" by reading primary literature and attending conferences in their field of specialization. Learning from primary sources poses a daunting task for a nonspecialist, but it is nevertheless essential since most recent results exist only in primary research articles and rarely, if ever, trickle down to secondary sources. Chapter 9 explores in much more detail how to effectively approach primary sources.

Research in applied mathematics

Problems in natural science (physics, chemistry, biology, ecology, epidemiology, etc.) or social science (psychology, sociology, economics, management science, political science, etc.) drive research in applied mathematics. All natural and social sciences progress by observation, formulating a hypothesis, designing an experiment, collecting qualitative and quantitative data, and using the data to either support or reject the hypothesis. This process, called the *scientific method* stands as the hallmark of any area of human inquiry bearing the designation of a science. Though not a necessary component of the scientific method, mathematics often plays three different roles. First, the formulation of a hypothesis often occurs in the context of a *mathematical model*, which consists in describing the behavior of an observed system as a mathematical equation. *Mechanistic models*, based on differential equations, can confirm or reject hypotheses concerning the underlying mechanisms that cause observed data patterns. Second, statistics provides tools to analyze the empirical data. Among other things, statistics allows the scientist to measure the likelihood that the conclusions pertaining to a limited sample can be inferred as valid for the entire population. Third, analysis of the mathematical model may allow the scientist to make predictions. For instance, management science (or decision science) uses models from *operations research* (OR) to determine the optimum value of an objective function and the values of the control variables that achieve the optimum.

Mathematical models can help gain insight into a system that goes beyond what is revealed by raw experimental data alone. Models can also simulate experiments quickly and inexpensively to help predict future experimental results, thereby guiding decisions about the best investments in money, time, and effort toward future experiments. For these reasons, many fields of experimental science have turned to mathematical modeling as a powerful complement to traditional experimental research.

Research in pure mathematics

Unlike the other sciences, research in pure mathematics does not rely on experiments or data. Instead, pure mathematics progresses by proving theorems within the context of a *mathematical theory*. A mathematical theory begins with precise definitions about the objects or phenomenon of interest in that theory and explores as much as can be known about those objects. Sometimes, we call parts of these definitions *axioms*, but in modern mathematics the term "axiom" does not mean the same thing as it does in philosophy. For example, "the axioms of a vector space" mean the various properties we require of the set and binary operations involved in the definition of a vector space. Consider also Peano's definition of the integers given in Section 4.1. Modern mathematics no longer considers the properties of integers as absolutely true but accepts Peano's Axioms as the definition of nonnegative integers.

Just as applied mathematics finds inspiration for problems of interest arising outside of mathematics, problems or topics of interest within the particular branches motivate the research. The process often begins by stating a *conjecture* (an unproved statement believed to be true – usually by intuition), and proving it deductively from other previously proved theorems and definitions, thereby converting it into a *theorem* within the context of a mathematical theory. Unlike scientific theories, which are continually revised as new data becomes available, the deductive nature of mathematics reasoning is such that a theorem established via a valid proof becomes a permanent extension of a mathematical theory and can be used as the basis for further theorems.

Not all conjectures can be proved immediately. In fact, there are hundreds of open conjectures in all fields of mathematics; some are famous and centuries-old. For example, the *Goldbach Conjecture* formulated in 1742, which states that every even integer greater than 2 is the sum of two primes, remains open to this day.

Fields Medal and Abel Prize

As in other academic fields, there are prizes in mathematics awarded every year to mathematicians who produce exceptionally good original research and important new results. Most of these awards are given by national or international mathematical associations, and many of them seek to reward contributions in a specific field of mathematics or by a certain type of mathematician. For example, the American Mathematical Society awards the *Frank Nelson Cole Prize in Algebra* every year for outstanding achievements in the field of algebra. Similarly, the National Academy of Sciences awards the *Maryam Mirzakhani Prize in Mathematics* biennially for exceptional contributions to mathematics by a mid-career mathematician, and the Mathematical Association of America awards the *Morgan Prize* every year for outstanding mathematical research by an undergraduate student.

There is no "Nobel Prize" in mathematics. Instead, mathematicians regard the *Fields Medal* and the *Abel Prize* as the highest awards in mathematics, comparable to the Nobel Prize in other fields. The Fields Medal is awarded to two, three, or four mathematicians under 40 years of age[3] at the International Congress of the International Mathematical Union, a meeting that takes place every 4 years. Famous recent Fields Medalists include the American mathematician *Maryam Mirzakhani* (1977-2017), the first female recipient of the award (in 2014), and Russian mathematician *Grigori Perelman*, who was awarded the Fields Medal in 2006 for solving a famous 100-year-old problem in topology (and one of the Millennium Prize Problems) called the *Poincaré Conjecture*.[4]

---

[3]The purpose of the age limit is to encourage early-career mathematicians.

[4]However, Perelman declined to accept the Fields Medal, as well as the $1 million Millennium Prize.

The Abel Prize, named after Norwegian mathematician *Niels Abel* (1802-1829), is modeled directly after the Nobel Prizes. The Abel Prize is awarded yearly by the King of Norway to an outstanding mathematician.[5] Famous recent Abel Prize Medalists include *Karen Uhlenbeck* in 2019, who was also the first female recipient, and *Andrew Wiles*, who in 2016 received the Abel Prize for solving Fermat's Last Theorem, a famous problem in number theory conjectured in 1637.

Millennium Prize problems

In 2000, the *Clay Mathematics Institute* of Cambridge, Massachusetts established a list of seven important open problems called the *Millennium Prize Problems*. The list was compiled in the spirit of a similar list of 23 open problems presented by *David Hilbert* (1862-1943) in 1900, and which guided much of the mathematical research during the 20th century. The Clay Institute offers a prize of $1 million for a correct solution to any of these problems, which represent some of the most difficult mathematical problems of today.

Millennium Prize Problems

1) P versus NP problem,

2) Hodge conjecture,

3) Poincaré conjecture,

4) Riemann hypothesis,

5) Yang-Mills existence and mass gap,

6) Navier-Stokes existence and smoothness,

7) Birch and Swinnerton-Dyer conjecture.

Russian mathematician Grigori Perelman solved the Poincaré Conjecture in 2006, but the other six problems remain open.

7.1.3   Open fields of research in mathematics

Mathematics subject classification

Two indexing services for mathematical documents, *Mathematical Reviews* and *Zentralblatt MATH*, collaborated to devise a classification scheme called the *Mathematics Subject Classification* (MSC) as an attempt to organize all the open fields of mathematics. This hierarchical scheme assigns a 5-digit alphanumeric code to every field: the

---

[5] The Nobel Prizes are awarded by the King of Sweden.

first two digits describe the general field (for example, 35 for Partial Differential Equations), followed by a letter to describe a more specific subfield (for example, 35L for hyperbolic PDEs), followed by two more digits for an even more specific area in the subfield (for example, 35L51 for second-order hyperbolic systems of PDEs). These indexing services continually update the MSC as mathematicians develop new fields and subfields. The latest update, called MSC2020, contains more than 7000 subfields.

Nine fields stand out as the most important: (1) number theory, (2) algebra, (3) topology, (4) analysis, (5) combinatorics, (6) differential equations, (7) statistics, (8) operations research, and (9) numerical analysis. Although researchers can approach any of these fields from a theoretical or applied point of view, the first four generally carry the label of "pure," the latter four fall on the "applied" side of the pure-applied spectrum, while combinatorics can be regarded as either pure or applied. There are of course many other fields, but a large portion of current mathematical output falls under one or more of these nine fields. We briefly describe each one below.

Number theory

German mathematician *Carl Friedrich Gauss* (1777-1855) dubbed *Number theory* the "queen of mathematics." It studies properties of integers and of integer-valued functions. Though one of the most ancient branches of mathematics, it remains a large open field of modern mathematics. Among other things, it studies patterns in the distribution of prime numbers, divisibility and factorization of integers, and integer solutions to polynomial equations (called *Diophantine equations*). Many open problems in number theory date back hundreds of years. Though simple to state and understand, these problems are extremely difficult to prove. For example,

1) Are there infinitely many primes of the form $n^2 + 1$?

2) (*Twin Prime Conjecture*) Are there infinitely many pairs of *twin primes*, i.e., primes that differ by 2, like 5 and 7, or 41 and 43?

3) (*Goldbach Conjecture*) Can every even integer be expressed as a sum of two primes, like $100 = 41 + 59$?

4) (*Legendre's Conjecture*) Is there always a prime between two consecutive squares, like $9 < 11 < 16$?

Diophantine equations are polynomial equations to which integer solutions are sought. For example, when $n = 2$, the equation

$$x^n + y^n = z^n$$

admits many integer solutions (x, y, z) called *Pythagorean Triples*. Interestingly, this equation admits no positive integer solutions for integer values of $n > 2$. Pierre de Fermat conjectured this in 1637. Famously, when he first jotted down the conjecture in the margin of the journal *Arithmetica*, he added that he had a proof but that it was too large to fit in the margin. Though often called a theorem, it remained only a conjecture until Andrew Wiles proved it in 1995.

In recent years, computers help to unlock many open problems in number theory, sometimes simply by a *brute force* verification of millions of test cases. For example, consider the open problem of determining which integers n can be expressed as the sum of three cubes

$$x^3 + y^3 + z^3 = n \qquad \text{with } x, y, z \in \mathbb{Z}. \qquad (7.1)$$

It is easy to show that n cannot have a remainder of 4 or 5 when divided by 9. (The reader familiar with modular arithmetic should try this.) However, it is an open problem whether there is an integer n with a remainder different from 4 or 5 when divided by 9 that cannot be expressed as in (7.1). Using supercomputers, Eisenhans and Jahnel (2009), then Huisman (2016), and then Booker and Sutherland (2019) found solutions for all $n \leq 100$.

British mathematician *G.H. Hardy* (1877-1947) claimed number theory as the purest and most beautiful form of mathematics, possessing supreme elegance, but devoid of any application [43]. However, number theory has recently become an applied field in business and government as a method to design encryption schemes, which rely on the prime factorization of large integers.

Algebra

Algebra is a broad field of mathematics divided into many subfields. *Elementary algebra* refers to the type of algebra studied in secondary school; it is an abstraction and generalization of arithmetic, as it replaces specified numbers with quantities without fixed values, known as variables. In all types of science, we use elementary algebra to express quantitative relationships as algebraic equations. Elementary algebra also considers methods to solve equations.

With the advent of set theory in the second half of the 19th century, mathematicians began considering different types of mathematical objects (numbers, vectors, functions, etc.), classifying them into sets, and developing algebraic techniques for them. Because the techniques involved were similar, mathematicians focused on the sets, rather than on their individual elements. Thus, in its more general sense, algebra refers to the study of sets, together with the properties of different types of operations between elements of the set. The different types of elements and properties of the operations (such as commutativity, associativity, etc.) give rise to sets with different *algebraic structures*. The study of each of these leads to different subfields of algebra.

At its beginning, *linear algebra* deals with structures that can be described by systems of linear equations (i.e., equations of the form $a_1x_1 + \cdots + a_nx_n = b$). Inspired by the algebra of geometric vectors, we call n-tuples of real numbers vectors and open the door to describe phenomena involving multiple variables at once. Natural problems from applications and geometry lead to the notion of matrix and matrix algebra. Linear then develops in two different but interlaced directions. Taking a step toward abstraction, we define a *vector space* as a structure having similar properties as addition and scalar multiplication. This abstraction leads to fruitful applications throughout all of mathematics. At the same time, the advent of computing raised linear algebra as the foundational framework for many applied fields, especially numerical analysis, differential equations, operations research (linear programming), and computer science.

Abstract algebra (also called *modern algebra*) refers to the study of generalized algebraic structures that arise in the process of solving problems in elementary algebra, number theory, modular arithmetic, and geometry. One of the central algebraic structures studied in abstract algebra is that of a *group* (specifically, in *group theory*), which finds many applications in science, especially in physics, chemistry, and computer science. Other algebraic structures studied in abstract algebra include rings, fields, and modules.

Abstract algebra is a very large and active field of research today and influences almost every other branch of mathematics. Abstract algebra divides into a large number of subfields, including algebraic geometry, Galois theory, Lie theory, representation theory, and category theory. Many of these fields have highly specialized applications in areas of advanced mathematics, information security, physics, and computer science.

As in most branches of mathematics, computational methods starte presenting new avenues of exploration since the late 20th century. Building on the foundation of algorithms that implement performing manipulation of algebraic expressions, computational algebra focuses on theorems that naturally lead to algorithms that answer otherwise intractable questions.

Topology

The name *topology*, in Greek, means the "study of place." As a field of mathematics, topology generalizes objects generally viewed as geometric (called *topological spaces*) and explores properties that are preserved under a continuous deformation, i.e., by stretching and

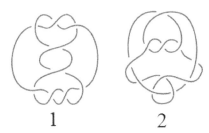

1   2

Figure 7.1: Is it possible to deform knot 1 into knot 2 without cutting the string?

deforming without cutting or tearing. For example, a circle is *topologically equivalent* to an ellipse, square, or triangle. Similarly, a sphere is topologically equivalent to an ellipsoid, and a doughnut (i.e., a torus) to a coffee mug. One of the central ideas in topology is that spatial objects can be treated as independent objects in their own right, regardless of how they are embedded or represented in space. Thus, topology studies how the points of geometrical objects are connected to each other, while disregarding the actual shape of the object itself.

Topology branches into *algebraic topology*, *differential topology*, and *point-set topology*. Algebraic topology uses algebraic structures such as groups and rings to study structures that contain "holes" in space. Differential topology studies differentiable (smooth) topological spaces called manifolds. Point-set topology studies the generalized abstract nature of continuity, connectedness, compactness, and dimension of topological spaces.

Another subfield of topology called *knot theory* studies knots, which are defined as closed, simple (not self-intersecting) curves embedded in three dimensions. Knot theory seeks to describe and to determine, for example, if a given knot can be continuously deformed, tangled, or untangled into another without cutting it (Figure 7.1). Although topologists developed knot theory without specific applications in mind, its principles have recently been used with great success to model the tangling of DNA strands in molecular biology, thereby connecting to a field in applied mathematics.

Analysis

At first brush, *mathematical analysis* (or simply *analysis*) studies the theoretical foundations of calculus. Analysis provides a rigorous foundation for the concepts initially encountered in an elementary calculus course (such as derivatives, integrals, infinite series, the fundamental theorem of calculus) and reframes them in terms of limits, convergence, and infinitesimals. For example, derivatives arise as the limit of a difference quotient when the denominator becomes infinitely small. Similarly, a definite integral is defined as a converging Riemann sum.

The concepts of analysis extend to the study of functions and their properties (continuity, smoothness, etc.). Functions may be either real-valued or complex-valued, leading to *real analysis* and *complex analysis*, respectively.

Functional analysis explores the analytical properties of *function spaces*, infinite-dimensional vector spaces where the functions play the role of vectors. Transformations between function spaces (called *functionals* or *operators*) are the main focus of study in a related field called *operator theory*, which is also particularly useful in the study of differential equations.

Combinatorics

Combinatorics is primarily concerned with counting, such as a number of combinations, permutations, configurations, arrangements, selections, partitions, etc. The means of such enumerations tie combinatorics to number theory. Combinatorics also finds many applications in probability theory, optimization, and decision theory.

Graph theory, a mathematical field related to combinatorics, studies mathematical graphs, defined as a set of points (called *nodes* or *vertices*) and lines (called *edges* or *arcs*) connecting some subset of the points. Graphs offer an ideal framework to model real-world problems that require describing which points of a network are connected to each other, while ignoring their geometric positions or the distance between the points.

A famous graph problem conjectured in 1852, called the *Four Color Problem*, states that, given any separation of a plane into contiguous regions, no more than four colors are required to color the regions so that no two regions that share a boundary have the same color. In 1976, Kenneth Appel and Wolfgang Haken announced they proved this conjecture with the aid of a computer. Their method did not immediately gain wide acceptance since it required checking far too many cases for a human to grasp. With a verification of the algoritm and its implementation, the mathematics community ultimately acknowledged it, and the proof became the first computer-aided proof.

Differential equations

Differential equations are equations that contain one or more functions and their derivatives. The study of differential equations consists mainly in the study of their *solutions* (the set of functions that satisfy the equation) and their properties. Another central problem consists in studying the existence and uniqueness of solutions of a given differential equation.

The first differential equations arose with the invention of calculus by Newton and Leibniz, providing the spark for early progress in calculus. The laws of motion set forth by *Isaac Newton* (1642-1726) define the velocity $v(t)$ of an object as the rate of change of its position $x(t)$ with respect to time (t):

$$v(t) = \frac{dx}{dt}.$$

Given a velocity function $v(t)$, solving this differential equation gives the position $x(t)$ of the object at any time. A similar differential equation arises with the definition of the acceleration $a(t)$ as the rate of change of the velocity. We commonly call these equations "laws" of motion, because until the late 19th century, physicists used them to describe the mechanisms that govern the motion of all objects in the universe. Many mathematicians and scientists followed Newton's example and used differential equations to describe other laws of nature, giving us for example the *Navier-Stokes equations* for fluid dynamics or the *Maxwell equations* for electromagnetism.

The trend to describe phenomena in nature using differential equations, though previously confined to physics, spread to all fields in the natural and social sciences. These *mechanistic models* allow scientists to understand which variables and parameters drive the behavior of a complex system, to simulate different scenarios, and to determine how to control a system toward desired outcome.

Differential equations come in several types, which depend on the properties of the equations and methods for obtaining solutions. *Ordinary differential equations* (ODEs) involve an unknown function of a single variable and its derivatives. We commonly use ODEs to study time-dependent systems, called *dynamical systems*, where the independent variable t represents time. *Partial differential equations* (PDEs) involve an unknown multivariable function and its partial derivatives. We typically use PDEs to study systems that vary in time and space, such as diffusion (of heat, sound, light) and transport phenomena (fluid flow, traveling waves).

Until the mid-20th century, the study of differential equations focused almost exclusively on linear differential equations, as these are the only ones for which there exist well-developed analytic methods. The recent advent of computing opened the door to the vast field of nonlinear differential equations, which play a role in describing natural phenomena like explosions, shock waves, and wave fronts. Although the large majority of nonlinear equations do not present exact analytical solutions, a computer makes it possible to obtain numerical (approximate) solutions and visualize them graphically.

Statistics

Statistics concerns itself with organizing, representing, interpreting, and modeling variability in data. Though a rich field of mathematics in its own right, statistics comes to the aid of the scientific method for its power to analyze any dataset collected by experiment in the natural and social sciences. Scientists of all stripes use statistical methods to characterize the data, to determine whether the data support or reject a scientific hypothesis, and to make informed decisions that take into account the inherent variability, uncertainty, and risk associated with an outcome.

Statistical methods rest upon the theoretical framework of probability theory, which studies chance events. Although it is not possible to perfectly predict random events, probability theory provides a framework for modeling randomness in a way that is useful for inference and prediction.

Descriptive statistics refers to methods of visualizing, describing, and summarizing the quantitative and qualitative features of a dataset, for example by calculating the mean, quartiles, standard deviation, or correlation coefficient. These methods are, in turn, used to study *inferences*, which is about understanding the relationships between measures and how they vary, and to make predictions.

Operations research

Operations research (OR), a mathematical subfield of *decision science* (or *management science*), uses mathematical tools to help make better decisions, especially in complex logistical situations involving many moving parts and large amounts of human and material resources. The central objective of operations research is optimization, i.e., identifying the extreme values (the maxima or minima) of an *objective function*, as well as the variables (called *decision variables* or *control variables*) which a decision-maker can control to achieve the optimum. Some situations may require that a solution also satisfy one or more *constraints*. As in other fields of applied mathematics, OR mathematicians regularly must make certain simplifying assumptions to model a real-life system in order to solve the mathematical problem either analytically or numerically.

A simple OR problem encountered in every elementary calculus course consists in finding the dimensions x and y that maximize the area of a rectangle with a given perimeter P. In OR terms, x and y are the decision variables, the area is the objective function, and the fixed perimeter represents a constraint.

It is not unusual for modern OR problems to involve millions of variables and thousands of constraints. For example, a factory manager may face a *resource allocation problem* to determine the quantities $x_1, ..., x_n$ of n products to produce and sell at unit profit $p_1, ..., p_n$, with the goal of maximizing the total profit

$$P(x_1, ..., x_n) = p_1 x_1 + \cdots + p_n x_n = \sum_{j=1}^{n} p_j x_j,$$

while remaining constrained by limited quantities of m production materials of the form

$$a_{i1} x_1 + \cdots + a_{in} x_n = \sum_{j=1}^{n} a_{ij} x_j \leq a_{i,max}, \quad i = 1, ..., m.$$

Operations research originated during World War II as a scientific method to provide a quantitative basis to guide military decisions. After World War II, the scope of OR expanded to civilian industries, especially in government, economics, finance, manufacturing, transportation, and logistics. For example, any major airport uses large amounts of OR to optimize flight arrival and departure schedules, fuel operations, on-ground aircraft maintenance and repairs, baggage handling, passenger transfer, food and beverage preparation, staffing of gate agents, ground personnel, and flight crews. The complex logistics involved keeping everything running smoothly while maximizing efficiency in these types of situations would not be possible without OR tools.

Not unlike models involving differential equations and statistics, early OR models were almost all linear; early OR problem-solving methods and numerical algorithms were not equipped for the difficulty of solving nonlinear problems. Recent progress in numerical analysis has made it possible to expand the scope of OR modeling to more sophisticated systems that require nonlinear equations.

Game theory, a mathematical field closely related to operations research, studies *mathematical games*, defined as situations involving two or more parties with conflicting interests. Though the principles of game theory apply to strategy games such as chess, checkers, and cards, they also analyze any form of competition such as those encountered in politics, warfare, economics, or property division.

Numerical analysis

Numerical analysis studies methods to solve mathematical problems by numerical approximation as opposed to symbolic manipulation. A numerical method is usually defined as an *algorithm*, namely a finite sequence of rules used to carry out a computation (i.e., a "recipe").

Some elementary numerical algorithms are centuries-old, such as *Newton's method* for finding the roots of a function, the *Trapezoid*

Rule for estimating a definite integral, *Euler's Method* for solving a differential equation, or *Gaussian elimination* for solving a linear system. Most of these algorithms involve *iterative methods*, meaning that they begin with an initial guess, and proceed by successive approximations that converge to the exact solution, but only as a limit and after an infinite number of steps. In practice, these algorithms *truncate* the sequence, i.e., terminate after a finite number of steps, thereby producing an approximate solution. Consequently, error estimation, error propagation, and convergence of the algorithm to the exact solution become matters of primary concern when designing a numerical method to solve a specific problem.

Numerical analysis has become a central topic of mathematical research with the advent of computing. Many numerical algorithms developed centuries ago drew little interest because of the large volume of computing required to implement them by hand. As if they simply were waiting for the right time, these methods now enjoy a sudden revival thanks to computers that can easily perform these computations at great speed. For this reason, numerical analysis associates closely with computer science, as it involves recasting mathematical problems in ways that a computer can solve.

Recent advances in numerical analysis open the door to study nonlinear problems in differential equations, statistics and operations research. Largely ignored until the mid-20th century because of their complexity, these problems play a central focus in modern applied mathematics research. Furthermore, numerical methods have become ubiquitous in the industry, where almost all mathematical models are solved numerically.

Research in numerical analysis progresses hand-in-hand with advances in other fields of mathematics: as new problems emerge, mathematicians develop new numerical methods to solve them by computer.

EXERCISES FOR SECTION 7.1

1. Choose one of the following fields of open mathematical research: Algebraic Geometry, Differential Geometry, Combinatorics, Complex Analysis, Fourier Analysis, Graph Theory, Group Theory. Write a short paragraph about your chosen field, which includes:

   (a) a general description of the field in 2-3 sentences,

   (b) approximately when the field was created,

   (c) the name of a mathematician who contributed to the creation or development of the field,

   (d) some of the applications of the field, if any.

2. The International Congress of Mathematicians (ICM) was first held in the late 19th century and is now a regular event. Do some research

and write a report about the history of the ICM. Include some information about the most recent ICM and the next one.

3. The first African American to be awarded a Ph.D. in mathematics by an American university was Elbert F. Cox, who received his doctorate from Cornell University in 1925. Find out more about Elbert Cox and at least two other African Americans who have earned Ph.D.s in mathematics. Write a report including comments about the contributions of African-American mathematicians to 20th-century mathematics.

4. *Ethnomathematics* is a relatively new field that studies the relationship between mathematics and culture. Do some research about ethnomathematics and write a report explaining the type of issues that are studied in this field.

5. A famous open problem in number theory is the *Collatz Conjecture*. Do some research about this conjecture and write a report explaining what the conjecture states and illustrating it with a few examples.

7.2 Collaboration, Associations, and Conferences

Popular media often depicts mathematicians as introverted, solitary recluses who work alone and come up with groundbreaking new theorems without ever sharing their ideas with anyone. Though maybe useful for interesting character arcs in fiction, this is a gross misconception of reality. There are a few cases where this is true. Andrew Wiles, for example, solved Fermat's Last Theorem in 1996 after working in near-total isolation for several years. However, he recounts that he did so simply to avoid unwanted publicity. Furthermore, his first attempt at a proof required corrections, and the feedback he received from other specialists helped him over the last hump. However, these are exceptional cases. Most progress in mathematics arises as the result of intense, fruitful collaborations, in which mathematicians teach and learn from each other, share ideas, compare approaches, and comment on each other's work.

Until the early 20th century, logistical issues and communication barriers (different languages, terminology, and notations) greatly hindered collaboration, but this did not deter mathematicians. They would travel great distances to meet with each other in person, or they would communicate by mail. The history of mathematics offers countless examples of great mathematical achievements emerging out of a fruitful correspondence between individuals who, in some cases, never met in person.

Research in mathematics, and science in general, evolved greatly during the 20th century, becoming increasingly collaborative with the ease of travel and long-distance communication. Moreover, notation and terminology in many fields has become standardized, and nowadays all mathematical research is discussed, presented, and published in English, regardless of the authors' mother tongues. However, it is important to note that English only recently (in the last few decades) became the *lingua franca* (common language) of mathematics and science. Study in many fields of mathematics still requires the ability to read other languages (especially French, German, and Russian) to access primary source documents published before the mid-20th century.

In the last section, we discussed how applications of mathematics reached into all fields of science by the late 20th century. Any mathematician who develops models or other tools to assist the progress in another science must necessarily collaborate with scientists in those areas. Even mathematicians who make advances in their own fields must borrow tools, ideas, and concepts from others.

Interdisciplinary collaboration presents several challenges. Working effectively in a team is a learned skill that requires practice. Subsection 7.2.1 describes the benefits of learning upper-level mathematics in a collaborative setting and offers some strategies for effective collaboration that students can practice within the context of their studies and assignments. Subsection 7.2.2 focuses on the challenges of distance collaboration and the tools to overcome them. Thanks to the internet and specialized distance collaboration tools, it is now possible for mathematicians on different continents to collaborate daily and communicate results in real time, making mathematical research a truly global endeavor. Subsection 7.2.3 describes the roles of mathematical associations, which have existed for almost 200 years to facilitate communication, collaboration, and networking among mathematicians. Finally, Subsection 7.2.4 focuses on mathematical conferences, where participants meet with each other, listen to each other's research, and collaborate in person.

7.2.1 Collaboration in mathematics

Barriers and benefits

Engaging in collaboration presents many logistical, technical, social, and linguistic barriers that we must overcome:

1) **Time.** It takes time and effort to organize a collaboration. Coordinating a joint approach to a common problem requires a substantial amount of teaching, learning, communicating, and making sure that all collaborators are "on the same page."

2) **Proximity and logistics.** Meeting with collaborators in person poses considerable challenges related to travel, scheduling,

and availability. Distance collaboration solves many logistical challenges, but it also requires that collaborators be equipped with appropriate communication tools and software.

3) **Personality differences.** Collaboration requires navigating and adjusting to diverse personalities and working styles. Team members must be willing to sacrifice doing things their way for the benefit of the team.

4) **Confidentiality.** In very competitive fields of active research, mathematicians and scientists may be reluctant to collaborate or prevented from openly collaborating because of copyright or competitive publication practices.

5) **Fear.** Mathematicians may fear that a collaboration may require an unmanageable amount of responsibility or that it may expose their ignorance (a phenomenon called the *impostor syndrome*).

Because of the fruitfulness of collaboration, it is well worth the effort and courage to strive to overcome these barriers. Approaching open-ended problems benefits from a discussion between collaborators who bring different viewpoints and work together to overcome roadblocks.

Research in pure mathematics requires grappling with new, difficult, and abstract concepts. Verbal processing, dialogue, and idea sharing facilitate understanding. Moreover, the interdisciplinary nature of research in applied mathematics provides an opportunity for mathematicians to learn about a field of application, and conversely for nonmathematicians to learn about the benefits of a mathematical approach to a problem in their area of specialty.

Collaboration among students

Students often discover the benefits of collaboration in upper-level mathematics courses, where the focus involves replicating existing techniques to building models and proving theorems. Collaboration among students provides a valuable platform for learning through mutual teaching and comparing approaches. Unfortunately, these benefits are lost to a student who chooses to do his or her work alone.

Upper-level mathematics students often naturally and spontaneously form informal *study groups* with classmates and friends and realize that an essential part of their learning occurs through these groups. Consequently, many upper-level mathematics courses require *group projects* and other built-in opportunities for students to practice working with others.

Effective collaboration is a learned skill that requires practice and does not come overnight. The first group project may turn out to fuel some frustration or may feel discouraging. However, with time and practice, students can eventually learn to implement strategies that promote an effective collaboration and a positive experience. Here are some factors related to effective teamwork:

1) **Skills inventory.** The most successful teams are those whose members have a wide array of complementary skills. Often, group projects require skills in various areas of mathematics as well as familiarity with technology and different software applications (such as Excel, MATLAB, LaTeX, or some programming language). Furthermore, if the project involves a written or oral presentation, the team should include members skilled with presentation software, public speaking, and English writing. In an ideal team, every member can take ownership of an aspect of the project that aligns with his or her strengths.

2) **Roles.** Effective teamwork requires that members identify their role within the team and define it in relation to other team members. Depending on their role, each member may be called upon to lead, follow, teach, or learn in different aspects of the project. The roles themselves may be dictated by the members' personalities and skill sets.

3) **Logistics and teamwork style.** Team members must communicate their level of commitment, availability, and schedule to other team members at the onset of a joint project. As long as all members agree on a common strategy, there are many ways to approach a group project. Some adopt the "divide and conquer" strategy (i.e., dividing the tasks and combining the results at the end), while others do everything together. Some groups prefer to meet in person; others are comfortable doing everything via email, text messages, and shared documents.

Learning to collaborate is an essential skill for success in studying upper-level mathematics, and few students are able to navigate the coursework without it. It is also one of the most important *transferable skills* linked to success in almost any profession. We discuss vocational preparation in more depth in Section 7.4.4, but we already emphasize here career payoff for students who learn to collaborate during their studies. Students learn a strong work ethic, flexibility, coordination, leadership, and conflict resolution. Moreover, these experiences provide specific examples that students can use to strengthen a résumé and letters of recommendation.

7.2.2 Distance collaboration

Many mathematical researchers are academics employed at a college or university. Some build a collaboration with other academics at the same institution, but most collaborations occur between academics at different institutions. This situation poses its own unique challenges. Distance and the logistical difficulties of traveling to meet in person imply that collaborators at different institutions may only meet face to face once or twice a year, often during a conference. Between these meetings, collaborators use a variety of distance collaboration tools to remain in contact and share their progress with each other.

The internet offers a large variety of tools to facilitate distance collaboration. These fall into three main categories: (1) distance communication, (2) document sharing, and (3) scholarly collaboration networks (SCNs). Like all computer applications, these tools are constantly changing, with new ones emerging on the market every year. Familiarity with the current tools and how they work is an essential skill for any career, whether in academia or in industry.

Distance communication applications

Email is the universal way to communicate over the internet, but there are much better applications designed specifically for distance collaboration. For example, the platforms *Slack* and *Trello* allow collaborators to communicate in writing while organizing and archiving conversations by project, phase, theme, date, and author. Videoconferencing applications such as *Skype*, *Zoom*, and *Google Meet* make virtual face-to-face communication possible in real time. Finally, *GitHub* has emerged as one of the leading collaboration platforms for software development.

Document sharing and file management applications

Effective collaborations require using a file management application that allows team members to generate, organize, and store documents on a centralized, accessible, and secure cloud-based platform. *Google Drive*, *Microsoft One Drive*, and *Dropbox* are common document sharing and file management applications that organize and store documents and also allow users to edit documents and track changes made by others. *Overleaf* combines these features with an online LaTeX editor and compiler, allowing users to create, edit, and share documents in LaTeX.

Scholarly collaboration networks

The term *Scholarly Collaboration Network* (SCN) refers to a social media platform for scholars to connect and share their research with each other. SCNs such as *ResearchGate* and *Academia* offer several features to facilitate collaboration and document-sharing:

- **Profile.** Users have a profile that identifies them, along with their affiliations, background, and research interests.

- **Follow.** Users can "follow" other users to stay up-to-date with their work.

- **Newsfeed.** Users can personalize their newsfeed to receive alerts when new content relevant to their interests becomes available.

- **Messaging.** Users can reach out to each other either via public or private communication.

- **Public groups.** Users can join public groups for discipline-specific news and topic discussions.

- **Private groups.** Collaborators on a common project can create a private group, where they can share full-text research articles along with their annotations. As discussed in Subsection 9.4.4, access to journal articles can be a major barrier for doing research. However, American copyright law allows article sharing between research collaborators within a private group hosted by a SCN. This feature is a great advantage to collaborators who belong to different institutions, and who may otherwise not have access to the same articles if their respective institutional libraries do not subscribe to the same journals.

EndNote, *Mendeley*, and *Zotero* are SCNs that also function as *reference manager* applications, where users can store and organize the references relevant to their research (articles, books, webpages, etc.) by topic, author, title, or journal. These systems offer a desktop version that syncs with the website and offers additional features, such as automatically creating a BibTeX file to manage a bibliography in a LaTeX document. See Subsection 9.4.3 for more information on reference managers.

Cost

All document sharing and file management applications offer basic features for free, but they require upgrading to a paid account to unlock advanced features, more storage, or collaboration with more users.

7.2.3   Mathematical associations

Mathematical associations are nonprofit, private organizations that exist to promote the interests of people whose work involves some aspect of mathematics, as well as people who are simply interested in mathematics. Their main purpose is to facilitate collaboration, networking, and presentation and discussion of new research results by holding regular conferences and publishing or sponsoring scholarly journals. Mathematical associations are similar to professional and academic associations (also known as *learned societies*) in other fields. Most mathematical associations allow anyone interested in mathematics to become a member for a yearly fee, regardless of their academic or professional credentials. These fees fund a large portion of the association's activities and allow members to benefit from discounts and access to events. Although some of the larger mathematical associations employ a paid staff, volunteers organize and manage most of the associations' activities.

There exist hundreds of mathematical associations of all sizes, from national or international associations with thousands of members to smaller ones with only a few dozen members from a handful of institutions in the same geographical region. An association's scope of topics may stretch to all of mathematics or may remain restricted (for example, only statistics). The larger associations break down further into smaller *interest groups* that focus on narrowly defined areas. Applied mathematicians who specialize in a specific field may belong to a nonmathematical association related to their area of application. For example, applied mathematicians who specialize in mathematical models for disease transmission may be members of the American Public Health Association.

History

The earliest mathematical associations began in England during the 19th century to promote collaboration, accelerate research, and prevent overlap. Established in 1833, the *Manchester Statistical Society* was the first one, closely followed by the *Royal Statistical Society* (1834) and then the *American Statistical Association* (1839) in the United States. All three remain vibrant societies and continue to focus on statistics, serving quantitative scientists and users of statistics across many academic areas and applications. A few decades later saw the creation of the *London Mathematical Society* (1865) with a broader scope including all areas of mathematics. It served as a model for the creation of the *American Mathematical Society* in 1888.

Nowadays, over 80 countries boast their own national mathematical association for members within their borders. Currently in the United States, several different national mathematical associations function. The four largest ones are:

1) The American Mathematical Society (AMS),

2) The American Statistical Association (ASA),

3) The Mathematical Association of America (MAA),, and

4) The Society for Industrial and Applied Mathematics (SIAM).

Each of these national associations claims many regional chapters and numerous interest groups.

Several other mathematical associations operate at a national level with a slightly narrower focus. For example, the *Association for Women in Mathematics* (AWM) dedicates its efforts to encouraging women in the mathematical sciences. The *Association of Christians in the Mathematical Sciences* (ACMS) focuses on the integration of mathematics with the Christian faith. In addition, *Mu Alpha Theta* (ΜΑΘ), *Kappa Mu Epsilon* (KME), and *Pi Mu Epsilon* (ΠΜΕ) gear their activities toward undergraduate students in the United States, with local chapters in many colleges and universities.

At the international level, the *International Mathematical Union* (IMU) devotes its efforts to international cooperation in mathematics across the world. Its members include the national mathematics associations of over 80 countries. Every four years, the IMU hosts the *International Congress of Mathematicians*, the largest mathematics conference in the world. It is during the opening ceremony of this event that the organizers present the Fields Medal and other prestigious awards.

Another important international association, the *National Council for Teachers of Mathematics* (NCTM) strives to improve mathematics education worldwide, especially at the primary and secondary levels. With approximately 100,000 members, the NCTM is the largest mathematics education association in the world.

Activities

The activities of a mathematical association fall into three main categories: (1) publications, (2) conferences, and (3) advocacy. Each one is briefly described below.

1) **Publications.** Publications constitute a large portion of the activities and budget of a mathematical association. Many associations manage one or more peer-reviewed journals, and many also publish books. All of these publications are made available to anyone regardless of their membership status, but members benefit from discounts on journal subscriptions and books published by the association and also receive additional information via magazines and newsletters.

2) **Conferences.** Every association organizes at least a general conference every one to four years for all members of the association. In addition, associations often organize several additional regional or topic-specific conferences. These events are usually open to anyone regardless of their membership status (as long

as they pay the registration fee), but members benefit from discounts on registration and attendance costs.

3) **Advocacy.** Associations dedicate a portion of their budget to various activities that promote or reward mathematics research and education. This includes awarding research grants, fellowships, and scholarships, as well as prizes for excellence and contributions in mathematical research, such as finding the solution to a difficult open problem or proving an important conjecture.

Costs and benefits

The membership cost for mathematical associations ranges from $20 to $500 per year. Associations with expensive membership fees typically offer discount prices for students and early career mathematicians to encourage them to join. Some associations allow institutions to purchase an institutional membership, which extends to all faculty and students affiliated with the institution.

For professional mathematicians and students, the many benefits of belonging to an association make the expense of membership worthwhile. Most importantly, membership facilitates sharing and disseminating news and information, and helps members to remain up to date about current developments and recent advances in research. It also allows members to find other mathematicians in their field to ask questions, get feedback on their work, and build collaborations. Finally, associations help members connect with potential employers and remain informed of career opportunities.

Membership also affords many financial advantages to academics and researchers, as the discounts to journal subscriptions, books, and conference costs outweigh the membership fees several times over. In addition, members are eligible to apply for research grants, fellowships, and scholarships awarded by the association.

American Mathematical Society

Claiming over 30,000 members, the *American Mathematical Society* (AMS) is the largest mathematical association in the United States. It publishes over ten peer-reviewed journals, including the *Notices of the American Mathematical Society*, which appears monthly and is one of the most prestigious mathematical journals in the world. The AMS also publishes many topic-specific journals as well as English translations of some of the most widely read foreign mathematical journals. The AMS manages the *MathSciNet* database, which is currently the most complete online bibliographical database for mathematical publications and original articles.

The AMS hosts the *Joint Mathematics Meetings* every year in January. This event is the national general conference of the AMS and the largest mathematics gathering in the world. Each of the four geographic sections (western, central, eastern, and southeastern) holds two sectional conferences every year.

The AMS sponsors many prestigious prizes for excellence in mathematics and selects an annual class of Fellows for outstanding contributions to the advancement of mathematics.

Mathematical Association of America

The *Mathematical Association of America* (MAA) focuses on undergraduate mathematics. It boasts about 14,000 members, many of which are university, college, and high school teachers, as well as graduate and undergraduate students. There are several special interest groups that bring together members who share a common interest, such as *Mathematical Biology*, *History of Mathematics*, or *Mathematics and Sports*.

The MAA publishes multiple peer-reviewed journals, which focus mostly on expository research accessible to a broad audience of mathematicians. Its primary journal, the *American Mathematical Monthly* (often simply called the *Monthly*) is the most widely read mathematics journal in the world. Other popular journals include the *College Mathematics Journal* and *Mathematics Magazine*. The MAA's book publishing program, *MAA Press*, has transferred its operation to the AMS, but it retains its name.

The MAA sponsors the annual summer meeting *MathFest*. MAA members also have a large presence at the Joint Mathematics Meetings, co-sponsored by the AMS and MAA. The MAA is divided into 29 geographical regions, each one holding at least one regional conference every year.

The MAA distributes many prizes and sponsors numerous competitions for students every year. Among them, most people consider the *William Lowell Putnam Mathematical Competition* (often simply called the *Putnam*) the most prestigious university-level competition in the world.

Society for Industrial and Applied Mathematics

The *Society for Industrial and Applied Mathematics* (SIAM) is the world's largest association devoted to applied mathematics. Members include applied and computational mathematicians, engineers, and scientists, whether employed in academia or in industry. About a third of its approximately 14,000 members reside outside of the United States. Like other large associations, SIAM includes a number of special activity groups related to different areas of application (such *Financial Mathematics*, *Geosciences*, and *Life Sciences*), or to mathematical theories and tools used in applications (such as *Dynamical Systems*, *Optimization*, *Orthogonal Polynomials*, and *Partial Differential Equations*).

Like the other national associations, SIAM publishes books and journals. Its primary journal, *SIAM Review*, contains articles in all areas of applied mathematics. SIAM also publishes about 20 peer-reviewed journals focusing on various areas of application, for example, the *SIAM Journal on Imaging Sciences* or the *SIAM Journal on Numerical Analysis*.

SIAM organizes several themed conferences throughout the year on various topics in applied mathematics and computational science, as well as a general meeting every two years for all its members.

SIAM awards prizes every year to recognize applied mathematicians for outstanding contributions to their fields. The society also distributes several research grants and student scholarships. In particular, it organizes the *MathWorks Mega Math (M^3) Challenge*, a mathematical modeling competition for high school students that awards more than \$150,000 annually in scholarships.

American Statistical Association

The *American Statistical Association* (ASA) is the oldest mathematical association in the United States and the second oldest continuously operating professional society in the United States (only the Massachusetts Medical Society, founded in 1781, is older). Its approximately 18,000 members include mathematicians, statisticians, quantitative scientists, and users of statistics across many academic fields and applications. Like SIAM, the ASA includes about 40 interest groups covering various subject-area and industry-area sub-disciplines such as *Statistics in Epidemiology*, *Mental Health Statistics*, *Statistics in Sports*, and *Transportation Statistics*.

In addition to its primary journal, the *Journal of the American Statistical Association*, the ASA publishes about 10 subject-specific journals and several magazines for a general audience, such as *Chance* and *Significance*.

The *Joint Statistical Meetings* is the yearly general meeting of the ASA. The association is divided into 78 local chapters arranged geographically into 6 districts. Each chapter organizes its own local activities and conferences.

The ASA awards several fellowships, scholarships, and research grants every year. It also organizes competitions for students at the high school and college level.

7.2.4 Conferences

Description

A mathematica *conference* (sometimes called a *colloquium*, *congress*, or *symposium*) is a meeting that mathematicians attend primarily to meet other mathematicians in person, to present their research, and to hear about the latest findings in mathematics. Mathematical conferences come in all shapes and sizes, from small local meetings for a handful of people to global events with thousands of attendees from many countries. They may last only one day or stretch up to four days. Smaller conferences often take place on a university campus, while larger ones gather at a large hotel or convention center. Some conferences have a specialized scope, while others are more interdisciplinary, welcoming perspectives from across several disciplines and bringing together people from within and outside of academia.

For professional mathematicians, attending a conference is especially important, as it represents a unique opportunity to present and discuss their latest results and to receive feedback in person, an invaluable opportunity that most cannot get anywhere else. Furthermore, conferences allow mathematicians to forge new connections, build collaborations, learn about ongoing or finished projects (often months before the same results appear in peer-reviewed articles), discover new fields of research, and converse with the people who are actually doing the research. Conferences are intense, intellectually invigorating events, and, for many mathematicians, the high point of the year.

Presentations, in the form of plenary sessions, oral sessions, and poster sessions, form the core of a mathematical conference's activities. However, several other events may take place at a conference, including workshops, business meetings, working groups, and other events that take advantage of the simultaneous presence of all attendees.

Plenary sessions

Every conference features a small number of keynote speakers who present at *plenary sessions*. Organizers schedule these events during prime-time slots that do not conflict with other conference events, thereby allowing everyone to attend. Keynote speakers are usually guests of honor, eminent mathematicians who cast a vision for the conference and discuss important themes accessible and of interest to all attendees.

Oral sessions

Most conferences structure the bulk of the schedule around oral sessions, with a program packed full of short presentations across a variety of topics. Oral sessions are open to all attendees, and some are specifically designed to be accessible to undergraduate students. At large conferences, organizers group oral talks by topic into parallel "streams" that run simultaneously, with speakers in separate rooms speaking at the same time. An oral session tends to run 2-3 hours and to feature 8-12 presenters speaking on a common topic. However, the presentations do not follow a specific sequence, so listeners can come in and leave at their leisure. Speakers typically use a set of presentation slides (less frequently, a chalkboard or whiteboard) as a visual support for their oral presentation.

Some presentations are *contributed talks*, meaning that conference organizers specifically select speakers to present their research and thereby to contribute to the conference's offerings. Contributed talks are designed to be short and concise (10-15 minutes), followed by a few minutes of questions. These are not complete lectures; they are snapshots of the speaker's research, just enough to give the audience the gist without the details, to pique their interest and to invite them to find out more later, on their own. For this reason, talks usually end with speakers sharing their contact information and inviting interested listeners to follow up in person or by correspondence.

Organizers select the contributed talks in the months leading up to the conference in a process that begins with the conference organizers issuing a *Call for Abstracts* (CFA). The CFA outlines the scope and themes of the conference and invites prospective presenters to submit an *abstract* describing their proposed presentation. Similar to a peer-reviewed journal, a program committee reviews the abstracts (or commissions a panel of peers to review them) and ultimately decides which contributed talks are invited to present at which session. However, unlike journals, program committees do not review all the details of every presentation, nor do they expect results to be necessarily complete or in polished form.

At the end of a conference, the organizers may compile and publish the set of contributed talks as *conference proceedings*, thereby enabling those who did not attend the oral sessions to read about what was presented.

Poster sessions

A *poster session* offers an alternative setting to present one's research. It involves creating a physical poster that summarizes a person's work on a specific topic. At a poster session, which typically lasts 2-3 hours, all presenters present their posters at the same time and in the same room. During this time, interested people browse the posters at their leisure. Many poster sessions offer food or beverages, thereby becoming informal social events with a lot of interaction.

Poster sessions offer many advantages to attendees, as they can quickly browse through a large number of posters in a few minutes and decide which ones to spend more time on. Then they can spend as much time as desired studying the selected posters and discussing the work with the presenters in detail. The presentation itself takes the form of a two-way discussion, where the listener can interrupt the presenter to ask questions and request clarification; similarly, the presenter can customize his or her presentation to the listener's background and interest level. Many presenters prefer this more flexible format, which has several advantages over a traditional oral talk:

- A poster presentation is more interactive, allowing the presenter to get real engagement from the audience (one person at a time), answer questions with no set time limit, receive criticism and suggestions, build relationships, and even spark future collaborations.

- Although a poster session lasts 2-3 hours, the posters themselves often remain on display much longer, sometimes for the entire duration of the conference. This allows interested viewers, or attendees who were unable to attend the poster session, to come back to the poster later.

- For many people, especially those whose first language is not English, an informal face-to-face conversation offers an easier venue than a formal talk. This makes a poster session a less intimidating experience for both the presenters and the audience.

Poster sessions are an increasingly popular trend, and many conferences are replacing traditional oral sessions by poster sessions every year. For this reason, most colleges and universities, whether in a specific department or in a science division, own a printer larger enough to produce a poster.

Panel sessions

Panel sessions and round tables usually involve multiple researchers discussing a common topic in front of an audience. By design, these events elicit an exchange of viewpoints among experts on a topic; they usually include a moderator who guides the discussion, sometimes drawing questions from the audience. Though less common at mathematical conferences than in the social sciences or humanities, they do occur regularly among groups that discuss the history and philosophy of mathematics or mathematics education.

Workshops and working groups

Conferences often offer short, intensive classes on a variety of topics called *workshops* or *minicourses*. The limited space requires attendees to sign up for these events ahead of time. Workshops are an excellent way for students and early-career mathematicians to learn a cutting-edge topic taught by experts and to interact with the other interested attendees.

Existing research groups who collaborate at a distance may take advantage of their attendance at the same conference to schedule a *working group*, i.e., a time and place where they can work together in person at the conference. For many collaborators, these working groups are the only time of the year that they meet face to face.

Social events

Conference organizers purposefully craft the schedule to foster a large amount of interaction among attendees, allowing for chance encounters and conversations in hallways, cafes, bars, and restaurants. Many conferences further enhance social opportunities by organizing banquets, sightseeing tours, music, theater, games, art shows, as well as informal on-site competitions and athletic events. Conference attendees make the most of these opportunities to follow up with speakers whose presentations caught their attention, to meet with colleagues over a meal or coffee to have a relaxed conversation about work, or to discuss the presentations that had the most impact. Most conferences also hold events targeted at event first-timers such as students and young researchers, to help them make connections with other attendees with common interests and enter into established circles.

Conferences provide a whirlwind of intense social activity, but the interactions do not end with the conference. Attendees return to their home institutions with lists of new contacts with whom they can follow up or to whom they can introduce themselves, using the conference as a point of reference for ongoing conversations or collaborations by correspondence.

Other events

Besides talks and presentations, workshops, and social events, conferences provide a venue for many side events that take advantage of the large number of attendees gathered in one place.

- **Interviewing.** Large conferences host a career center for the benefit of employers and those seeking employment, where employers and attendees may interview and be interviewed for many positions at the same time.

- **Vendors.** Every conference provides a space for publishing companies, software designers, and merchants of all types of products of interest to mathematicians (calculators, t-shirts, etc.) to exhibit their wares and interact with potential customers.

- **Award ceremonies.** The various awards for excellence in research and education sponsored by the organizing association are usually presented at an award ceremony at a conference.

- **Business meetings.** The governing boards of associations and special interest groups usually take advantage of the presence of all board members in attendance to conduct meetings concerning the business of the association.

Costs

Conferences are expensive events to attend. Between the $100-$500 registration fee (which gives access to conference events), travel, lodging, and meals, the costs can add up quickly. However, the fact that conferences attract thousands of participants every year gives evidence that the benefits clearly outweigh the costs.

When a researcher receives grant money to fund their work, the granting agency may require that person to attend a conference and present their research. In these cases, the grant sets aside a certain amount of money to cover conference costs. Academic budgets at many academic institutions make funds available to sponsor the attendance of faculty and students, particularly when these intend to present talks or posters. Furthermore, the association that sponsors the conference may offer discounts and travel grants to members, students, early-career mathematicians, or attendees from developing countries to help defray their attendance costs. Finally, because conference organizers rely on a large number of volunteers to run the conference, a participant may sometimes obtain a discount on the registration fee in exchange for service as a volunteer.

EXERCISES FOR SECTION 7.2

1. Create a personal profile on a Scholarly Collaboration Network such as ResearchGate, Endnote, Mendeley, or Zotero. Indicate your institution and your research interests. Build a personal network by connecting with your professor, other students at your institution, and/or following mathematicians who share your research interests.

2. Do some research on the previous International Congress for Mathematicians (ICM) and write a report that explains (a) when and where the last ICM took place, (b) what notable results and awards were presented at the last ICM, and (c) when and where the next ICM will take place.

3. Do some research on the AMS, MAA, SIAM, ASA, or another national mathematical association of your choice. Write a report that
   explains:

   (a) when and where the next general conference will take place,
   (b) what regional chapter, district, or section your college or uni-
       versity is in, and when and where the next regional/sectional
       conference will take place,
   (c) whether your college or university has an institutional mem-
       bership to the association, and/or whether you can obtain a
       discounted or free membership as a student at your institution,
   (d) some of the notable special interest or activity groups within
       the association that match your personal interests,
   (e) some of the notable prizes awarded by the association.

4. Determine whether your college or university has a mathematics club,
   or a student chapter of MAΘ, KME, and/or ΠME.

5. Determine whether your college or university participates in the Putnam competition, the Mathematical Contest in Modeling (MCM), or
   other national or regional association-sponsored mathematics competitions. Determine what mathematics courses would help you prepare
   for these competitions.

6. Do some research on the Putnam competition and write a report
   that explains the format, scoring system, awards, and the date of
   the next Putnam competition.

7.3 Studying Upper-Level Mathematics

Upper-level mathematics differs dramatically from elementary mathematics in the way we teach and learn it. In most academic institutions, the lower-division courses emphasize methods and applications, as a wide variety of academic programs across the natural and
social sciences, engineering, economics, computer science, and mathematics require these courses. In contrast, upper-division courses,
aimed primarily at mathematics students, carry a heavier emphasis
on abstraction, writing, and solving problems. Students who transition from lower- to upper-division courses must adjust their learning
strategies and study habits to navigate the different nature of the
content and pedagogy of upper-level courses.

Section 7.3.1 outlines some of the distinctive aspects of upper-level
mathematics, such as the increased emphasis on abstraction, writing,
problem-solving, and learning from books. Section 7.3.2 offers some
advice for students on how to approach studying for upper-division
courses. Finally, Section 7.3.3 centers on the professors and instructors who teach upper-level mathematics and their roles within academic institutions.

7.3.1   Distinctive aspects of upper-level mathematics

Abstraction

One of the most distinctive aspects of upper-level mathematics is the increased depth of abstraction, even in applied mathematics. Elementary arithmetic and geometry exhibits a close connection to problems encountered in everyday life, starting with real numbers, concrete quantities, and finite lengths, areas, and volumes. Elementary algebra takes a first step into abstraction, replacing lengths and quantities by abstract "variables" represented by symbols such as "x" and "y." Linear algebra takes the intuitive concept of a "vector," i.e., an "arrow" that represents a force or velocity, and extends it to an "ordered n-tuple of real or complex numbers." Similarly, the concept of the "sine," defined in elementary trigonometry as the opposite-over-hypotenuse ratio in a right triangle, is later redefined as an abstract transcendental function.

This process of delving deeper into abstraction, of distancing oneself from concrete connections to the material world, is ubiquitous in upper-level mathematics. There are good reasons for this trend. Abstraction allows us to generalize mathematical theories, to solve larger classes of problems, and to cross boundaries between fields of mathematics. Progress in all areas of pure and applied mathematics requires the student become comfortable with high levels of abstraction.

For example, an elementary differential equations course introduces a differential equation (say, $y' + 2y = 0$) as a problem that consists in finding a function $y = y(x)$ that satisfies the equation.[6] A more abstract view of the same problem consists in finding the kernel of the linear transformation $\mathcal{L}(y) = y' + 2y$ between function spaces. Recasting the problem in the abstract framework of linear algebra and set theory may seem more complicated than simply separating the variables and integrating, but it gives more insight on the underlying theory of linear differential equations than merely learning to apply methods. Furthermore, all the theorems about linear transformations between vector spaces apply to this particular $\mathcal{L}$ operator.

Boolean logic and set theory sit as the common foundation for many different fields of mathematics, including algebra, analysis, combinatorics, probability theory, and topology. The rigorous definitions that articulate each of these theories utilizes the common framework of set theory to define their objects of study and organize them into sets. Upper-level mathematics views numbers as mathematical objects belonging to different sets (integers, rational, real, complex,

---

[6]We can solve this differential equation by the method of separation of variables.

etc.). As we illustrate in other chapters of this book, set theory offers a natural construction to pass from set of numbers to sets of other mathematical objects such as vectors, matrices, and functions, as well binary operations between them.

Mathematical writing

Elementary mathematics classes do not focus much on writing; instead, they prioritize learning mathematical methods and executing calculations that lead to getting the "right answer" to a given problem.

In upper-level mathematics, the focus shifts away from merely *doing* the math, to *explaining* how it is done. A larger part of assessments concerns the clarity of mathematical writing and the ability to use the correct terminology, style, and syntax. In theoretical courses, proof writing becomes a central component of the coursework, and applied mathematics courses require extensive writing within the context of building mathematical models. Students do not acquire overnight the ability to express oneself with clarity, accuracy, correct terminology, and formal mathematical style. It takes practice, mostly developed over the course of the upper-level mathematics curriculum.

Like all fields of study, mathematical writing carries its own distinctive style and uses its own specialized terminology. Mathematical speech also has its own informal expressions and jargon that mathematics students pick up gradually over the years, by spending time listening to professors, guest lecturers, or reading mathematical literature (books and journals).

Section 10.1 further explores the distinctive aspects of mathematical writing and suggestions on how to approach practicing this type of writing.

Applying methods versus solving problems

A large portion of the pedagogy in elementary mathematics revolves around learning to solve problems by identifying an appropriate procedure, method, technique, algorithm, or recipe, and implementing it to obtain a correct answer. The methodology for learning these procedures often requires studying a *worked example*, provided in class or in the textbook, that illustrates how the method works, followed by a series of *exercises* to practice replicating the template of the worked example to other examples. The ability to study and emulate worked examples and proofs is valuable, especially since the assessments often consist in identifying, implementing, and properly adjusting the correct method or creative idea to a specific example. Because of this, students often consider the theoretical concepts underlying the methods to be of secondary importance. For instance, a first-semester calculus course may present the theoretical foundations of the product rule, but the main skill to acquire at that level is the ability to apply the rule to find the correct derivative of a variety of functions. Most exercises involving the product rule are aimed at increasing the students' comfort level with the application of the rule, rather than proving it.

The pedagogy of upper-level mathematics coursework shifts away from merely doing exercises to solving open-ended problems for which there is no ready-made method or unique way to approach the solution. In theoretical courses like abstract algebra and analysis, the typical problem consists in proving a theorem, a task that requires creativity, intuition, experience, and thinking outside the box. Applied mathematics courses often require building a mathematical model, a task that involves collecting experimental data, deciding which variables to include, arguing which simplifying assumptions are justified, running simulations, analyzing alternate scenarios, refining the model, and writing a report. In either case, the end goal of working through these types of problems is no longer to obtain a single correct answer, but rather to explain a thought process clearly and convincingly. Ultimately, solving problems of this type carries the invaluable benefit of sharpening the intuition and deepening mathematical maturity. In the end, these skills are more meaningful and marketable than the ability to apply methods, especially since the task of implementing a mathematical method, such as solving differential equations, is nowadays increasingly delegated to computers.

Learning from books

Many college-level courses dispense knowledge via a combination of lectures and textbooks. Ideally, these two means complement each other, so that listening to a professor's lecture and reading books expose students to different aspects of the material.

In elementary mathematics courses, students often consider the in-class lectures and activities as the primary locus of learning, while textbooks play an accessorial role, a backup to clarify material presented in class and a source of exercise problems. In upper-level mathematics courses, these roles are reversed: the textbooks gradually become the main source of learning, and in-class lectures and activities merely complement and highlight the important points. The main reason for this shift is that students of all stripes only fully apprehend complex mathematical concepts through *intensive reading*, a slow process that requires time and focus. Because of the high level of abstraction and complexity, full understanding does not come immediately. Understanding only sinks in after continual exposure and an extended time of mental rumination, during which the mind slowly processes information.

For this reason, students who transition into advanced mathematics may feel unaccustomed to learning from books. Many advanced courses in the humanities or social sciences may require multiple books, sometimes upward of 10. However, the intensive reading in mathematics is such that advanced courses usually only draw on one textbook or just portions of one textbook. Consequently, mathematics and most of the humanities involve a very different type of reading. Mathematics textbooks are not merely *descriptive*, but *prescriptive*, where the author expects the reader to be actively engaged in intensive reading, working through the problems as they are presented, actually doing the mathematics in real time.

As students progress into areas of cutting-edge mathematics, textbooks become scarce, so it becomes necessary to learn from journal articles and other forms of primary literature. Learning from journal articles can be challenging (even more than learning from textbooks), but it is a valuable skill to develop when transitioning into upper-level mathematics. In Chapter 9, we explore strategies to learn from primary literature.

7.3.2   Study habits

Reading

There are different ways of reading academic texts. Two common ways are called *extensive reading* and *intensive reading*.

1) **Extensive reading.** The primary goal of this type of reading, also called *reading for gist*, is to gain exposure to the main ideas and get a general feel for the cadence and structure of a narrative. Extensive reading occurs at a relatively fast pace, in large swaths at a time, without pausing to look up every unfamiliar term (unless it recurs frequently).

2) **Intensive reading.** The goal of an intensive read is to perform a complete analysis of the text in detail to absorb as much information from it as possible. This requires fully engaging with the text, attending to every detail, analyzing every word in every sentence, and pausing to look up every term that is not completely understood. Depending on the complexity of the text, the pace might feel very slow. It may take an hour or more to read and digest a single page. Literature classes often call this *close reading*.

Extensive reading occurs frequently in the humanities and social sciences, where an author may take several pages to make a point, state a nuanced opinion, or present a complex argument. The main ideas shine through because of regular repetition and culminating points in the narrative. Deep understanding in those fields comes gradually as the reader allows the narratives to wash over him or her while reading through hundreds of pages. The author often requires redundancy and paraphrasing to effectively communicate the essence of an argument or to capture shades of meaning. Therefore, even if a reader misses a few details, he or she can still walk away with the gist.

Learning mathematics from a textbook cannot happen by extensive reading. It requires intensive reading. Mathematical language is designed to capture ideas perfectly, so concepts and results require only one statement. Like poetry, every word is saturated with meaning and every symbol is important. Every sentence presupposes comprehension of every previous one; missing even a single word can compromise the understanding of everything that follows. When reading mathematics, half-understanding does not gradually evolve into full understanding by merely plowing through unclear passages. In mathematics, partial understanding is no understanding at all.

Intensive reading requires active participation from the reader. Full understanding may require rereading a passage multiple times, pausing to look up unclear words or symbols in previous chapters, the glossary, index, or table of contents. It also may require verifying calculations, re-working examples, finding additional examples or counterexamples, drawing diagrams, or plotting graphs. Authors of mathematics textbooks often explicitly solicit readers to do this work by stating "the details are left to the reader." Finally, doing the exercises becomes an important part of the learning process, as deep understanding only sets in by solving problems. Furthermore, most assessments in mathematics classes measure students' understanding of the material by their ability to solve problems.

Intensive reading requires mental effort and focus. It is taxing on the mind, so it can only be done well when mentally well-rested and should be limited to 30-60 minutes between breaks. Intensive reading also requires frequent note-taking, so the responsible reader will not do it reclining on a couch or laying in bed, but sitting upright at a desk, without distractions, and with tools at hand: paper, pencil, and computer.

Recitations

Unlike other fields where extensive reading alone suffices to gradually build knowledge and understanding, learning mathematics requires actively *doing* mathematics. Not unlike developing musical skill or a sport, merely reading about mathematics, or watching others do

it, does not promote sufficient experience or understanding. In the
same way that learning upper-level experimental sciences (physics,
chemistry, biology, etc.) requires a significant component of hands-on
"lab" work, learning upper-level mathematics requires a significant
amount of "hands-on" work that involves doing exercises and solving
problems. For this reason, many mathematics courses alternate lectures with *recitations* (also called *lab sessions, drill sessions, exercise
sessions*, or *problem-solving sessions*), which are an important part
of the learning process. The process of solving problems clarifies concepts and solidifies true understanding. To successfully progress in
mathematics, it is essential to dedicate a significant amount of study
time to actually solving problems, rather than merely rereading lecture notes and books.

Learning definitions

Because of the heavy emphasis on written and oral expression in
mathematics, the student should approach learning upper-level mathematics like learning a language, where communication and understanding are impossible without first knowing the words. Thus, the
student must learn mathematical terms like the vocabulary of a foreign language or the specialized jargon of a field like medicine or
law. However, though some terms exhibit a Greek or Latin etymol-
ogy, mathematicians often borrow terms from everyday English. This
adds a deceptive difficulty in that a reader may not immediately realize that a certain word has a very different mathematical definition
than its descriptive definition in a nonmathematical dictionary.

Mathematical definitions play a central role in axiomatic mathematics. Every one is carefully worded to unequivocally assign a
mathematical term to a highly abstract concept, object, operation,
or property. An anthropologist once joking said of his own field that
there are as many definitions of "culture" as there are anthropologists
[78]. Not so for mathematics. Though intuition might shape the definition of a mathematical object, one cannot prove anything and say
anything of depth without a precise definition. Consequently, student
and mathematician alike must memorize definitions word for word.

Unlike a *descriptive* definition in a typical English dictionary,
a mathematical definition is *prescriptive*, a mathematical statement
that provides a method for solving a problem or a roadmap for proving a theorem. For example, consider the following definition:

A set of vectors $\{\vec{v_1}, ..., \vec{v_n}\}$ is *linearly independent* if and
only if the equation $c_1\vec{v_1} + \cdots + c_n\vec{v_n} = \vec{0}$ has a unique
solution.

This definition ties the abstract concept of *linear independence* to the number of solutions of an equation, which can be determined mathematically. Thus, the definition does not attempt to describe the concept, but rather relate it to a mathematical problem, in this case, solving an equation and counting the solutions.

It is important to realize that knowing the definition of a term is separate from knowing a mathematical method related to that term. For example, knowing a *method* for finding the eigenvectors of a given matrix differs from knowing the *definition* of an eigenvector. As valuable as it is, the method to find the eigenvectors will not always help with proving theorems about eigenvectors, and we must return to the definition. Knowledge of definitions becomes increasingly important in proof-based mathematics courses, where problem solving shifts from applying methods to proving theorems.

Learning theorems and proofs

Besides definitions, theorems are the building blocks of mathematical theories. A theorem is any mathematical statement for which someone has provided a proof relative to the theory's definitions. Combining, extending, and generalizing them can lead to new theorems. Unlike mathematical methods designed to solve specific problems, theorems deepen understanding by revealing important key properties, structures, relationships, and patterns. We also use theorems to confirm the validity of a method, or as a starting point to design new methods. Like definitions, a student should strive to commit theorems to memory, especially theorems bearing a specific name.

The reader may opt to study the proof of a theorem separately from learning the statement of the theorem itself. Studying a theorem's proof is especially useful if the proof illustrates a common technique, as one of the primary goals of this type of study is to gradually build a personal repertoire of tools that can be used to prove other theorems. If there exist multiple different ways to prove a theorem, the reader may find it an instructive exercise to compare proof strategies and to decide which is more intuitive, insightful, elegant, or beautiful.

One of the ultimate goals of studying proofs is eventually to be able to write one's own proofs. As a difficult skill to develop, some liken proof writing more to an art than a science. Because almost every proof requires some unique insight, no fail-proof recipe or algorithm exists to prove any but the most basic proofs. Proof writing takes much practice and comes slowly as mathematical experience increases and maturity deepens.

The student should note that not all proofs are useful to study. Sometimes proofs are straightforward and uninteresting, but the author will provide them to satisfy due diligence. Other proofs are difficult to follow, even if the theorem itself is easy to apprehend. Moreover, proofs are often presented in a polished, final form, concealing the multiple false starts and dead ends that someone attempting to prove the theorem from scratch would likely encounter.

Examples and counterexamples

Euler [47] said: "Some facts can be seen more clearly by example than by proof." Because many students learn elementary mathematics primarily by following worked examples, elementary textbooks illustrate every theorem and method by numerous examples. In contrast, upper-level textbooks usually lack in examples. This deficiency can be seen as an invitation for a student reading intensively to create his or her own. Both definitions and theorems benefit immensely from illustrative examples and counterexamples. Ideally, every student should imagine at least one example and one counterexample for each definition and theorem and commit them to memory.

Example 7.3.1. The following definition of a *subspace of* $\mathbb{R}^3$ can be illustrated by an example and a counterexample. For the counterexample, it is further necessary to explain why the set is not a subspace of $\mathbb{R}^3$, i.e., to demonstrate that it is not closed under addition or scalar multiplication.

- **Definition.** A *subspace of* $\mathbb{R}^3$ is a nonempty subset of $\mathbb{R}^3$ that is closed under addition and scalar multiplication.

- **Example.**

$$\left\{ \begin{bmatrix} a \\ b \\ c \end{bmatrix} \in \mathbb{R}^3 \middle| c = a + 2b \right\}.$$

- **Counterexample.**

$$\left\{ \begin{bmatrix} a \\ b \\ c \end{bmatrix} \in \mathbb{R}^3 \middle| c = a + 2 \right\}. \qquad \triangle$$

For a theorem stated as an "if (hypothesis) ..., then (conclusion) ..." implication, we can imagine an example that verifies the implication, as well as a counterexample that demonstrates that the converse is false:

Example 7.3.2. Consider the following theorem:

- **Theorem.** Let $f : D \to \mathbb{R}$ be a differentiable function. If f has a local minimum or maximum at a point $x_0 \in D$, then $f'(x_0) = 0$.

- **Example.** Consider the differentiable function $f(x) = x^2$. The function f has a local minimum at 0. We verify that $f'(0) = 0$, as predicted by the theorem.

- **The converse is not true.** Consider the differentiable function $f(x) = x^3$. We observe that $f'(0) = 0$, yet f does not have a minimum or maximum at 0. $\triangle$

Example 7.3.3. Consider the following theorem:

- **Theorem.** If $T : \mathbb{R}^2 \to \mathbb{R}^2$ is a linear transformation, then $T\left(\vec{0}\right) = \vec{0}$.

- **Example.**

$$T\left(\begin{bmatrix} a \\ b \end{bmatrix}\right) = \begin{bmatrix} a + b \\ a - b \end{bmatrix}$$

The transformation T is linear. We verify that $T\left(\vec{0}\right) = \vec{0}$.

- **The converse is not true.**

$$T\left(\begin{bmatrix} a \\ b \end{bmatrix}\right) = \begin{bmatrix} a^2 \\ b^2 \end{bmatrix}.$$

We observe that $T\left(\vec{0}\right) = \vec{0}$. Yet the transformation T is not linear. $\triangle$

Delayed Understanding

In elementary mathematics courses, concepts are often simple enough that students can grasp them at the same pace as the instructor presents them in class. Many students benefit from *immediate understanding*: their minds mentally digest the new material in "real time" and they can apply new methods immediately. Gifted students often leave a mathematics class with the satisfying feeling of having learned something new that makes sense and is solidly anchored in their mind.

In upper-level mathematics, immediate understanding occurs rarely; students must grapple with abstract concepts, complex methods, and intricate chains of reasoning that require extensive "rumination" before understanding sets in. Therefore, students often experience the unpleasant and unsettling feeling of leaving a class in a blur of confusion. This feeling is perfectly normal and is not a reflection of a student's background, academic preparation, or ability to maintain focus in class. Advanced mathematical knowledge simply takes time to sink in, a phenomenon called *delayed understanding*. This delay is normal, and experienced mathematicians are accustomed to it.

The length of time varies from person to person. Although students may develop helpful strategies in this area, not much that can be done to substantially accelerate the process. Studying upper-level mathematics requires a combination of active learning (reading, writing, studying, doing exercises, asking questions), and times during which the mind unconsciously engages in passive learning. Giving the mind time to rest between study sessions is as important as the quality of active study time, as the mind can continue to subconsciously ruminate on a mathematical problem even while one is eating, exercising, or showering. The history of mathematics recounts many examples of mathematicians receiving insight for a problem during a time of passive learning.[7]

Students who work as tutors or teaching assistants often experience delayed understanding when explaining mathematical concepts a few years after learning them themselves. They often report that true understanding came only after allowing the concepts to coalesce in the mind over the course of a few semesters.

Being stuck on a problem

Whether engaged in original research or working through a weekly homework set, no student of mathematics will escape the experience of getting "stuck on a problem." Some may find the sensation unpleasant, unsettling, stress-inducing, aggravating, or depressing. Gifted students, in particular, who rarely faced this challenge in elementary mathematics courses may view getting stuck as alarming and abnormal.

In upper-level mathematics, the state of being stuck on problems is common and normal. It is not a reflection of a person's intelligence or background preparation. Instructors expect students to get stuck, and remain stuck for long periods of time. Wrestling with difficult problems forms part of the nature of being a mathematician, and experienced mathematicians realize that most challenging problems take time to solve. This goes hand-in-hand with the *delayed understanding* concept: difficult problems need time to mature and are best approached by alternating periods of active and passive problem-solving, realizing that the mind still processes the problem subconsciously during passive periods. Professional mathematicians

---

[7]Archimedes discovered the now-famous *Archimedes' Principle* while taking a bath. Henri Poincaré describes receiving a flash of insight for a problem he had been stuck on while boarding a bus on a geological excursion [64]. W.R. Hamilton came up with the fundamental concept of *quaternions* while taking a walk with his wife. Hamilton did not have anything to write with at that moment, but did not want to lose his idea, so he carved the now-famous formula for quaternions ($i^2 = j^2 = k^2 = ijk = -1$) with his pocket knife into the stone of a nearby bridge.

keep multiple unsolved problems marinating in the back of their mind at any given time.

This aspect of active engagement contrasts with the modern expectation of always finding solutions to most questions at one's fingertips. In order to learn well, the contemporary student must resist the instinctual urge to simply look up how someone else "did the problem." Doing so only short-circuits the learning process by which perseverance leads to insight.

Foundational self-study

In mathematics, new knowledge builds on previous knowledge, so, to some degree, students must learn concepts in a specific order. For example, multivariable calculus builds on single-variable calculus, which builds on concepts of analytic geometry learned in precalculus, which themselves rely on concepts learned in algebra, geometry, and trigonometry. Students may already have realized the necessity of relearning the material of past courses in order to successfully apprehend advanced concepts that build on them in subsequent courses.

In upper-level mathematics courses, students often need to engage in a significant amount of personal self-study to fill knowledge gaps, i.e., to learn or relearn foundational knowledge from previous courses. For example, students in a differential equations course may encounter a problem that features a *hyperbolic trigonometry* function, *integration by partial fractions*, a *trigonometric substitution*, or some other concept presented in an elementary calculus course. If a student does not remember, or never learned, these concepts, it is essential to take the time to (re)learn them on his or her own. This is common and normal; instructors realize this and expect it. Most upper-level mathematics courses list one or more *prerequisite* courses to provide some advisory guidance about the type of prerequisite knowledge expected to successfully navigate the course. However, merely taking a prerequisite course does not dispense the need for foundational self-study, as students enter upper-level mathematics courses with widely different backgrounds.[8]

The necessity of solidifying one's own foundational knowledge forms an essential component of any upper-level mathematics course, so students should expect to invest time and effort for this, in addition to the time required to study the concepts of the course itself. This self-study often requires consulting books and notes from previous courses, re-doing exercises, and reaching out to classmates, teaching assistants, or instructors for guidance and clarification.

---

[8]For example, two students who have taken different Linear Algebra courses may have been exposed to different sets of concepts or may remember different things.

Learning from others

Whereas many students experience elementary mathematics as a solitary activity, learning mathematics alone becomes exceedingly difficult when transitioning into advanced coursework. Studying upper-level mathematics benefits immensely from interactions with others. Wrestling with a difficult problem with one or more other people can unlock situations where a single person would remain stuck indefinitely, even after extensive and intensive reading and rereading lecture notes and books.

Although instructors and teaching assistants can certainly provide guidance when a student is stuck on a problem, most students find that working alongside other students offers an ideal setting to promote learning and sharpen one another's problem-solving skills. When students study together, they can answer each other's questions, clarify concepts, quiz each other, and get each other unstuck. Many minds working together can achieve better results than the sum of them working independently. For this reason, upper-level mathematics courses often assign group projects, where several minds are expected to work in concert to obtain a solution.

7.3.3 Professors

The job of a professor consists of various duties, many of which are not immediately visible to students. In addition to teaching classes and interacting with students, professors get involved in mentoring, advising, researching, writing, publishing, going to conferences, and performing service to their institution and to the community.

The duties of a professor fall roughly into three categories: (1) teaching, (2) research, and (3) institutional service. Depending on the professor and the institution, the time dedicated to each of these three areas may differ widely. Some professors spend most of their time teaching, whereas at a research university, doing research occupies the majority of a professor's time and energy.

In addition to employment at their primary institution, professors are also members of professional associations, such as the Mathematical Association of America, the American Mathematical Society, or the Society for Industrial and Applied Mathematics. Professors may hold administrative or executive roles within these organizations, and their involvement may require traveling to attend regional or national conferences.

As in many academic fields, teaching mathematics at the college level requires at least a master's degree. This type of degree certifies that the recipient sufficiently "masters" the existing body of knowledge in his or her field to be qualified to teach it. In addition to a master's degree, many professors hold a Ph.D. (Doctor of Philosophy) degree. This degree requires more than merely mastering existing knowledge; it certifies that the recipient has demonstrated the ability to create new knowledge, to engage in original research.[9]

In addition to professors, many universities employ other faculty members with the title of *lecturer*, *instructor*, or *adjunct professor*. Their primary role is teaching; these positions usually do not require doing research or institutional service.

Teaching

The teaching role of a professor is the most visible to students. It involves three main facets: lecturing, assessing, and advising.

In many colleges and universities, departments hire *teaching assistants* (TAs) to support professors in their teaching duties. In mathematics, TAs might grade assessments, administer quizzes, and occasionally give lectures. Grading large assessments (midterm and final exams) and assigning final grades at the completion of the course remain the purview of the professor.

In addition to day-to-day teaching, professors who advise students hold a long-term perspective on their advisees' education, a role that becomes more important as students progress into upper-level coursework. This type of advising includes giving recommendations on course sequencing and scheduling, as well as providing guidance on vocational and career preparation, summer research programs, internships, and scholarships.

Research

Many professors view their role as researchers as their primary identity. It is within this facet of their work that they have license, granted by their institutions, to pursue their own research interests. Furthermore, many colleges and universities view a professor's scholarly research as the major, even the sole, criterion for tenure, promotion and compensation. Because research output is often measured by the number of papers published, their length, and their quality, many professors feel a continuous pressure to publish ("publish or perish"). "Research" may mean different things to different professors: it may be original or expository, solitary or collaborative. The research topics often vary widely from the topics of the courses they teach. Research requires a great deal of writing, as all research results must appear

---

[9]The word "philosophy" in Ph.D. is to be understood in its general sense of "knowledge." Therefore, one can obtain a Ph.D. in any field of academia by creating new knowledge in that field. The Ph.D. differs from other types of "professional" doctorate degrees such as M.D., J.D., Psy.D., or Pharm.D., which do not require producing original research.

in published form (often in peer-reviewed journals) to "count." In addition, professors often give presentations related to their research at conferences.

Mathematics professors often do research in collaboration with other researchers, who may themselves be professors of mathematics or scientists in other fields. These collaborators often work at different institutions, which requires that professors travel regularly to interact with their collaborators in person.

Many researchers actively seek out students to assist them in their research as *research assistants* (RAs). Often, professors prefer to give these positions to graduate students, but, at some institutions, professors or departments regularly involve undergraduate students in research activities.

Finally, conducting research of any kind requires monetary funds to cover experimental equipment, literature resources, publication fees, travel to conferences, meetings with collaborators, and stipends for research assistants. Because professors are themselves responsible for securing the necessary funds to pursue their research, they must allocate a portion of their time to soliciting money and grant writing.

Institutional service

Many colleges and universities give administrative duties and a significant amount of executive power to professors on matters concerning the curriculum, admissions, finances, academic policies, recruiting, hiring, promotion of faculty, and other aspects of college administration. In addition, colleagues and collaborators often call upon each other to peer review their scholarly work. These tasks, though not directly related to teaching or research, fall under the rubric of "institutional service" and can consume a significant amount of a professor's time and energy. Some institutions may reduce the teaching load of a professor assigned a particularly heavy administrative role, such a department chair or divisional dean.

EXERCISES FOR SECTION 7.3

1. State the definition of a *subspace of* $\mathbb{R}^3$. Find an example and a counterexample, i.e., a subset of $\mathbb{R}^3$ that is a subspace of $\mathbb{R}^3$, and one that is not.

2. A *binary operation* $*$ on set S is *commutative* if and only if $x*y = y*x$ for all elements x and y in S. Find an example and a counterexample of a commutative binary operation.

3. State the definitions for an *invertible matrix* and a *diagonalizable matrix*. Find examples of a matrix that is (a) invertible but not diagonalizable, (b) not invertible but diagonalizable, (c) both invertible and diagonalizable, and (d) neither invertible nor diagonalizable.

4. Consider the theorem: "A square matrix is invertible if and only if all of its eigenvalues are nonzero." Give an example that illustrates the theorem and its converse.

5. Consider the theorem: "If x is a real number such that $x > 0$, then $x^2 > 0$." Give an example that illustrates the theorem and a counterexample that its converse is not true. You do not have to prove the theorem.

6. Consider the theorem: "If the series $\sum_{n=1}^{\infty} x_n$ converges, then the limit $\lim_{n \to \infty} x_n = 0$." Give an example that illustrates the theorem and a counterexample that its converse is not true. You do not have to prove the theorem.

7. Do some research on the mathematics professors at your college or university. For each one, determine:

   (a) which mathematics courses they regularly teach,

   (b) what type of research they do, and

   (c) what type of institutional service they perform.

7.4 Mathematical Vocations

If you are reading this book, you are probably at least considering pursuing a degree in mathematics. You may even have already committed yourself by declaring a mathematics major. There are many excellent reasons for choosing to study upper-level mathematics: you may be aiming for a well-rounded education in a quantitative field, or maybe mathematics comes easily and naturally to you, or perhaps you simply enjoy doing mathematics. You may have also made the decision with an eye toward the future and considered the benefits of earning a degree in mathematics in terms of preparation for life in general. Maybe you have already identified a specific vocation for which a degree in mathematics is advantageous or necessary.

The following sections explore how the study of advanced mathematics can intersect with a student's plans and desires for the future. Subsection 7.4.1 focuses on the high value of a bachelor's degree in mathematics in today's marketplace. Such a degree is a flexible and versatile credential that opens many doors leading toward a wide variety of vocations. However, because a mathematics degree is an academic degree (as opposed to a professional degree like business or engineering), it is important that a student learn how to articulate the value of the degree and to showcase the distinctive character of a mathematics education in a way that other people, especially potential employers, will appreciate.

Subsection 7.4.2 explains the meaning of the term *vocation* and how a mathematics student can begin the process of defining a post-college vocation and building a vocational pathway that leads to it.

College offers an excellent environment to start taking proactive steps toward planning a first vocational pathway. Proper planning will ease the transition from student life to early professional life. Subsection 7.4.3 lists some of the more common pathways available to people who study upper-level mathematics. Finally, Subsection 7.4.4 offers some practical advice on what a student can do beyond the classroom to prepare for various mathematical vocations.

7.4.1 Value of a bachelor's degree in mathematics

Employability

In popular culture, there are two common perceptions about mathematics. On one hand, many see mathematics as a difficult, esoteric, and daunting subject. Even well-educated people often openly admit to possessing little aptitude for mathematics; even fewer know anything about advanced mathematics. On the other hand, countless people readily acknowledge mathematics as an essential part of everyday modern life, from economic prosperity, to national security and daily comfort. For these reasons, much of society views the mathematically adept as extremely valuable human resources [10]. A bachelor's degree in mathematics holds a reputation as a high-level credential that carries prestige and historical pedigree. The curriculum of an undergraduate program in mathematics is well established, standardized, demanding, and rigorous. Those who can successfully navigate its challenges are generally seen as having the talent, ability, grit, and intelligence to successfully tackle the world's most difficult problems.

Many studies conducted over the past 20 years reveal that a bachelor's degree in mathematics remains one of the most valuable undergraduate degrees, ranking high in terms of employability, salary earnings, and job satisfaction:

- A yearly study conducted by the website CareerCast.com compares 200 professions according to 25 factors such as job environment, income, growth potential, employment outlook, and stress. Over the past 20 years, the professions of Mathematician and Statistician have consistently ranked in the top 10. Furthermore, most of the other professions appearing in the top 20 also require substantive mathematical reasoning ability. These include Accountant, Actuary, Computer Systems Analyst, Data Scientist, Database Administrator, Financial Analyst, Logistician, and Operations Research Analyst [1].

- Employment data compiled by the National Association of Colleges and Employers [3], the Georgetown University Center on Education and the Workforce [2], and the Michigan State

College Employment Research Institute [56] reveal that a bachelor's degree in mathematics stands among the most lucrative of undergraduate degrees, exceeding those of the other sciences, business, economics, humanities, and arts.

- The *Occupational Outlook Handbook*, compiled by the U.S. Bureau of Labor Statistics, predicts a 30% increase in the demand for mathematicians and statisticians between 2018 and 2028 [5].

Generality and flexibility

The demand for professionals who are well versed in mathematics has steadily increased in the past 50 years, owing to the increased use of computation, statistics, data analysis, and mathematical modeling in virtually every sector of industry, business, and government. The rapid emergence of new technologies in the past few decades brought with it a proliferation of new professions (such as Data Scientist or App Developer) that did not exist 20 years ago. These have since become very common and lucrative. This dramatic shift in the employment market poses a deep challenge to university programs, which must not only keep up with the current trends and technologies but also anticipate the skills that will be in demand when students graduate. Former U.S. Secretary of Education Richard Riley summarized this challenge as follows: "We are currently preparing students for jobs that don't yet exist, using technologies that haven't been invented, in order to solve problems we don't even know are problems yet [40]."

Mathematics curricula by design remain general in nature. This gives the degree a great deal of flexibility to be relevant to a wide range of current and future vocations and careers. The general and flexible nature of an education in upper-level mathematics is especially well suited to respond to the demands of a rapidly changing employment market, as it trains mathematicians to approach new and complex problems for which no clear solution path yet exists. Although few job postings ask for a "mathematician" (and positions which do call for this job title often require an advanced degree), an increasing number of openings with titles such as "analyst," "engineer," "investigator," or "scientist" require quantitative problem-solving skills. As a future mathematics graduate, the student should recognize these positions as especially well suited for him or her and realize that an analytical skill set offers a uniquely competent edge to thrive in these types of emerging professions.

However, the general nature of skills developed for a degree in mathematics may pose some challenges for selling oneself to potential employers as compared to professional degrees such as accounting or engineering. For this reason, it is important to realize that the strength of a mathematician's training does not necessarily reside in specific skills, such as being able to solve a second order differential equation by hand, but rather in the more abstract abilities and attributes acquired during the course of an entire undergraduate mathematics curriculum. We describe some of these attributes below, so that the reader may be aware of them and learn to articulate them to potential employers.

The general background gained with an undergraduate degree in mathematics also constitutes a solid foundation to pursue graduate studies or a vocation that requires a graduate degree, such as law or medicine. A student can supplement a bachelor's degree in mathematics with a variety of master's degrees in business, computing, data science, economics, finance, or management. A mathematics degree offers an excellent preparation for a career as a lawyer: the training in critical, logical thinking, combined with the ability to write proofs and solutions clearly, sequentially, and convincingly, is directly applicable to any legal profession. Admission into a graduate program in business or law requires scoring well on the GMAT or LSAT standardized tests, respectively. A study by the National Institute of Education shows that mathematics majors generally perform better on these tests than students from other majors (even business, accounting, political science, or pre-law) [4].

Attributes and skills

The employment market has shifted in recent decades toward an environment where the demands for professionals with specific skills evolve rapidly, causing new fields to emerge in response to advances in technology. Because it is impossible to anticipate what specific skills employers will demand in 5 to 10 years, the key to continued employability in today's economy lies in an undergraduate training that cultivates *attributes* and *transferable skills*, in addition to specific skills, which may be obsolete a few years after graduation.

Attributes, also called *soft skills*, consist in the general way one approaches work. These qualities increase a professional's worth regardless of the field. In today's marketplace, the four most important professional attributes are:

1) **Self-learning.** Ability to learn new ideas, concepts, and technologies independently and quickly.

2) **Problem solving.** Ability to solve new problems creatively, methodically, and sequentially. This includes the ability to apply methods used to previous problems to new problems, and the ability to approach a complicated problem by breaking it down into smaller pieces, solving those individual pieces separately, and reassembling them to solve the original problem.

3) **Attention to detail.** Ability to work with precision and reliability in small details, without losing sight of the big picture.

4) **Perseverance.** Ability to face difficulties with determination, grit, and tenacity.

These attributes are all naturally developed throughout the course of any undergraduate mathematics curriculum. By virtue of their training, mathematics students come to excel at them. Although cultivated in the process of developing specific mathematical skills, they are transferable to virtually any vocation. Being self-aware of these unique qualities will allow a student to highlight them effectively to potential employers.

Skills (or *hard skills*) consist in the ability to effectively perform certain tasks and can be broken down into *specific* (or *contextual*) skills and *transferable* (or *portable*) skills. Specific mathematical skills, such as the ability to solve equations by hand and prove theorems, are important skills to master to obtain a mathematics degree and constitute marketable skills for a career in academia, research, or math education. But specific mathematical skills are not directly applicable outside of those vocations. In contrast, transferable skills are those which are learned in one context and effectively applied in another [56]. In today's rapidly changing marketplace, transferable skills are very important, especially the following three:

1) **Computing.** Proficiency with computing tools and computer programming. This is an essential skill to master, as nowadays virtually every quantitative task in any career involves a computer.

2) **Communication.** Proficiency in expressing oneself verbally and in writing with clarity, precision, and professionalism. This is another essential skill, as the use of specific mathematical skills for analyzing and solving problems typically come into play *after* one has extensively discussed the problems themselves in order to understand them. Similarly, one is typically asked to communicate results in some way after solving a problem.

3) **Collaboration.** Proficiency in working effectively with other people is essential, as almost every vocation requires coordinating your skills with those of other people in a teamwork environment.

A mathematics curriculum may help develop these three essential transferable skills to some extent. But not all mathematics programs emphasize them because of the difficulty in assessing them academically. For this reason, unaware students may graduate with a deficiency in these areas. In reality, computing, communication, and collaboration skills are as important, if not more, as specific mathematical skills for any mathematical vocation. However, students are usually expected to develop them on their own.

Comparison with other undergraduate majors

Majoring in mathematics typically leads to a bachelor's degree in mathematics, which affords the reputation of a respected, prestigious, and marketable degree while remaining general and flexible. Students interested in mathematics might pursue the following majors instead of, or in addition to, mathematics:

- **Applied Mathematics and Statistics.** Some institutions offer a separate major in applied mathematics and/or statistics. These programs usually sacrifice some of the coursework in theoretical, proof-based mathematics and require instead courses in mathematical modeling, numerical methods, computing, and applied projects. Applied mathematics and statistics majors help develop the same attributes and transferable skills as a standard mathematics major and may offer a slight advantage for vocations in the industry, but they may also put students who later decide to pursue graduate studies in theoretical mathematics at a slight disadvantage.

- **Data Science.** These are relatively new programs created in the past decade in response to the recent increase in the number of professions that involve working with data. An undergraduate degree in data science opens many doors in today's economy. Because of its relative novelty as a field, there is no broadly accepted curriculum for a data analytics degree. Every program approaches the content and education in these fields differently.

- **Engineering.** Majoring in engineering provides a direct vocational pathway to a career as an engineer, a well-established and lucrative profession. Engineering programs require the same lower-level mathematics coursework as a mathematics program (calculus, differential equations, linear algebra) but do not require any further training in theoretical mathematics, focusing instead exclusively on numerical methods and computing tools specific to the needs of the engineering profession. As a professional degree, an engineering degree has a much narrower focus than a general mathematics degree, but it is usually required to gain entrance and acceptance into the engineering profession.

- **Natural Sciences.** Majoring in physics, chemistry, biology, geology, ecology, and similar majors requires some mathematics, but the coursework is limited to mathematical and computing tools, to the exclusion of theoretical mathematics. As experimental sciences, these majors emphasize the scientific method, requiring a significant amount of lab work, which standard mathematics programs do not require.

- **Business and Economics.** As social sciences, these programs focus on applying the scientific method and empirical modeling to the development of economic theories. The more theoretical of these programs provide training in mathematical methods applied to economics (mostly regression tools used in econometrics), while others follow a narrower, professional focus, leading to specific careers in accounting or actuarial science.

Selling oneself

The student entering the professional world will need to "sell" himself or herself, meaning that he or she will need to convince other people, especially potential employers, of his or her value as a professional. Holding a bachelor's degree in mathematics is an excellent first step that sets the applicant apart, but it is only a piece of the puzzle that occupies a single line on a résumé. Successfully selling oneself requires creating a narrative that highlights personal attributes and skills and supports claims with specific examples.

The goal of the application process is to obtain an interview, where the applicant is invited to sell himself or herself in person. Until that crucial moment, one must sell oneself via three types of documents: (1) résumé, (2) transcript, and (3) letters of recommendation. Each of these documents offers a different perspective of a person's worth as a potential employee or graduate student. Employers and graduate program directors use the information contained in all three documents together to gauge the value of a candidate and decide whether to pursue him or her. Developing the content of these documents takes time, so it is important to start the process early and to make incremental progress on them as a student.

1) **Résumé.** The résumé describes credentials and experience. A bachelor's degree in mathematics is a strong centerpiece for a résumé, but a job applicant should complement this with work experience in industry, research, and/or teaching. Mathematics programs usually do not require any experience of either type to grant a degree, so students must be proactive in building experience as they progress through their studies. The résumé may also mention extracurricular activities or volunteer work that contributed to developing one's attributes or transferable skills, but these items cannot replace actual industry, research, or teaching experience.

2) **Transcript.** The transcript records a student's courses and grades. Students often focus exclusively on grades, but employers rarely look at them; many do not even require a transcript at all. Employers who do require a transcript are more interested in the scope of your coursework than grades. For this reason, a student should carefully and strategically choose elective upper-level courses in accordance with vocational plans and in consultation with his or her academic adviser and/or career counselor. A specific course on the transcript may make a significant difference in employability, regardless of the grade obtained.

3) **Letters of recommendation.** Employers and graduate programs frequently request letters of recommendation to help build a more complete portrait of potential candidates. Letters come from people who know the student well enough to be able to speak authoritatively on the student's attributes, skills, and other intangible qualities such as emotional maturity, professionalism, attitude, engagement, motivation, perseverance, and integrity. Because grades cannot effectively measure these aspects, their best assessment comes from recommendation letters. Ideally, students who need letters of recommendation for graduate school or a job application should request them from professors, work supervisors, or colleagues who can comment on the student's work ethic and attitude, and who can support their observations with specific examples and anecdotes. Strong letters require that the student establish a personal rapport with potential writers ahead of time. It is wise to take steps to make oneself known to potential recommendation letter writers early on in one's studies, so that they have ample time to observe one's performance before being asked for a letter.

7.4.2 Vocational discernment

The undergraduate experience offers countless opportunities to grow and mature, to expand knowledge, to develop attributes, and to sharpen skills. All of these aspects of a student's education play a role in defining his or her identity, which includes discerning a first vocation after graduation. Growing in self-awareness about what constitutes the facets of one's identity involves a long-term, introspective process by which we define what motivates us and gives purpose to our lives. For students, this includes building a vision for the type of work they wish to pursue after college.

A *vocation* is a type of job or occupation that one desires, that gives satisfaction, and that provides a sense of meaning to one's life

[18]. Some people possess a clear sense of vocation from their childhood, settling on a specific profession early on. Other people do not develop a clear sense of vocation until much later in life. For some people, a unique vocation guides them throughout their whole life, while others experience different seasons throughout their life, each one defined by a different vocation. A vocation may also include other important aspects not necessarily connected to a money-making activity, such as getting married, raising children, caring for elderly parents, or engaging in a meaningful volunteer activity. Furthermore, a vocation encompasses not only one's desires and talents, it is also heavily influenced by family, friends, community, and social environment. A *vocational pathway* is a series of steps that a person follows along the road toward defining and fulfilling a vocation. This may include taking courses, developing your attributes and skills, obtaining a credential, gaining experience, or interacting with certain people.

A student's choice of undergraduate major may find a vocational pathway more or less clearly built into the curriculum. A program leading to a professional degree, such as engineering, business, economics, pre-med, or pre-law, presents a clearly defined vocational pathway. A student need simply follow the requirements of the major to emerge from college with the necessary coursework, credentials, skills, and experience needed to be immediately employable in a specific profession. On the other hand, programs leading to a mathematics degree do not have a specific built-in vocational pathway, thereby giving students the flexibility to creatively design their own.

Mathematics programs are excellent for students who do not already have a specific vocation, do not know which vocational pathway to follow, or have a vision for an unconventional vocational pathway. The vocational flexibility of a mathematics degree is one of its greatest strengths, as it opens a wide variety of doors and puts the choices into students' hands. However, on the flipside of this strength, a student who does not exploit the pathway well runs the risk of graduating with a general degree and no clear plan on how to effectively use it. Designing a personal vocational pathway takes time and effort. The mathematics student should start this process early, ideally 2 to 3 years before graduating. Forging one's own pathway requires that the student be proactive, assertive, curious, and willing to take risks. Despite the occasional discouragement, the reward is worth the effort.

Identity and vocation

Defining one's vocation fits into the more general framework of defining one's *identity*, which is complex and multifaceted, influenced by factors such as personality, cultural environment, spirituality, and current vocation. A person's cultural identity is the part shaped by the culture in which one grew up and includes cultural reference points, values inherited from family, upbringing, and education. This part of our identity also encompasses how we interact with other people in society and what role we play in our community.

Defining vocation(s) is a life-long journey that, for many, begins in college. Therefore, the first vocation for most people is that of being a student, which is a vocation in and of itself, in addition to being a vocational pathway toward the next vocation after graduation. Co-curricular activities, such as job, hobby, sport, or instrument, and other commitments such as family obligations may significantly influence a student's vocation. Students should live out their vocation to the fullest but keep in mind that it is a temporary vocation, designed to lead to the next season of life, which will carry a different vocation. It is important to prepare for this inevitable transition, so that it does not take one by surprise.

Landing a good job after graduation is an important and necessary component of a first vocational pathway. However, the job search process, which occurs shortly before graduation, should fit into a longer journey punctuated by conscious vocational decisions. The actual job search should arise as the culmination of a season of vocational discernment, rather than a last-minute thoughtless process guided by pure randomness. Taking strategic steps to define one's first vocation after college early will give the student time to carefully craft a pathway that informs and shapes choices when time comes to look for a job.

A student who has never given any thought to vocation may start the journey by asking the following four questions:

1) What do I love to do?

2) What am I good at?

3) What does the world need?

4) What can I be paid to do?

Answers to these questions will provide a starting point to begin researching possible vocations. However, they may also require some preliminary research, consulting an adviser or career coach, or speaking informally with people who know us well, such as friends and family. An ideal vocation lies at the intersection of the answers to these questions, i.e., a vocation where passions and talents coincide with the things that the world needs and is willing to pay for. It may be hard to imagine a single realistic vocation that fulfills all four of these requirements, but someone may be able to think of one or more that fulfill three out of the four.

Vocational pathways

After a student identifies one or more possible vocations, he or she can start planning vocational pathways that lead to them, perhaps in terms of which courses to take or what type of experience to acquire to reach those goals. Besides earning a degree in mathematics, there are certainly other vocational steps the student can work on at the same time, such as learning a software tool or networking with professionals in a certain line of work. Some steps of a vocational pathway require a sequencing that delays them until later. For example, to pursue a career in mathematical ecology, the student may decide to focus on mathematical studies first and plan to deepen knowledge of ecology later.

Most people develop parallel vocations or a sequence of vocations during their life. Consequently, a student should not feel pressured by the false notion that planning a vocational pathway means planning one's entire future. Mathematics students may desire to explore vocations that have nothing to do with mathematics, such as pursuing a career in the performing arts, teaching English in a developing country, or getting married and raising children. It is perfectly possible to pursue more than one vocation either in parallel or sequentially, putting one on hold and returning to it at a later season of life. Planning vocational pathways does not entail abandoning dreams, although it may involve prioritizing them, or imagining strategic ways to combine them. In any case, most students need only to focus on the pathway leading to the first post-college vocation, which will give a sense of direction and a goal other than graduation.

Above all, undergraduate students should approach the process of vocational discernment with an open mind. With maturity and life experiences, dreams and desires may change, especially as life surprises us with unexpected challenges and opportunities. Most people change vocations multiple times throughout their lives, so planning vocational pathways will probably be a life-long engagement.

Guidance

The process of defining one's vocation(s) and imagining strategic pathways leading to them is a journey that requires a great deal of introspection, but students do not have to navigate it alone. Many people can provide guidance along the way:

- **Career counselors.** All colleges and universities employ career counselors who specialize in guiding students along the process of defining vocations. These counselors can assist in the beginning stages by administering personality tests or prompting self-reflection questions like the ones mentioned in the previous section. In later stages, they can help identify specific professions and strategically plan every step of a vocational pathway, including searching and applying for internships, writing a résumé, and creating strategies to sharpen vocation-specific skills that are not naturally integrated in the college curriculum.

- **Friends and family** are an excellent resource in the first stages of vocational discernment, as they know us well and can often objectively identify skills and talents better than we can ourselves.

- **Professors and teaching assistants** who evaluate class performance can give very specific feedback about skills and attributes. Unable to accurately gauge themselves objectively, students can benefit immensely from a professor's honest assessment for how the student compares to others in the field. Asking professors or teaching assistants for vocational advice is a natural way to initiate a personal rapport with them, which may be useful later on, especially if that professor could serve as a good recommendation letter writer. Finally, if the student considers pursuing an academic career, professors can give pertinent advice from their personal experience.

- **Books and online tools.** There are many resources to consult at all stages of the process, whether just beginning to think about one's identity or looking for insight into a specific career. The larger mathematical associations, in particular MAA, AMS, ASA, and SIAM, also provide a wealth of vocational resources on their websites.

7.4.3   Mathematical vocations

Some undergraduate students delay thinking of vocation as long as possible, while others commit to pursuing a specific career, whether mathematical in nature or otherwise, from long before college. We expect that anyone reading this book enjoys or is good at upper-level mathematics, and hence may consider a mathematical theme for post-college vocation. The following sections lay out various typical vocational pathways that undergraduate mathematics students pursue in parallel with their coursework. Even if a student changes direction later, it is a good idea to pursue one of these; doing so gives a goal other than merely graduating, and provides direction for vocational decisions while in college.

Vocations in industry

Working in industry encompasses any for-profit or not-for-profit business, in the private or public sector, as an employee or business owner. A degree in mathematics can lead to a great variety of different jobs in many areas, including accounting, computer programming, engineering, finance, manufacturing, medical, merchandising, purchasing, software development, sales, supply chain, and transportation. Because a mathematics degree is so general, it opens doors in almost every sector of the economy, even those for which there exist specialized degrees (such as business, economics, and engineering). Most of the jobs available to mathematicians in these fields involve analytical, quantitative, problem-solving, and computational skills, and often carry job titles with the words "analyst," "engineer," "investigator," or "scientist."

In most cases, these careers do not require an advanced degree beyond a bachelor's degree in mathematics. However, obtaining a master's degree may open additional doors later, a few years into a career. Many large universities offer professionally oriented master's degrees with a specific focus that develops the necessary skills for advancement in a specific career. Some examples include a Master's in Business Administration (MBA), Data Analytics, Scientific Computing, Biomathematics, Biostatistics, Mathematical Finance, Operations Research, or Public Policy. Admission into most of these programs requires a bachelor's degree and a good score on the Graduate Management Admission Test (GMAT) or the Graduate Record Exam (GRE). A bachelor's degree in mathematics is a strong credential that facilitates entrance into these specialized programs, and mathematicians usually score higher than average on the GMAT and GRE.

Vocations in finance or law

Careers in finance and law usually require an advanced credential or degree beyond a bachelor's degree.

It is possible to start a career in finance with a bachelor's degree, but career growth remains limited without an additional credential. Actuaries and accountants need a certification, financial advisers and stock brokers need to be licensed to sell investment products and financial securities, similar to how real estate and insurance brokers need to be licensed to sell their products. These certifications often require having a certain number of years of experience on the job and passing one or more exams.

A career in law requires going to law school to obtain a *Juris Doctor* (J.D.) or another graduate-level professional law degree.

Admission into law school requires a bachelor's degree and a good score on the Law School Admissions Test (LSAT). A bachelor's degree in mathematics is a strong credential that facilitates entrance into law school, and mathematicians usually score higher than average on the LSAT.

Vocations in secondary education

Working as a mathematics teacher at the secondary level in the public school system requires a bachelor's degree in mathematics as well as a secondary teaching certification. The requirements for obtaining the certification vary by state (and country), but generally require passing a basic skills exam, a content area exam, a teaching assessment, and working a certain number of hours as a student teacher under the observation of a mentor. Certification is not a requirement for teaching in a private school, although many private schools require that their teachers be certified.

Another pathway consists in working on the certification by enrolling in a one- or two-year Master's conversion program after obtaining a bachelor's degree. In addition to granting the necessary credential to teach, these programs arrange for placement into a school for student teaching and award a Master of Arts in Teaching (MAT) or Master of Education (M.Ed.), which also garners a higher salary in public schools.

Vocations in academia

Academia refers to a career that combines teaching (usually at the college level and above) and scholarly research at an institution of higher education. Academics occupy faculty positions at colleges and universities and carry the title of *Professor* (or variations thereof). In addition to their permanent, full-time faculty, academic institutions rely on several *adjunct professors* (also called *instructors* or *lecturers*) to teach additional classes. Adjunct positions are often temporary or part-time in nature and do not require any research or scholarly work outside of teaching. Most institutions require that mathematics adjuncts hold at least a master's degree in mathematics or a related field.

The job of a full-time mathematics professor depends on the relative amounts of teaching versus research that their institution requires. On one extreme are those professors who spend all of their time teaching and performing teaching-related duties (such as curriculum development, advising, preparing lectures, grading assessments, and managing teaching assistants). On the other extreme, research professors teach very little and spend most of their time engaged in research and research-related duties (such as writing grants, publishing articles, presenting papers, and attending conferences). Depending on their institution and employment contract, mathematics professors can occupy any position between these extremes and must split their time between teaching and research accordingly. Professors who focus primarily on research usually work at research universities, which include many large public and some private universities that largely delegate the task of undergraduate teaching to graduate students. At smaller institutions, especially those that lack graduate programs, professors spend a larger proportion of their time teaching.

An academic position that involves any amount of research requires a *Doctor of Philosophy* (Ph.D.). Ph.D. programs involve taking courses, passing exams, doing research, and writing and defending a dissertation. Academic positions with a heavy research component require, in addition to a Ph.D., a certain number of years of research experience, usually obtained by working as in full-time research in a post-doctoral research program (called a *postdoc*) at a research institution.

Anyone considering an academic vocation in mathematics must apply to a Ph.D. program after the bachelor's degree. Applicants should start the process early, as some programs have a deadline as early as November for programs starting in August of the following year. The application must include a transcript, a résumé, a statement of purpose, several letters of recommendation, and the score obtained on the standardized Graduate Record Exam (GRE). In addition, many programs require that the applicant provide the score obtained on the mathematics subject GRE, a comprehensive exam that covers the entire undergraduate mathematics curriculum (including calculus, differential equations, linear algebra, modern algebra, real and complex analysis, discrete math, geometry, topology, probability and statistics). It can take several months to study for this exam and a few more months to obtain the results, so it is important to start early.

Embarking on a Ph.D. program is a daunting, long-term journey that takes at least 5 years of full-time work. During this time, the student will immerse himself or herself in advanced mathematics coursework and pursue cutting-edge research mentored by faculty. In most mathematics Ph.D. programs, all students also work in positions as teaching or research assistants, which provide stipends or tuition reductions. Consequently, it is often possible to obtain a Ph.D. in mathematics without getting into debt.

Vocations in research

Besides becoming an academic at a research institution, there exist other types of work that involve full-time mathematics research. These pathways start out similarly to vocations in academia, as most of these vocations require a Ph.D. in mathematics.

There are several research institutes in the U.S. and abroad that focus exclusively on increasing and disseminating new mathematics. The most famous of these is the *Institute for Advanced Study* (IAS) in New Jersey (famous for having been the academic home of Albert Einstein and Kurt Gödel, among others). Other examples include the *Courant Institute* (New York), the *Clay Mathematics Institute* (New Hampshire), and the *Mathematical Sciences Research Institute* (California).

The U.S. government sponsors mathematics research in several areas. The *National Security Agency* (NSA) is the largest employer of mathematicians in the United States, promoting research activities in computing and encryption (some of this research is classified). The *National Science Foundation* (NSF) promotes the progress of science, advances national health, prosperity and welfare, and secures the national defense. Some of its budget goes to support mathematics research done by faculty at universities. The *National Institutes for Health* (NIH) dedicates a significant part of their budget to researching mathematical applications in medicine and public health.

Finally, many large corporations in the private sector see mathematics research as an investment. Companies such as AT&T, IBM, Microsoft, and Xerox maintain a mathematics research department that employs a permanent staff of mathematicians who do research in commercial applications.

7.4.4 Vocational preparation in college

There are many things a student can do to develop a first vocational pathway. Earning good grades in class certainly matters, as these appear on the transcript. However, merely getting good grades does not suffice for a successful foray into the competitive post-graduation job market. As we mentioned previously, potential employers measure candidates' worth initially via three types of documents: résumé, transcript, and letters of recommendation. We mention below several strategies to develop each one.

Skill development

Developing attributes and skills form the key to a strong vocational pathway. Any mathematics curriculum naturally develops attributes such as problem-solving and attention to detail, but some programs do not emphasize computing, communication, or collaboration as much. Furthermore, these types of skills are difficult to assess by grades, so mathematics grades may not accurately reflect a person's skills in these areas. Developing them will strengthen any vocational pathway and increase employability just as much as the mathematical skills do, if not more. As the student develops skills, he or she should keep a list of concrete examples to illustrate and showcase them to a potential employer.

1) **Computing.** Any mathematician must be proficient with basic computation tools, including at least one spreadsheet (Microsoft Excel or Google Sheets), one computer algebra system (Mathematica or Maple), and one programming language (Python, Java, R, MATLAB, etc.). If the student's coursework does not develop these tools, he or she should spend some time self-learning them.

2) **Communication.** Regular assignments develop verbal and written communication skills. Students should use every written assignment as an opportunity to develop effective writing skills. As students progress through a mathematics program, they could make every assignment reflect professionalism by writing it as a formal mathematical document in LATEX. Professors will certainly notice the effort and comment on this positively in their recommendation letters.

3) **Collaboration.** Section 7.2.1 highlights the importance of collaboration in mathematics and offers suggestions on how to develop effective teamwork techniques. Students should take advantage of opportunities built into their classes to experiment with different collaboration strategies to determine their natural collaboration dynamic through personal experiences. Professors will take note if a student proactively embraces the challenges of collaborative work, and they may offer concrete examples to support their comments in their recommendation letters.

Professionalism

Professionalism refers to the way people act, dress, and interact with each other in a professional environment. These behaviors are not innate or learned overnight; they take time and must be practiced actively before they become natural. Most workplaces require a much higher level of professionalism than a typical college environment, so students will benefit from gradually cultivating a sense of professionalism in order to facilitate the transition to a first post-college vocation. The first job after graduation will certainly require the acquisition of new specific skills in a short time, so already possessing habits of professionalism will make the transition easier. Furthermore, professors notice when a students adopts a professional attitude and will comment positively on this in their recommendation letters.

The very informal nature of some college environments imply that it requires intentionality to develop professionalism. Here are some professional habits that students can practice:

- **Behavior and attitude.** A professional attitude includes arriving to class on time, adopting a physical posture that communicates interest and engagement, refraining from multitasking (for example, checking one's phone).

- **Social skills.** Professional social skills include showing courtesy to peers and respect toward persons in authority.

- **Quality and presentation of work output.** A student should submit written assignments as professional, formal documents, showing concern not only for the content, but also the presentation, handwriting, and paper quality.

- **Personal communication.** This includes the ability to communicate professionally and effectively by email and verbally.

- **Attire.** Students may wish to experiment with styles of casual and business wear that feel comfortable but look professional.

Networking

Vocational development involves not only developing knowledge and skills, but also contacts. Only about 15 to 20% of available jobs are formally advertised; the rest are part of the *hidden job market* and do not appear in published channels [56]. For this reason, the key to landing a job often consists more in "*who* you know" than "*what* you know." Consequently, building a strong personal professional network is essential to access the opportunities of the hidden job market.

College offers an excellent time to meet people who can help beyond the practical aspect of merely getting a job. Established professionals can help discern one's aptitude for a vocation, give advice on how to develop skills, put people in contact with key employers, and perhaps even eventually hire young people in their network.

Building a personal network requires being proactive and making the initial contact. Here are some concrete steps to start networking:

- **Online.** This involves creating a profile on LinkedIn and on a scholarly collaboration network such as Mendeley or Zotero (see Subsection 7.2.2 for more information about SCNs), or joining a discussion group in a field of interest and reaching out to other members in the group.

- **Alumni networking.** Many institutions facilitate networking with alumni in various professions who have indicated their desire to mentor students in vocational matters.

- **Professional associations.** The large mathematical associations, like MAA, AMS, SIAM, ASA, AWM, and others, enable younger members to connect with more seasoned ones. Many offer substantial career resources and student membership at a heavily discounted price.

- **Conferences.** Attending a conference provides many opportunities to meet and interact with a large number of professionals in person.

Work experience

If a student wishes to pursue vocations in the industry, he or she should take steps to gain as much industry work experience as possible. Actually working in a field related to a prospective vocation, even for a short duration, allows someone to test the waters and experience the field of work from an insider's perspective, giving vocational insight, as well as a valuable addition to a résumé.

Summer *internships* offer an ideal setting to gain solid work experience while in college. This kind of environment gives the student sufficient time to acquire specific training, become proficient in performing specific tasks, and get a good feel for the overall work and dynamics of the workplace. Many businesses treat summer interns as regular short-term employees, providing on-the-job training, expecting a certain quantity and quality of work output, and sometimes paying a decent salary. Applying for an internship feels like applying for a regular job, so just going through the application process already offers a valuable experience, even if the student does not get the position. Students who consider summer internships should start the application process in January or February, when many businesses make plans for summer projects and staffing needs.

Aside from a formal internship, many businesses welcome students to experience their workplace by offering a tour or a day visit where one can *job-shadow* a professional (also called an *externship*). These short-term, low-commitment experiences can provide valuable insight into the day-to-day aspect of a certain type of work.

Teaching experience

Students considering a vocation in education or academia, should gain some form of teaching experience before graduation. This is especially true when applying to a Ph.D. program, since mathematics graduate schools often require their students to teach courses starting in their first year as part of their stipend agreement. Among many other things, applications to graduate school ask recommending faculty to evaluate the applicant's ability to teach.

Most institutions employ undergraduate mathematics students as teaching assistants or tutoring lab employees. It is also possible to gain teaching experience while in college as a private tutor, or by getting a part-time tutoring job with a tutoring academy.

Research experience

Students considering a vocation in academia or research should strive to get some research experience. Most undergraduate mathematics students do not pursue vocations that involve any research, and most mathematics programs do not require students to engage in research at the undergraduate level, so those wanting to do research must be intentional in seeking out research opportunities.

As a requirement of their position, most mathematics professors engage in research, and some may decide to take on one or two students as research assistants for the duration of a semester or summer. Often these positions are preferentially given to graduate students, but professors may employ undergraduate students as well, especially at baccalaureate colleges or when graduate students are unavailable. In any case, faculty do not usually advertise these positions in regular channels. Instead, professors individually approach students who exhibit a promising aptitude for research with a personal offer to appoint them as a research assistant. For this reason, students interested in this type of position should strive to make themselves known to professors and inform them of their interest in participating in the professor's research.

The National Science Foundation and various other granting agencies fund *Research Experiences for Undergraduates* (REU) and *Summer Institute in Biostatistics* (SIBS) programs at many colleges and universities. These 8 to10-week-long summer programs offer an intensive research experience in a collaborative setting, mentored by eminent professors. Furthermore, they provide participants with a generous stipend, and cover room, board, and travel expenses. There are over 100 such programs offered every summer at various institutions; each one welcomes about 15 to 20 students to delve into one or more cutting-edge research topics in pure or applied mathematics. REU and SIBS programs offer a unique opportunity to explore emerging fields of mathematics not encountered in courses or available in books, while making new friends and future professional contacts. Furthermore, research conducted in these settings often lead to a journal publication or presentation at a conference.

Applications must be completed in January or February and typically require a transcript and one or more letters of recommendation. It is important to read all application conditions carefully, as some programs require prerequisite coursework and some are reserved for special groups, such as U.S. citizens, women, or minorities.

Study abroad

Studying abroad for a semester or during the summer offers many advantages, such as increasing cultural awareness, adaptability, self-reliance, and foreign language skills. Because mathematics curricula are fairly standardized worldwide, it is not difficult to find upper-level mathematics courses offered in foreign universities that are equivalent to those required in U.S. institutions. Students can continue their mathematics coursework and not fall behind while abroad.

Several university programs around the world specialize in undergraduate mathematics and cater to U.S. students looking for a study abroad experience. One notable such program is the *Budapest Semesters in Mathematics* (BSM), which offers American and Canadian undergraduate students the opportunity to study mathematics in Budapest, Hungary for a semester or a summer. The program offers an extensive list of upper-level mathematics courses taught in English, in small classes, by eminent professors from various prestigious Hungarian universities.

EXERCISES FOR SECTION 7.4

1. Write a personal essay about your current vocation as a student and your vision for a potential future vocation (after graduation). Your essay should answer the following questions:

   • What are some elements of your cultural identity (upbringing, family, extracurricular activities) that shape your thoughts about your vocation?

   • What are your reasons for deciding to study upper-level mathematics and/or pursue mathematics as a major?

   • How much thought have you given to life after college? Do you already have a vision for a post-college vocation?

   • Besides doing well in your classes, are you taking steps to develop a vocational pathway?

CHAPTER 8

History and Philosophy of Mathematics

Katherine Johnson, one of the first African American women to work as a NASA mathematician and who appeared in the book and the movie entitled, *Hidden Figures*, stated about mathematics, "Some things will drop out of the public eye and will go away, but there will always be science, engineering, and technology. And there will always, always be mathematics." Johnson's sentiment highlights the stunning aspect that mathematics boasts ancient roots reaching back to the very beginnings of civilization.

This chapter provides a broad overview of the history of mathematics from its origins in antiquity until today. Along the way, we present key philosophical ideas inspired by developments in mathematics. Students will also recognize many names of mathematicians presented here in their social, historical and geographical context. Studying the history of any field of inquiry helps us frame the trends leading to its present state. The history of mathematics is singularly rich because of its deep connections with virtually all areas of human life and thought, from the practical to the philosophical.

Section 8.1 sketches the history of mathematics from its ancient roots to the Scientific Revolution, a momentous era that dramatically transformed mathematics, elevating its trajectory to become the prominent academic field it is today. Section 8.2 focuses on the connection between mathematics and science that originated during the Scientific Revolution. Section 8.3 discusses the axiomatic method, from its origins in ancient Greece to its universal application to all fields of modern mathematics. Section 8.4 outlines historical movement in modern mathematics since the Scientific Revolution. Finally, Section 8.5 presents philosophical issues related to the epistemology of mathematics, the relationship between mathematics and the physical world, and the ontology of mathematical objects.

8.1 History of Mathematics Before the Scientific Revolution

Some authors claim that math could only exist in an established civilization, often a stable empire [15]. It comes therefore as no surprise that ancient mathematical history falls into distinct periods of intense activity that usually coincided with the period between the rise and fall of an empire. Societies with sufficient peace, prosperity, and social stability formed a fertile environment for its citizens to engage in scholarly activities like mathematics. Up until the early 17th century, mathematics progressed by short, intense bursts of localized activity, separated by long periods of stagnancy during which mathematical development slowed to a standstill, sometimes for centuries. For this reason, we divide this broad, 6000-year overview of ancient mathematics into separate sections, each one focusing on a specific geographical area and time period.

History marks the beginning of the Scientific Revolution as the early 17th century. For many reasons, this critical period presents a natural dividing point for this historical overview. Before the Scientific Revolution, societies across the world did not widely regard mathematics as a serious scholarly activity. Builders or scholars primarily saw mathematics as a collection of methods for solving certain practical problems. In Europe, learned citizens sometimes viewed it as a set of mildly interesting "mind-puzzles" that members of upper-class society would engage in as a leisurely recreational pastime. A few scholars specialized in mathematics, dedicating their lives to studying patterns arising in numbers and geometrical figures, but most viewed their field of study as obscure and esoteric. Moreover, natural philosophy, the systematic study of the physical world that we now call "science," rarely drew on mathematics until the 17th century.

Ancient mathematics was not an organized or unified field of study. With the exception of ancient Greece, mathematical developments in ancient cultures existed as a set of disparate recipes for solving practical problems, with little concern for theoretical foundations. Furthermore, scholars penned ancient mathematical texts entirely written in words, devoid of symbols and numerals, presenting only specific examples with no easy way to generalize them to other cases. The first appearance of mathematical symbols dates back only to the 15th century, and the symbols themselves did not become standardized until after the development of calculus. For these reasons, mathematics progressed slowly prior to the Scientific Revolution.

Subsection 8.1.1 describes the characteristics of mathematical developments in several ancient civilizations. Subsection 8.1.2 focuses specifically on mathematics in ancient Greece, as it differed dramatically in quality from mathematics in other ancient civilizations and ultimately became an essential foundation that shaped the way we structure mathematics today. Subsections 8.1.3 and 8.1.4 focus on the important contributions of India and the medieval Islamic empire, respectively. Finally, Subsection 8.1.5 describes the evolution of mathematics in Europe during the Middle Ages and Renaissance, setting the stage for the major changes in the nature of mathematics that emerged during the Scientific Revolution.

8.1.1 Ancient civilizations

Features of ancient mathematics

The history of mathematics is as old as civilization itself, originating with humans building cities and developing a writing system. Many early civilizations devised writing systems to facilitate the administrative duties associated with a central government. The earliest forms of writing date back to about 5000 BC in the Ancient Near East, specifically the regions of Mesopotamia (known as the *Cradle of Civilization*) and Egypt. Ancient Mesopotamians wrote in cuneiform characters on clay tablets, which once baked, remained intact indefinitely. This format preserved their writing for millenia. Archeologists have deciphered many of them, giving us good insight into those ancient cultures. The everyday necessities of government administration and trade required a special class of professional civil servants called "scribes" to maintain records of production, taxation, and other quantitative data, and to perform mathematical calculations with these quantities. Interestingly enough, vestiges of Mesopotamian mathematics have survived to present day in our use of 60 minutes in an hour and 60 seconds in a minute: Ancient Mesopotamians' number system used base 60. We also see this influence in our use of 360° in a circle; Egyptians, who inherited the base-60 number system from Mesopotamia, considered the equilateral triangle a particularly pleasing figure and counted 60° in one interior angle of an equilateral triangle. Since six equilateral triangles fit around one point, we get 360° degrees around a point [11].

As other civilizations arose across the world, each eventually developed its own writing systems, complete with methods of keeping numerical records and doing mathematical calculations with them. The civilizations of Ancient Egypt, China, India, the Minoan civilization of Crete, and the civilizations of Ancient Mesoamerica all developed their own number systems and methods of performing calculations with them. However, very little information about the details of these ancient forms of mathematics exists, either because most of the records produced by these cultures have not survived, or because archaeologists have not been able to decipher the writing. For example, the Ancient Egyptians wrote with ink on papyrus, a material that does not easily survive for thousands of years (unlike Mesopotamian clay tablets). Furthermore, the key to deciphering Egyptian hieroglyphics, an artifact called the Rosetta Stone, was not discovered until 1799. We now know that the Egyptians were the first to divide a day into 24 hours and a circle into 360 degrees [11].

It is interesting to note that not all civilizations invested in promoting mathematics. For example, the Roman Empire, known for its great developments in law, literature, philosophy, and engineering, did not produce any notable progress in mathematics. We might speculate that the awkward Roman number system (called *Roman numerals*), though well suited for record keeping, might not have served well for abstract thinking [15]. This example suggests that the nature of a civilization's writing system may have had a great impact on the extent of its mathematical progress.

Practical mathematics

The mathematics developed in ancient civilizations such as Mesopotamia or Egypt was almost entirely driven by the necessity to solve practical problems related to everyday life in an organized, urban society. This includes

- production of food and other natural resources required to sustain a population,

- trade and commerce, including measuring quantities in units of weight or volume,

- accounting and taxation, including keeping records of what is owed to whom,

- defining property lines and measuring the areas of fields,

- timekeeping and calendars,

- surveying, construction, and engineering of public works,

- astronomy and navigation.

All of these aspects of daily life require keeping track of numerical quantities and performing calculations with them, giving rise to elementary arithmetic, geometry, and algebra. The users of mathematics in these cultures could perform arithmetic operations, solve linear (and, in many cultures, quadratic) equations, compute areas and volumes of geometrical shapes, and generally apply a method used in a specific instance to different examples.

Scholarly mathematics

The first evidence of mathematics pursued as a scholarly activity, independent of any specific problem or application, dates back to between 1900 and 1600 BC in Mesopotamia, called the Old Babylonian period. During that time, the scribes began devising techniques to generalize their practical mathematical methods, discovering interesting patterns in the numbers themselves. These discoveries gave rise to the first general rules of elementary number theory and geometry. Other civilizations also eventually developed a culture of pursuing mathematical knowledge as an intellectual endeavor, not necessarily motivated by the need to solve a practical problem. In many cultures, this scholarly aspect of mathematics originated with the development of astronomy and astrology. As humans in every culture had leisure to observe the sun, moon, stars and planets, they sought to develop mathematical tools to study the patterns of their motion in time and space.

8.1.2   Ancient Greece

Ancient Greek mathematics

Of all the ancient civilizations that developed their own form of mathematics, the Greek civilization stands apart because it is the only one that incorporated logic and reasoning as a fundamental part of its mathematics. This unique aspect of how the Greeks structured their mathematical ideas fundamentally influenced the progress of modern mathematics more than any other culture's methods. The word "mathematics" itself derives from the Ancient Greek *mathema*, which means "subject of learning."

The Ancient Greek civilization flourished between 800 and 300 BC. The empire encompassed a wide geographical area surrounding the Mediterranean and Black Seas, including modern-day Greece, but also many parts of modern-day Turkey, Italy, France, and northern Africa. Although this geographical area counted people from many different ethnic groups, all of them shared a common culture (called the *Hellenistic* culture) and used the Greek language as their *lingua franca* (common language). Founded in northern Egypt by Alexander the Great in 331, the city of Alexandria quickly became the main center of Hellenistic intellectual activity and for many centuries remained the Mediterranean nucleus for knowledge and learning in all fields, including literature, philosophy, mathematics, art, and architecture. Alexandria boasted the famous Great Library, which at the time housed the largest book collection in the world.

As in many other ancient civilizations, Greek mathematics began with arithmetic, number theory, and geometry but soon branched off into several directions not directly connected with any specific application. For example, an early school of Greek mathematics called *Pythagoreanism*, created around 500 BC by *Pythagoras*, started assigning mystical properties to numbers, claiming that "God is number," extending number theory to a form of number-worship. Similarly, Greek mathematicians extended geometry in an abstract direction, separating it from its practical uses, and applying it instead to philosophy, as a method of organizing patterns of thought and distinguishing truth from falsehood by deductive reasoning. One distinctive aspect of Greek geometry involved *straightedge and compass* construction problems: questions where the solution involves a sequential method to build geometric lengths, angles, and figures using only a straightedge (a ruler with no markings) and a compass.

Greek mathematicians favored the pursuit of mathematical knowledge for its own sake, generalizing practical mathematical methods to formulate statements called *theorems*. The Greeks also developed a method, called the *axiomatic method*, to prove these statements by relating them to a set of statements assumed to be true, called *axioms*. Instead of relying on intuition and observation, the axiomatic method relies on logic and deductive reasoning to derive theorems from a set of axioms. Although our perspective on the axioms has changed in last 150 years or so, this method of organizing mathematics as a consistent network of theorems centered on a common set of axioms has provided a foundation for mathematics and its philosophy for millenia.[1]

Euclid and *The Elements*

The most important Ancient Greek mathematician is arguably *Euclid of Alexandria*, known primarily for his monumental work called *The Elements*, which he wrote around 300 BC. Organized into 13 "books," *The Elements* provides a comprehensive compilation of all known mathematical results of Euclid's day, most of which had been developed by previous mathematicians. The books cover mostly plane and solid geometry (known today as *Euclidean geometry*), as well as some number theory.

The defining feature of Euclid's accomplishment resides not only in the content itself, but the way in which he presented the material. Far more than a mere compilation of haphazardly listed results, Euclid organized them into a single, logically coherent framework. Although

---

[1] Morris Kline [52] asserts that this approach derived from a Platonic disdain for the physical/material world. Reality was found in a world of ideas, not empirical evidence. The reliability of an argument and its conclusions depended completely on the infallibility of the reasoning steps, and not upon some validation in a material application.

the actual material did not originate with him, the wording of the theorems and their proofs do. Euclid used a systematic development, starting with a small set of axioms and definitions, then proceeded to build one theorem upon another, each one proved by appealing only to previously proved theorems and axioms, every theorem fitting perfectly into the overall structure, like pieces of a puzzle.

The first book of *The Elements* begins by establishing the axiomatic foundation, which consists of:

- 23 *definitions* of geometrical terms (for example, "line," "circle," and "triangle"),

- 5 *common notions* (for example, "Things which are equal to the same thing are also equal to one another"), and

- 5 *postulates* (for example, "Given a point O and a point A not equal to O, there exists a circle centered at O with radius OA").

The 5 common notions and 5 postulates constitute a set of 10 axioms, assumed to be true without proof. Then, the first book continues with a sequence of theorems (also called *propositions*) and straightedge and compass constructions. Each proposition comes with a rigorous proof in which Euclid justifies every step by referencing a previously proved theorem, a common notion, or a postulate. The other 12 books of *The Elements* follow the same strict structure.

Greek mathematics contains no numbers, calculations, or measurements. Euclidean geometry compares distances to one another as ratios and uses the right angle as the only unit by which to compare angles. Furthermore, Euclid provides no motivation or examples in *The Elements*, presenting only the bare facts and their proofs. For these reasons, the writing style of *The Elements* often feels cold, terse, minimalist, and dispassionate to modern readers accustomed to learning from texts that strive to engage the student.

However, for over two millennia, scholars across the world aware of its existence have considered *The Elements* a masterpiece in the application of logic to mathematics. Many have been struck by the sheer perfection of its structure, which gives the work an austere beauty that has stood the test of time. *The Elements* is one of the most influential books ever written, one of the earliest books to be printed after the invention of the printing press, and second only to the Bible in the number of editions published. Abraham Lincoln kept a copy of *The Elements* in his saddlebag and studied it late at night by lamplight, eventually memorizing large portions of it. Albert Einstein referred to it as the "holy little geometry book [27]."

When early universities shaped their curricula around the *quadrivium* (see Subsection 8.1.5), Euclid's *Elements* became one of the classical works that constituted a fundamental part of anyone's education. This requirement persisted until the 20th century, when schools started to include some of the contents of *The Elements* in other courses (such as geometry) instead. However, the axiomatic, deductive structure of *The Elements* became the universally adopted system for building and organizing mathematical knowledge. The impact of *The Elements* transcends mathematics, we can apply its structure as a general template for the logical organization of ideas and arguments. As such, students in philosophy, law, theology, science, and politics also study *The Elements*.

8.1.3   India

Mathematics emerged in the Indian subcontinent as early as 1200 BC, with roots strongly influenced by the mathematics of Mesopotamia and Ancient Greece. Indian mathematics flourished between 400 and 800 AD, a period when Europe and the Middle East experiencing great upheavals caused, in part, by the fall of the Western Roman Empire in the 5th century and the rise of Islam in the 7th century. While all mathematical progress came to a halt in Europe and the Middle East, Indian mathematics enjoyed its Golden Age.

As with many other ancient systems of mathematics, the early problems studied by Indian mathematicians emerged from astronomical questions. The Indians soon developed an interest in mathematics for its own sake and began studying relationships between numbers and discovering geometrical patterns by generalizing earlier practical mathematical tools. Indian mathematics made large contributions to number theory, algebra, combinatorics, and trigonometry (a field inherited from Greek mathematics), which the Indians used to describe the motion of planets.

The most famous feature of Indian mathematics is its number system in base 10. Indian mathematicians devised a place-value system based on powers of 10, which allowed them to represent an arbitrarily large number by recycling a finite number of digits. In contrast, most ancient number systems (like Roman numerals) required the invention of new symbols for larger and larger numbers. Indian mathematicians used 9 symbols for the numbers 1 through 9, and a dot as a placeholder for the empty space used to symbolize "nothing." This dot later became a small circle, and the Sanskrit word *sunya* for this symbol eventually gave rise to the word *zero*. This counterintuitive idea of representing nothing by something greatly increased the potential of this numeration system. Eventually, this system spread to the whole world, gradually superseding all others.

8.1.4   The medieval Islamic world

The rise of Islam in the 7th century marks the beginning of the Islamic Golden Age. By 800 AD, Islam had spread from western India to modern-day Spain, including the Middle East and northern Africa. The city of Baghdad was founded during this time and became the capital of this civilization [65]. Baghdad quickly rose in size and influence to become the largest city in the world, as well as a significant cultural, commercial, and intellectual center. The city earned the name of "Center of Learning" because of its numerous academic institutions and its great library, called the *House of Wisdom*.

Scholars from all over the world received welcome at the House of Wisdom, regardless of their ethnicity or religion. Like the Greek civilization, many ethnic groups made up the Islamic world, but all of them shared a common culture and used Arabic as their *lingua franca*. The House of Wisdom grew as scholars from all over the world arrived; by 850 AD, it housed the largest book collection in the world. Scholars translated works from many sources into Arabic, facilitating an easy exchange of ideas. Arab mathematicians eventually compiled all the mathematics of ancient Greece and India and built upon it, promoting considerable progress until the 14th century.

Muhammad Ibn Musa Al-Khwarizmi, the "Father of Algebra," stands as one of the great mathematicians of his time. Al-Khwarizmi authored a book explaining the benefits of the Indian number system in base 10. Arab mathematicians adopted this system, with slight modifications, giving rise to "Hindu-Arabic" numerals. Al-Khwarizmi's work was later translated into Latin and became the main source leading to the adoption of this number system throughout Europe. The Latin translation of this book begins with the words *Dixit Algorismi* ("thus spoke Al-Khwarizmi"), giving rise to the English words *algorism*, the technique of performing arithmetic using Hindu-Arabic numerals, and *algorithm*, any finite sequence of steps to perform a computation. In another book, Al-Khwarizmi explained how to solve linear and quadratic equations, under the Arabic title *Al-jabr wa'l-muqabala*. Later translated into Latin, his book took the latinized title *Algebra*.

Arabic mathematics made great progress in geometry, trigonometry, combinatorics, and algebra. An interesting area of geometry developed during this period concerns the study of tilings, scaling, and symmetry, motivated by the frequent use of repeating shapes and abstract geometrical patterns in Islamic art and architecture.[2]

---

[2]This feature of Islamic art is a consequence of the Islamic prohibition against graven images and artistic representation of the human body.

8.1.5   Medieval and Renaissance Europe

The Dark Ages and Roman legacy

At the height of its power, the Roman Empire encompassed most of western Europe and the British Isles. When the western Roman Empire fell to "barbarian" tribes in the 5th century AD, it plunged all of western Europe into a long period of political and social instability, commonly called the *Dark Ages*, that lasted until the 10th century. Literacy and learning declined, halting any possibility for progress in mathematics. During this period, western European society retained many of the ancient structures of the Roman civilization, including Latin as the *lingua franca*, the use of Roman numerals, and the Christian religion. The Church became the only institution that retained a measure of stability in those troubled times of intellectual darkness, preserving a small spark of learning in cathedrals and monasteries.

The use of mathematics in the West did not extend much beyond elementary arithmetic. Adding and subtracting numbers required the use of an *abacus* or *counting board*, a table on which little pebbles were placed in grooves.[3] Roman numerals, ill-suited for any form of mathematics, were only used for noting the final result of these basic calculations and for record keeping.

Contacts with the Islamic world

Islamic control of the Iberian peninsula, modern-day Spain and Portugal, spared that region of the intellectual decline that western Europe suffered during the Dark Ages. Because of this connection with the more intellectually advanced Islamic world, much of mathematical knowledge that later spread to Europe came from the Islamic world via Spain, as early European scholars discovered and translated into Latin the Arab mathematics texts housed in Spanish libraries.

Commerce offered another important source of contact between medieval Europe and the Islamic world. Avid travelers and navigators, European merchants established trade routes to distant regions by land and sea bringing back exotic goods, such as silks, spices, and mathematical ideas. *Leonardo of Pisa* (1170-1240), better known by his nickname *Fibonacci*, was one of these merchants, exposed to Arabic mathematics while accompanying his father, a successful Italian merchant, on his travels to the Middle East. Around 1200 AD, Fibonacci wrote a book titled *Liber Abaci* ("Book of Calculation"), in which he discussed several practical mathematical methods for trade and accounting, gave an explanation of the Hindu-Arabic number system, and explored the now famous *Fibonacci Sequence*. Widely

---

[3]The name of these pebbles, in Latin, is *calculus*, from which we get the English word *calculate*.

translated and circulated in Europe, the *Liber Abaci* prompted the gradual replacement of Roman numerals by the more convenient Hindu-Arabic numerals, better-suited for mathematics. Fibonacci wrote two other influential mathematical texts: *Practica Geometriae*, a manual for practical geometry (also inspired by Arabic mathematics), and the *Liber Quadratorum* ("Book of Squares") containing number theory and algebra.

Universities

The concept of a *university* has its roots in medieval Europe, originating with the *cathedral schools* established by the Church in major cities to provide education for children of noble families desiring to pursue a career in the Church. Cathedral schools eventually became centralized education centers for teaching and learning called *universities*, admitting students destined for any profession and attracting a community of scholars. The first such universities were established in Bologna (1088), Paris (1150), Oxford (1167), and Cambridge (1209). By the end of the Middle Ages, many other major cities boasted a university.

Like other early centers of learning such as the *Great Library* of Alexandria and Baghdad's *House of Wisdom*, universities compiled knowledge and promoted the exchange of ideas among scholars. Beyond this, early universities strove to provide their students a balanced and well-rounded education. Every university crafted a carefully designed curriculum, approved by the Church. With courses given in Latin, all students followed the same course sequence with no electives. The Church also provided detailed instructions on how to teach the various topics, ranging from literature to mathematics. The curriculum itself was structured around seven fundamental subjects, called the *seven liberal arts*, themselves grouped into a set of three basic subjects consisting of grammar, rhetoric, and logic (called the *trivium*), and a set of four more advanced subjects consisting of arithmetic, geometry, astronomy, and music (called the *quadrivium*). Mathematics played a central role both in the trivium and quadrivium and played an essential role in preparation for pursuing studies in the higher disciplines of philosophy, theology, law, or medicine.

Many medieval universities adopted a teaching method called *scholasticism*, based on logical reasoning, the formulation of arguments and counterarguments, and the extension of knowledge by inference. *Thomas Aquinas* (1225-1274), an influential philosopher and theologian at the University of Paris, was one of the principal proponents of scholasticism. Known particularly for his natural theology, he strove to reconcile the principles of Aristotelian philosophy with Christian theology, and approached the analysis of biblical passages with reason. Many thinkers consider Aquinas's masterpiece *Summa Theologica* as the pinnacle of scholastic, medieval, and Christian philosophy. Another influential medieval philosopher in the scholastic tradition, *Roger Bacon* (1220-1292) at the University of Oxford promoted the study of nature through *empiricism* and emphasized the study of mathematics and *natural philosophy* as an important part of the curriculum of medieval universities.

The Renaissance

The *Renaissance*, which in French means "rebirth," refers to the period of European history marking the end of the Middle Ages. A renewed interest in classical art, architecture, literature, values, and ways of thinking sparked a sudden cultural, political, and economic rebirth. This widespread cultural revival began in Italy during the 14th century and quickly spread to all of Europe. The invention of the printing press in 1450 catalyzed the rapid spread of new ideas, making it possible for the masses to have access to ancient texts for the first time. Euclid's *Elements* became one of the first texts printed and disseminated in 1482. European mathematicians, now armed with translated copies of mathematical texts on Greek, Indian, and Arabic mathematics, began to progress again.

Algebra witnessed many advances during the 14th and 15th centuries, mostly thanks to the introduction of *symbolic notation*. Unlike Arabic algebra, written entirely in *rhetorical syntax* (i.e., in words), European mathematicians introduced symbols to represent operations (such as addition, subtraction, multiplication, and division), making it possible to write equations in a more efficient and compact syntax. However, mathematicians in different countries utilized different symbols. It took several more centuries for the notation to become fully standardized. Italian algebraists *Scipione del Ferro* (1465-1526) and *Niccolò Tartaglia* (1499-1557) are credited with finding the general solution of the cubic equation, a feat that surpassed Greek and Arabic mathematics. The result was published by *Girolamo Cardano* (1501-1576) in his book *Ars Magna* ("The Great Art"), a work that also includes the solution to the quartic equation developed by his student *Ludovico Ferrari* (1522-1565).

In France, *François Viète* (1540-1603) made another decisive step forward in algebra when he started using symbols to represent unknown quantities and parameters. This pushed algebra deeper into abstraction, but it also allowed writing general forms of equations, instead of relying on wordy explanations illustrated by specific numerical examples. Historians of mathematics give credit to the French *René Descartes* (1596-1650) for the algebraic notation we use today, first appearing in his work *La Géometrie*. Another decisive step, credited to both Descartes and *Pierre de Fermat* (1601-1665), is the unification of algebra and geometry into a common framework, which we now call *analytic geometry* (or *coordinate geometry*). This framework became an essential foundation for the invention of calculus.

Natural philosophy

Philosophy, in its broadest sense, simply refers to the human practice of thoughtfully engaging with the world. A subfield within philosophy, *natural philosophy* formerly referred to the study of nature, including all aspects of the material world perceived by human senses, in contrast to abstract ideas or the spiritual world.

Natural philosophy has its roots in Greek antiquity and remained an established field of study throughout the Christian era and Middle Ages. It constituted an important discipline studied in European medieval universities alongside other streams of philosophy and theology. Aristotelian philosophy, which claims that all knowledge begins with the senses, shaped how medieval universities taught natural philosophy. The Aristotelian approach consisted in building knowledge about the world through qualitative and descriptive analysis of natural phenomena initially, but then proceeded by deduction and reasoning to find natural causes.

In the 16th century, the Scientific Revolution started to change the ways of establishing knowledge about the natural world. Natural philosophers adopted the *scientific method*, which eventually distanced them from other forms of philosophy. However, the term "natural philosophy" remained the most common way to describe a systematic study of the material world until the 19th century. The term eventually gave way to the term *natural science* in the 19th century and later divided further into the separate fields of physics, chemistry, and biology. The *social sciences*, which refer to the study of man-made structures and institutions (including economics, psychology, political science, and sociology), now fall under the general umbrella term *science* because knowledge in these fields also progresses by the scientific method.

EXERCISES FOR SECTION 8.1

1. Most of what we know today about the mathematics of Ancient Egypt comes from the *Rhind Papyrus*. This famous mathematical document, which dates back to 1550 BC, was discovered in Egypt by the Scottish archaeologist *Alexander Henry Rhind* in 1858. Do some research on the Rhind Papyrus and its contents and describe the important features of Ancient Egyptian mathematics and its number systems.

2. Do some research on the mathematics of Ancient Mesopotamia and Ancient Egypt. Compare them by describing their similarities and differences.

3. The most important text of early Chinese mathematics is the *Jiuzhang Suanshu* (also known as the *Nine Chapters on the Mathematical Art*), which contains material dating back to the 11th century BC. Do some research on the *Nine Chapters* and its contents and describe the important features of Chinese mathematics.

4. In many ancient cultures, astronomy provided the first mathematical questions that were not merely related to an everyday, practical problem. Describe what these specific questions were and what areas of mathematics were developed to answer them.

5. A defining feature of Greek mathematics (which is evident in Euclid's *Elements*) consists in devising *geometrical constructions*, i.e., establishing a sequence of instructions to obtain a geometrical result. The following are some geometrical constructions studied by the Greeks:

   (a) dividing a line segment into n segments of equal length,
   (b) building an equilateral triangle,
   (c) building a square,
   (d) bisecting an angle,
   (e) trisecting an angle,
   (f) squaring a circle,
   (g) doubling a cube.

   Greek mathematicians discovered that some of these constructions can be accomplished using only a straightedge and compass (called *straightedge-and-compass* constructions), and some cannot. Do some research to determine what these constructions consist in and which are straightedge-and-compass constructions.

6. Euclid's *Elements* contains several results that are now considered fundamental in many areas of mathematics, such as the *Pythagorean Theorem* (*The Elements* 1:47). Here are a few others:

   (a) The *isosceles triangle theorem* (*The Elements* 1:5). State this theorem and explain why it has the nickname *pons asinorum* (Latin for "bridge of asses").
   (b) The *inscribed angle theorem* (*The Elements* 3:31). State this theorem, as well as the special case of this theorem called *Thales's Theorem*.
   (c) The *fundamental theorem of arithmetic* (*The Elements* 7:30-32). State this theorem.
   (d) The *infinitude of primes theorem*, also known as *Euclid's Theorem* (*The Elements* 9:20). State this theorem.

7. Book 13 of Euclid's *Elements* is entirely devoted to the properties of five three-dimensional shapes known as *Platonic solids*. These solids have captured the interest of countless mathematicians, scientists,

and philosophers throughout the ages. Do some research and explain what a Platonic solid is and provide the names of the five Platonic solids.

8. European merchants were experienced at long-range navigation by sea, a skill that requires a good knowledge of astronomy and certain types of mathematics. Do some research and explain what types of mathematics are used in navigation and include an example.

9. The discovery of the general solution for the cubic equation is an exciting story of competition, intrigue, and broken promises. Do some research on this story, detailing the roles of Del Ferro, Tartaglia, Cardano, and Ferrari.

8.2 Mathematics and Science

Nowadays, a deep relationship exists between mathematics and science. Many people think of mathematics primarily as a field closely associated with science, and some mathematicians even refer to mathematics itself as a science. Similarly, many scientists consider mathematics an essential component of what they do.

This close relationship between mathematics and science, however, is relatively recent. Historians generally trace it back to a period in 16th-century Europe called the *Scientific Revolution*, which defined the modern concept of "science." Before the advent of science, the study of natural phenomena fell under the purview of *natural philosophy*, which shared the same methods employed in other branches of philosophy, based on a thoughtful qualitative and descriptive analysis of nature. The Scientific Revolution saw the rise of a different method adapted to natural philosophy called the *scientific method*, based on experimentation and quantitative analysis of experimental data.

The adoption of the scientific method to study natural phenomena had the effect of distancing natural philosophy from other forms of philosophy. Concurrently, the natural philosophers started borrowing mathematical tools to analyze experimental data, thereby initiating the relationship between natural philosophy and mathematics and elevating the status of mathematics from an ancillary subject to a central field of study in its own right. The invention of calculus, motivated by the desire to describe natural phenomena using mathematics, greatly accelerated the advance of natural philosophy, as this new form of mathematics provided the necessary mathematical syntax to write the laws that govern natural phenomena as differential equations.

The inclusion of mathematics in the scientific method, a process dubbed the *mathematization of science*, initially only applied to physics. In the 19th century, scientists used the same mathematical process designed to describe physical laws to also describe hypothetical mechanisms behind the patterns observed in other fields of science. This process of *mechanistic modeling* allowed mathematics to infiltrate biology, chemistry, ecology, epidemiology, and meteorology, among others. The rise of probability theory and statistics in the 19th century brought with it another type of modeling, namely *empirical modeling*, which takes into account the randomness of nature and experimental procedures. Empirical modeling occurs widely in the social sciences and informs the experimental testing of mechanistic models in the natural sciences. All fields of science were gradually mathematized during the 20th century, to the point that mathematical methods now arise in every natural and social science.

Subsection 8.2.1 describes the shifts in science and mathematics initiated during the Scientific Revolution. Subsection 8.2.2 describes the process of mechanistic mathematical modeling and provides an illustrative example. Subsection 8.2.3 outlines the nature and use of empirical modeling. Finally, Subsection 8.2.4 focuses on the current uses of mathematical modeling in science and the practices encompassed by the now commonly used phrase *applied mathematics*.

8.2.1   The Scientific Revolution

The *Scientific Revolution* encompasses a series of events that took place in Europe in the 16th century, at the end of the Renaissance.

These ideas of the Scientific Revolution shaped the views of society about nature and the physical universe. It saw the rise of an inductive method for advancing knowledge in natural philosophy based on experimentation, called the *scientific method*. This method contrasted sharply with the deductive methods rooted in Greek philosophy that had dominated scientific inquiry for almost 2000 years. The Scientific Revolution arguably influenced not only the sciences but also the social movements of the 18th century, called the *Age of Reason* or *Enlightenment*.

The scientific method

The *scientific method* finds its roots in a philosophical movement called *empiricism*, a term derived from the Greek word *empeiria*, which means sense-based experience. Empiricism flows from an Aristotelian idea that knowledge comes only, or at least primarily, from a sensory observation of the world. The traditional medieval approach consisted in determining the causes of what was observed by deductive reasoning, in order to formulate an explanation about how the world works. In contrast, the scientific method uses an inductive approach, which promotes observing with an open mind, being willing to question assumptions, and placing a greater emphasis on empirical evidence, rather than on intuitive ideas, preconceived notions, dogma, tradition, or revelation.

Whereas previous methods focused on a qualitative study of the world, the scientific method distinguishes itself by its use of quantitative measurements of physical phenomena. Thus, for the first time, scientific inquiry became a quantitative endeavor, separating it from other forms of philosophy and bringing it closer to mathematics.

The scientific method proceeds by experimentation, which requires following carefully planned procedures and systematically isolating the hypothetical causes of a phenomenon through inductive elimination. To this effect, experimental design and the use of observation instruments are a central part of the scientific method.

The philosophical ideas behind the Scientific Revolution can be attributed to *Francis Bacon* (1561-1626), an English philosopher and scientist known as the "Father of Empiricism." He thought that traditional natural philosophy concerned itself too much with words (discourse, debate, etc.), rather than actually observing the material world. He championed the inductive approach for scientific inquiry, proceeding from fact to theory to law. His ideas about the methods for scientific inquiry, initially known as the *Baconian method* but later dubbed the "scientific method," eventually became the universal standard method for scientific progress in all natural and social sciences to this day.

The scientific method proceeds iteratively, repeating the following steps in sequence as needed:

1) **Observation.** The first step consists in observing a "real-world" phenomenon. This implies that the scientific method only applies to the parts of the physical world that are perceivable by the human senses.

2) **Hypothesis.** Formulating a hypothesis consists in proposing an explanation regarding the observed phenomenon.[4] A scientific hypothesis must be testable by experiment, so it is usually formulated with a view toward an experiment. The hypothesis is usually accompanied by a prediction of what we expect the experiment will reveal.

3) **Experiment.** This step requires designing and carrying out an experiment (or series of experiments) and collecting data that enable us to determine whether the phenomenon agrees or conflicts with the behavior predicted in the hypothesis.

---

[4]It is important to distinguish this definition of a scientific "hypothesis" from the concept of a "hypothesis" in a mathematical theorem stated within the context of the axiomatic method.

4) **Analysis.** The results of the experiments are recorded and interpreted. The conclusion serves as a starting point for a new hypothesis. The method proceeds recursively by returning to step 2 (hypothesis).

Scientific results are only valid if the scientist discloses the experimental data and if other scientists examine the experimental procedure for quality control. To be valid, an experiment must be repeatable and generate the same results every time someone repeats it under the same conditions. Therefore, the scientist must describe the experimental procedure in full detail to allow others to replicate the results and confirm their validity.

Knowledge produced by the scientific method progresses by tentative steps and involves false starts, continued revision, and falsification. However, when scientists find that the results of several experiments support a hypothesis are confirmed valid by peer review and gain general acceptance by the scientific community, they may establish them as a semi-permanent *scientific theory*.[5] Like any scientific result, a scientific theory can serve as a starting point for further hypotheses that may refine, revise, supplant, or overthrow it.

The modern person inclined to view the experimental approach as obvious and therefore to think of the earlier forms of natural philosophy as backward must understand the perspective of the era. To ascribe to empiricism required someone to abandon the irrefutable nature of conclusions obtained by a deductive method and involved what scholars of the time probably considered a lower form of reasoning.

The mathematization of science

The advent of the scientific method caused natural philosophy to distance itself from theology and higher forms of philosophy, which employed deductive reasoning. The inclusion of measurement-based methods and the analysis of experimental results required mathematics. Thus, the natural philosophy became closely associated with mathematics, initiating a historical phenomenon dubbed the *mathematization of science*. Mathematics underwent a fundamental shift from an esoteric and relatively useless field of abstract knowledge to a branch of study playing a major role in the progress of science.

One of the first natural philosophers who turned to mathematics to describe his observations of the physical universe was the astronomer and mathematician *Nicolaus Copernicus* (1473-1543) from

---

[5]Again, it is important to distinguish this concept of a "scientific theory" from a "mathematical theory," such as number theory or group theory, which is built deductively within the context of the axiomatic method.

Prussia (modern-day Poland). In his work *De Revolutionibus Orbium Coelestium* ("On the Revolutions of the Heavenly Spheres"), Copernicus proposed a heliocentric model for the universe, in which the earth and other planets move along circles on concentric spheres around the sun.

The German *Johannes Kepler* (1571-1630) followed in Copernicus' footsteps and eventually formulated three laws of planetary motion. Kepler's laws, written as mathematical equations, refine Copernicus' model by defining elliptical (instead of circular) planetary trajectories to account for the observed varying velocities. This model later became one of the foundations for Isaac Newton's theory of universal gravitation.

Italian astronomer and mathematician *Galileo Galilei* (1564-1630), often called the "Father of Modern Science," solidified the scientific approach of combining observation and mathematics, going as far as claiming that all laws of nature were mathematical. In his work *Il Saggiatore* ("the Assayer"), generally considered to be one of the pioneering works of the scientific method, Galileo put forth the idea that the book of nature should be read with mathematical tools rather than those of scholastic philosophy, as generally held at the time. He writes:

> [Natural] philosophy is written in this grand book—I mean the universe—which stands continually open to our gaze, but it cannot be understood unless one first learns to comprehend the language and interpret the characters in which it is written. It is written in the language of mathematics, and its characters are triangles, circles, and other geometrical figures, without which it is humanly impossible to understand a single word of it; without these, one is wandering around in a dark labyrinth [26].

Laws of nature and faith

Galileo's statement described the novel idea, born out of the Scientific Revolution, that the patterns observed in the physical universe are governed by laws, called *laws of nature* (or *laws of physics*), expressed in the language of mathematics. It is interesting to note that Galileo, Kepler, Copernicus, and later Newton were all devout Christians; they viewed the assumption that the universe is governed by laws in harmony with the existence of a divine lawmaker, not in opposition to it. These men saw their work of discovering the laws of nature within the context of their Christian faith. Their approach was consistent with the biblical "cultural mandate" that commands humans to care for God's creation, which first requires that humans understand how it works.

Galileo's famous statement also illustrates the rising status of mathematics during the Scientific Revolution. Under the medieval scholasticism, despite serving as an essential element of the trivium and quadrivium, mathematics took a secondary role to the more elevated fields such as philosophy or theology. Thomas Aquinas famously dubbed theology the "queen of the sciences." The mathematization of science and the accompanying belief that God had written the laws of nature in the language of mathematics elevated mathematics to the rank of "God's language." By the 19th century, mathematics had proven so successful at uncovering the nature of physical reality that German mathematician Carl Gauss (1777-1855) dubbed mathematics the new "queen of the sciences," thereby reassigning the phrase previously attributed to theology.

Calculus, mechanism, and determinism

The development of calculus by Newton and Leibniz in the late 17th century came from their desire to invent a method capable of describing mathematically the evolution of natural phenomena in time and space. Both Newton and Leibniz achieved considerable strides by creating the concept of a *derivative* to describe a rate of change. For example, Newton's second law of motion, $F = ma$, relates the force F acting on an object with its mass m and acceleration a, where the acceleration is defined as the rate of change of the velocity (i.e., the time-derivative of the velocity), which is itself the rate of change of the object's position. Thus, calculus made it possible for Newton to write the law that governs the patterns of motion as a mathematical equation containing derivatives, i.e., a *differential equation*.

Newton's laws are just one example in a long list of natural phenomena described using differential equations. In fact, most laws of nature involve this syntax, effectively making differential equations the veritable "language of God." Table 8.1 shows a sample of natural phenomena along with the differential equations that describe them.

Calculus allowed natural philosophers to describe the mechanisms of nature and to understand how and why natural systems behave the way they do, removing much of the mystery behind patterns observed in nature. Armed with differential equations, they confirmed their understanding of natural patterns by predicting future outcomes by simply solving those equations. Furthermore, the mathematical form of these laws provides clues about which parameters we can control to produce a specific, desired outcome.

In its extreme form, this view of the world leads to *mechanism*, a philosophy that claims that all natural systems behave like predictable machines, whose mechanisms can be explained in terms of mathematical laws. Mechanism goes hand-in-hand with

Natural phenomenon	Law (differential equation)
Motion of an object	Newton's laws of motion
Electricity, magnetism, light	Maxwell equations
Fluid dynamics	Navier-Stokes equations
Diffusion	Fourier's heat equation
Wave-like phenomena (sound, light, etc.)	Wave equation
Vibrations in solid matter	Euler-Bernoulli beam equation
Quantum mechanics	Schrödinger equation

Table 8.1: Natural phenomena described by differential equations.

determinism, the belief that all events are completely determined by previously existing causes. Under this belief, the laws of nature become all-powerful and universal, with nothing able to escape their scope. Knowing them becomes the key to understanding the past and predicting the future of any system. French mathematician *Pierre-Simon de Laplace* (1749-1827) sums up this belief as follows:

> We may regard the present state of the universe as the effect of the past and the cause of the future. An intellect which at any given moment knew all of the forces that animate nature and the mutual positions of the beings that compose it, if this intellect were vast enough to submit the data to analysis, could condense into a single formula the movement of the greatest bodies of the universe and that of the lightest atom; for such an intellect nothing could be uncertain and the future just like the past would be present before its eyes [58].

Many Christian natural philosophers attempted to reconcile this view with their belief in an all-powerful, sovereign God. As a compromise, they adopted *deism*, the belief that God exists as a divine "clock maker" who created the universe in the beginning and set it in motion by establishing the laws of nature, but who does not intervene in present times.

8.2.2   Mathematical modeling

Mechanistic models

By the 19th century, scientists had found natural laws expressed as differential equations to successfully describe many natural phenomena whose patterns had remained mysterious for centuries. However, several phenomena still resisted all attempts to be described in this way, especially in biology, the life sciences, and the social sciences. Although these systems followed some general mechanistic principles in certain cases and under special circumstances, they did not appear to follow universal laws. Furthermore, many such "laws" initially considered as universal failed to apply outside of certain scopes. For example, near the end of the 19th century many physicists believed that only a few small corners of their science needed full resolution, especially in regards to the very small objects or to objects moving close to the speed of light. These few small corners led to special relativity and quantum mechanics, which so changed physics that philosopher of science Thomas Khun called it a *paradigm shift* [55].[6]

This realization did not deter scientists from describing observed patterns using differential equations, but they did so cautiously, with the understanding that their mathematical description was not a perfect representation of reality but rather an approximation that was only reasonably accurate in certain circumstances. These imperfect descriptions of natural phenomena are called *mathematical models*. Scientists no longer viewed laws as describing nature in an absolute sense but recognized models as inherently imperfect yet capable of providing a reasonably good approximation of reality in certain circumstances. For example, although the classical laws of Newtonian mechanics have been demoted from "laws of nature" to a mere "model," we still commonly refer to them as "laws" with the understanding that they have certain limits of applicability. Even though these models do not claim to describe a universal law, they still aim at describing an underlying mechanism. For this reason, models of this type are called *mechanistic models* to differentiate them from empirical models that we discuss below. Nowadays, most applied mathematicians no longer ascribe to the lofty goal of discovering a universal law of nature; they settle instead for the more modest goal of building mathematical models that provide a mechanistic explanation for some observed pattern and a reasonably good approximation of reality in certain cases.

Building a mechanistic model requires that the modeler make certain assumptions about the mechanisms at play. Deciding upon simplifying assumptions of reality constitutes striking a balance between tractability and accuracy, i.e., obtaining an equation that is mathematically solvable without sacrificing too much accuracy. Thus, an important element of mechanistic modeling consists in defining the limits of the model's applicability outside of which the model no longer faithfully represents reality. Mechanistic modeling fits squarely

---

[6]Quantum mechanics describes the motion of very small objects, below the radius of an atom, and the theory of relativity describes the dynamics of objects with velocities approaching the speed of light.

Natural phenomenon	Mathematical model (differential equation)
Competitive species in ecology	Lotka-Volterra equations
Population dynamics with harvesting	Beverton-Holt equation
Gene propagation	Fisher's equation
Shock waves	Burgers' equation
Spread of diseases	Kermack-McKendrick equations
Action potentials in neurons	Hodgkin-Huxley equations
Electric transmission lines	Telegraph equations
Age-structured populations	Leslie equation
Options pricing in finance	Black-Scholes equation

Table 8.2: Mathematical models that describe natural phenomena.

within the framework of the scientific method, as it requires careful observation of a system, based on experiments, to derive sufficient knowledge about the system's behavior. The model's solution, when compared to experimental data, gives insight into the mechanisms that drive the system and helps decide which simplifying assumptions are warranted to describe them mathematically. Table 8.2 shows a sample of natural phenomena along with the mathematical models that describe them.

Example

The field of ecology makes extensive use of mechanistic models to describe the interactions between different species of plants and animals in an ecosystem. These interactions, which may be cooperative, competitive, predatory, parasitic, or symbiotic, drive the evolution in population levels. Managers of protected ecosystems often turn to mathematical models to understand the underlying mechanisms and to predict the possible consequences of any human influence. This example concerns a state park that supports a population of wild hares growing out of control and threatening to drive to extinction several species of protected plants and animals. Suppose that park managers or ecologists propose to introduce a population of wild bobcats to act as predators. To predict the consequences of this course of action, we build a mathematical model describing the scenario.

Model. The model aims to simplify the complexity of the situation by ignoring features that we might judge to be of little influence, while retaining the primary mechanisms that drive the evolution of

the hare and bobcat populations. The decisions about which influences to include and ignore follow a period of observation, consultation with experienced field specialists who can give a qualitative
description of the main mechanisms at play and field experimentation to collect quantitative data.

Consider the following mechanistic model (8.1). It consists of two
ordinary differential equations, which describe the rates of change of
the hare $\left(\frac{dh}{dt}\right)$ and bobcat $\left(\frac{db}{dt}\right)$ populations, respectively:

$$\frac{dh}{dt} = Rh\left(1 - \frac{h}{K}\right) - \gamma bh \qquad\qquad \frac{db}{dt} = \alpha bh - \mu b \qquad (8.1)$$

In these equations, $h(t)$ and $b(t)$ represent the populations of hares
and bobcats as time-dependent functions. Each equation contains
two terms, one positive and one negative, to model the increase and
decrease mechanisms of each species. The terms are based on existing
laws that approximate the dynamics of predator-prey interactions
under a set of simplifying assumptions:

- $Rh\left(1 - \frac{h}{K}\right)$ is the growth term for the hare population, based
  on the *logistic growth law*. It assumes that the per capita rate
  of increase decreases linearly with the population and that, in
  the absence of predators, the population grows asymptotically
  to maximum size K.

- $-\gamma bh$ and $+\alpha bh$ describe the interactions between hares and
  bobcats. Predation contributes to a decrease of the hare population and an increase in the bobcats. The terms are based on
  mass action law, which assumes that the populations are well-
  mixed and uniformly distributed in space. Furthermore, this
  law assumes that the growth and decrease rates are proportional
  to the probability of encounter between hares and bobcats, and
  that the probability of predation given an encounter is constant.

- $-\mu b$ is a linear decrease term, which describes the bobcats' natural death rate. It assumes that the per capita death rate is
  constant regardless of the population size.

The model makes further simplifications of reality by ignoring other
species, age distributions, gender ratios, seasonal mating cycles, alternate food sources for bobcats (other than hares), and alternate death
causes for hares (other than predation).

In addition to the variables $h(t)$ and $b(t)$, the equations also contain the following parameters, assumed constant, which represent
quantitative features of the system's dynamics:

- K: carrying capacity (the maximum hare population that the
  ecosystem can sustain).

- R : maximum per capita growth rate of hares.

- α and β: rates of predation and bobcat increase.

- μ: rate of bobcat mortality.

The numerical values of these parameters are *a priori* unknown. Some can be determined by direct experimentation. Others can be determined by finding those that provide the best fit of the model's solution to experimental data collected on the variables $h(t)$ and $b(t)$.

Analysis. Once we define the differential equations, we must integrate them to obtain the *solution*, i.e., the values of functions $h(t)$ and $b(t)$ for any time t. Although it is not possible to solve the system 8.1 analytically, we can obtain an approximate solution using a numerical method. For example, Figure 8.1 shows the solution curves for a scenario that consists in introducing 10 bobcats into a population of 4500 hares, with parameter values $R = 0.018 \, \mathrm{day}^{-1}$, $K = 9500$, $\alpha = 3.5 \cdot 10^{-6} \, \mathrm{day}^{-1}$, $\gamma = 0.0003 \, \mathrm{day}^{-1}$, $\mu = 0.004 \, \mathrm{day}^{-1}$.

The goal in designing a mathematical model for a situation like this one is not to obtain a one-time "correct" answer, but rather to build a tool that can probe the system repeatedly and reveal various features about its dynamics that are not apparent to the naked eye or by experimental observation alone. For example, the model can:

- simulate a variety of scenarios, for example to determine what would happen if more or less bobcats are introduced, or on different dates;

- predict the long-term behavior of the system. For example, the solution curves extended to 10 years, shown in the side panel of Figure 8.1, reveal that the two populations oscillate out of phase with decreasing amplitude, reaching an asymptotic steady state in the long run;

- determine the maximum and maximum hare and bobcat population levels and the corresponding dates;

- reveal the system's sensitivity to a parameter, for example, how the maximum number of bobcats responds to a variation in the carrying capacity of the hares, all other parameters remaining constant;

- determine the value of parameters that cannot be determined by a direct experiment, such as the carrying capacity or the predation rate. We can achieve this by fitting the model's solution to time-series experimental data collected on the variables $h(t)$ and $b(t)$.

Model refinement. Because of the large number of simplifying assumptions, this model is obviously very rudimentary. However, we can progressively refine it to take more complex mechanisms into

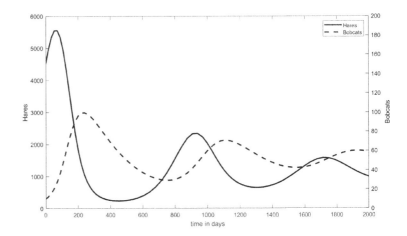

Figure 8.1: Solution of the model for the introduction of 10 bobcats into a population of 4500 hares, with parameter values $R = 0.018\,\mathrm{day}^{-1}$, $K = 9500$, $\alpha = 3.5 \cdot 10^{-6}\,\mathrm{day}^{-1}$, $\gamma = 0.0003\,\mathrm{day}^{-1}$, $\mu = 0.004\,\mathrm{day}^{-1}$.

account and thereby better reflect the reality of the ecosystem. We can relax the simplifying assumptions by taking into account more predators, more food sources, other causes of death, seasonal effects, or delayed feedback effects. Another possible modification consists in converting this deterministic model into a stochastic model that takes randomness into account. In this version, we no longer assume that parameters and variables have a uniquely determined numerical value; we model them instead as random variables distributed according to a probability density function.

8.2.3   Empirical models

The realization that many natural phenomena are inherently random, or exhibit random-like behavior, led to the emergence of a different type of model in the 20th century, one that tries to describe the variability and random nature of these systems. Models of this type, unlike traditional mechanistic models, do not assume that observed patterns are the consequence of a law of nature written as a differential equation. Instead, these models rely on the experimental observations alone. Consequently, we call them *empirical models*, or *data-driven models* (although "data-driven" leads to confusion, as mechanistic models are also driven by experimental data). Empirical models assume that the experimental data exhibit some variance caused by errors inherent to the experimental procedures or by the natural variability of the system itself. These models employ statistical tools, themselves based on probability theory, which take into account the "messiness" of real data and help interpret results based on them. Empirical models provide a guide to support or reject hypothetical cause-and-effect relationships quantifying the uncertainty in conclusions.

One of the main aspects of creating an empirical model consists in formulating a hypothesis in terms of measurable variables and causal relationships between them (i.e., defining what causes what). An empirical model uses a mathematical equation to define a hypothesis for the way certain *explanatory* (or *predictor*) variables produce an observed effect in the *outcome* variables. The scientist then uses statistical tools to compare experimental data to the model's predictions and obtain a *coefficient of determination*, which quantifies the degree to which the model succeeds at explaining the variance in the outcome data.

Another use of empirical models is *uncertainty quantification*, i.e., quantifying the uncertainty in the conclusion of whether experimental results support or reject a hypothesis. Empirical models do not answer this question by a simple "yes" or "no," but rather as a probability (called a *p-value*) that the observed patterns in the data arise from random variability rather than from the causal relationships set forth in the hypothesis. Finally, empirical modeling tools can also quantify the uncertainty arising from flawed data that may not accurately represent reality, either because of flawed experimental procedures, or because the data proceeds from a nonrepresentative sample of a larger population.

Social sciences (especially economics and psychology) use empirical models extensively, so much that social scientists generally refer to this type of modeling when they use the term "mathematical modeling." Empirical modeling has also permeated the "hard" natural sciences to some extent. For example, biology commonly employs both empirical and mechanistic models, which are the main tools studied in the fields of *biostatistics* and *biomathematics* (or "mathematical biology"), respectively. Similarly, weather prediction modeling combines both types of models: empirical models exploit the large quantities of weather data collected over decades and often available in real time, while mechanistic models use the laws of thermodynamics and fluid dynamics that govern the diffusion of thermal energy, air flow, and moisture through the atmosphere.

The importance of empirical models has risen in recent decades with the "big data" trend, fueled by the increasing computational power available, leading businesses and governments to collect increasing amount of data. Empirical modeling is the primary tool used in the emergent fields of *data science* and *data analysis*, whose goal is to manage, organize, and reveal useful information contained in large amounts of raw data.

8.2.4 Applied mathematics

Mathematics in the service of science

Since the mid-20th century, the term *applied mathematics* has come to describe the mathematical methods used in service of modern science. The scientific method has become the standard method by which new knowledge is produced in all natural and social sciences, so we conceive all applied mathematics within this framework. Though the scientific method does not make explicit reference to the use of mathematics, by the 21st century it has become standard practice for almost every field of natural or social science to include some form of mathematical modeling or statistical analysis into the standard process of scientific inquiry.

Some of the advantages of mathematization for the progress of science include the following characteristics:

- Models are powerful experimental tools themselves. Solvable for different initial conditions, they can predict the outcomes of countless alternate scenarios by simulation easily and inexpensively.

- Models are powerful time-saving and cost-saving tools, useful for deciding how to best invest money, time, and effort toward future experiments.

- The insight gained by a mathematical model's solutions is valuable in its own right and often impossible or cost-prohibitive to obtain experimentally.

- Mathematical models can reveal patterns in experimental data that are invisible to the naked eye, and thus would go undiscovered by experimental science alone.

- Overall, mathematics lends the strength and clarity of mathematics to the process of scientific inquiry. As American mathematician Glenn Ledder said, "The muscles of mathematics are connected to the bones of science by the tendons of mathematical modeling [59]."

Mathematical modeling, whether mechanistic or empirical, has become the central activity of applied mathematicians in the 21st century. University programs in mathematics reflect this new reality, emphasizing coursework in differential equations, statistics, data science, and mathematical modeling, to prepare future professional mathematicians to step into a world where modeling skills are valued.

Collaboration in applied mathematics

Applied mathematics is a uniquely collaborative field. The whole process of mathematical modeling requires extensive qualitative knowledge of the phenomenon under investigation, a good grasp of experimental design and data collection techniques, mathematical skills in differential equations and/or statistics, and computational skills to solve the models numerically. It is rare for a single person to possess all these skills. For this reason, mathematical modeling requires that experimental scientists, mathematical modelers, and computational scientists come together, each one contributing to the overall project with their specialized skills, while learning about the other parts in the process.

8.3   The Axiomatic Method

As we mentioned in the introduction to Chapter 1, every area of human inquiry that makes any form of claims to knowledge possesses a distinctive notion of truth and methodology to ascertain it. All sciences employ the scientific method; the methods of justification in history carry the label of historiography; philosophy utilizes yet another more deductive form of reasoning; political science, law, theology, literature, and other areas use yet different methods. Each methodology only applies with a limited scope. For example, science starts with observation and proceeds by experimentation, so can only apply to that which we can perceive by our senses. Other areas of inquiry deal also with questions involving ideas, abstract concepts, and unperceivable objects, placing them outside the scope of the scientific method. The various methods for acquiring knowledge and discerning truth in different fields constitutes an interesting field of study in its own right, called *epistemology*.

Mathematics exhibits a dual nature. On one hand, it provides the tools employed by the scientific method to determine truth in science, while on the other, mathematics need not be limited to its use within the scientific method. Unlike science, mathematics is not constrained to describe things that are perceivable by human senses, dependent on human experience, or even consistent with physical reality. Outside the bounds of science, mathematics uses the *axiomatic method* to determine whether to ascribe to a statement the label of true or false. With roots in Greek philosophy, the axiomatic method determines the truth of a statement by relating it to a set of true axioms via a proof, that is, a valid argument starting from a set of premises and using rules of inference to establish each intermediate conclusion.

Not only do the various methods for discerning truth (scientific, axiomatic, and others) differ in how they discern truth, but they also vary in how they define truth. For example, science assigns truth to facts that agree with objective reality, are consistent with the real world, and can be supported by experimental data. In contrast, the axiomatic method discerns truth in the absence of empirical data.

Part I of this book aims to provide a modern introduction to proofs along with an overview of some of the fundamental objects that appear throughout mathematics. In the first two sections of that part, we discuss the notion of truth in mathematics and the concept of a logical proposition. The examples and the exercises invite the reader into active participation with the subject. In this section, we offer a broad overview of the axiomatic method. Subsection 8.3.1 outlines the main differences between the axiomatic and scientific methods. Subsection 8.3.2 provides a brief history of the axiomatic method, from its origins in ancient Greece to its current status as the universally adopted method in mathematics. Subsection 8.3.3 describes the concept of a mathematical theory, and Subsection 8.3.4 describes the important role of mathematical terms and definitions in the axiomatic method. Finally, Subsection 8.3.5 focuses on axioms and axiomatic systems, and Subsection 8.3.6 offers some insight into how the axiomatic method facilitates progress in mathematical research.

8.3.1   Axiomatic versus scientific method

As previously discussed, the dual nature of mathematics implies that mathematics can be involved within the scientific method or the axiomatic method. Although both methods endeavor to discern truth, they proceed differently, involve different contexts, and implicitly use different definitions for what it means for something to be "true." Furthermore, some of the common terms, such as *hypothesis* and *theory*, mean different things depending on the method. We outline some of the main differences between the two methods below.

1) The scientific method applies to objects in the empirical world, i.e., the part of the world that can be probed by experiments and perceived by human senses. In contrast, the axiomatic method revolves around *mathematical objects*, which may or may not be perceivable. For example, a vector is a common type of mathematical object encountered in many mathematical theories. With the exception of Euclidean vectors in $\mathbb{R}^2$ and $\mathbb{R}^3$, vectors are unperceivable, abstract objects.

2) The scientific method considers a scientific hypothesis to be true if that hypothesis can be supported by experimental data. The axiomatic method considers a statement to be true if it is consistent with a set of axioms, that is, if it can be logically

derived from the axioms by deductive reasoning.

3) In the scientific method, a true hypothesis must agree with objective reality in the physical world. In the axiomatic method, a true statement must be consistent with a set of axioms, but the axioms themselves need not be consistent with objective reality. For example, it is possible to build an entire system of consistent mathematics on an axiomatic foundation that claims that the angles of a triangle do not sum to 180°. For this reason, Albert Einstein observed, "As far as the laws of mathematics refer to reality, they are not certain, and as far as they are certain, they do not refer to reality [30]."

4) A *scientific theory*, established by the scientific method, is a conclusion confirmed by repeated experiments, verified by peer review, and generally accepted by the scientific community. A *mathematical theory*, established by the axiomatic method, is a collection of theorems consistent with a common set of axioms. Because the axiomatic method always establishes the truth of a theorem in relation to axioms, "truth" is always to be understood as relative (to the axioms), i.e., not absolute. For this reason, mathematician Bertrand Russell once quipped, "Mathematics may be defined as the subject in which we never know what we are talking about, nor whether what we are saying is true [68]."

5) Facts about the physical world deduced from experimental results via the scientific method can never claim to be true or false with 100% certainty. Scientists often state their findings along with a measure of significance (for example, a confidence interval or p-value) that quantifies the probability that the experimental data supports the wrong conclusion. Furthermore, scientific theories are continually at risk of being refined, revised, or overthrown by new evidence. Because of this, progress in science often experiences fundamental shifts and revolutions as new theories emerge to replace old ones. In contrast, progress in axiomatic mathematics does not experience the sudden reversals seen in science. Unless a reader discovers an error in a proof, the truth of a theorem proved within the context of a mathematical theory stands immutable.

6) Scientific knowledge established by the scientific method progresses by doing new experiments and analyzing new data. Mathematical knowledge established by the axiomatic method progresses by proving new theorems that abstract, extend, generalize, or connect existing concepts.

8.3.2   History of the axiomatic method

The axiomatic method and its application to mathematics has its roots in ancient Greece. Early Greek mathematics started similarly to the mathematics of other ancient civilizations (Egypt, Babylon, India, and China, among others), which developed sophisticated mathematical methods without any axiomatic structure. However, the Greek civilization is unique in that Greek mathematicians abstracted the concepts of geometry into a system of carefully worded statements, using them as building blocks to determine the truth of more complex and less intuitive statements. Soon, Greek geometry in particular distanced itself from its utilitarian roots and became primarily a tool to determine truth by deductive reasoning and logic. Thus, Greek mathematicians became more interested in philosophical implications than in practical applications. Historians generally attribute many early Greek mathematical developments to the *Pythagoreans*, an order of mathematicians established by *Pythagoras of Samos* (6th century BC). This order promoted the pursuit of both mathematics and philosophy as a way to provide a moral basis for the conduct of life. To the ancient Greek mind, mathematics and philosophy were inseparable. It is this relationship that gave rise to the axiomatic method, which has become the universal framework for modern theoretical mathematics.

Around 300 BC, Euclid of Alexandria compiled all the known mathematical results of his day in his work *The Elements*, organizing them according to an axiomatic structure. Starting with a set of axioms and primitive terms, Euclid proceeded to prove every result by relating it only to previously proved results by logical deduction. *The Elements* is the earliest example of a work structured as mathematical theory and stands as a timeless masterpiece that perfectly displays the beauty of logic. This structure has been used ever since as a template for other mathematical theories.

However, the axiomatic structure of *The Elements*, which was adopted by Greek mathematicians, did not immediately take hold in other parts of the world. The mathematics of India and the Islamic world consisted primarily in collections of methods for solving various problems and never developed according to an axiomatic structure [53]. Neither did the great mathematical achievements of the Scientific Revolution in Europe. The invention of calculus, created primarily as a method for solving practical problems in physics, sparked an explosion of new mathematical concepts and methods developed haphazardly, lacking an axiomatic foundation or the organized structure of interlocking theorems and proofs.

Despite these advances, no one completely forgot Euclid's *Elements* during this time. Educators considered the study of Euclidean geometry an essential tool, specifically for the development of logical reasoning skills. For a while, newer mathematical developments did not attempt to frame themselves in the context of an axiomatic structure. A major shift occurred at the beginning of the 19th century, when mathematicians decided to axiomatize mathematics, turning to the structure of Euclid's *Elements* as a template to organize the existing concepts of mathematics into mathematical theories, thus providing axioms, precise mathematical definitions, and theorems. The 19th century saw the axiomatization of calculus, algebra, number theory, probability theory, and many others.

For many centuries earlier, mathematicians attempted to prove Euclid's fifth postulate, the parallel postulate, from his other four, and always came up short. Lobachevsky in 1829 and later but independently Bolyai in 1832 proved that the postulate is independent of the other axioms. Rephrasing the parallel postulate to Playfair's version, "Given any line L and a point P not on L, there exists a unique line parallel to L that falls on P," Lobachevsky and Bolyai developed non-Euclidean geometries by modifying "a unique line" to either "no lines" or "more than one line." Each of these modified axioms led to a fully consistent geometry, namely elliptic and hyperbolic, respectively.

The discovery of non-Euclidean geometries shook the foundations of the mathematical world. Because of the central role of *The Elements* in teaching logic, for many people it shook the notion of truth itself. Nevertheless, the 19th century efforts to axiomatize the different fields of mathematics continued to search for a universal, consistent, and complete foundation for all of mathematics. This project, called the *Foundations of Mathematics*, lasted into the early 20th century, but ultimately failed when Austrian mathematician *Kurt Gödel* proved two *Incompleteness Theorems* in 1931, which imply that there does not exist a formal system of logic robust enough to perform arithmetic and in which it is possible to prove all the true statements expressible within that system.

The discovery of non-Euclidean geometry and later the proof of the Incompleteness Theorems, approximately coincided with the postmodern trend of relativism, which encouraged abandoning the concept of an absolute truth.[7] For mathematicians, nothing could be further from accurate. Mathematicians slowly realized that their error came from conceiving of an axiom as a self-evident truth that describes our reality. People formerly viewed points, lines, and circles as objects that could exist in our physical reality and conceived of the axioms as truths about that reality. After all, the theorems

---

[7]Many amateur scientist-philosophers also muddied the waters by interpreting Einstein's theory of relativity to imply the vacuous comment "everything is relative."

proved via the deductive reasoning of Euclidean geometry matched so well with the corresponding physical constructions that countless students, architects, and engineers performed with their rulers and compasses.

Nowadays, we view mathematical axioms as definitional, meaning that they describe the properties of a given structure. For example, Peano's axioms do not describe the integers as they are but rather define the set of integers. As in applied mathematics, we recognize any application of mathematics to reality as a model. Euclidean geometry models our physical reality well at our scale. The axiomatic method remains the method by which we establish all mathematical theories with the one difference that all results are contingent on definitions. Research consists in extending existing theories or creating new theories via the axiomatic method.

8.3.3   Mathematical theories

In mathematics, a *theory* (or *mathematical theory*) refers to a consistent, self-contained body of knowledge formulated as *mathematical statements*. The statements of the theory consist of a set of *axioms*, in effect definitions, and all the *theorems* (or *propositions*) that are derived logically from them using deductive reasoning. Consequently, all mathematical truths are *contingent* truths in that they depend on certain definitions. A theorem is true because it follows from the axioms by a sequence of logical *rules of inference* called a *proof*.

At the foundation of a theory, the statements involve *primitive terms*, words given without a definition that we can effectively think of as place holders. Axioms make statements about how these terms interact. For example, the Euclidean axiom, "On any two distinct point falls one line," is a quantified statement that connects these undefined terms. In order to refer to patterns concisely, every theory introduces *mathematical definitions* that attach a new term to some mathematical object or some property.

Consequently, a mathematical theory is a collection of theorems that fit together precisely, like interlocking pieces of a puzzle or bricks of a wall. The collection extends by stacking and combining existing theorems to form new theorems.

The vast majority of mathematics rests on the foundation of Boolean logic. After that, most of modern mathematics exists in the context of set theory. Much of the content of Part I in this book introduces set theory, defining the notion of a set, an element, a Cartesian product, and then subsequently defining functions, cardinality, relations, equivalence relations and so on. However, we only introduced what most people call *naïve set theory* as opposed to the formal axioms of set theory, Zermelo-Fraenkel with the Axiom of Choice, labeled as **ZFC**. These axioms give a precise definition of a set from logic. Many other theories build entirely on set theory. For example, group theory introduces no primitive terms; the axioms of group theory define the concept of a group using primitive terms of set theory.

There are hundreds of different theories in mathematics; some are large and ancient, and others are relatively new. Some theories branch off into separate subtheories, and sometimes people discover "bridges" between separate theories. Here are some examples of well-known mathematical theories: set theory, Euclidean geometry, bifurcation theory, chaos theory, graph theory, group theory, knot theory, number theory, operator theory, probability theory, representation theory, category theory, model theory, topos theory, and so on. Some theories carry the name of their inventor/discoverer; for example, Fourier theory, Galois theory, and Hodge theory.

Mathematical statements

A mathematical theory is formulated in terms of *mathematical statements*, technically called *logical propositions*, which are declarative sentences that must be either true or false. This concept of truth, which undergirds Boolean logic, implies the principles of *noncontradiction* (it cannot be both true and false) and *excluded middle* (it must be either true or false). As we discuss in Section 1.1, logical propositions can be negated or combined to form *compound statements* by use of *quantifiers* (for all..., there exists...) and *logical connectives* (and, or, if..., then...). Logical rules of inference determine how the truth value of these more complex statements depends on the truth value of their individual components.[8]

The mathematical statements that make up a theory can be of different types:

1) **Axioms** are definitional statements that provide the framework for a given theory. They do not require a proof. These basic statements provide a starting point from which flow all the theorems in the theory.

---

[8] A handful of mathematicians utilize another form of logic called, *intuitionistic logic*. This logic involves a different notion of "truth." In contrast to Boolean logic, a proposition cannot have a truth value inherently and is only considered true once a valid proof exists. This implies that there does exist an "excluded middle" possibility and that the logical equivalence of double negative does not apply in intuitionistic logic. From a practical perspective, this means that proofs by contradiction are not a valid proof strategy. Because it is weaker than classical logic, every statement proved with intuitionistic logic holds in Boolean logic, but the converse is not true. Mathematicians who use intuitionistic logic exclusively often believe this form of logic to be more consistent with their constructivist/empirical view of mathematics (i.e., that humans create mathematics), whereas those who hold to a more Platonicist view favor Boolean logic (See Section 8.5.3).

2) **Theorems** are true statements by virtue of their proof, which tie them to the axioms and/or previously proved theorems.

3) **Definitions** are equivalence statements (if and only if...) that associate a mathematical term to an abstract concept, object, or property.

4) **Conjectures** are statements that appear to be true, but for which there is no proof (yet). As soon as someone produces a valid proof, the conjecture becomes a theorem.

Primitive terms and axioms

The rules of inference of the axiomatic method forbid *self-referential* statements (statements that assert what they seek to prove, or definitions that contain the term they seek to define), as well as *circular* statements and definitions (for example, defining a *hill* as "elevated landmass lower than a mountain," and *mountain* as "elevated landmass higher than a hill"). Therefore, at the heart of every theory lies a set of *axioms* and *primitive terms*. Together, these provide a foundation for the theory and a starting point for the logical structure of the axiomatic method.

The primitive terms are a set of foundational notions that are not defined in terms of previously defined concepts. We typically either assume the terms to be sufficiently self-evident, so that there is no possibility of ambiguity, or they simply remain undefined terms so that the term serves as a place holder. In Euclidean geometry, for example, the primitive terms include "point," "line," and "plane." These terms are not defined but their meaning for how they model our ambient space is inferred from properties stated in the axioms.

Axioms are true by definition, usually stating a fundamental relationship between the primitive terms. For example, Euclidean geometry has ten axioms (which are divided into five postulates and five common notions in Euclid's *Elements*). One of them states that "a straight line segment may be drawn from any given point to any other." Many theorems of Euclidean geometry hold because of this assumption, and their proof is justified by invoking this axiom.

The axioms that constitute the heart of a theory, called the *axiomatic system* of the theory, are usually carefully chosen to avoid redundancies and the possibility of contradictions.

Logic

Whereas mathematical terms constitute the vocabulary, logic constitutes the grammar of the language employed in axiomatic mathematics. Logic forms the "cement" that holds an axiomatic structure together. It defines the rules of inference that govern the deductive reasoning steps that determine the truth value of mathematical statements and validate the proof of mathematical theorems.

In its mathematical sense, logic is a subfield of philosophy concerned with discerning truth using criteria of validity of inference. Like the axiomatic method itself, the formal logic used in mathematics has its roots in ancient Greece. *Aristotle* (4th century BC) in particular, studied the principles of reasoning by employing variables to show the underlying logical form of an argument. He sought relations of dependence that characterize necessary inference from the truth of the premises, and he was the first to deal with the principles of contradiction and excluded middle in a systematic way [76]. Euclid systematized and mathematicized this form of logic in *The Elements*. Another school of ancient Greek logic is that of *Chrysippus of Soli* (3rd century BC) and his successors, the Stoics, who studied propositional logic and conditional (if..., then...) statements. Proponents eventually refined this form of logic in the use of symbols.

Modern philosophy departments offer courses in formal logic that introduce symbolic logic and its rules, which function very similarly to mathematics. However, although logic is the backbone of the axiomatic method, mathematical proofs typically only make use of relatively simple rules of inference. It is not necessary to be an expert in symbolic logic to be a good mathematician.

8.3.4 Mathematical definitions

The statements of a theory involve different types of abstract concepts, including mathematical objects (numbers, vectors, manifolds, graphs, vector spaces, etc.), operations between objects (addition, scalar multiplication, matrix multiplication, dot product, cross product, etc.), and properties of objects (convergence, continuity, commutativity, etc.). Communication in mathematics requires that every abstract concept be associated with a mathematical term. The semantic equivalence is given by a *mathematical definition*.

Unfortunately, no language can perfectly describe abstract concepts accurately in a concise way. In the humanities and social sciences, abstract ideas often require lengthy descriptions that take up entire paragraphs, chapters, or even books to capture their essence. In contrast, a mathematical definition assigns a highly abstract concept unequivocally to a single word. For this reason, we word definitions carefully and precisely. Anyone learning a particular field must memorize them verbatim since the essence of a mathematical term cannot be grasped by either intuition or deduction. An incorrect or incomplete knowledge of a term's definition makes it impossible to fully understand any statement that contains it or to prove anything related to that term. Unfortunately, even professional mathematicians often trace their errors to a misunderstanding of a definition.

Establishing definitions

Like all mathematical statements, every definition must satisfy the principles of noncontradiction and excluded middle, i.e., it must be either true or false. Furthermore, every definition is worded as an *equivalence* (if and only if) statement that associates a term (usually highlighted in bold or italics) to an abstract mathematical concept.[9] For example,

> A 4-sided polygon is a *parallelogram* if and only if both pairs of opposite sides are parallel.

In the example above, the noun "parallelogram" describes the properties of a certain type of polygon. Other examples of nouns include: vector, matrix, slope, gradient, category, ideal, group, ring, and field. In mathematics, we often name objects after the person who invented them, such as an abelian group (named after *Niels Abel*), a Gorenstein ring, a Cauchy sequence, a Laplace transform, or a Hilbert space.

We may describe mathematical properties by a qualifying adjective; for example,

> A matrix is *square* if and only if it has the same number of rows and columns.

In this case, "square" is an adjective that describes a property of a matrix. Other examples of adjectives that define matrix properties include: invertible, singular, skew, and triangular. Similarly, a function may be continuous, differentiable, smooth, or analytic.

A definition may describe an operation; for example,

> The *determinant* of a square matrix $\begin{bmatrix} a & b \\ c & d \end{bmatrix}$ is the scalar $ad - bc$.

A definition may use previously defined terms; for example,

> A square matrix is *singular* if and only if its determinant is zero.

Understanding this definition requires knowing the terms "square," "matrix," and "determinant." Definitions are often presented in sequences, stacked upon each other, each one using previously defined

---

[9]Although some texts informally use "if" instead of "if and only if," a definition should always be understood as a two-way equivalence.

terms to define a new one. For example, in Euclidean geometry, the definition of a circle uses the terms "point," "line," "straight," "shape," and "boundary" (previously established as primitive terms):

A circle is a shape whose boundary is a single line, and such that straight lines from all points on the boundary to one point inside the shape are equal.

A mathematician may introduce a mathematical theory by providing all the necessary definitions upfront, as Euclid did at the beginning of *The Elements*. However, it is more common nowadays to introduce definitions gradually, immediately preceding the theorem where they first appear. For this reason, upper-level mathematics texts usually organize their presentation as a sequence of statements that alternates between theorems and definitions. Examples, whether before or after a definition, motivate and illustrate the concept, while proofs, usually given immediately after the theorems provide the logical justification.

Precision versus clarity

Mathematical definitions require being concise without unambiguity. Thus, every word is important and serves a specific purpose. Consider this example:

A nonzero vector $\vec{v}$ is an *eigenvector* of a square matrix A if and only if $A\vec{v} = \lambda\vec{v}$ for some scalar λ called the *eigenvalue*.

In this definition, the adjective "nonzero" is essential. Without it, the zero vector would satisfy the definition, leading to errors. This definition also contains a statement whose truth value can be verified, thereby providing a test: given a square matrix A and a nonzero vector $\vec{v}$, the equation $A\vec{v} = \lambda\vec{v}$ constitutes a clear and unequivocal mathematical test to determine whether or not a given $\vec{v}$ is an eigenvector of A. (If the equation holds, then yes; if not, then no.) The definition also provides a road map for its use in a theorem, as the equation $A\vec{v} = \lambda\vec{v}$ constitutes a key step in the proof of any theorem involving eigenvectors. Thus, students must develop the habit of learning definitions word-for-word and know how to use them (which is a different skill than knowing how to use a *method* for finding eigenvectors).

Furthermore, mathematical definitions are written to facilitate the logical mechanisms of the axiomatic method. As such, they may not provide a clear and intuitive vision of the term, as their primary function is not to clarify or explain a term as would a descriptive definition in a nonmathematical dictionary. Unfortunately, in an effort to be precise, a mathematical definition often loses intuitive clarity.[10] For example, consider this definition from real analysis:

A real-valued function $f(x)$ is *continuous* at x_0 if and only if for every $\varepsilon > 0$, there exists a $\delta > 0$ such that if $|x - x_0| < \delta$, then $|f(x) - f(x_0)| < \varepsilon$.

This definition does little to help gain an intuitive understanding of what it means for a function to be *continuous*. A more intuitive definition would say, for example, "A function is continuous at x_0 if it does not exhibit any abrupt change (discontinuity) at x_0." Continuity did indeed originate from this more intuitive notion, which professors often teach in an elementary calculus course and was sufficient for the development of calculus up through the end of the 18th century. As the imprecision of this definition led to contradictions, mathematicians later settled on the formal definition above, in the 19th century, when they formalized and axiomatized calculus. Although less intuitive, this new definition provides precise mathematical statements whose truth value can be verified, allowing it to play an important role in any theorem involving this concept.

Sources of confusion

Describing abstract concepts using any human language presents many difficulties, compounded in mathematics by the simultaneous need for precision and conciseness. We can often trace a great deal of confusion in mathematics back to the treacherous, but unavoidable, use of language. Here are some common sources of confusion and errors:

- Mathematics often borrows words that have an everyday definition and assigns them a different, mathematical definition. For example, the mathematical terms *group, ring, field, gradient, slope, trace, derivative,* and *integral* all have a common definition that appears in any standard nonmathematical dictionary. It is the same way with the adjectives *defective, diagonal, inconsistent, invertible, similar, singular, skew, symmetric,* and *triangular.* Because these words look familiar, unaware readers may not realize that, in a mathematical context, they refer to a very specific (and often very nonintuitive) mathematical concept. For example, a casual reader may rush across the word *similar,* not realizing that in matrix algebra, this adjective has the following precise definition:

Two matrices A and B are *similar* if and only if $B = PAP^{-1}$ for some invertible matrix P.

---

[10]Physicist Richard Feynman remarked on this when he said, "Mathematicians sacrifice clarity for precision [32]."

- A mathematical term may have different mathematical definitions in different theories. For example, the term *similar* (defined above for matrix algebra) has a different definition in Euclidean geometry, where two shapes are *similar* if we can obtain one from the other by a uniform scaling, translation, rotation and/or reflection.[11] Similarly, there is a definition for continuous functions between metric spaces (a set equipped with a generalized concept of distance) and also between topological spaces (a further generalization of metric spaces). In each of these contexts, the definition of a *continuous function* differs from the definition stated in the previous context but each time generalizes it.

- We may assign a mathematical concept to multiple synonymous mathematical terms. For example, in dynamical systems, the terms *fixed point*, *equilibrium point*, *stationary point*, *critical point*, and *steady state* all mean the same thing.

- Doing mathematics in more than one language, such as learning from a mathematical text in German or French, raises additional difficulties linked to inconsistent translations: different terms in one language are sometimes translated as the same term in English, or vice versa.

8.3.5   Axiomatic systems

Every mathematical theory is centered around a set of axioms called an *axiomatic system*.

Historically, many mathematical developments coalesced into various bodies of mathematical knowledge long before the concerted effort to *axiomatize* them into formal mathematical theories. Thus, the process of *axiomatization* of mathematical knowledge tended to occur, and in some instances still occurs, in reverse. This requires organizing existing mathematical results into a consistent axiomatic structure, formulating the necessary definitions and axioms to support the theory. Before the 19th century, axioms were typically chosen to be self-evident truths that were consistent with objective reality, whose truth could simply be verified by observation and supported by scientific evidence. Many mathematical theories nowadays are based on this type of axiom. Applied mathematicians in particular, who are concerned with applying mathematical methods to problems of the real world, work with mathematical theories based on axioms that reflect physical reality.

---

[11] In this example, the use of the same word suggests that the two concepts are related. Indeed, similar geometrical shapes may be obtained algebraically by a similarity transformation.

Unlike contexts that depend upon the scientific method, axiomatic systems need not be constrained by physical reality, a mathematician may coin undefined terms and set an arbitrary statement with those primitive terms as an axiom. Therefore, it is possible to purposely define an axiomatic system that does not appear to model anything observed within the real world or perceived by human senses. The discovery of non-Euclidean geometries inspired 19th century mathematicians to explore this freedom afforded by the axiomatic method to build entire mathematical theories upon axiomatic systems defined more or less arbitrarily, unconcerned with applications to the real world or even coherence with the real world. This movement and ensuing mathematical progress helped initiate the separation between pure and applied mathematics.

Properties of axiomatic systems

For an axiomatic system to be valid, it must possess one fundamental property: *consistency*. There are two additional properties that we may desire in an axiomatic system: *independence* and *completeness*. We describe these three properties below.

1) **Consistency.** An axiomatic system is *consistent* if the axioms cannot be used to simultaneously prove a statement and its negation. In other words, the statements built upon the axioms may not lead to a contradiction. An axiomatic system that lacks this property is of no interest in mathematics.

2) **Independence.** An axiomatic system is *independent* if all axioms are fundamental truths that do not rely on each other for existence. If one of the axioms can be proved by the others, it is technically redundant and should be considered merely a theorem.

3) **Completeness.** An axiomatic system is *complete* if the axioms are sufficient to prove every statement containing the primitive and defined terms of the system as being either true or false. This means that whatever anyone attempts to prove with the axioms will either be proved, or its negation will be proved.

As noted above, independence and completeness are not absolute requirements for a functioning axiomatic system. Independence is a desirable property when seeking to eliminate redundancies and reduce the number of axioms to the bare minimum. As we mentioned earlier, for over 2000 years mathematicians suspected that the *parallel postulate* of Euclidean geometry followed from the other four, in other words that Euclid's axioms were not independent. Many mathematicians tried, in vain, to prove the parallel postulate from the other axioms. Lobachesky's and Bolyai's work introduced non-Euclidean geometries but it was not until 1868, when *Eugenio Beltrami* (1835-1900) finally proved the independence of the parallel postulate from the other axioms.

In 1931, Austrian mathematician *Kurt Gödel* proved two famous *Incompleteness Theorems*. The first one states that even in elementary arithmetic, there exist *undecidable statements* that cannot be proved or disproved within the system. The second one states that the axioms of a system cannot prove that the system itself is consistent (assuming that it is indeed consistent). These theorems had a profound impact on the philosophy of mathematics and the search for the foundations of mathematics.

College geometry courses often introduce very simple axiomatic systems. (See for example [17, 39, 70].) However, we list below some examples of axiomatic systems of well-known mathematical theories.

Euclidean geometry

Euclidean geometry is the mathematical theory compiled and organized by Euclid in *The Elements*. Its axiomatic system consists of 10 statements, divided into 5 *common notions* and 5 *postulates*. Early Greek mathematicians made a distinction between common notions (assumptions common to all sciences) and postulates (assumptions specific to a particular science). In modern times, we make no distinction between the two types of axioms.

1) **Common notions**

   (a) Things which are equal to the same thing are also equal to one another.

   (b) If equals are added to equals, then the wholes are equal.

   (c) If equals are subtracted from equals, then the remainders are equal.

   (d) Things which coincide with one another are equal to one another.

   (e) The whole is greater than the part.

2) **Postulates**

   (a) A straight line segment can be drawn joining any two points.

   (b) Any straight line segment can be extended indefinitely in a straight line.

   (c) Given any straight line segment, a circle can be drawn having the segment as radius and one endpoint as center.

   (d) All right angles are equal to one another.

(e) If two lines are drawn which intersect a third in such a way that the sum of the inner angles on one side is less than two right angles, then the two lines inevitably must intersect each other on that side if extended far enough.

The fifth postulate is equivalent to the *parallel postulate*, which states: "Given any straight line and a point not on it, there exists one and only one straight line which passes through that point and never intersects the first line, no matter how far they are extended." From the beginning, this postulate was problematic to many mathematicians, if only because it reads differently than the others: whereas the first four are clear and simple assertions, the fifth is an awkward sentence, difficult to state succinctly. Euclid himself probably disliked using it, as he avoided it for the first 28 propositions of *The Elements*, before being forced to invoke it on the 29th. Many people throughout history suspected that the fifth postulate was not an independent axiom and attempted to prove it as a theorem from the other axioms. Efforts in this direction produced more general results than standard Euclidean geometry, which led to theorems as soon as mathematicians formally defined elliptic and hyperbolic geometry. For example, first studied by Omar Khayyam in the 11th century and later rediscovered by Sacchieri in 1733, the Sacchieri-Khayyam quadrilateral is defined as a quadrilateral with two equal sides perpendicular to its base. It is easy to show in Euclidean geometry that this shape must be a rectangle, but it is possible to prove a number of interesting properties about this figure that apply in any geometry involving the first four of Euclid's axioms.

Natural numbers

In 1889, Italian mathematician *Giuseppe Peano* (1858-1932) developed a set of axioms for the set of natural numbers (denoted $\mathbb{N} = \{0, 1, 2, 3, ...\}$), which became the standard axiomatization of the natural number system used in number theory. The system uses the concept of a *successor* as a primitive term.

1) Zero (0) is a natural number.

2) If a is a number, then the successor of a is a natural number.

3) There is no natural number whose successor is 0.

4) Two natural numbers that have the same successor are themselves equal.

5) If a set S of natural numbers contains 0 and also the successor of every number in S, then S contains every natural number.

Section 4.1 of this book describes in some detail how all the properties of the natural numbers follow from the five Peano axioms.

Real numbers

The axiomatic system that defines the set of real numbers (denoted $\mathbb{R}$) and its properties is called the *real number system*. This system constitutes the foundation for real analysis, which includes elementary arithmetic and calculus. The axioms are divided into three sets: the field axioms, the order axioms, and the completeness axiom.

1) **Field axioms**

   (a) $a + b = b + a$, for all $a, b \in \mathbb{R}$.

   (b) $a + (b + c) = (a + b) + c$, for all $a, b, c \in \mathbb{R}$.

   (c) There exists an additive identity element denoted $0 \in \mathbb{R}$ such that $a + 0 = a$, for all $a \in \mathbb{R}$.

   (d) Every element $a \in \mathbb{R}$ has an inverse denoted $(-a)$ such that $a + (-a) = (-a) + a = 0$.

   (e) $a \cdot b = b \cdot a$, for all $a, b \in \mathbb{R}$.

   (f) $a \cdot (b \cdot c) = (a \cdot b) \cdot c$, for all $a, b, c \in \mathbb{R}$.

   (g) There exists a multiplicative identity element denoted $1 \in \mathbb{R}$ such that $a \cdot 1 = a$, for all $a \in \mathbb{R}$.

   (h) Every element $a \in \mathbb{R} \backslash \{0\}$ has an inverse denoted a^{-1} such that $a \cdot a^{-1} = a^{-1} \cdot a = 1$.

   (i) $a \cdot (b + c) = a \cdot b + a \cdot c$, for all $a, b, c \in \mathbb{R}$.

2) **Order axioms**

   (a) (Trichotomy) For all $a \in \mathbb{R}$, exactly one of $a < 0$, $a = 0$, $a > 0$ is true.

   (b) If $a, b > 0$, then $a + b > 0$ and $a \cdot b > 0$.

   (c) If $a > b$, then $a + c > b + c$, for all $c \in \mathbb{R}$.

3) **Completeness axiom.** If a nonempty subset $A \subset \mathbb{R}$ has an upper bound, then it has a least upper bound in $\mathbb{R}$.

In the language of algebra, a set of elements, together with two operations ($+$ and $\cdot$), that satisfies the field axioms is called a *field*. So, if we already had the notion of a field at our disposal, we could summarize the first set of axioms by stating that "$\mathbb{R}$ is a field."

More specifically, the first four field axioms (a-d) state that $\mathbb{R}$, together with the addition operation, has the structure of an *abelian group*. Similarly, the next four (e-h) state that $\mathbb{R} \backslash \{0\}$, together with the multiplication operation, is also an abelian group. The last field axiom of distributivity ties the two operations together.

A field, together with a relation $<$ that satisfies the order axioms, is called an *ordered field*. An ordered field that additionally satisfies the completeness axiom is called a *complete ordered field*. So the entire set of axioms can be summarized by the statement: "There exists a complete ordered field, which we call the real numbers, and denote by $\mathbb{R}$."

Finally, in 1903, Edward Huntington, based on previous work by Dedekind and Hilbert, proved that there is a unique complete ordered field, which allows us to use the system thus defined as a definition of $\mathbb{R}$ [48].

These axioms suffice to prove all the algebraic properties of real numbers, the rules of elementary arithmetic, the existence of irrational numbers, the existence of decimal expansions, and any property that can be derived about the real numbers. It is important to note that there are alternate ways to define the real numbers. Section 6.4 in Part I uses set theory to define $\mathbb{Z}$ from $\mathbb{N}$, then $\mathbb{Q}$ from $\mathbb{Z}$ and finally $\mathbb{R}$ from $\mathbb{Q}$. The theorem that there exists a unique complete ordered field implies that $\mathbb{R}$ as constructed in this way is the same set as that described by the above axioms.

Probability theory

The study of probability originated with attempts to analyze chance events. Using combinatorics to study discrete events led to the ideas of a *probability* and a discrete *probability distribution*, a concept that mathematicians later extended to continuous distributions. These results were eventually organized into modern probability theory, which rests on an axiomatic foundation laid in 1933 by Russian mathematician *Andrey Kolmogorov* (1903-1987). Based on set theory and measure theory, this system formally defines the notions of a *random event*, *mutually exclusive events*, and an event's *probability* as taking a value between 0 and 1. The three Kolmogorov axioms are:

1) The probability of an event is a real number greater than or equal to 0.

2) The probability that at least one of all the possible outcomes of a process (such as rolling a die) occurs is 1.

3) If two events A and B are mutually exclusive, then the probability of either A or B occurring is the probability of A occurring plus the probability of B occurring.

The deep results of probability, such as the *law of large numbers* and the *central limit theorem*, follow from these axioms. Probability theory itself underlies statistics and data science, which offer many direct applications in the natural and social sciences.

Non-Euclidean geometries

Non-Euclidean geometries are mathematical theories based on axiomatic systems that are close to, but different than, Euclidean geometry. Most of these theories arise from modifying the fifth postulate (the *parallel postulate*), which states that, "Given any straight line and a point not on it, there exists one and only one straight line which passes through that point and never intersects the first line, no matter how far they are extended." In 1829 and 1831, respectively, the Russian *Nicolay Lobachevsky* (1792-1856) and Hungarian *János Bolyai* (1802-1860) independently created a consistent version of geometry that modifies the parallel postulate to admit infinitely many (instead of only one) distinct parallel lines through the given point. This new system was named *Bolyai-Lobachevsky geometry*, and later *hyperbolic geometry*.

For his habilitation[12] in 1854, the German *Bernhard Riemann* (1826-1866) prepared a lecture entitled, *Uber die Hypothesen die der Geometrie zu Grunde liegen* ("On the hypotheses which lie at the foundation of geometry"). His paper took Gauss' work on differential geometry and extended it beyond its bounds at the time. Defining differentiable manifolds, the metric tensor, curvature and many other deep notions, he ultimately produced a framework that subsumed both Euclidean and hyperbolic geometries, and gave a foundation for Einstein's theory of general relativity. As a simple example, Riemann modified the parallel postulate in a different way, such that no parallel line passes through the given point. As a result, he created *elliptic geometry*, which can be modeled as occurring on a sphere. Each one of these geometries has its own set of theorems that differ from those of Euclidean geometry. For example, the interior angles of a triangle (which sum to 180 degrees in Euclidean geometry) sum to less than 180 in hyperbolic geometry, and more than 180 in elliptic geometry.

Although non-Euclidean geometries were originally defined axiomatically as purely theoretical exercises, some of them later found applications in the real world. Spherical geometry, a special case of elliptic geometry, plays a key role in astronomy, cosmology, geodesy, and navigation. Hyperbolic geometry offers a model for the theory of special relativity in physics.

8.3.6 Research involving the axiomatic method

Ever since its emergence in Ancient Greece, the axiomatic method has served as a model for clear and logical reasoning. Euclid's *Elements* was an important part of the curriculum in all universities until the early 20th century; every scholar was familiar with its contents. However, the axiomatic method did not become widespread in mathematics until it experienced a revival in the 19th century. At that time, mathematicians made a concerted effort to axiomatize all

---

[12]A higher level than a doctorate, the habilitation is a qualification required in many European universities to obtain a professorship.

existing fields of mathematics using Euclid's *Elements* as a template. During this period, dubbed the *Age of Rigor* in mathematics, all existing mathematical results were gradually organized into formal theories, each one with its definitions, axioms, and theorems. Since then, mathematicians develop mathematical knowledge according to an axiomatic structure. This axiomatic method has a central place in the way mathematics is taught, learned, and researched in both applied and pure mathematics.

What defines the distinction between applied and pure mathematics is not the rigor or the method by which results are developed, but rather the original motivation of the research and the intended use of the work. Applications drive research in applied mathematics, but often they also spark extensions into mathematical theories that have no known application. Similarly, mathematical theories developed without external application in mind often result in applications to unexpected real-world situations, sometimes long after their development. Section 8.5.2 has a more detailed discussion of this phenomenon dubbed the *passive effectiveness* of mathematics.

Research in applied mathematics refers to mathematical developments motivated by an application to the real world. In searching for solutions to these problems, scientists and mathematicians develop mathematical methods, models, and algorithms. However, these tools are only mathematically sound and reliable insofar as they are firmly established on a theoretical mathematical foundation. To ensure the reliability of these methods, it is necessary that they arise from an existing mathematical theory, or that the applied mathematician create a new theory to support them.

Research in pure mathematics proceeds by extending, generalizing, or unifying existing theories. Results obtained in this fashion need not be motivated by a specific application, but this does not mean that applications will not emerge at a later date. Sometimes, results obtained entirely by axiomatic deduction can later provide the basis for mathematical methods useful in applications. Here are some examples:

- When Johannes Kepler needed to describe the motion of planets in 1609, he used a family of curves called *conic sections* (i.e., ellipses, hyperbolas, and parabolas), which the Greek Apollonius had studied around 200 BC with no specific application in mind. Isaac Newton used them again in his law of gravitation a few decades later.

- *Bolyai-Lobachevsky geometry*, a mathematical theory obtained axiomatically by modifying the axioms of Euclidean geometry, later became the mathematical framework for Einstein's theory of relativity in physics.

- *Matrix theory*, developed by the British Arthur Cayley in 1858, was considered to be a strange and awkward theory at the time of its development. But in the 1920s, matrix theory helped Werner Heisenberg solidify his explanation of quantum mechanics, and it went on to become a central tool in linear algebra, graph theory, combinatorics, and statistics.

- *Knot theory*, a subfield of topology inspired by physical knots, was axiomatized in the early 20th century with no specific application in mind. Several decades later, it generated several practical mathematical tools used in the study of knotting phenomena in DNA and other polymers.

- *Mathematical logic*, the foundation of the axiomatic method, has become the basis for the development of computing machines and computer languages.

- American mathematician Leonard Dickson (1874-1954) once famously quipped, "Thank God that *number theory* is unsullied by any application." Admittedly, a few applications already did exist for number theory but many mathematicians considered number theory the purest branch. However, less than 50 years after his quote, number theory became the essential tool for cryptography, information security, e-commerce and countless other applications in the digital era.

The following sections describe various processes by which mathematics progresses involving the axiomatic method: axiomatization, extension, generalization, and unification.

Axiomatization

Many mathematical theories arise from efforts to organize and formalize mathematical methods developed to solve practical problems. As a well-known historical example, Isaac Newton invented the concepts of a derivative, integral, and differential equation in his attempt to describe and formulate laws about how objects move under gravity or other forces. More than a century later, Cauchy and Weierstrass, among others, axiomatized these concepts and created the theory of analysis that establishes the methods of calculus on a solid axiomatic foundation. This process of *axiomatization* required working backwards, searching for a consistent axiomatic system that is capable of deriving a set of existing applied methods axiomatically. Axiomatization of an applied mathematical method also requires *abstraction*, i.e., stepping back from the intuition of the application to establish the necessary abstract concepts that underpin the axiomatic structure, along with their definitions and notations.

Once a new theory rests on a solid axiomatic foundation, mathematicians can explore this theory in various directions, no longer requiring any connection with the real world. For example, the axiomatization of methods developed to solve applied problems involving differential equations led to theoretical areas of dynamical systems, functional analysis, and others. Given a solid foundation, these areas can develop without connection to the original application. Mathematical research continues in those theories, and many of the extensions are not connected to solving problems in the real world. In some theories, research continues to create new abstract extensions long after the mathematical methods that inspired their foundation have been abandoned.

Extension and generalization

Extension refers to the process of applying the theoretical results of an existing mathematical theory to other types of situations or mathematical objects. The process often requires defining new terms, notations, operations, and incorporating them into new theorems.

Example 8.3.1. The algebra of Euclidean real-valued vectors and matrices is an extension of scalar algebra. For example, let us define a *two-dimensional real-valued vector* as an ordered pair of real numbers and denote it $\vec{u} = (u_1, u_2)$, where $u_1, u_2 \in \mathbb{R}$. Furthermore, let us define the *vector addition* operation as follows:

$$\vec{u} + \vec{v} = (u_1, u_2) + (v_1, v_2) = (u_1 + v_1, u_2 + v_2).$$

We are now in the position to prove the theorem that $\vec{u} + \vec{v} = \vec{v} + \vec{u}$, i.e., that this operation is *commutative*.

Proof. Let $\vec{u}, \vec{v} \in \mathbb{R}^2$. Then

$$\vec{u} + \vec{v} = (u_1, u_2) + (v_1, v_2) = (u_1 + v_1, u_2 + v_2)$$
$$\overset{*}{=} (v_1 + u_1, v_2 + u_2) = (v_1, v_2) + (u_1, u_2) = \vec{v} + \vec{u}. \qquad \square$$

This proof uses the commutativity of scalar addition (the first axiom of the real numbers) to justify the central equality (denoted by $*$).

Generalization is a similar process to extension, consisting in extending the range of applicability of an existing theorem, thereby making it more general. This process often involves *relaxing* a hypothesis of the existing theorem (i.e., making it less restrictive), and modifying the conclusion accordingly. By broadening the assumptions, the generalized theorem becomes stronger, and the original theorem becomes a mere special case of the generalized theorem.

Example 8.3.2. The Pythagorean theorem applied to a plane triangle with sides of lengths a, b, and c, states:

If the angle between a and b is $90°$, then $c^2 = a^2 + b^2$.

If we relax the hypothesis ("the angle between a and b is $90°$") to allow the angle to take on an arbitrary value $0 \leq \theta \leq \pi$, the theorem becomes:

If the angle between a and b is θ, then $c^2 = a^2 + b^2 - 2ab\cos\theta$.   △

This generalized theorem is called the Law of Cosines. Note that it reduces to the Pythagorean theorem in the special case when $\theta = 90°$. The theorem can be generalized further; for example, one may relax the requirement that it be a plane triangle, allowing it instead to be on the surface of a sphere of arbitrary radius.

Generalizing a theorem sometimes uncovers unsuspected connections with other existing theorems, in some cases revealing both theorems to be special cases of a single, unifying result.

Unification

Mathematical theories develop like the branches of a tree; some are short, while others are prolific, growing quickly and branching out themselves into multiple subtheories. A subtheory may even become a narrow specialized field in its own right, maintaining only a distant relationship with the main theory that spawned it.

Unification is the process of discovering connections between different existing theories or subtheories. Mathematicians who aim for unification build bridges between different parts of the mathematical world. Unification consolidates and solidifies theories, making it possible to apply theorems from one area to concepts in a newly connected field. A unifying bridge often unlocks research in areas that have been stalled for centuries.

A historical example of unification occurred in the 17th century, when the French mathematicians Fermat and Descartes developed the field of *analytic geometry*, which uses a coordinate system to unify algebra and geometry. Thanks to this bridge, we can enlist the powerful tools of algebra (such as solving equations) to solve problems in geometry (for example, finding the intersection of two curves). Conversely, this connection endows the abstract concepts of algebra (equations) with representations by geometrical objects (curves and surfaces).

A more recent example of unification is the *Erlangen program*, developed at the University of Göttingen (Germany) primarily by the mathematician *Felix Klein* (1849-1925). This program synthesizes the results of Euclidean and non-Euclidean geometry and provides methods for organizing the properties of the various types of non-Euclidean geometries using results from group theory in modern algebra.

We have mentioned *set theory* a few times in this book so far and offered an introduction to its basic concepts in Part I. The modern approach to set theory began in the 1870s with work by Georg Cantor and Richard Dedekind. Despite emerging so recently, it quickly became a unifying foundation for many areas of modern mathematics. Since its inception, number theory, algebra, analysis, combinatorics, topology, and probability theory, among others, have all been recast into the common framework of set theory, allowing mathematicians to apply the common concepts of set theory across the boundaries of these distinct fields. Set theory has become the *lingua franca* for the study and development of modern mathematics.

Many mathematicians view the *Langlands program* as one of the most profound projects in modern mathematics research of this day. A collection of conjectured connections between number theory and geometry, the Langlands program suggests a bridge between these two areas of mathematics that would provide the keys to solving a large number of open problems on either side. American-Canadian mathematician *Robert Langlands* (b. 1936) proposed the conjectures in 1967 and subsequently won the Abel Prize in 2018 for proving one of the elements of this connection. The British mathematician Andrew Wiles won the Abel Prize in 2016 for his proof of Fermat's Last Theorem, which he achieved in the process of proving another connection conjectured by Langlands. At least three other mathematicians have won Fields Medals in recent years for work related to the Langlands program.

EXERCISES FOR SECTION 8.3

1. The French philosopher Voltaire (who was not a mathematician) shared the widespread 18th century belief that universal agreement was a marker for truth. He illustrated his opinion by saying, "There are no sects in geometry. One doesn't say, 'I am a Euclidean,' or 'I am an Archimedean.' When truth is evident, it is impossible to divide people into parties and factions. Nobody disputes that it is broad daylight at noon. Demonstrate the truth, and the whole world will agree. There is one morality, just like there is one geometry." Based on what you learned about how truth is established in geometry and 19th-century developments in axiomatic mathematics, write an essay that explains how Voltaire's statement may be partially correct, but also partially incorrect.

2. Write the mathematical definition of the following emphasized terms. Then for each one, write a sentence that illustrates how the same word is used in everyday speech.

(a) Matrix A is *similar to* matrix B.

(b) Matrix A is *defective*.

(c) The two linear systems are *equivalent*.

(d) The linear system is *inconsistent*.

(e) The number x is *irrational*.

(f) The *trace* of matrix A is positive.

(g) The set G, together with the operation $\otimes$, is a *group*.

3. The *Mean-Value Theorem* states: "If the function f is differentiable on the interval (a, b) and continuous on $[a, b]$, then there exists a point $c \in (a, b)$ such that $f'(c) = \frac{f(b) - f(a)}{b - a}$."

(a) This theorem is the generalization of another well-known theorem that requires, as an additional hypothesis, that $f(a) = f(b)$. What is the name of this theorem? What is the conclusion in this special case? Which of the two theorems is stronger?

(b) The Mean-Value Theorem is itself a special case of the *Cauchy Mean-Value Theorem*. Write this theorem, and state the additional hypothesis that reduces it to the Mean-Value Theorem. Which of these two theorems is stronger?

8.4 History of Modern Mathematics

The invention of calculus in the late 17th century marked a turning point in the history of mathematics. The events that led up to this event started in the European Renaissance, a period of artistic, cultural, political, and economic "rebirth" that underscored the end of the Middle Ages. Europeans had inherited the mathematics of ancient Greece, India, and the Arab empire, including the axiomatic method, the Hindu-Arabic base 10 number system, methods for solving algebraic equations, and had started building upon them. Prior to the Renaissance, Europe generally lagged far behind other parts of the world in terms of intellectual achievements. After the Renaissance, Europe gradually started to occupy center stage in mathematical and scientific developments. European universities played an essential role in the progress and dissemination of new ideas, becoming centers of intense intellectual activity, and establishing a structured curriculum to educate the masses in an organized and effective way.

The Renaissance ushered in the beginning of the Scientific Revolution, which promoted the study of natural philosophy (which would later be called "science") and the quest to discover the inner workings of the physical world by observation and experimentation. The Scientific Revolution witnessed the rise of empiricism and the scientific method, based on inductive reasoning and observation, which set the goals to formulate scientific theories and to establish laws of nature. During this period, especially under the influence of Copernicus, Galileo, and Kepler, mathematics gradually came to play a central role in the quantitative study of nature and formulation of natural laws in mathematical form. The increasing importance of mathematics in science, a phenomenon called the *mathematization of science*, began during the Scientific Revolution and continues to this day.

This section explains the historical context surrounding the exponential increase in the rate of mathematical progress, starting with the last part of the Scientific Revolution and the invention of calculus. Originating in medieval Europe and refined throughout the following centuries, the European university model played a central role in this progress and was later adopted in all other parts of the world. This section focuses heavily on Europe, which led the world in mathematical and scientific progress from the late 17th until well into the 20th century. Unfortunately, the European society during these centuries heavily discriminated against women and minorities in academia, which explains the small number of female mathematicians mentioned in this section. Until the 20th century, university administrators barred most women from holding academic positions or even attending classes. A few independently wealthy women were able to access the education reserved for men and achieve success and recognition. Others, not so fortunate, had to fight for recognition, and were often forced to publish their ideas under the name of their (male) professor.

The 18th century, known as the *Age of Enlightenment*, was a period of intense mathematical progress, especially in calculus and its applications to science. Mathematical developments often consisted of methods to solve problems arising in scientific investigation, and were motivated by inductive, rather than deductive, reasoning. The 19th century saw the continued progress of mathematical methods applied to science, but also a greater concern for rigor. During this century, dubbed the *Age of Rigor*, the axiomatic method made a resurgence, with great efforts to axiomatize the mathematical developments of previous centuries. This led to the creation of new fields of mathematics, established entirely as mathematical theories from the ground up, thereby creating a gap between theoretical mathematical developments obtained in this fashion and applied developments motivated by problems in science.

Finally, the 20th century saw a continued acceleration in pure and applied mathematical progress, ushering in a Golden Era of mathematics that continues to this day. The rate at which mathematical results emerge currently stand at an all-time high, with an estimated 95% of today's mathematics created after 1900 [24]. Mathematicians expand theories or create new ones every day and the generally collaborative nature of mathematics has allowed it to infiltrate virtually every field of natural and social science.

Subsections 8.4.1, 8.4.2, 8.4.3, and 8.4.4 provide an overview of several important events and persons in the development of modern mathematics, from the 17th to the 20th century (roughly one section per century), leading to the current landscape of mathematics described in Section 7.1.

8.4.1 17th century

The impetus for scientific and mathematical progress that started during the Scientific Revolution continued throughout the 17th century. The invention of analytic geometry and calculus in the 17th century sparked an explosion of mathematical activity that gave mathematics a central place in the intellectual pursuits of the following centuries.

Analytic geometry

Analytic geometry refers to the study of geometry using a coordinate system. This approach unifies classical *synthetic geometry* (as developed in Ancient Greece) and algebra. French philosopher and mathematician *René Descartes* (1596-1650) first published the concept of a coordinate system and developed methods of analytic geometry. Working independently, *Pierre de Fermat* (1601-1665) devised a similar system. We still call the rectangular coordinate system *Cartesian* coordinates in honor of Descartes. Later, alternate coordinate systems (polar, cylindrical, spherical, etc.) emerged for other applications.

Analytic geometry provides a powerful unifying bridge that represents algebraic equations as geometrical objects, applies algebraic methods to geometrical problems, and assigns algebraic properties to geometrical objects. For example, analytic geometry reveals that the geometric curves known as parabolas, hyperbolas, and ellipses, which the Greeks called *conic sections* (because they are obtained by cutting a cone with a plane), can all be described algebraically by quadratic polynomial equations. This ability to visualize abstract equations as geometrical objects, or to describe geometrical objects by algebraic equations, gives rise to now-common expressions such as "the graph of an equation" or "the equation of a curve." Analytic geometry recasts compass-and-straightedge constructions in synthetic geometry (such as finding the intersection of two curves) into systems of equations, which we can study by algebraic methods instead.

Descartes and Fermat created analytic geometry independently in the same year. Descartes' work appeared in his book *La Géometrie*, which begins with geometric curves before introducing a coordinate system to produce their equations. Conversely, Fermat's work proceeds in the opposite direction, starting with algebraic equations and then using a coordinate system to describe the geometric curves that represent them. Thus, both men established the same bridge starting from opposite ends. Analytic geometry serves as an essential foundation for the invention of calculus by Leibniz and Newton a few decades later. Nowadays, the elementary concepts of analytic geometry are typically first taught in a precalculus course, while curricula tend to introduce other coordinate systems in the calculus sequence.

Calculus

Similar to the advent of analytic geometry, the invention of calculus is credited to two persons: *Gottfried Leibniz* (1646-1716) and *Isaac Newton* (1642-1726).

In 1684, Leibniz published *Nova methodus pro maximis et minimis* in the journal *Acta Eruditorum*. (Though a German peer-reviewed journal, all European mathematics and science used Latin as the *lingua franca* until the 19th century). In this publication, Leibniz describes a "calculus" (i.e., a method) for calculating minima, maxima, and tangent lines. It is because of his initial use of the word "calculus" in this article that the name "Calculus" came to be associated with this branch of mathematics. Leibniz's work introduces the concepts and notation still used in modern calculus, such as the fractional notation for derivatives (dy/dx), and the integral symbol $\int$ adapted from the now-archaic "long s" letter that symbolizes a *sum* (Latin *summa*). Leibniz uses the lowercase d notation for a *differential*, which he defined as an *infinitesimal* (infinitely small) change in some varying quantity. For example, a small change in x (Δx) becomes dx when the change becomes infinitely small. For this reason, we often use the more precise name of *differential calculus* or *infinitesimal calculus*.

Newton claimed to have started working on a form of calculus in 1666, while quarantined at Woolsthorpe Manor during the Great Plague of London. However, his first published ideas about calculus did not appear until 1687, in his work *Philosophiae Naturalis Principia Mathematica*. Like Leibniz, Newton used infinitesimal quantities to calculate an instantaneous rate of change with respect to time (i.e., a time derivative), which he called a *fluxion*. His concepts are similar to those of Leibniz, although Newton used the dot notation for the time derivative $(\dot{y})$ and a rectangle for the integral $\boxed{y}$. The now familiar "prime notation" y' appeared a few decades later, in work by

Joseph-Louis Lagrange (1736-1813). Newton's notations have largely fallen into disuse, except in physics where the dot notation still means a derivative specifically with respect to time.

Both Leibniz and Newton developed calculus in separate ways, Leibniz starting with integration, and Newton with differentiation. Both men eventually discovered the link between differentiation and integration, known today as the *Fundamental Theorem of Calculus*, which enables the algebraic closed-form computation of definite integrals by finding formulas for antiderivatives.

Modern historians who carefully examined the works of Leibniz and Newton generally conclude that both men developed calculus independently and more or less simultaneously. It is less of a coincidence if we consider that with the invention of analytic geometry a few decades prior and the general interest in mathematical applications to physics, the intellectual climate of the time was ripe for the discovery of calculus. However in those years, both Newton and Leibniz accused each other of stealing and plagiarizing each others' ideas. The bitter rivalry between the two men extended to their supporters in Germany and England, the German mathematicians siding with Leibniz, and the English with Newton. Before long, all mathematicians in the two countries were involved in the dispute, regularly publishing insults and slurs against each other in academic journals. The dispute persisted for over a century after the death of Leibniz and Newton. During this time, German and English mathematicians stubbornly refused to adopt each other's notation and collaborate with each other, ultimately hindering mathematical progress in both countries.

Newton's *Principia*

Newton's landmark treatise *Philosophiae Naturalis Principia Mathematica* (Latin for "Mathematical Principles of Natural Philosophy"), commonly referred to simply as the *Principia*, is considered one of the most important works in the history of science. The *Principia* contain not only the first concepts of calculus, but also Newton's three laws of motion, his universal law of gravitation, and his derivation of Kepler's laws of planetary motion, thereby laying the foundations of classical mechanics and astronomy. In his work, Newton carefully proceeds from observation, to theory, to law, explaining the workings of the physical universe in mathematical terms.

The *Principia* stands as a monumental masterpiece that represents the culmination of ideas brought forth during the Scientific Revolution. Newton's work confirms the supremacy of the scientific method and the central place of mathematics in the pursuit of science. In a few mathematical equations, Newton successfully captures the laws that govern the motion of the planets, which had baffled humans for millennia. His success encouraged the idea of a universe governed by mathematical laws that could be discovered by human reason and careful observation. The impact of Newton's *Principia* influenced science and philosophy so much that its publication is widely seen as marking the beginning of the era called the *Enlightenment*.

8.4.2   18th century

The Enlightenment

The *Enlightenment*, also known as the *Age of Enlightenment*, *Age of Reason*, or *Age of Inquiry*, is an intellectual, philosophical, cultural, and political movement that dominated the 18th century. It originated in Europe under the influence of several thinkers including *Francis Bacon*, *René Descartes*, and *Isaac Newton*. The movement gained momentum throughout the 18th century, eventually spreading to the United States. At the beginning of the 19th century, the Enlightenment gave way to the *Romantic Era*.

Principles driving the Enlightenment emphasized human reason as the primary source of knowledge, with the belief that everything could be explained, demystified, cataloged, and improved by simply applying an inductive method, as exemplified by Newton's *Principia*. Another project spurred by this ideal is the publication of the first *Encyclopedia* between 1751 and 1772. This massive undertaking, organized by the French *Denis Diderot* and *Jean le Rond d'Alembert*, sought to compile the entirety of human knowledge. Several European and American thinkers popularized the ideas of the Enlightenment, promoting skepticism and freedom of thought over blind faith in dogma and existing institutions.

The Enlightenment movement is a direct consequence of the Scientific Revolution and overlaps with it. Enlightenment thinking encouraged scientific progress in the vein inspired by Newton's *Principia*, claiming the sovereignty of the human senses, the scientific method, and mathematics as the way to increase scientific knowledge. As a consequence, science and mathematics blossomed during the 18th century and became fashionable subjects of discussion and study in all learned circles.

The 18th century also saw deep shifts in society, religion, and politics, as its ideas encouraged questioning long-held beliefs and demanded a justification for the order of society. Enlightenment thinkers denounced supernatural occurrences as superstition, gradually replacing the dogmatic teachings of the church with *deism*, which is the belief that although God created the world and set it in motion, he does not directly interact with his creation and chooses to let the universe proceed according to natural laws. This worldview undermined the authority of the church, eventually giving way to a new society structure based on the separation of church and state.

Similarly, the ideas of the Enlightenment undermined central government authorities and absolute monarchies, as people espoused the idea of a new society built on rational thinking, equality, tolerance, personal freedom, and human dignity. This worldview paved the way for the American Revolution (1775-1783) and French Revolution (1789-1793), which overthrew the authority of the monarchy and replaced it with constitutional governments.

Dissemination of mathematics

The ideas of the Enlightenment encouraged the pursuit of knowledge, prompting efforts to make the sources of knowledge more widely available and intelligible. As a result, the production of books, periodicals, and newspapers increased exponentially. Previously, mathematical ideas remained relegated to scholarly works in Latin, and therefore only accessible to a learned elite. The Enlightenment produced a large number of *expository* authors, who strove to organize, translate, explain, and popularize mathematical concepts, making them accessible to a wider audience. For example, several authors wrote expository commentaries on the ideas of Leibniz and Newton, complete with step-by-step explanations, examples, and diagrams. Thanks to these books, the ideas of calculus spread rapidly to the general population.

The textbook *Analyse des Infiniment Petits pour l'Intelligence des Lignes Courbes*, a textbook written in French by *Guillaume de l'Hospital* (1661-1704) in 1696, offered the first systematic expository work on calculus. In it, L'Hospital uses Leibniz's point of view and notation, but also mentions Newton's published work and acknowledges that it contains the same material. Widely published and translated into many languages, L'Hospital's book became a template for future treatments of calculus. Another well-known French expository author, *Emilie du Châtelet* (1706-1749), translated and commentated Newton's *Principia*. She was also an early advocate for women to have equal access to education as men, especially at the secondary level. In 1748, *Maria Gaetana Agnesi* (1718-1799) published in Milan, Italy, *Instituzioni Analitiche Ad Uso Della Gioventù Italiana*. Her textbook offered a comprehensive and systematic presentation of calculus, complete with illustrations of exceptional quality for its time period. Written in Italian, others subsequently translated it into French and English.

Euler

In 1999, mathematics popularizer William Dunham wrote a book entitled, *Euler: The Master of Us All*[28]." This title reflects how much *Leonhard Euler* (1707-1783) dominated 18th century mathematics. His massive accomplishments and immense productivity overshadow those of all other mathematicians of his century.

Originally from Basel, Switzerland, Euler became acquainted with the many mathematicians of the Bernoulli family, who were active in that same city. Later, Euler was appointed to positions at the Russian and Prussian academies of science, in St. Petersburg and Berlin, respectively. The father of 13 children, Euler enjoyed a thriving family life. Sociable but modest, he drew the friendship and respect of other great minds of his time. As a devout Protestant Christian, however, he did not subscribe to the deistic beliefs common to other Enlightenment thinkers.

Euler's most notable attributes were his photographic memory and his unusual ability for mental calculation.[13]. He could recall by memory virtually everything he had ever read, and could calculate in his mind "as easily as other people breathe." Thanks to these phenomenal abilities, Euler remained mathematically active until his death, even though blindness claimed his sight during the last 17 years of his life.[14]

By far the most prolific mathematician in history, mathematicians have not yet compiled the entirety of Euler's work, more than 200 years after his death. A massive project called *Project Euler* is currently underway to catalogue all of his writings. Described as a *polymath*, Euler also wrote prolifically in all areas of natural science, especially physics, solid mechanics, fluid dynamics, acoustics, optics, astronomy, and cartography. All these fields contain fundamental results named after Euler. For example, *Euler's formula* in engineering gives the critical load under which a thin, vertical column buckles. Euler derived this formula from the application of calculus of variations to elasticity theory:

$$P_{cr} = \frac{\pi^2 EI}{(KL)^2},$$

where P_{cr} is the critical load, E is the modulus of elasticity, I is the area moment of inertia of the column's cross-section, L is the length of the column, and K is the effective length factor. In engineering

---

[13]Perhaps a third remarkable attribute was his ability to concentrate despite the many distractions going on in his household. A contemporary wrote that Euler often wrote and did mathematics "with a child on his knees and a cat on his back" [36]

[14]Euler generated more than half of his scientific output after he was blind [36].

mechanics, *Euler-Bernoulli beam theory* uses a partial differential equation (PDE) called the *beam equation* to describe the load-carrying and deflection characteristics of a beam. In fluid mechanics, the *Euler equations* are a set of PDEs that govern adiabatic (zero thermal conductivity) and inviscid (zero viscosity) fluid flow.

Euler was well read and also wrote abundantly about music theory, philosophy, and theology. His phenomenal memory contributed to his extensive knowledge about many fields outside of his research subjects. It almost does not surprise that Euler could converse in or read at least eight languages: his native German and Swiss-German, but also French, Latin, Greek, Hebrew, English, and Russian.

Almost all areas of mathematics show some influence by Euler, from geometry to trigonometry, algebra, and number theory. He made significant developments in calculus, integrating Leibniz's differential calculus with Newton's method of fluxions into a form of calculus that would be recognizable today. He popularized modern terminology and notation, such as the concept of a *function* and its notation $f(x)$, the notation for trigonometry functions ($\sin x$, $\cos x$, etc.), the sum notation $\sum$, the Greek letter π for the ratio of a circle's circumference to its diameter, the notation i for $\sqrt{-1}$ (the imaginary unit), and the letter e for the basis of the natural logarithm.

In Euler's fertile mind, simple problems would often blossom into entire new fields of mathematics. After solving a problem, he would extend the problem in a variety of directions, ultimately creating powerful generalized methods and opening up multiple new areas of mathematical research. For example, Euler tackled and solved the famous *brachistochrone problem*, which consisted in finding the curve between two points on which an object slides downward under the effect of gravity in the shortest time. The analytic approach he used to solve this problem eventually gave rise to the *calculus of variations*, a broad new field created and developed by Euler together with Lagrange. Similarly, Euler seeded the ideas of *graph theory*. This field emerged soon after Euler published a mathematical method to solve the *Seven Bridges of Königsberg* problem, another famous problem that consisted in devising a walk through the city of Königsberg (in modern-day Russia) that crosses each of its seven bridges exactly once.

Euler popularized the use of infinite series and made several advances in this area. A notable result is his solution to the famous *Basel problem*, which consists in finding the sum of the series

$$\sum_{n=1}^{\infty} \frac{1}{n^2} = 1 + \frac{1}{4} + \frac{1}{9} + \frac{1}{16} + \cdots,$$

known today as a p-series with $p = 2$. Euler showed that the sum converged to $\pi^2/6$, but did not stop there. He generalized the problem and derived a formula for the sum of any p-series for even $p > 1$. Furthermore, he defined the *zeta function* $\zeta(s)$ to denote the sum of such a series and discovered a connection between the zeta function and the sequence of prime numbers:

$$\zeta(s) = \sum_{n=1}^{\infty} \frac{1}{n^s} = \prod_{p \text{ prime}} (1 - p^{-s})^{-1}.$$

The zeta function later became the subject of intense research after Bernhard Riemann (hence its name as the *Riemann* or *Euler-Riemann zeta function*) extended it by allowing the variable s to take on fractional and complex values.

Another notable result is Euler's proof that the harmonic series has a logarithmic rate of divergence. A comparison of the harmonic series and natural logarithm function gives rise to a constant, called the *Euler-Mascheroni constant*, denoted by γ:

$$\gamma = \lim_{n \to \infty} \left(1 + \frac{1}{2} + \frac{1}{3} + \cdots + \frac{1}{n} - \ln n \right) \approx 0.57721...$$

This constant appears frequently in analysis and number theory. Not unlike e or π, a web search on this constant reveals an enormous number of alternative expressions arising naturally from various fields of mathematics.

One of Euler's most famous results is the *Euler formula* that provides a link between exponential and trigonometric functions:

$$e^{i\theta} = \cos\theta + i\sin\theta.$$

Recasting this formula for $\cos\theta$ and $\sin\theta$ gives $\cos\theta = \frac{1}{2}\left(e^{i\theta} + e^{-i\theta} \right)$ and $\sin\theta = \frac{1}{2i}\left(e^{i\theta} - e^{-i\theta} \right)$. Substituting $\theta = \pi$ reduces it to *Euler's identity*:

$$e^{i\pi} + 1 = 0, \tag{8.2}$$

an elegantly simple identity that relates the five most important numbers of mathematics $(0, 1, i, \pi, e)$ together with the three basic operations of arithmetic (addition, multiplication, and exponentiation). Though mathphiles the world over pause and admire (8.2), it is just one of the countless applications of Euler's formula to complex analysis, differential equations, Fourier series and so on.

Another beautiful formula for its simplicity is *Euler's (polyhedral) formula*, which states that in any convex polyhedron (such as a tetrahedron or cube), the number of vertices and faces together is exactly two more than the number of edges (symbolically, $V - E + F = 2$). A vast generalization of this result to other shapes in any number of dimensions gives rise to the *Euler characteristic* associated with an arbitrary topological space.

Mathematization of science

The mathematization of science, a trend inherited from the Scientific Revolution, continued throughout the 18th century. Besides Euler, many mathematicians made great strides in this direction. One of the most notable ones is the Franco-Italian *Joseph-Louis Lagrange* (1736-1813), who worked closely with Euler, and whose name appears in many areas of 18th century mathematics and physics. Another French mathematician, *Jean le Rond d'Alembert* (1717-1783), one of the co-editors of the Encyclopedia project, did significant work in mathematics applied to oscillatory phenomena in vibrating strings and in fluid mechanics, paving the way for further advances in those areas during the 19th century.

8.4.3   19th century

The *Romantic Era* was a period following the Enlightenment that roughly overlaps with the 19th century. In the arts, literature, and music, this period is characterized by an emphasis on passionate feelings, emotions, and the awe of nature. In mathematics, we often label this same period as the *Age of Rigor*, because of the shift from intuitive methods motivated by applications to a deductive approach that emphasized logic, rigor, and internal consistency.

The ideals of the Enlightenment elevated the scientific method as the way to attain true knowledge. As a result, most mathematical advances of the 18th century were motivated by applications to science. Mathematics was considered science's faithful "handmaiden," providing the right tools at the right time to assist the progress of science. Mathematics grew rapidly, generating a large number of methods, techniques, algorithms, and recipes developed to solve various problems that arose in the process of scientific investigations. However, few scientists made efforts to organize these developments into formal mathematical theories. Instead, mathematics amounted to a relatively disorganized and amorphous collection of results that lacked firm theoretical foundations. Even Euler, despite the magnitude of his accomplishments, is not known for his rigor. Many of his results were never formally proved, and many of his proofs would not meet modern standards for acceptance.

Because of contradictions that arose as one attempted to generalize results, mathematicians of the 19th century saw the need to formalize the mathematical concepts of the previous centuries. They turned to Euclid's *Elements* as a template for the organization of existing mathematical concepts into an axiomatic structure. Prior to the 19th century, the axiomatic method in *The Elements* primarily served an educational role for the development of logical reasoning skills. The Age of Rigor saw the revival of the axiomatic method and its application to all areas of mathematics. Mathematicians proceeded to gradually axiomatize all existing mathematical results into formal mathematical theories, complete with axioms, definitions, and theorems. By the end of the century, algebra, calculus, probability theory and many other fields rested squarely on solid theoretical foundations. In addition, the focus on the axiomatic structure led to the emergence of new fields, such as non-Euclidean geometries, statistics, group theory, set theory, and symbolic logic.

The new emphasis on rigor and axiomatic mathematics did not slow down the progress of new mathematics applied to science. The term "applied" mathematics appears for the first time in the 19th century to distinguish this form of traditional mathematical research from research in "pure" mathematics not driven by applications, but rather proceeding by extending mathematical theories axiomatically.

Gauss

Just as Euler dominated the mathematical scene of the 18th century, many consider the German *Carl Friedrich Gauss* (1777-1855) as the greatest mathematical figure of the 19th century. He is sometimes referred to as the "greatest mathematician since antiquity" or *Princeps mathematicorum*, which is Latin for the "Prince (as in the foremost) of mathematicians."

Gauss was a child prodigy, born to poor, working-class parents. There are numerous anecdotes of Gauss surprising his teachers and parents at a young age with his mathematical ability, such as discovering an arithmetic error in his father's accounting records at the age of three. His illiterate mother never recorded his birth date, remembering only that it happened on the Wednesday of a week before the Ascension, a feast that occurs 39 days after Easter. With this information, Gauss was able to determine his own birth date, and then proceeded to derive a generalized formula to compute the date of Easter in any year, a complicated feat that requires taking into account the solar and lunar cycles, as well as the irregularities of the Gregorian calendar. Gauss made significant mathematical discoveries as a young university student; he studied at the prestigious University of Göttingen and later held an appointment there for many years.

Not only an eminent leader in mathematics, Gauss also contributed significantly to physics, astronomy, geodesy, geophysics, electrostatics, and optics. He rivals Euler with the number of important concepts, laws, formulas, and equations that bear his name. In physics, *Gauss's law* (or *Gauss's equation*) relates the electric charge to the resulting electric field, and *Gauss's law for magnetism* states that a magnetic field has zero divergence. These laws are two of the four *Maxwell equations*, which constitute the foundation of classical electromagnetism. The *Gauss*, a unit for magnetic flux density, is named after him. Collaborating with Wilhelm Weber, Gauss also constructed the first electromechanical telegraph.

Gauss called mathematics the "queen of the sciences," a phrase he coopted from theology, which had previously been considered the queen of the sciences since the foundation of the first medieval universities. In mathematics, Gauss explored nearly every field including algebra, analysis, differential geometry, and statistics. But his favorite area was number theory, which he called the "queen of mathematics." All of Gauss's mathematical work exhibits an absolute axiomatic rigor, performed in a style that had not been seen since the ancient Greeks. Though not as prolific as other mathematicians, everything he published is marked by extreme beauty in demonstration, a feature that amazed his contemporaries and contributed to reviving enthusiasm for the axiomatic method [9].

As a perfectionist, Gauss continually rewrote his notes until they met, in his mind, the highest level of quality. He wrote proofs in a pure and elegant style, but devoid of many explanations, as Gauss would erase the intuition behind his proofs, preferring to present them as appearing out of thin air.[15] Gauss regularly looked for alternative or more elegant proofs of known results. For example, after first publishing a proof of the *Law of Quadratic Reciprocity* in his *Disquisitiones Arithmeticae* (1801), he regularly sought other simpler or alternative proofs of the same result, providing at least eight significantly different proofs over the course of his life.

Because of this extreme perfectionism, Gauss only published a fraction of his work. Examination of his personal notes after his death revealed that he had in fact discovered many important results now attributed to others who discovered and published them later. For example, Gauss never published his results on non-Euclidean geometry, which he discovered around 1813. We now generally attribute this discovery to János Bolyai and Nikolai Lobachevsky, who published the first papers on hyperbolic geometry in 1829-1830.

Gauss offered the first proof of the *Fundamental Theorem of Algebra*. This important theorem states that every polynomial of degree n with complex coefficients has exactly n complex roots, counted with multiplicity. Gauss inaugurated the fields of differential geometry and complex analysis, providing the geometrical interpretation of a complex number in the form $a + ib$ as a point in the *complex plane* (also known as the *Gaussian plane*).

---

[15]Niels Abel famously said, "Gauss is like the fox, who erases his tracks in the sand with his tail" [71].

In probability theory and statistics, we encounter Gauss' mark with the density function of the *normal* (or *Gaussian*) *distribution*, also called the *Gaussian function*. We can represent this function as a bell-shaped *Gaussian curve* with the following equation:

$$f(x) = \frac{1}{\sigma\sqrt{2\pi}} e^{-\frac{1}{2}\left(\frac{x-\mu}{\sigma}\right)^2},$$

where μ is the *mean* and σ is the *standard deviation* of the distribution. Every undergraduate mathematics student also recognizes Gauss's name in the method of *Gaussian elimination* and other methods for solving linear systems of equations (Gauss-Jordan, Gauss-Seidel, Gauss-Jabobi).

Axiomatization

The 19th century English mathematician Oliver Heaviside once said, "Mathematics is an experimental science; definitions do not come first, but later on" [44]. Until they are "grafted" into a sound mathematical theory, mathematical results that are derived by intuition, and which may appear at first to be sound by intuitive examination alone, are on shaky ground, at risk of collapsing under contradictions and internal inconsistencies.

For example, the early ideas of calculus were based on intuitive concepts based on observation of nature, such as the motion of planets or the trajectories of falling objects. In his *Principia*, Newton defined the concept of a *fluxion* (which we would call a *derivative*) as a "quotient of two quantities, which in some cases are both zero, but in others are not." (Similar "intuitive" definitions sometimes appear in an elementary calculus course today as well.) Some people in Newton's day worried about the inherent contradiction in this type of definition, but many did not, being satisfied by the observation that the results seemed to agree with the behavior of the real world. For over a century, calculus was applied to the natural sciences with great success without a clear, rigorous, theoretical explanation for how it actually works.

The discovery of several logical gaps and inconsistencies led several 19th century mathematicians to try to formalize and axiomatize different areas of mathematics, including calculus. For example, on one occasion, the German mathematician *Richard Dedekind* (1831-1916), being asked to teach calculus at the Polytechnic School in Zurich, remarked,

> I feel more keenly than ever before the lack of a really scientific foundation for arithmetic. [...] For myself this feeling of dissatisfaction was so overpowering that I made

the fixed resolve to keep meditating on the question until
I should find a purely arithmetic and perfectly rigorous
foundation for the principles of infinitesimal analysis [51].

In this context, *analysis* refers to the analysis of how things change in
relation to one another, i.e., the more general theoretical framework
in which the methods of calculus are derived deductively.

The French *Augustin-Louis Cauchy* (1789-1857), in his textbook
Cours d'Analyse, stressed the importance of rigor in analysis, establishing for the first time the concepts of calculus in terms of infinitesimal increments, which became the unifying idea behind his definitions
for derivatives, integrals, the continuity of a function, and the convergence of infinite series. For example, Cauchy defined a function
$f(x)$ to be *continuous* on an interval if, on this interval, an infinitely
small increment in the variable always produces an infinitely small increment in the function. Later, the German *Karl Weierstrass* (1815-
1897) refined the definitions for continuity, derivatives, and integrals
in terms of *limits*. He introduced the "lim" notation for a limit, and
established definitions that compare distances, represented by the absolute value notation, to small increments represented by the letters
ε and δ. Thus, Weierstrass provided the following modern notation
and definition of a limit ($\lim_{x \to c} f(x) = L$):

A function $f(x)$ has a *limit* L at c if and only if for all
$\varepsilon > 0$, there exists a $\delta > 0$ such that if $0 < |x - c| < \delta$,
then $|f(x) - L| < \varepsilon$.

This type of "ε-δ terminology" remains the standard syntax for the
rigorous language of analysis.

Mathematicians also formalized algebra during the 19th century.
An important step forward in this process occurred when Gauss proved
the fundamental theorem of algebra in his doctoral dissertation in
1799. In 1824, the Norwegian *Niels Abel* (1802-1829), after whom the
Abel Prize is named, proved the impossibility of solving the general
quintic equation, a problem which had remained open for over 250
years. The French *Evariste Galois* (1811-1832) extended the results
on solvability of polynomial equations, thereby creating the foundations of *group theory* and *Galois theory*, which constitute two main
branches of modern algebra.

The 19th century also saw the creation of differential geometry,
which offered a coherent framework not only for Euclidean geometry
and Lobachevsky and Bolyai's non-Euclidean, but also for general
contexts with variable metrics. Manifold theory, symplectic geometry,
Lie theory, and several other branches of geometry blossomed during
the second half of the 19th century, quickly eclipsing the state of
geometry at the beginning of the century.

Set theory

Of all the mathematical theories that emerged in the 19th century, *set theory* deserves a special mention because it has since become woven into the fabric of virtually every area of modern mathematics.

The creation of set theory is attributed to the German *Georg Cantor* (1845-1918), and specifically to the publication of his 1874 article titled, *On a Property of the Collection of All Real Algebraic Numbers*. In this article, Cantor formalizes several ideas surrounding infinite sets of numbers. For example, he proved that, although the sets of real and natural numbers are both infinite, the real numbers are more numerous than the natural numbers. (See Theorem 5.4.6.) Thus, he distinguishes between two types of infinite sets: *uncountably infinite* (like the real numbers), and *countably infinite* (like the natural numbers). Moreover, Cantor proved that the set of rational numbers is also countably infinite by establishing a *bijection* (i.e., a one-to-one correspondence) between the natural numbers and the rational numbers. (See Theorem 5.4.5.)

As in Definition 1.2.1, a *set* is a collection of objects called *elements*. As applied to the usual objects in mathematics, sets might contain numbers, vectors, matrices, functions, or even other sets. The number of elements in a set (called the set's *cardinality*) may be finite or infinite (Definition 4.1.9) and, as Cantor demonstrated in his 1874 paper, certain types of "infinity" are more numerous than others. Set theory also studies the operations between elements, and the properties of the elements with respect to those operations, such as commutativity, associativity, and so on. Many areas of mathematics involve sets equipped with operations that satisfy some list of properties. Specific types of properties give rise to structures with names such as *group, ring, field*, or *vector space*.

Mathematicians slowly adopted this simple, yet powerful and universal framework of set theory as a foundation for many other mathematical theories, including algebra, analysis, differential equations, number theory, graph theory, combinatiorics, and probability theory. For example, in the modern theory of differential equations, solutions of differential equations are seen as elements in a *solution set*, and an important theorem states that the solution set of a linear, homogeneous differential equation has the structure of a vector space. In this sense, the solution set is the kernel of a linear operator between sets of functions, so solving a differential equation reduces to finding the kernel of a linear operator. Using a common framework in multiple theories allows for crossover between different areas of mathematics, as in this example, where theorems in linear algebra help us study a problem in differential equations.

Set theory also bears deep philosophical implications, especially regarding the internal logical structure of mathematical theories.

Applied mathematics

The renewed interest in axiomatic mathematics during the 19th century did not slow down the progress in mathematics driven by applications in science. Encouraged by the many successful applications of mathematics to science developed during 18th century, mathematicians continued researching along that vein in the 19th century, many of them with little concern for rigor. English mathematician *Oliver Heaviside* (1850-1925), a quintessential applied mathematician, remarked,

> Rigorous mathematics is so narrow. To have to stop and formulate rigorous proofs would put a stop to most physical inquiries. Am I to refuse to eat because I do not fully understand the mechanism of digestion? [29].

Heaviside, along with the American *Josiah Gibbs* (1839-1903), contributed significantly to *vector calculus* and is responsible for reformulating the *Maxwell equations* as a set of four partial differential equations in terms of electric and magnetic fields, which is their most common form today.

The 19th century witnessed enormous strides in applied mathematics. In France, *Pierre de Laplace* (1749-1827) published the most influential book on probability theory of his day, *Théorie Analytique des Probabilités*, and laid the foundations for statistics. He also made numerous contributions to mathematical physics and astronomy. *Joseph Fourier* (1768-1830) developed important mechanistic models for heat transfer, diffusion, and oscillatory phenomena, and *Sophie Germain* (1776-1831) pioneered the mathematical formulation of elasticity theory. Meanwhile in England, Sir *George Stokes* (1819-1903), in collaboration with the French *Claude-Louis Navier* (1785-1836), developed a set of partial differential equations known as the *Navier-Stokes equations* to describe fluid motion, which are used today to model the weather, ocean currents, water flow in a pipe, airflow around a wing, and other phenomena.

This form of mathematical research came to be called "applied mathematics," to distinguish it from the newer type of research that emphasized axiomatic developments not directly driven by applications. Thus, the divide between "applied" and "pure" mathematics appears for the first time in the 19th century. By the early 20th century, the French mathematician Henri Poincaré (1854-1912) had already noticed these two types of mathematical progress, which he describes in his book, *The Value of Science* (1905):

It is impossible to study the works of the great mathematicians, or even those of the lesser, without noticing and distinguishing two opposite tendencies, or rather two entirely different kinds of minds. The one sort are above all preoccupied with logic; to read their works, one is tempted to believe they have advanced only step by step, after the manner of a Vauban who pushes on his trenches against the place besieged, leaving nothing to chance. The other sort are guided by intuition and at the first stroke make quick but sometimes precarious conquests, like bold cavalrymen of the advance guard [38].

Universities

Nowadays, universities play two fundamental roles: education and research. This dual role, a model adopted by all universities worldwide, results from the evolution of universities throughout the 19th century, especially in regards to the locus and methods of building and disseminating mathematical and scientific knowledge.

Ever since the appearance of the first medieval universities in Europe, mathematics had been an important part of the curriculum, playing an ancillary role for the branches of the trivium and quadrivium. During the Renaissance, the ideas of the Scientific Revolution elevated mathematics and natural philosophy to a central place in university curricula. Concurrently, the study of mathematics and natural philosophy (the precursor of "science") went beyond mere exposure to classical texts and mastery of existing knowledge. The pursuit of scientific knowledge required original research, so universities became centers of research activity, while still maintaining their traditional role as educational institutions. At the same time, several kings and emperors of Europe began sponsoring scientific and mathematical research as well by establishing their own *royal academies*. These academies grew in importance and prestige throughout the 17th and 18th centuries, competing against each other to attract the best scholars. Euler, for example, did most of his work while appointed to positions at two academies: the Imperial Russian Academy of Sciences (also known as the St. Petersburg Academy) and the Royal Prussian Academy of Sciences (also known as the Berlin Academy). During this time, universities operated in parallel with the royal academies, housing research scholars and fulfilling their educational role. Universities, however, remained under the strict control of the church, which exerted a heavy influence on the curriculum until the end of the 18th century.

During the 19th century, as the emphasis on original research continued growing, universities began combining research and education, gradually placing research at the center of the curriculum and primary means of instruction. This evolution originated at the University of Berlin, soon followed by other German universities. Traditional classroom instruction based on lectures gave way to a variety of hands-on instructional styles based on labs, seminars, and discussions, which promoted students' engagement in original research instead of merely learning existing material. The *University of Göttingen*, under the leadership of German mathematician *David Hilbert* (1862-1943), rose to fame as the center for cutting-edge mathematical research activity, inspiring the universities of France, England, and the rest of Europe to follow this example and adopt the German university model.

In the United States, the first colleges and universities, established in the late 17th century and 18th century, imitated the educational institutions of England in curriculum, structure and even architecture. At the time, these institutions primarily provided education for the upper class and clergy. The curriculum in these early American colleges contained very little science and mathematics, instead focusing almost exclusively on studying the classics and humanities. The mid-19th century saw the creation of the first *polytechnical institutes* with a focus on science and mathematics, mostly to provide training for professions in technology and engineering. Although some of these institutes were associated with universities, the degrees they offered carried little prestige, as society generally considered applied sciences as an inferior program for inferior students. For example, the students enrolled at the Sheffield Scientific School, associated with Yale University, remained segregated from the other students in chapel. Americans who wanted to pursue research or an academic career in science or mathematics had no choice but to study in Europe. Many early American scholars attended German universities, and in particular, the University of Göttingen [41].

In 1876, *Johns Hopkins University* became America's first university established according to the German model, blending teaching and research. Within a few years, Johns Hopkins earned the reputation as the leading research institution in science and mathematics in the United States, hosting numerous scholarly journals and conferences, and setting the trend for doctoral programs in those fields. Soon, many other American university programs followed this trend and adopted the German university model. Among the earliest of these is the mathematics department of the *University of Chicago*, established in 1892, which became a pioneer in the development of modern mathematics programs, graduating many mathematicians who went on to establish mathematics programs of their own according to the same model throughout the United States [41].

8.4.4   20th century

Many historians of mathematics dubbed the 20th century the discipline's "Golden Age." The exponential rate of mathematical progress of the 17th to the 19th centuries continued throughout the 20th century and continues still. Scholars of the discipline itself estimate that over 95% of the mathematics we have today was produced after 1900, with hundreds of mathematical journals in existence and thousands more scientific journals devoting a large share of their space to mathematical methods and applications [10].

Foundations of mathematics

The *foundations of mathematics* refers to an early 20th-century movement that investigated the philosophical and logical foundations underlying all of mathematics and sought to define a universal logically consistent axiomatic framework to prove the consistency of mathematics. Several mathematicians across Europe undertook this monumental task. Two of the leaders were *David Hilbert* (1862-1943) and *Felix Klein* (1849-1925), both German mathematics professors at the University of Göttingen. Of crucial import in this program, philosopher and logician Gottlob Frege (1848-1925) sought to show that mathematics flows out of logic. In his work, *Begriffsschrift: Eine der Arithmetischen Nachgebildete Formelsprache des Reinen Denkens* ("Concept-Script: A formal language for pure thought modeled on that of arithmetic") [8], while providing a rigorous logical analysis of functions and variables, he developed the concept of predicates and quantification, a central concept in modern logic. (See Section 1.5 for an introduction.) Though generally overlooked during his lifetime, logicians of the early 20th century championed his efforts to remove all appeals to intuition from any proof.

Without a natural consensus on the approach to this project, various competing schools of thought emerged. They bore such names as formalism, intuitionism, logicism, set-theoretic Platonism, structuralism and others, all pursuing the same goal but using different approaches. One by one, these schools ran into difficulties, such as the emergence of various mathematical *paradoxes*, i.e., mathematical statements that appear to contradict themselves while simultaneously seeming completely logical. One of the most succinct examples is *Russell's paradox*, named after the English *Bertrand Russell*. In set theory, consider a set A defined as "the set of all sets that do not contain themselves," $A = \{ S \mid S \notin S \}$. If A is not a member of itself, then by definition, it must contain itself ($A \notin A \Rightarrow A \in A$). However if A contains itself, then it would contradict its own definition; therefore, $A \in A \Rightarrow A \notin A$. The conclusion is that the set A seems to simultaneously contain and not contain itself, a statement which violates the *rule of noncontradiction*.[16]

The foundations of mathematics movement came to a sudden halt in 1931 when a young Austrian mathematician named *Kurt Gödel* (1906-1978) published two *Incompleteness Theorems*. The first one states that in any axiomatic system, there exist propositions which cannot be proved or disproved, i.e., for which it is impossible for a proof to exist within the system. In other words, the theorem states that no mathematical system can be *complete*.[17] Gödel's second theorem states that every axiomatic system contains *undecidable statements*, such as the question about the consistency of the system itself. Therefore, if an axiomatic system is consistent, the axioms of that system are incapable of proving it.[18] Together, these theorems demonstrate the nonexistence of the type of mathematical foundation that the various schools of mathematics were seeking to formulate [41].

Although this crisis ended the hope of defining a unifying, consistent framework for all of mathematics, it did not discourage mathematicians from trying to resolve several of the paradoxes they encountered. In particular, German mathematicians *Ernst Zermelo* and *Abraham Fraenkel* developed the *ZF axiomatic system* which resolves most of the paradoxes of set theory. For example, one of the ZF axioms (called the *axiom of specification*) forbids recursive definitions for a set in the way that leads to Russell's paradox. An additional axiom, the *axiom of choice*, was later added to the ZF axioms, resulting in *ZFC set theory*, which constitutes the most common foundation for many mathematical theories of today [15].

Separation of pure and applied mathematics

The separation between applied and pure mathematics, which began in the 19th century, continued throughout the 20th century. Furthermore, though the applications remained largely confined to physics and engineering through the 19th century, mathematics started to infiltrate other fields of natural science in the 20th century, including chemistry, biology, ecology, epidemiology, and pharmacology. This mathematization process gradually extended to the social sciences, starting with psychology and economics. Scientists of all stripes now regularly employ mathematical models, confirming the ubiquity of mathematics in virtually every field of science.

---

[16]A puzzle derived from this paradox, the *barber's paradox*, supposes a barber who shaves all those, and those only, who do not shave themselves. The question is, does the barber shave himself?

[17]See Franzén's book [34] for a survey of Gödel's Incompleteness Theorems.

[18]French mathematician André Weil famously said, "God exists since mathematics is consistent, and the Devil exists since we cannot prove it" [41].

With solid logical foundations, untethered to applications, theoretical mathematics progressed rapidly in the 20th century, with
a constant focus on extension, generalization, and unification. Of
particular note, the large collection of textbooks known as *Eléments
de Mathématique* (*Elements of Mathematics*), published in 1939 in
French under the name *Nicolas Bourbaki*, laid out with precise logic
the foundations and modern formulation of set theory, modern algebra, analysis, geometry, and topology in a generalized, unifying
framework. In a twist of drama, it was revealed several years after
its publication that Bourbaki was not a real person, but instead the
pseudonym for a group of about a dozen French mathematicians. Although the original members of the Bourbaki group have all died, its
legacy lives on as the *Bourbaki seminar*, still holds regular conferences
attended by the top mathematicians from around the world. Their
work established a standard of rigor that remains in effect today.

Computers

The appearance of computers in the 20th century profoundly impacted both theoretical and applied mathematics. In theoretical
mathematics, the computation speed of a computer makes it possible
to test a conjecture by simulating millions of cases in a few minutes.
Similarly, if the proof of a theorem can be broken down into a finite
number of cases, no matter how large, then a computer can test them
all, thereby proving the theorem by *brute force*. For example, the
Four Color Theorem, which had been conjectured in the mid-19th
century, was finally proved in 1976 by brute force with the help of a
computer.[19] Similarly, computers help with finding large prime numbers or prime number pairs (primes p such that $p + 2$ is also a prime
number).

The advent of computers has promoted the development of mathematical methods specifically designed to be implemented by a computer, a new and growing field of mathematics called *computational
science*. Thanks to advances in computational science, it is now possible to build a numerical method capable of approximating the solution of virtually any mathematical model, no matter how complex,
thereby removing the barriers that previously restricted mathematical modeling to those few models that could be solved by analytical
methods alone. This has led to the creation of numerous fields of
applied mathematics that require complex models. Concurrently, the
field of *numerical analysis*, which develops mathematical techniques
for numerical approximation, has risen to one of the fastest-growing

---

[19]This theorem states that, given any separation of a plane into contiguous
regions, no more than four colors are required to color the regions so that no two
adjacent regions have the same color.

fields of modern mathematics. Interestingly enough, because numerical methods often rely on linear algebra, the effort to find algorithms that minimize the time of computation spark countless interesting theoretical questions in matrix analysis, tensor geometry and representation theory [57].

The graphical capabilities of computers, which allow the representation of data and the visualization of complicated geometrical objects, have opened up new avenues for mathematical research and education. For example, the mathematical field of *chaos theory*, which studies chaotic dynamical systems, experienced a rapid growth when computers revealed the interesting numerical behavior of chaotic solutions and the mesmerizing graphical beauty of basins of attraction and strange attractors.

World War II

With the mathematization of so many areas of science in the 20th century, it is almost not surprising that mathematics played a critical and ultimately a decisive role during World War II. The applied fields of *cryptography*, *operations research*, and *statistics*, among others, experienced a rapid growth as the war increased the demand for mathematical methods that would provide a military advantage. At its height, the war influenced so much of the mathematical and scientific research worldwide that "applied mathematics" came to be understood as mathematics applied to the wartime effort [6]. Many of the mathematical methods used nowadays in business and civilian applications emerged from declassified military information, originally developed for military applications.

World War II deeply impacted the rise of mathematical research in American and British universities in another profound way. In 1933, the Nazi government conducted a great purge of German universities, dismissing hundreds of academics who were openly opposed to the Nazi regime; many of whom claimed Jewish heritage. A large number of Germany's finest mathematicians and scientists fled their country, causing the downfall of the great universities of Germany, especially the University of Göttingen, once considered to lead the world in mathematical research. These eminent mathematicians found a welcome in the United States, Canada, and Great Britain, where they became leaders of academic institutions and thereby greatly contributed to elevating the mathematical prestige of their adoptive countries.

Women in mathematics

Before the 20th century, women who wanted to pursue mathematical and scientific studies faced insurmountable difficulties, both in Europe and in America. Both the royal academies and universities barred women from attending or even auditing classes, and scientific journals refused to publish female authors. An academic career was simply impossible for nearly all women. The few exceptions mentioned in previous sections (Maria Gaetana Agnesi, Emilie du Châtelet, Sophie Germain) were born into noble or wealthy families, and therefore had access to books and private tutors to educate themselves. These few women pursued their work outside of universities and royal academies, although they engaged in extensive correspondence with several men at these institutions and earned their respect.

As the first woman to obtain a doctorate in mathematics and the first woman to land a professorship, the 19th-century Russian mathematician *Sofia Kovalevskaya* (1850-1891) became a pioneer for women in mathematics. To gain access and then achieve success in a male-dominated field, she faced and overcame immense social obstacles. Educated by private tutors during her youth in Russia, she later moved to Germany to pursue further studies, after entering into a marriage of convenience that gave her the freedom to travel abroad alone. At the University of Berlin, she impressed Karl Weierstrass and earned his support. The university did not allow Kovalevskaya even to audit classes, but Weierstrass agreed to teach her privately for several years, eventually becoming her doctoral adviser. Weierstrass arranged for the University of Göttingen to allow her to defend her doctoral dissertation, where she earned the high distinction of *summa cum laude*. For several years, Kovalevskaya was unable to secure a professorship anywhere in Europe, as university administrations summarily turned down her applications because of her gender. Finally, in 1889, thanks to her friend Gösta Mittag Leffler, a fellow mathematician she had known as a fellow student of Weierstrass, the newly founded Stockholm University appointed her Professor, making her the first woman in modern times to hold such a position. Kovalevskaya made significant contributions to the fields of partial differential equations and analysis.

Despite the pioneering example of Sofia Kovalevskaya, heavy discrimination against female mathematicians in academia persisted well into the 20th century, with very few (less than 10%) female mathematics professors until the 1960s. Only in the past few decades have women started to enjoy the same level of privilege and recognition previously reserved to men. In the United States, the Association for Women in Mathematics (AWM) was founded in 1971 to support and encourage women in mathematics. In France, the French Academy of Sciences inaugurated the *Sophie Germain Prize* in 2003 in honor of female mathematician Sophie Germain. In 2014, Maryam Mirzakhani became the first woman to win a Fields Medal, and Karen Uhlenbeck was the first woman to win the Abel Prize in 2019.

EXERCISES FOR SECTION 8.4

1. During the 17th and 18th centuries, many members of the Bernoulli family of Basel (Switzerland) pursued mathematics and science, and achieved several important advances. Well acquainted with the family, Euler frequently collaborated with various members. Do some research and write an essay on the Bernoulli family, focusing especially on the mathematical achievements of brothers Jakob and Johann, and of Johann's son, Daniel.

2. The *gamma function* is a mathematical function that has captured the interest of several mathematicians throughout the last three centuries. Daniel Bernoulli, Euler, Gauss, and Weierstrass, among others, researched the properties, extensions, and applications of this function. How is this function defined? What other well-known function is it related to?

3. Partial differential equations (PDEs) are differential equations that contain an unknown multivariable function and its partial derivatives. They are extensively studied today because of their many applications, especially in physics and engineering. The following three PDEs: the *heat equation*, the *wave equation*, and *Laplace's equation* (collectively known as the "big three"), are particularly important because of their ability to model a wide array of physical phenomena. Provide the mathematical form of these three PDEs and explain what types of physical phenomena they describe.

4. Besides the "big three" PDEs, the *Navier-Stokes equations* and the *Maxwell equations* are also considered important systems of PDEs. Provide the mathematical form of these PDEs and explain what types of physical phenomena they describe.

5. As mathematical theories became established in the 19th and 20th centuries, mathematicians identified a "fundamental" theorem in every theory, one which they considered to be the most central theorem of the theory. Nowadays, over thirty "fundamental theorems" exist in different theories. State the following well-known fundamental theorems:

   (a) fundamental theorem of arithmetic,
   (b) fundamental theorem of algebra,
   (c) fundamental theorem of calculus,
   (d) fundamental theorem of linear programming,
   (e) rank-nullity theorem (sometimes known as the fundamental theorem of linear algebra).

6. As mathematics grew and expanded throughout the 19th and 20th centuries, the various mathematical theories began to lose sight of each other and the separation between pure and applied mathematics widened. Mathematics became a fragmented body of knowledge and mathematicians were forced to choose an area in which to specialize.

The French mathematician *Henri Poincaré* (1854-1912), often nick-named the "Last Universalist," is considered to be the last person to have been at ease in all areas of mathematics of his day, both pure and applied. Do some research on Poincaré and write a short essay about his life and contributions to mathematics.

7. German mathematician *David Hilbert* (1862-1943) was one of the most influential mathematicians of his day. Among other things, he is famous for having proposed a list of 23 problems in the year 1900. Do some research on Hilbert and write a short essay about his famous 23 problems and his other contributions to mathematics.

8. Indian mathematician *Srinivasa Ramanujan* (1887-1920) is considered one of the most influential mathematicians of the 20th century. Do some research on Ramanujan and write a short essay about his life and contributions to mathematics.

9. Hungarian mathematician *Paul Erdös* (1913-1996) is one of the most famous and prolific mathematicians of the 20th century, publishing around 1,500 mathematical papers and engaging in more than 500 collaborations with other mathematicians. Do some research on Erdös and write a short essay about his life, his contributions to mathematics, and the concept of an *Erdös number*.

10. Besides Sonia Kowalevski, several women are well known for their important mathematical contributions and their efforts to promote gender equality in mathematics. Do some research on *Emmy Noether* (1882-1935), *Julia Robinson* (1919-1985), or another female mathematician and write a short essay about her life and contributions to mathematics.

8.5   Philosophical Issues in Mathematics

The nature of mathematics and mathematical knowledge touches upon many philosophical questions. Mathematical epistemology and ontology, in particular, have deep ramifications that are the subject of much debate in the philosophy of science and even in theology. This section focuses on three issues. Subsection 8.5.1 presents the roles of intuition and deduction in mathematical *epistemology*, a branch of philosophy that concerns itself with the nature of mathematical knowledge, and in particular the methods by which humans can acquire it. Subsection 8.5.2 focuses on the complex and mysterious relationship between mathematics and science, and in particular the questions raised in Eugene Wigner's seminal article, *The Unreasonable Effectiveness of Mathematics in the Natural Sciences*. Finally, Subsection 8.5.3 presents issues concerning mathematical ontology, a branch of philosophy that seeks to answer questions regarding the nature of mathematical objects and concepts.

8.5.1   Mathematical epistemology

Intuition and deduction

Epistemology, from the Greek *episteme*, which means "knowledge," refers to the branch of philosophy that studies the methods by which humans build and acquire knowledge. One of the key questions in epistemology concerns the extent to which humans depend upon sense experience in their effort to gain understanding. On one hand, *intuition* refers to knowledge gained through sense experience (i.e., proceeding by what "makes sense"); on the other, *deduction* refers to knowledge gained by logical inference (i.e., proceeding by what "follows logically"). Intuition and deduction lead to different epistemological views: *empiricism* favors intuition and claims that knowledge comes primarily from a sensory observation of the world, while *rationalism* favors deduction and claims that we primarily gain concepts and knowledge independently of sense experience.

The scientific and axiomatic methods rely on intuition and deduction, respectively. Both methods play important roles in mathematics. Both have been alternately emphasized throughout the history of mathematics, rising and falling as their limitations became apparent.

Intuition and its limitations

A well-known epistemological viewpoint since antiquity, the epistemological perspective of empiricism came to the forefront during the Scientific Revolution. At that time, Francis Bacon (the "Father of Empiricism"), among others, promoted an inductive approach for scientific inquiry, proceeding from observation, hypothesis, and experimentation, to theory, and to law. His Baconian method, which was later dubbed the *scientific method*, became the universal method for the pursuit of knowledge in science. Galileo and Kepler promoted the use of mathematics within this method, paving the way for the mathematization of science. In his magnum opus, *Principia*, Isaac Newton exemplified the power of mathematics combined with this inductive approach, providing an impetus for the rapid progress of both science and mathematics during the Enlightenment. Throughout the 18th century, the scientific method reigned supreme, widely considered the key to understanding all aspects of the physical universe. The belief that the world could be entirely apprehended by observing it with our senses and modeling it with mathematics led to the worldviews of mechanism and determinism, which claim that the consistency of mathematics perfectly reflects all orderly patterns observed in the world.

However, the view that mathematics exists primarily to provide tools for science, although common, falls far short. In reality, mathematics is not subservient to science. Unlike the other sciences, because mathematics is not confined to the physical world, its scope reaches much further than its application to an empirical approach within the scientific method. Mathematics obeys its own universal laws that transcend human experience and often reveals that the world is not as it seems to our human perceptions, that our senses are not trustworthy, and that our intuition can lead us astray. The Ancient Greeks had already discovered the ability of mathematics to expose the fallacies of our intuition; for example, when they discovered the existence of irrational numbers by mathematical deduction. All previous Greek theories relied on the intuitive, but false, assumption that all numbers were rational. According to legend, the discovery of irrational numbers so troubled the Pythagoreans that they kept it secret and killed the mathematician who proved that $\sqrt{2}$ was not rational. Many centuries later, Georg Cantor determined more interesting and counterintuitive facts about numbers by mathematical deduction, such as the fact that the set of rational numbers is countably infinite, but that the set of irrational numbers is uncountable. This discovery led to the even less intuitive fact that between any two arbitrary irrational numbers, there is a rational one, and vice versa.

We humans are not only limited by our perceptions, but also by our scale. Formerly, people believed that the laws of Newtonian physics, which Newton derived by empirical observation, perfectly reflected the reality of the physical world. However, in the early 20th century, physicists discovered that they do not. Although Newtonian mechanics provides a good *approximation* of reality at human scales, they break down when an object's velocity approaches the speed of light (for which the effects of special relativity become nonnegligible) and at the scale of atoms and subatomic particles, where quantum physics takes over. Interestingly, in order to support his law of special relativity mathematically, Einstein turned to Lobachevskian geometry (a non-Euclidean geometry), instead of the more intuitive Euclidean geometry.

Deduction and its limitations

The 19th century brought forth the Age of Rigor in mathematics, with the revival of the axiomatic method. Although the scientific method persisted as the universal method to make progress in empirical science and related areas of applied mathematics, it no longer drove mathematical research as a whole. The weaknesses and limitations of a purely intuitive approach urged mathematicians to turn instead to the axiomatic method, which proceeds deductively, by human reasoning, instead of appealing to the senses and intuition. This deductive approach led to the discovery of several counterintuitive results, such as the existence of non-Euclidean geometries, the summability of infinite series, or the uncountably infinite cardinality of the set of real numbers. The nonintuitive nature of these results fueled the opinion that human intuition was unreliable and untrustworthy. As a result, this new perspective led mathematicians to revisit all previous methods, supplementing their intuitive foundations by axiomatic ones, providing sharply worded definitions and rigorous proofs for even the most basic mathematical concepts. By the end of the 19th century, mathematics viewed the axiomatic method as the universal method for progress in pure mathematics.

Then, in the early 20th century, the limitations of a purely deductive approach began to appear. The foundations of mathematics movement stalled in its attempt to rely on the power of deductive reasoning and axiomatic mathematics to achieve a universal formulation to unify all mathematical theories. Gödel's two incompleteness theorems sounded a death knell to Frege's hope that all of mathematics could grow directly out of logic. This does not mean, however, that mathematics is inherently imprecise but rather that all mathematical knowledge is contingent. It means that the axioms of mathematics are not true in some absolute sense as a consequence of Boolean logic; instead, mathematical knowledge depends on the definitions. Currently the vast majority of mathematics rests on the foundation of ZFC set theory (Zermelo-Fraenkel definitions with the Axiom of Choice) and the definition of $\mathbb{N}$ using Peano's Axioms. However, various mathematicians, especially logicians, may study mathematics without the Axiom of Choice, or they may consider the implications of adding another axiom to ZFC. Then depending on the application, no matter how abstract, one is free to choose a system of axioms that most accurately model what we are studying.

Intuition/deduction thesis

A common view today is that in mathematics, intuition and deduction work in concert, both playing essential roles in the progress of mathematical knowledge. This epistemological view, called the *intuition/deduction thesis*, argues that mathematical knowledge begins with intuition, but must then proceed by deduction. Gottfried Leibniz (one of the co-inventors of calculus) claims that mathematics is "intuitively rational." He describes it this way:

> The senses, although they are necessary for all our actual knowledge, are not sufficient to give us the whole of it, since the senses never give anything but instances, that is to say particular or individual truths. Now all the instances which confirm a general truth, however numerous

they may be, are not sufficient to establish the universal
necessity of this same truth, for it does not follow that
what happened before will happen in the same way again.
[...] From which it appears that necessary truths, such as
we find in pure mathematics, and particularly in arithmetic and geometry, must have principles whose proof
does not depend on instances, nor consequently on the
testimony of the senses, although without the senses it
would never have occurred to us to think of them [60].

Intuition provides a starting point and a guide, illuminating the pathway along which deduction may proceed. Thus, mathematical conjectures may appear to be true by intuition, but must be proved
deductively to become theorems. This duality often brings together
two mathematicians in collaboration, one of them provides the intuition, while the other establishes the deductive proof. A famous
example is that of Srinivasa Ramanujan, a mathematical genius who
had an uncanny intuition about numbers, and G.H. Hardy, an expert
in rigorous, axiomatic deduction. Together, Ramanujan and Hardy's
fruitful collaboration produced many stunning results in number theory and mathematical analysis.

8.5.2   Mathematics and science

What we call *science* in the 21st century is a modern extension of the
traditional field of *natural philosophy*, which focuses on studying phenomena that take place in space and time within the physical world,
in contrast to studying concepts and ideas that pertain to the human
mind or the spiritual realm. The process by which we do science today
finds its roots in the Scientific Revolution, with the establishment of
the scientific method and the mathematization of science. Since then,
mathematics and science have been inextricably linked. However, the
relation between the two is complex, and the progress of science and
mathematics in recent decades has raised even more questions about
this mysterious relationship. Although many mathematicians, scientists, and philosophers have attempted to explain the nature of this
relationship, many questions remain unanswered to this day.

Unreasonable effectiveness of mathematics in the natural sciences

In 1960, the Hungarian American mathematician and physicist (and
Nobel Prize winner) *Eugene Wigner* (1902-1995) published an article
entitled, *The Unreasonable Effectiveness of Mathematics in the Natural Sciences* [77]. In this article, Wigner highlights the central and
mysterious role of mathematics in the progress of science. He raises

several questions about the "passive" effectiveness of mathematics and revives a centuries-old debate about the nature of mathematics itself. Instead of suggesting answers, Wigner leaves the questions open, pointing instead to the inexplicable and miraculous nature of the mysterious interplay between mathematics and science. Drawing examples from the laws of planetary motion, quantum mechanics, and quantum electrodynamics, Wigner reflects on the fact that the laws of physics are all written as mathematical equations, recalling Galileo's statement that the "laws of nature are written in the language of mathematics." Wigner states,

> The enormous usefulness of mathematics in the natural sciences is something bordering on the mysterious and there is no rational explanation for it. [...] The miracle of the appropriateness of the language of mathematics for the formulation of the laws of physics is a wonderful gift which we neither understand nor deserve [77].

Furthermore, Wigner considers that the existence of laws of nature in the first place, and the human ability to apprehend them, are also miracles in and of themselves:

> It is difficult to avoid the impression that a miracle confronts us here, comparable to the miracles of laws of nature, and of the human mind's capacity to divine them [77].

Since its publication in 1960, Wigner's article has been quoted thousands of times, provoking and inspiring many articles and books written in response from across a wide range of disciplines, including philosophy, mathematics, physics, biology, computer science, and economics. Each one echoes Wigner's questions and offers some answers, rebuttals, or further questions. Wigner's article has become a seminal paper in the philosophy of mathematics and science, launching an intense scholarly debate that continues to this day. "Unreasonable effectiveness" is now a well-known phrase to describe the set of philosophical discussions surrounding these topics, some of which we outline below.

Dual nature of mathematics

The dual nature of mathematics refers to the fact that mathematics interacts with the physical world but is not confined to it. Primarily nonempirical, mathematics does not require any application or physical support. It can be done entirely within the bounds of the human mind, not requiring any perception of the material universe. Furthermore, only a small portion of the large body of mathematical knowledge we possess now in the 21st century was developed with a specific application in mind. The vast majority of theories and theorems in pure mathematics was created for purely academic or aesthetic reasons.

On the other hand, mathematics is the undisputed *lingua franca* of empirical science; no other language has been found to describe the physical world better than mathematics. All laws of physics are written as mathematical equations. Philosopher Mark Steiner writes,

> How does the mathematician – closer to the artist than the explorer – by turning away from nature, arrive at its most appropriate descriptions? [73].

Mathematics is also the language of mechanistic and empirical mathematical modeling, used extensively in all natural and social sciences. In many cases, scientists invent the mathematical concepts needed to accurately describe newly observed patterns in nature, such as Newton inventing calculus to write the laws of motion. However, in other cases, scientists borrow a mathematical theory already developed long ago, one that seems to have been mysteriously custom designed to fit the description of a newly observed phenomenon. Physicist Steven Weinberg writes,

> It is very strange that mathematicians are led by their sense of mathematical beauty to develop formal structures that physicists only later find useful, even when the mathematician had no such goal in mind. Physicists generally find the ability of mathematicians to anticipate the mathematics needed in the theories of physics quite uncanny. It is as if Neil Armstrong in 1969, when he first set foot on the surface of the moon, had found in the lunar dust the footsteps of Jules Verne [75].

For example, Einstein utilized the differential geometry of manifolds, first introduced by Riemann to generalize hyperbolic and elliptic geometry, as a perfect framework to describe his general theory of relativity. Similarly, in recent decades, computational molecular biologists have discovered that knot theory, a particular subfield of topology developed in the 1920s with no direct application in mind, offered an ideal tool to apprehend the invariant properties of complex tangles in DNA molecules. These two cases are examples of what Wigner calls the *passive effectiveness* of mathematics.

Passive effectiveness

On one hand, the passive effectiveness of mathematics refers to cases where abstract mathematical theories, developed with no applications in mind (often out of a purely academic or aesthetic motivation), turn out decades, or even centuries later, to offer powerfully predictive models. Another instance of passive effectiveness is when scientists find that a mathematical concept developed for a specific application has applicability far beyond its context. Thus, the mathematical description of a natural phenomenon can reveal a connection with the behavior of another, seemingly unrelated, phenomenon. In this sense, mathematics plays the role of a unifier of physical theories. As the French physicist and applied mathematician Joseph Fourier stated, "Mathematics compares the most diverse phenomena and discovers the secret analogies that unite them [23]."

In the previous statement, Fourier was speaking in particular of the *law of diffusion*, also known as *Fourier's Law*, which he discovered in 1822 through his study of heat conduction. Fourier's observations led him to write his *heat equation* (8.3), which governs the diffusion of heat through a solid body in space and time. If k is the *thermal conductivity* constant, a property of the body that can be determined experimentally, then the heat energy density $u(x,t)$ satisfies

$$\frac{\partial u}{\partial t} - k\nabla^2 u = 0. \tag{8.3}$$

Soon after, scientists discovered that this same equation also models other diffusion phenomena, such as the diffusion of a chemical through water, nutrients through the intestinal tract, a disease through a population, or an animal species through an ecosystem.

Similarly, in 1746, the French Jean d'Alembert initially wrote the *wave equation* (8.4) to describe the motion of a vibrating string or membrane. In this equation, $u(\boldsymbol{x},t)$ represents the vertical displacement and c is the wave propagation speed:

$$\frac{\partial^2 u}{\partial t^2} - c^2\nabla^2 u = 0. \tag{8.4}$$

Scientists later found that this equation also described the motion of water surface waves, seismic waves, sound waves, and electromagnetic waves.

In 1865, the Scottish *James Clerk Maxwell* (1831-1879) formulated the laws of electricity and magnetism in terms of a set of 20 differential equations. In doing so, he discovered that the two phenomena were related, i.e., that electricity and magnetism were in fact two different manifestations of the same phenomenon. Thus, his mathematical equations unified the physical theories of electricity and magnetism. The English *Oliver Heaviside* (1850-1925) later reformulated the equations using the syntax of vector calculus, which allowed him to reduce them to just four equations in terms of the magnetic field ($\boldsymbol{B}$), electric field ($\boldsymbol{E}$), current density ($\boldsymbol{J}$), and charge density (ρ)[20]. Together, the four *Maxwell equations* (8.5) represent the foundation for classical electromagnetism.

$$\nabla \cdot \boldsymbol{E} = \frac{\rho}{\varepsilon_0} \qquad \nabla \cdot \boldsymbol{B} = 0$$
$$\nabla \times \boldsymbol{E} = -\frac{\partial \boldsymbol{B}}{\partial t} \qquad \nabla \times \boldsymbol{B} = \mu_0 \left(\boldsymbol{J} + \varepsilon_0 \frac{\partial \boldsymbol{E}}{\partial t} \right) \tag{8.5}$$

Maxwell's equations also reveal other unexpected facts. For example, in vacuum, they can be decoupled, giving rise to the following two equations:

$$\frac{\partial^2 \boldsymbol{B}}{\partial t^2} - \frac{1}{\varepsilon_0 \mu_0} \nabla^2 \boldsymbol{B} = 0 \qquad \frac{\partial^2 \boldsymbol{E}}{\partial t^2} - \frac{1}{\varepsilon_0 \mu_0} \nabla^2 \boldsymbol{E} = 0.$$

These equations have the exact mathematical form of the wave equation (8.4), thereby revealing that electric and magnetic fields propagate as waves, leading to the now common term "electromagnetic waves." Furthermore, according to these equations, the wave propagation speed is constant and equal to $c = 1/\sqrt{\varepsilon_0 \mu_0}$, which is the speed of light. So Maxwell's equations suggest that light is itself an electromagnetic wave. This example illustrates the passive effectiveness of mathematics: a mathematical description of nature ultimately explains far more than was ever bargained for, unifying multiple physical theories into a remarkably simple set of equations.

8.5.3   Mathematical ontology

The questions raised by Wigner's article strike at the heart of the nature of mathematics itself, raising questions that have occupied the minds of mathematicians and philosophers for over 2000 years and continue to be highly debated to this day.

Ontology, from the Greek *ontos*, which means "being," is a branch of philosophy that studies being and existence. Mathematical ontology seeks to answer questions such as, "What is mathematics?", "In what sense are the abstract objects and concepts of mathematics (such as numbers, shapes, functions, sets, theorems, and models) real?", "Do they exist in the sense of being part of objective reality, or do they only exist in the human mind?". Another related question concerns the nature of the system of logic itself, the mortar that holds all of mathematics together. Is it a human invention, or does it exist outside of human thought?

At its core, mathematics studies abstract entities. Mathematical theories use different mathematical objects, e.g., numbers, lines,

---

[20]In addition, the equations contain two constant parameters, ε_0 (the *permittivity*) and μ_0 (the *permeability*), which can be determined experimentally.

circles, polygons, scalars, vectors, matrices, groups, rings, fields, functions, and sets. These abstract objects constitute the vocabulary of
mathematical thought; statements assemble them according to natural grammar, and logical proofs establish some statements as theorems. A fundamental question arises regarding the existence of these
mathematical objects, statements, and theorems, as well as the methods by which humans perceive them. As mathematicians, we are able
to impart existence to abstract entities by our words. For example,
by saying, "Let A be a square matrix," we now have a matrix that
has a name (A) and a property (being square). Did we *create* this
matrix out of nothing? Or did we *invoke* it from elsewhere? A similar
question arises when we prove a theorem, or develop a mathematical
theory. Did we invent the theorem, or merely discover it? In a similar
vein, did Newton and Leibniz invent calculus, or discover it?

Ultimately, this dilemma concerns the *locus* of mathematics, i.e.,
the place where the essence of mathematics itself originates. Does it
originate in the human mind or does it come from elsewhere? If it
originates in the mind, by what process do we create it? If it originates
elsewhere, how does it ultimately reach the mind? Philosopher Morris
Kline summarizes the dilemma this way:

> Is then mathematics a collection of diamonds hidden in
> the depths of the universe and gradually unearthed, or is
> it a collection of synthetic stones manufactured by man
> [53]?

Mathematicians are divided in their answers to these questions. We
outline below two schools of thought: mathematical Platonism and
empiricism.

Mathematical Platonism

Mathematical Platonism, a school of thought named after the Greek
philosopher *Plato* (429-347 BC), claims that mathematics has a real
existence independent of human thought. Although not a mathematician himself, Plato's cosmology and philosophy (called Platonism) form the basis for mathematical Platonism. Plato claimed that
there exists an invisible reality underlying the physical world, and the
essence of reality lies in patterns called *Forms*, which are unperceivable by human senses, but are more real than the perceptible objects
of the physical world. Forms are eternal, timeless, unchanging, perfect blueprints of reality, existing outside of space and time in a place
Plato called the *realm of being* (or *Platonic realm*). According to
Plato, because we humans are trapped in our physical bodies, we
have a limited perception of reality. What we perceive in the physical
world are not the forms themselves, but mere reflections (or shadows)

of them. Although we cannot perceive the forms, we can form ideas of them in our minds. In his work *Republic*, Plato illustrates his point of view with an allegory (the *Allegory of the Cave*), where humans are chained prisoners facing the back wall of a dark cave, unable to see the objects passing in front of the cave's entrance, but able to see the shadows they cast on the cave's wall. Thus, Platonism instills a mistrust in the material, perceivable world. It claims that human perceptions are limited, unreliable, and deceiving, and that the true source of perfect knowledge comes from the mind instead.

Mathematical Platonism extends Plato's ideas and applies them to mathematics. According to this worldview, mathematical truths are prefect, immutable, and unchanging, existing outside of time and space in a realm of their own. Thus, mathematics has its own objective reality and exists independently of human thought, language, and practices. However, we humans are able to access this realm of mathematical ideas with our minds, explore it and discover its treasures. So although mathematics itself does not evolve, human knowledge of mathematics does. The contemporary English mathematician Roger Penrose (b. 1931) says,

> I imagine that whenever the mind perceives a mathematical idea, it makes contact with Plato's world of mathematical concepts. [...] When one sees a mathematical truth, one's consciousness breaks through into this world of ideas, and makes direct contact with it. [...] When mathematicians communicate, this is made possible by each one having a *direct route to truth*, the consciousness of each being in a position to perceive mathematical truths directly, through this process of 'seeing'. Since each can make contact with Plato's world directly, they can more readily communicate with each other than one might have expected. The mental images that each one has, when making this Platonic realm contact, might be rather different in each case, but communication is possible because each is directly in contact with the *same* eternally existing Platonic world! [63].

Several mathematicians throughout history have adopted this point of view. Charles Hermite states, "I believe that the numbers and functions of analysis are not the arbitrary product of our spirits; I believe that they exist outside of us with the same character of necessity as the objects of objective reality; and we find or discover them and study them as do the physicists, chemists, and zoologists" [53]. Similarly, Godfrey Hardy says, "I believe that mathematical reality lies outside us, that our function is to discover or observe it, and that the theorems which we prove, and which we describe grandiloquently as our 'creations', are simply our notes of our observations [43]." Paul Erdös imagined that somewhere in this realm of mathematical truths, there was a grand book containing the most beautiful proofs of all theorems. So when anyone produced a particularly beautiful proof, he would exclaim, "This proof is from the Book!". Other mathematical Platonists include Gottfried Leibniz, Georg Cantor, and Kurt Gödel.

Some mathematical Platonists claim that the unreasonable and passive effectiveness of mathematics evoked by Eugene Wigner can be explained in that what we perceive in the material world are reflections of the perfect truths of mathematics. Heinrich Hertz states, "One cannot escape the feeling that these mathematical formulae have an independent existence and intelligence of their own, that they are wiser than we are, wiser even than their discoverers, that we get more out of them than was originally put into them [21]." Physicist Max Tegmark says, "Physics is so successfully described by mathematics because the physical world is completely mathematical, isomorphic to a mathematical structure, and we are simply uncovering this bit by bit [74]."

Mathematical empiricism

An opposing point of view regarding the nature of mathematics also traces its roots back to ancient Greek philosophy, in the work of *Aristotle* (384-322 BC). According to Aristotle, human knowledge comes from perception of the world by our senses. The mind is initially a blank slate, on which experience leave marks by sense perception and thus contributes to building our knowledge. Abstract entities do not exist in any transcendent way; rather, they are convenient linguistic conventions that humans use to communicate their thoughts to one another. Aristotle's philosophy became very popular during the Middle Ages, when medieval universities adopted a method of teaching called *scholasticism*, largely based on the principles of Aristotelian philosophy. It gained traction again during the Scientific Revolution, when the scientific method required that conclusions be empirically based on the evidence of the senses. This aspect of Aristotelian thought is called *empiricism*, and we call its application to mathematics, *mathematical empiricism*.

According to this worldview, mathematics does not exist independently of human thought. Rather, as the human mind abstracts knowledge gained by sensory experience, it creates mathematics as a mental tool to describe and communicate what it perceives. So ultimately, mathematics is invented, not discovered. This worldview agrees with the modern practice of mathematical modeling, which considers mathematical models to be mere imperfect descriptions of natural and social phenomena. Furthermore, unlike Platonism, mathematical empiricism does not appeal to the existence of a nonperceivable extra-human realm, which many mathematicians find problematic. For example, mathematician Reuben Hersch says,

> There is only one universe, one real world, which is physical reality, including its elaboration into the realm of living things, and elaborated from there into the realm of humankind with its social, cultural, and psychological aspects. I have argued that these social, cultural, and psychological aspects of humankind are real, not illusory or negligible, and that the nature of mathematical truth and existence is to be understood in that realm, rather than in some other independent 'abstract' reality [14].

Then, in regards to the effectiveness of mathematics, he says,

> There is no way to deny the obvious fact that arithmetic was invented without any regard for science, including physics, and that it turned out (unexpectedly) to be needed by every physicist [21].

Wigner himself seems to hold to an empiricist view in his statement that "mathematics is the science of skillful operations with concepts and rules invented just for that purpose [77]." Several other mathematicians agree with the empiricist view, including Karl Weierstrass, who compared a true mathematician to a poet, and Richard Dedekind, who said, "We understand by number not the class itself, but something new...which the mind creates. We are of a divine race and we possess...the power to create [53]."

Mathematical empiricism is a strong current nowadays, especially with the discovery that many "laws of nature" are not as universal as once thought and have been downgraded to the status of mere mathematical models. Similarly, the failure to unify the theories of mathematics has also swayed several mathematicians to reject the lofty place that mathematics occupies in the Platonist view. As philosopher Morris Kline states, "The present conflicts about the nature of mathematics itself and the fact that mathematics is not today a universally accepted, indisputable body of knowledge certainly favor the view that mathematics is man-made [53]."

EXERCISES FOR SECTION 8.5

1. Pure mathematicians work primarily with the axiomatic method, whereas applied mathematicians and experimental scientists work with the scientific method. Does this have an influence on the philosophical positions taken by pure and applied mathematicians in regards to their view of the nature of mathematics (mathematical Platonism or empiricism)?

CHAPTER 9

Reading and Researching Mathematics

In a 2014 interview with *The Guardian*, Maryam Mirzakhani gave the following advice for those who want to know more about mathematics: "I can see that without being excited mathematics can look pointless and cold. The beauty of mathematics only shows itself to more patient followers." Reading, research, exploration, and discovery in mathematics requires perseverance and, as Mirzakhani emphasizes, patience.

This chapter focuses on *scholarly communication*, the process by which scholars and researchers share their work with the rest of the world. Scholarly communication encompasses several aspects related to the dissemination of scholarly knowledge, including the cycle of writing, peer review, publishing, disseminating, organizing, and preserving of knowledge for future use. The scholarly practice of peer reviewing each other's work and disseminating it via journals originated during the Scientific Revolution, and has since become the standard process for scholarly communication in many academic fields, including mathematics. Scholarly communication in mathematics has undergone major changes in the past few decades with most journals offering their content exclusively online, and an increasing number of journals functioning as open access resources.

Peer-reviewed journal articles containing original research results are considered *primary sources*. The most significant original results may appear later in other forms of *secondary literature*, such as review articles, expository articles, books, encyclopedias, and websites. Some may even be translated into other languages or become incorporated into the curriculum of a mathematics course. Approaching mathematical knowledge via secondary literature, when available, is often preferable. Compared to a primary source, secondary sources often present ideas more clearly, in a more organized fashion, using standardized definitions and notations, making connections with other existing ideas, and correcting or clarifying concepts that may have been unclear when they were first developed. For example, although the original 17th-century publications of Newton and Leibniz contain all the concepts of elementary calculus, it is much easier to learn calculus from a 21st-century textbook that contains the same material.

However, only a small amount of original results ever make it beyond the stage of primary literature. Because of the smaller readership, textbooks in specialized fields of upper-level mathematics are rare. Furthermore, secondary literature emerges months or years after the original results appear in primary sources. Thus, the extent of mathematical knowledge available to learn from secondary sources represents only the tip of the iceberg. For this reason, accessing the vast number of undeveloped treasures and latest results of upper-level mathematics requires becoming comfortable with reading and learning from primary sources. This essential skill of *information literacy* is necessary when transitioning to advanced mathematics.

Section 9.1 introduces mathematical journals and the peer review process. Section 9.2 presents the overall structure of original research articles in pure and applied mathematics. Section 9.3 offers strategies to read and learn from primary sources and introduces the useful skill of expositing specialized content to a broader audience. Finally, Section 9.4 focuses on how to search for primary and secondary sources and highlights the unique role of libraries.

9.1   Journals

A *scholarly journal* (or *academic journal*, or simply a *journal*) is a specialized, periodical publication where scholars share their knowledge. Nowadays, journals are the primary vehicle for scholarly communication in mathematics, science, and many other fields of study. Currently, approximately 30,000 journals are active in the world, together publishing over 1 million articles each year.

Research journals publish primarily, but not exclusively, articles of original research, i.e., articles containing novel results, never published before, authored by the researchers themselves. Thus, journals contain the most accurate and up-to-date results. Furthermore, journals carefully monitor the quality of their content and the correctness of published results by requiring that every submitted manuscript undergo a stringent *peer review process* before it is published. Some journals are very selective, publishing less than 1% of submitted manuscripts.

Journals also represent an important platform for collaboration in a narrow field, as scholars build their own research on results published by others, citing one another's articles and being cited in return by future scholars building upon their work. It is often possible to trace the development of an idea across many scholars through a chain of articles, each one citing a previous article in sequence. In mathematics, where knowledge progresses quickly, journals are essential to remain on the cutting edge of a rapidly evolving field.

Virtually every new advance in science and mathematics made after the 17th century can be traced back to its appearance in a journal. Created during the Scientific Revolution, some of the first journals remain active to this day. For example, the English journal *Philosophical Transactions of the Royal Society*, established in 1665, continues to publish regular issues in the 21st century.[1] Over the course of its long history, this journal has published discoveries from many great scientists, including Isaac Newton, Benjamin Franklin, Charles Darwin, Michael Faraday, James Clerk Maxwell, Alan Turing, and Stephen Hawking.

Nowadays, many specialized journals exist for specific areas of mathematics, such as the *Journal of Number Theory* or the *Journal of Nonlinear and Convex Analysis*. In addition, many advances in applied mathematics appear in journals related to their field of application. For example, the *Journal of Molecular Biology* and *Computational Modeling and Epidemiology* contain cutting-edge mathematics developed in applications to molecular biology and epidemiology, respectively. Finally, some journals focus on mathematics education, for example the *Journal of Statistics Education*.

Subsection 9.1.1 describes the contents and main features of a journal, and Subsection 9.1.2 outlines the process and purpose of peer review. Subsection 9.1.3 provides guidelines for determining the value of a journal relative to others in its field. Subsection 9.1.4 offers some perspective on the price and access of journal articles. Finally, Subsection 9.1.5 lists a few other types of scholarly sources.

9.1.1 Content and features

Articles

The majority of a journal's content consists of *original research articles*. Articles of this type are considered *primary literature*, meaning that they contain new research results that have never been published before and whose authors are the researchers themselves. Some journals may also publish *letters*, *notes*, or *communications*, which are shorter articles containing smaller or partial results. Editors process these quickly in order to keep the readership aware of the latest results with minimal delay.

---

[1]The word *Philosophical* in the title refers to *natural philosophy*, which is the equivalent of what is now called *science*.

Journals may also publish pieces of *secondary literature* such as expository or review articles, which do not contain original results, and whose authors typically are not the same people who discovered the original results. For example, an *expository article* may present known alternative methods for proving a theorem. Other expository articles aim at explaining the content of an original research article to a broader audience by providing additional background, highlighting key steps, and clarifying terms. *Review articles* are secondary sources that summarize, organize, compile, and compare results from recent primary sources. These articles are particularly prevalent in rapidly expanding fields, where many researchers are making progress in similar directions simultaneously, but using different terminology and notations. These articles help readers get a bird's-eye view of recent advances in a field, determining who is doing what and who is collaborating with whom. Review articles often contain large lists of recently published primary articles with comments to differentiate and compare them.

Some journals publish other items besides articles, such as letters from readers, book reviews, open questions, and announcements.

Features of a journal

Not all periodical publications are scholarly journals, and not all journals are alike. Journals differ in intent from magazines or newspapers, which do not involve the peer-review process. In many cases, the appearance and style of the publication contain clues to identify it as a journal or magazine. Magazines, even serious ones containing technical material, often have an inviting style, with colorful, glossy pages containing numerous pictures, interspersed with advertisements. In contrast, journals are short, stark booklets with a formal, professional appearance, usually devoid of color and advertisements. Nowadays, most periodical publications offer all their content online, and many have discontinued their print format.

The best way to determine whether a publication meets the criteria of a journal is to turn to the internet. Every journal's website clearly identifies it as such and provides the following additional information:

1) **Peer-reviewed status.** Peer review is the primary feature that sets apart a journal from other types of scholarly publications. Peer-reviewed journals typically state this fact explicitly and prominently on their website.

2) **Frequency.** Depending on the journal, issues may appear weekly, biweekly, monthly, bimonthly, quarterly, or less frequently. Some journals publish at irregular intervals; others may occasionally publish a "special issue" on a specific topic in

addition to their regular issues. If the website does not explicitly state its issue frequency, the reader can easily determine it by looking at the dates of previous issues.

3) **Scope.** The *scope* of a journal refers to the range of topics covered in its articles. The scope of a journal may be very narrow and specialized, or sometimes quite broad.

4) **Readership.** The intended audience of a journal varies. Some cater to narrow specialists, whereas some are more interdisciplinary and address a variety of scholars at the intersection of several fields. Some journals are for professional researchers and scholars, while others target students, such as the *College Mathematics Journal* and *Involve*.

5) **Types of articles.** Most journals publish primarily original research articles. Some may also contain expository articles, review articles, book reviews, letters, notes, methods, and announcements.

6) **Articles.** Since most journals no longer print issues on paper, their websites contain links to all their articles. Some articles may be *open access*, i.e., accessible to view and download for free. Others may require a subscription or be available for purchase.

7) **Metrics.** Many peer-reviewed journals publicize a variety of *citation-based metrics* that rank the overall quality and prestige of the journal compared to other journals. Citation-based metrics calculate an article's score on the basis of how many times other articles cite the original article in other peer-reviewed publications.

8) **Board of editors.** Each journal operates under the guidance of a *board of editors* whose names are listed on the journal's website. The board of editors, under the direction of the Editor-in-Chief, is composed of eminent experts in the journal's field. Editors solicit and review manuscript submissions, determine the journal's scope and target audience and provide advice and feedback to authors. If the journal operates under a double-blind peer-review system, the editors act as intermediaries between authors and peer reviewers, transmitting all communication between them without revealing names.

9) **Indexing.** An *index* is a specialized database for journals. A journal's website often states the indexes where it is listed. See Subsection 9.4.3 for more information on indexes.

Example 9.1.1. The *Journal of Number Theory* is a monthly, peer-reviewed journal that publishes original research articles in number theory and allied areas, including computational number theory. The journal aims its content at professional, established researchers. It also encourages junior researchers and shorter submissions for its "general section." Every issue (called a *volume*) contains 20-30 articles; some are open access. The journal's website lists several metrics, including the journal's impact factor and CiteScore. The journal is indexed in eight indexes, including Scopus, Web of Science, Zentralblatt MATH, and five others.   △

Example 9.1.2. The *Journal of Mathematical Biology* is a monthly, peer-reviewed journal that publishes original and expository research articles. Its scope includes mathematical ideas, methods, techniques, and results that are useful to gain understanding or explain phenomena in cell biology, physiology, neurobiology, genetics, population biology, ecology, evolution, epidemiology, immunology, molecular biology, DNA and protein structure and function. Every issue contains 10-20 articles; some are open access. The journal's website lists the journal's impact factor and the names of all (approximately 30) members of the board of editors.   △

9.1.2   Peer review

The peer-review process

Peer review refers to a system of evaluation of a scholar's work by *peers*, i.e., people who are as competent as the author in his or her field of research. A peer-reviewed journal submits all *manuscripts* that the editors consider potentially worthy of publication for evaluation by a team of *peer reviewers*, also called *referees*. Unlike the journal's editors, peer reviewers are not associated with the journal, but they work closely with the editors to ensure that articles are of sufficient quality before they are published.

Not all manuscripts submitted to a journal are automatically forwarded to peer reviewers. The journal's editors, who are experts themselves, are usually able to judge the overall quality of a manuscript and decide whether it fits the journal's scope and target readership. In some journals, the editors reject over 95% of submissions outright without going through peer review. Journal editors may also give preliminary feedback to authors and solicit their own revisions before submitting a manuscript to peer review. Only if the editors are enthusiastic about a manuscript will they initiate the peer review process.

In specialized fields, editors may find it difficult to identify potential peer reviewers. In some cases, only a handful of people in the world might be qualified to judge a scholar's work, check it for correctness and originality, and provide feedback for improvement.

A journal typically tries to find at least two peer reviewers for each manuscript, and it may spend months searching for them. Peer reviewers are busy people with full-time jobs of their own, so it may be several more months before they can review a manuscript.

Editors task reviewers to carefully read and check a manuscript, making sure that everything is correct and, in the case of an original research manuscript, original. Therefore, peer reviewers must have a deep and comprehensive knowledge of all past and current literature in their topic so that they can confirm that the contents of a manuscript consist of new, unplagiarized results that have never been published anywhere before. Finally, peer reviewers make comments to improve the content by suggesting the inclusion of additional resources, clarifications, alternative solution methods, insightful extensions, or valuable corollaries.

In many cases, the peer review process is *double-blind* (i.e., the peer reviewers do not know the authors' identities and vice versa) to preserve anonymity and promote impartiality. All communication passes back and forth through the journal's editors.

A manuscript may go through multiple rounds of review and resubmission, with the peer reviewers making suggestions and authors revising the manuscript accordingly at each round. Because each round takes several weeks or months, the whole peer-review process may take several months or years from beginning to end. At any point in the process, the editors can choose to reject the manuscript, and the authors can choose to retract their submission.

If a manuscript reaches a state that the peer reviewers judge to satisfy the standards for the journal, they may recommend it to the editors for publication. If the editors agree and proceed with publication, the manuscript ultimately becomes an *article*. The term *archived* is used to designate an article that has successfully completed the peer-review process and become a permanent part of the scholarly literature.

Advantages and disadvantages

Peer review improves the quality of a journal's articles. By submitting a author's work to scrutiny by experts in the field, the journal reduces the likelihood of errors and, in the case of original research, of inadvertently publishing duplicated (and therefore nonoriginal) results. Because the peer-review process establishes communication between the authors and other specialists, articles effectively become the result of a collaboration between peers, benefiting from a form of consensus and a less subjective point of view.

Unfortunately, peer review is time-consuming and slow. Peer reviewers are difficult to find and may not be available to review a

manuscript in a timely manner. Manuscripts can become stuck in a
cycle of review and resubmission for months or years, rendering the
results obsolete by the time they are finally published. Peer review
also often opens up arguments between authors and peer reviewers,
which further delay publication. Finally, the process is not perfect:
occasionally errors slip through, only discovered later by readers. In
this case, the journal may decide to publish corrections retroactively
or retract the article altogether.

9.1.3 Prestige and ranking

Citation-based metrics

A citation-based *metric* is a quantitative method of ranking the relative prestige, importance, and significance of a journal compared to
others in its field. Such a metric is a number calculated by a formula based on the average number of times a journal's articles are
themselves cited in other peer-reviewed articles. Thus, it is not sufficient for an impactful article to be merely widely downloaded, read,
and circulated; it must generate citations. Its content must provide a
new, fertile platform on which future researchers may base their own
original results.

   Several different metrics exist, each one using a slightly different
formula. Some of them take into account the size of the journal, the
narrowness of the field, or how quickly the articles are cited or become
obsolete (i.e., the article's *half-life*). The following are some of the
more common metrics:

1) **The CiteScore (CS).** Average number of citations in a calen-
    dar year divided by all articles published in that journal in the
    preceding three years.

2) **The Impact Factor (IF).** Number of citations received in
    a given year of articles published in that journal during the
    two preceding years, divided by the total number of articles
    published in that journal during the two preceding years.

3) **The SCImago Journal Rank (SJR).** A complex formula
    that takes into account the importance of the journals from
    which the citations came (as measured by those journals' SJR).

4) **The Source-Normalized Impact per Paper (SNIP).** A
    complex formula that compares journals within the same spe-
    cific field. A SNIP greater than 1 indicates that the journal is
    cited more than average for that field.

In each case, an *indexing service* calculates the metric on the basis of the journals included in the index's database. Because not all

journals appear in every index, the citation counts used in the various formulas may be slightly different for each metric. Furthermore, many indexing services are not free and only reveal metric information to subscribers. For example, the index *Web of Science*, which calculates the Impact Factor, only discloses the IF of its journals to paid subscribers. However, many journals proudly disclose their own metrics on their webpages, especially if they are high.

Metrics are primarily used to compare and rank journals within a field. Scientists and mathematicians deem a journal with a higher score as more important than those with lower scores. For example, an IF less than 1 indicates a low-impact journal, whereas an IF greater than 5 is high-impact. It is important to realize that metrics only give an average view of a journal's articles. In reality, the citation count per article in a journal is not uniformly or normally distributed, so a metric based on the average number of citations may be misleading.

A journal's metric is not to be confused with an author's metric, which is based on the average number of citations garnered by articles from the same author.

Indexing

An *indexing service* (or *index*) organizes and facilitates finding information in peer-reviewed journals. Hundreds of different indexes exist; some are field-specific, others are broader and may include thousands of journals and millions of articles. The most common indexes for mathematics and science are: MathSciNet, ScienceDirect, Scopus, and Web of Science. To appear in these indexes, a journal must comply to the index's strict quality standards in terms of publishing frequency and timeliness, copyright policy, archiving of past issues, budget, and searchability of information. Thus, an indexed journal signals quality in terms of its administrative reliability.

Indexing services also promote the accessibility, availability, and searchability of information contained in their journals. Each one typically manages a sophisticated search engine allowing users to find specific information through all the journals in the index, a valuable tool enabling researchers to find what they are looking for. Therefore, journals can increase their prestige and visibility by being indexed with multiple indexing services.

A journal's website often lists the services where it is indexed, confirming that the journal has been vetted and maintains a certain level of quality. Conversely, a journal that is not indexed may signal that it has fallen into disuse, or that it is one of the many disreputable *predatory journals* that plague the publishing world. These fraudulent journals masquerade as legitimate journals to trick authors into publishing with them, charging high publication fees without providing any peer review or editorial services.

More information on indexes can be found in Subsection 9.4.3.

9.1.4 Price and access

The costs associated with publishing and accessing articles vary by journal.

Journals operate on a tight budget, and their very small readership prevents them from benefiting from advertisement income or economy of scale to reduce costs. Small journals are entirely managed by scholars who are already fully employed as researchers or professors and who do not collect any additional pay for managing the journal.[2] Furthermore, many journals rely on volunteer peer reviewers and operate entirely online to save the cost of printing paper issues.

A large portion of a journal's operating expenses covers the administrative work of managing the journal, including proofreading, copyediting, indexing, and archiving articles. Although some of the larger journals have a paid staff to do this work, most journals turn to third-party journal publishing companies, such as *Elsevier* or *Springer*, to handle these administrative tasks for a fee. Other journals are managed by mathematical associations (see Subsection 7.2.3 for more information on these associations).

Journals recover their costs by charging the authors or the readers, or both. Journals may charge authors an *article processing charge* (APC) when a manuscript is accepted for publication. Or they may charge readers a *subscription* fee or a *pay-per-view* (PPV) fee to access published articles. Academic institutions may help pay for some of these fees; in other cases, researchers must allocate a part of their research budget to cover them.

Access paid by readers

Most journals only reveal the abstract of their articles for free, hiding the rest of the content behind a *paywall*. An interested reader can bypass the paywall and access the full article by paying a one-time PPV fee, which is typically between $20 and $100 per article, although academic institutions may obtain a volume discount on PPVs. Academic institutions may also maintain yearly subscriptions to commonly accessed journals via a *site license*, which allows unlimited access to articles in that journal for any user affiliated with the institution. Depending on the journal, a yearly subscription can cost between

---

[2]Their involvement with the journal falls under the purview of their research activities or institutional service.

$100 and $10,000 per journal. See Subsection 9.4.4 for more information on how institutional academic libraries manage their journal subscriptions.

In particularly active fields of research, journals may only require payment to access articles within a 1-5 year *embargo period* after publication. After the embargo period, access to the article becomes open to the public free of charge.

Open Access

A journal may offer the option of having the authors (instead of the readers) cover the publication cost of an article by paying a $100 to $3000 *article processing charge* (APC). In this case, the article becomes *open access* (OA), i.e., permanently accessible to the public for free. Authors may choose this option to increase the visibility of their article, or they may be required by their research grant to publish their research as OA articles.

Some journals are strictly *open access journals*, requiring OA publication of all their articles, while others (*hybrid journals*) offer authors the option to decide to publish under OA or to allow subscription and PPV fees to cover the publication costs.

9.1.5 Other publications

Besides peer-reviewed journals, mathematicians often turn to other primary and secondary sources: conference proceedings, dissertations, preprints, working papers, and books. Some indexing services may also include these sources in search listings.

Conference proceedings

Conferences are important opportunities for researchers to present their research and get feedback from other researchers in their field. Researchers will often present papers containing their latest results at a conference, even before submitting them for publication in a journal. Conference organizers may compile all the papers presented during the conference and publish them as *conference proceedings*. These publications are useful for those unable to attend the conference, or a particular presentation, in person.

Papers presented at conferences often contain original results and are presented by the researchers themselves. However, not all are peer reviewed; some are still at the stage of unfinished "working papers" written in draft form or in a less formal format than a journal article. Many of these papers will go on to be published in peer-reviewed journals later, but non peer-reviewed papers in conference proceedings may contain errors and go through substantial revision before they are accepted as formal journal articles. In particular, the ideas may not be fully developed, as some researchers use conference presentations to solicit feedback from attendees or to connect with potential collaborators.

Conference proceedings are easy to identify as such, because their title contains the name, date, and location of a specific conference. Some indexing services will include conference proceedings in their database alongside related peer-reviewed journal articles.

Dissertations

Obtaining a Ph.D. degree (in mathematics or any field) requires that the candidate write a doctoral *dissertation* (also known as a doctoral *thesis*), which must contain substantial original research results. A dissertation represents the culmination of many years of full-time research, so it is typically much longer than a journal article.

A dissertation is not peer reviewed in the same sense as a journal article is. However, Ph.D. candidates are closely followed throughout their dissertation by a committee of 2 to 5 professors led by the candidate's primary doctoral adviser. These professors give continual feedback to the candidate, soliciting revisions, corrections, and extensions, thus playing a similar role to peer reviewers. In order for the candidate to obtain the Ph.D., all professors in the committee must approve the dissertation, and the candidate must defend it orally in public. For these reasons, the content of a dissertation is usually as reliable as a peer-reviewed article, even though it is still considered supervised student work. After approval of the dissertation, the candidate often chooses to submit its contents for publication in one or more peer-reviewed journals.

A dissertation can be identified as such by its title, which indicates it to be a requirement for some candidate to obtain a Ph.D. at a specific university. Most academic libraries at institutions who grant Ph.D.s make all their dissertations accessible to the public, and indexing services may include them in their databases.

In some universities, students are required to write a *master's thesis* to obtain a master's degree. A master's thesis is much shorter than a doctoral dissertation, and its content is usually not required to be original.

Preprints and working papers

Researchers will occasionally publicly share a paper that contains original results before it has been peer reviewed. These papers, called *preprints* or *working papers*, can be at any stage of completion. Some are little more than informal drafts containing preliminary or incomplete results, while others are complete manuscripts close to their final form. Some preprints may even have been already submitted to a journal, where they await completion of the peer-review process.

In 2001, Cornell University created *ArXiv*, pronounced "archive," an index and online repository for preprints in mathematics, physics, astronomy, electrical engineering, computer science, quantitative biology, statistics, mathematical finance, and economics. This database (arxiv.org) holds several million preprints in these fields and receives more than 10,000 new submissions every month. ArXiv posts all submitted preprints the same day they are received and offers free open access to all preprints in the database. ArXiv has become the worldwide central repository for preprints in mathematics.

Researchers often post preprints early in the research process to receive feedback from colleagues or to search for collaborators. Many preprints are eventually published as formal peer-reviewed journal articles. A preprint, however, has not completed peer review, so it may contain errors and undergo revisions before publication as an article.

Books

Books are perhaps what first comes to mind when considering written publications containing mathematical knowledge. There are certainly many excellent mathematics books, and many schools rely heavily on mathematics books to teach elementary mathematics. However, books only contain a small fraction of current mathematical knowledge, as only a small number of original mathematical results ever makes it beyond the stage of primary literature. Because the audience for upper-level mathematics is small, it is often not worth the time and energy to write a book that merely compiles mathematical results that already appear in journal articles. The only mathematical ideas that appear in books are those that have become important enough to interest a wider audience, i.e., significant enough to have been accepted as well-established theories in the mathematical community. Even in these cases, the concepts presented in a book lag several years behind their original appearance in primary sources, an unacceptable delay for mathematicians who strive to remain on the cutting edge of their field.

However, mathematics books are excellent resources for learning new areas of mathematics; some are excellent expository works written with a broader audience in mind and are more approachable than journal articles. Book authors are usually passionate about expositing complex material to facilitate learning and understanding. They strive to present material in an organized, sequential manner, using standardized definitions and notations, making connections with other existing ideas, and correcting or clarifying concepts that may have been unclear when they were first developed.

EXERCISES FOR SECTION 9.1

1. Determine whether the following publications are peer-reviewed journals. (a) *Annals of Mathematics*, (b) *Components in Electronics*, (c) *Discover Magazine*, (d) *Facilities*, (e) *International Journal of Science and Mathematics Education*, (f) *Journal of Functional Analysis*, (g) *Mathematical Biosciences*, (h) *National Geographic*, (i) *Nature*, (j) *Networks and Spatial Economics*.

2. For the publications in Exercise 1 that are peer-reviewed journals, determine their (a) frequency, (b) scope, (c) target readership, (d) Impact Factor or CiteScore. Then rank the journals by IF or CS.

3. Suppose you have identified the following three candidate journals to publish an article describing your original mathematical model for climate change caused by emissions of carbon dioxide. You are not allowed to submit your manuscript to all three at once; you must choose one. Determine the scope, frequency, and impact factor for each journal. Then, in light of this information, write a paragraph arguing which of the three journals would be your first choice for your manuscript.

    (a) *Applied Mathematical Modeling*
    (b) *Climatic Change*
    (c) *Journal of Advances in Modeling Earth Systems*

4. Nonmathematical journals often publish the *results* of a mathematical model but relegate the mathematical details pertaining to the model itself to an appendix or online supplementary material. Suppose you have identified the following four candidate journals to publish an article describing your original mathematical model for the spread of a new infectious disease. Research the scope and target readership of each journal and decide which one(s) are likely to publish the complete mathematical details of the model, and which may publish the results but relegate the mathematical details to an appendix.

    (a) *International Journal of Epidemiology*
    (b) *Mathematical Biosciences*
    (c) *Journal of Public Health*
    (d) *BMC Bioinformatics*

5. Some journals have been criticized for establishing editorial policies aimed at artificially increasing their citation-based metrics. Explain what some of these policies are and how they contribute to raising these metrics.

6. Search the *arXiv* (arxiv.org) for preprints in the fields of (a) algebraic topology, (b) differential geometry, and (c) probability. Count how many preprints were submitted today in each of these fields.

7. Although it is a common practice in mathematics, it may be risky for an author to publicly share a paper as a preprint before it has been peer reviewed. Explain what these risks may be and how posting a preprint may turn out to be detrimental to a researcher.

9.2 Original Research Articles

Reading a journal article is a skill acquired over time. It always requires a lot of effort, even for experienced mathematicians, but it becomes easier with practice. Because most recent mathematics only exists in journal articles, being able to confidently approach a journal article and decipher its contents is an essential skill that all mathematicians must develop. Whereas a student typically acquires elementary mathematical knowledge primarily from textbooks, as the level progresses, it becomes increasingly necessary to learn directly from primary sources.

Journal articles can seem very intimidating at first glance. Written by specialists for specialists, their authors assume that readers are familiar with specialized terminology, concepts, and notations, as well as recent advances in the field. Entering into such an article often feels like stepping into an ongoing conversation and having to catch up on what was said before in order to understand what is going on. Furthermore, authors use a formal mathematical style (described in Subsection 10.1.1), which is purposely cold and dispassionate, very different than the warm inviting style encountered in mathematics textbooks.

Before delving into the *content* of an original research article, it is important to spend some time understanding its *structure*. These two aspects are covered separately in this book: the current section (9.2) outlines the overall structure of original research articles in applied and pure mathematics, and the following section (9.3) offers some guidance on deciphering the content.

Research in *applied* mathematics is always connected to research in a field of experimental (natural or social) science. For this reason, articles that feature applied mathematical results follow the standard article structure used in experimental science, which divides the narrative into standard sections: Introduction, Methods, Results and Discussion (often abbreviated IMRAD), which align with the way research is conducted according to the *scientific method*. The authors typically present a mathematical model for a system that they also study experimentally. They may combine qualitative observation with experimental data to build a model, which in turn serves as an additional method to gain additional insight into the behavior of the system under investigation.

Mathematical models and other applied mathematics research can appear in a variety of journal types. Some address readers who are primarily interested in the mathematical details related to building and using a mathematical model. These types of articles may skim over the nonmathematical explanations and even relegate them to an appendix or supplementary material. Conversely, others address primarily nonmathematicians and focus instead on data collection and results, relegating the mathematical details to an appendix or supplementary material.

Conversely, research in *pure* mathematics proceeds according to the *axiomatic method*. Researchers obtain original results by extending or generalizing a mathematical theory axiomatically, building new theorems, and proving them by logical deduction based on previously proved theorems and axioms. This process often requires creating new terms and notations for new mathematical concepts, objects, and operations. Unlike applied mathematics, which often appears in nonmathematical journals, research articles in pure mathematics primarily appear in specialized journals written by mathematicians for mathematicians.

Subsection 9.2.1 presents the preliminary steps when approaching any mathematical article, whether pure or applied. Subsection 9.2.2 outlines the IMRAD structure common to articles in experimental science and applied mathematics, and Subsection 9.2.3 outlines the structure of pure mathematics articles. We conclude this section with a discussion of the final sections of mathematical research articles in Subsection 9.2.4.

9.2.1   Preliminary information

Authors

Authors of current articles, whether professional researchers or students, are people with real lives and interests. It is often possible to glean a great deal of information about them on the internet, which may even reveal photos and details about authors' personal lives. Professional researchers often have a personal webpage with their current contact information, employment, position, education, past and current research interests, and details about how their career has evolved. It is common for academics to develop interests in specialized fields that are substantially different than what they studied as students. Researchers are generally quick to respond to questions and open to feedback. They are usually thrilled to learn that someone is interested in their research and reading their articles.

If there are multiple authors, the order in which they are listed matters. If all authors are equal contributors to a publication, they appear alphabetically by last name. Otherwise, they appear in their order of relative contribution. The *first author* is the primary contributor; usually the one who provided the main insight, did the most work, and/or wrote the bulk of the manuscript. It is a coveted position, as some citation styles only cite the first author and combine all other named authors into "et al." One of the co-authors (not necessarily the first author) may also play the role of *corresponding author*, who acts as a spokesperson and handles all communication on behalf of the others.

The authors of an applied mathematics article often form an interdisciplinary team of researchers with different areas of expertise; for example, a sociologist, economist, psychologist, computer scientist, and applied mathematician. In this case, the article may list the co-authors in their order of contribution and explicitly describe the nature of each one's contribution to the project. The *last author* position is sometimes reserved for the team supervisor (also known as *principal investigator*) who directed and funded the research.

Journal

All peer-reviewed journals have a website that provides detailed information about their scope, readership, frequency of publication, types of articles, and citation-based metrics. It is useful to look up all of this information before reading any article in the journal.

Date of publication and current impact

Mathematical research is an ongoing, collaborative effort, where researchers refer to each other's results and build on them. Eventually, a "chain" of articles emerges, each one extending a previous result, incrementally expanding a mathematical theory, refining a model, or enhancing a method.

Research progresses fast in certain fields, with new results emerging every few weeks or months. In these cases, a publication that is a few years old will not have the most up-to-date results, although it may provide useful background to better understand a more recent article. When comparing multiple articles in the same field, looking at the publication dates allows one to order the articles chronologically to recreate the sequential narrative of ongoing developments.

In fields where research progresses very quickly, journals publish new results online immediately after the peer-review stage but before going through copyediting and final formatting. For this reason, some articles have an *online publication date* that precedes the official publication date by a few weeks or months.

The citation count of an article measures its impact in terms of providing a fertile platform that generates further original results. Since the citation count increases with time, the true impact of an article must consider the citation count in light of the publication date. Some indexing services also provide the average citation half-life of articles in different fields to help evaluate their impact.

Title and abstract

Unless the article is open access, only the title and abstract are viewable free of charge. Journals impose strict word count limits for both title and abstract, so authors carefully craft them to provide enough information for readers to decide whether to invest time and money to read the rest of the article. For this reason, the title is often several lines long and explicitly describes the main features of the article. Similarly, the abstract summarizes all aspects of the article in one or two paragraphs.

9.2.2   Original research in applied mathematics

Applied mathematics articles typically feature a mathematical *model*, which usually falls into one of the following three broad categories:

1) **Mechanistic models.** These models make assumptions on the underlying mechanisms that govern an observed behavior. Differential equations, or their discrete cousins, difference equations and recurrence relations, form the heart of this type of model. See Subsections 7.1.3 and 8.2.2 for more information on the role of differential equations in mechanistic modeling.

2) **Empirical models.** Empirical models assume that the experimental data exhibit some variance caused by errors inherent to the experimental procedures or by the natural variability of the system itself. These models employ statistical tools to interpret results and to support or reject a hypothetical cause-and-effect relationships between variables. See Subsections 7.1.3 and 8.2.3 for more information on the role of statistics in empirical modeling.

3) **Operations research models.** Used in the context of decision science (or management science), these models describe the logistics of complex systems, usually involving many moving parts and large amounts of human and material resources. The goal of this type of model is to guide a decision maker toward the optimum (minimum or maximum) of an objective function and to identify the value of the decision variables to achieve the optimum. See Subsection 7.1.3 for more information about operations research.

Because mathematical modeling fits within the context of investigating a system by the scientific method, research articles that feature a mathematical model follow the standard article structure used in experimental science, with the sections: Introduction, Methods, Results, Discussion (often abbreviated IMRAD), which we discuss below.

Introduction

The introduction of a research article in the applied sciences serves four main purposes: (1) educating readers, (2) describing the scope of the research, (3) highlighting the practical relevance of the research, and (4) situating the research in relationship to other research results and articulating the originality of the current research. Each one is further discussed below.

1) **Education.** Applied mathematics articles are the fruit of a collaboration between researchers in different fields. Since the topic of research itself usually lies at the intersection of various mathematical and nonmathematical fields, the article must begin by situating the field of research. Furthermore, because the readers are likely to come from different backgrounds, the authors must educate target readers in all concerned fields to a certain extent. In particular, this information should allow those whose primary background is in mathematics to acquire enough knowledge in the field of application to understand the article, and conversely, nonmathematician readers to gain mathematical knowledge relevant to the article. This process of cross-education may be several paragraphs long and define specialized jargon, acronyms, ongoing research trends, and other useful information about the research topic. Here, the authors often provide references to books, websites, or other outside sources of learning, so that readers can consult them to fill their knowledge gaps as needed.

2) **Scope.** The introduction continues with a description of the scope of the research, which is often formulated as a scientific *hypothesis*, which the authors attempt to answer by a combination of experimental evidence and mathematical modeling. However, not all research articles have an explicitly stated hypothesis. Some develop a mathematical model that reveals a variety of interesting features about a system, without focusing on a unique research question.

3) **Practical relevance.** Although research involving applied mathematics can be motivated by mere curiosity, i.e., purely for discovery's sake, this type of research is usually motivated

(and funded) because it has some practical goal beyond mere scientific discovery. The introduction provides a rationale for the time and effort dedicated to the research by highlighting what the results seek to accomplish in practical terms: increasing profit, curing a disease, saving the environment, saving lives, or enhancing some aspect of the quality of life.

4) **Literature review and originality.** Finally, the introduction provides a comprehensive *literature review* to situate the article in relation to other research efforts. Here, the authors cite all the previous results used and describe how each one contributed to their research. The authors identify out a *research gap* they seek to fill, thereby laying a *claim of originality* for their own contribution.

Methods

Mathematical articles often divide the traditional "Methods" section into "Data," "Model," and "Solution Methods," to separate the data collection, mathematical modeling, and solution phases. We discuss each one below.

1) **Data.** If the authors conducted their own experiment(s) and collected their own data, they provide here a clear and precise description of their experimental procedures (sample size, materials used, etc.). If, instead, they used data collected by someone else, they provide here the reference(s) for the data source(s). The data itself, usually provided in the form of tables and illustrated by graphs, may be partially relegated to an appendix or supplementary material, especially if it is lengthy.

2) **Model.** This section is the mathematical heart of the article, often the only part that contains any mathematics at all. Here, the authors explain and justify the steps leading to the mathematical model, which may include

- the equations, along with comments regarding the choice for their mathematical form, i.e., linear vs. nonlinear, discrete vs. continuous, deterministic vs. stochastic,
- the type of model, i.e., mechanistic, empirical, operations research, or other,
- the variables, parameters, and effects in the model and those that are simplified or ignored altogether, with a justification supported by quantitative data, qualitative observation, or experience,
- hypothetical underlying mechanisms and interactions between variables,

- the assumed distribution of population data from which data samples were drawn.

3) **Solution methods.** Once the authors have defined their mathematical model, they must explain which mathematical methods are employed to solve it. Some models can be solved *analytically*, leading to an exact, closed-form solution. Others require a *numerical method* instead. If the authors used a computer (which is usually the case for numerical methods), they often mention the software and hardware used.

One of the essential requirements of valid science conducted under the scientific method is that experimental results be *repeatable*, meaning that anyone reproducing the experiment using the same methods must obtain the same results. The same requirements hold for the solutions of mathematical models, so the Methods section must contain sufficient mathematical and computational details so that any reader can replicate the model, solve it, and obtain the same results.

Results and discussion

The purpose of a mathematical model is not to be solved once to obtain a unique "correct answer," but rather to shed light on a real life phenomenon as a whole under a variety of scenarios. As such, researchers will use a model as an experimental tool to probe the system in different ways and analyze it under different angles, thereby generating additional computational results to complement the data collected from traditional experiments. Therefore, the authors will need to solve their model multiple times to simulate different experiments, scenarios, or case studies, each time changing parameter values, the number and combination of variables, or the form of the equations.

In the Results section, the authors provide the raw output of these analyses without commentary, usually in the form of tables and graphs. If they are lengthy, only the most significant results may appear here, the rest being relegated to an appendix or supplementary material.

In the Discussion section, which immediately follows, the authors comment on the results. Here, they interpret the results and explain how they respond to the research questions and goals set forth in the introduction by:

- drawing attention to which results are particularly interesting, unexpected, surprising, or counterintuitive,

- explaining how the results support or reject the hypotheses posed as research questions,

- explaining the practical implications of the results, and

- explaining how the results reveal original findings, thereby filling a gap in the research literature.

The discussion may also defend the model's validity by presenting the results of a *validation study*, where the authors "test drive" the model by simulating a previous traditional experiment and verifying that the model's results agree with the experimental results. Finally, the discussion may also include a *self-criticism*, where the authors evaluate the strengths, weaknesses, and limits of applicability of their own model.

Conclusion

Much like the Abstract, the Conclusion summarizes the main points of the article, from the Introduction to the Discussion of results, with a special focus on the practical implications and accomplishments. For this reason, readers will often read the conclusion of an article first to get the overall context and a broad overview of the narrative before reading from the beginning.

In the conclusion, authors typically include a list of suggestions for possible future research directions related to their research. For example, they may encourage interested readers to extend or generalize their model by suggesting which simplifying assumptions to revisit or which variables to add to make the model more realistic or more widely applicable.

9.2.3   Original research in pure mathematics

In comparison to research articles involving applied mathematics, the structure of pure mathematics articles is less rigid and exhibits more variability. As we discuss in Subsection 8.3.6, authors may obtain original results in pure mathematics by axiomatization, extension, generalization, or unification of existing results. Although each method gives rise to a different narrative structure, in all cases, the authors carefully lay out their results in a logical sequence, consistent with an axiomatic structure, defining evey term before using it and justifying every step of every proof by previously proved theorems and axioms.

Introduction

The Introduction of an original research article in pure mathematics shares some of the same goals as for applied mathematics, in particular (1) educating readers, (2) describing the scope of the research, and (3) situating the research in relationship to previous results and articulating the originality of the current research. Because research in pure mathematics is not necessarily driven by a specific application, these articles usually do not state a practical motivation behind the research, although some authors may mention applications, if they exist.

As another notable difference, the target audience of pure mathematics journals is typically much narrower than the audience of an interdisciplinary scientific journal. Authors who publish in specialized mathematical journals often assume that readers already have a solid background in their area of specialty and provide minimal background information before diving into complex details. Others take more time to situate the field of research, providing context and explanations regarding existing theorems and open problems. As in applied mathematics articles, the introduction contains a *literature review* and a claim to originality. Some authors may provide references to books or other sources of additional background information for interested readers to consult as needed in order to understand the article.

Main body

In a pure mathematics article, the structure of the main body reflects the axiomatic structure of its contents. Because new results in pure mathematics must be grafted into an established *mathematical theory*, the authors typically begin with the existing definitions and theorems, on which they build their own definitions and theorems.

The definitions and theorems of a new mathematical development are presented in a logical order dictated by how they relate to each other, ultimately fitting together like pieces of a puzzle. Some authors present a sequence of theorems, each one used as a step to justify the proof of the following one. Others present a set of independent small *helping theorems* (or *lemmas*) as "stepping stones" to prove a larger result. In this case, a set of lemmas culminates in a large important theorem, whose proof invokes the results of the previously proved lemmas.

The syntax of theorems, i.e., the way they appear in an article, is fairly standardized across all mathematical journals. Every theorem is numbered and introduced by the heading "Theorem" in bold, which may also mention its author and date. The statement of the theorem itself appears after this heading, followed by a proof. The proof begins with the word "Proof" in bold or italics, and ends with QED (which stands for *Quod Erat Demonstrandum*; "what was to be shown" in Latin), or a symbol like $\square$, called a *tombstone, dingbat,* or *Halmos.*[3]

---

[3]In honor of the 20th century Hungarian-born American mathematician Paul Halmos.

Although the proof is essential to justify a theorem, a reader often skips over it in a first reading in order to get a clear overall view of the article's structure and the connection between the various lemmas and theorems.

Corollaries and examples

A *corollary* of a theorem is a direct consequence or an interesting special case of the theorem that is self-evident or that requires only a simple proof. After proving a theorem, authors may provide one or more corollaries to illustrate an application of the theorem or to show how a generalized theorem reduces to a familiar, special case.

For example, the *law of cosines* states that, in a triangle with sides of lengths a, b, and c, with angle θ between a and b, the following holds:

$$c^2 = a^2 + b^2 - 2ab\cos\theta.$$

A corollary of this theorem is that for a right triangle, i.e., when $\theta = 90°$, then the conclusion of the theorem simplifies to

$$c^2 = a^2 + b^2,$$

which is the familiar Pythagorean theorem. Thus, this corollary illustrates that the law of cosines is a generalization of the Pythagorean Theorem.

Authors will often illustrate a definition or a theorem with one or more examples. In this case, they may end examples with $\triangle$ to clarify where they end and the regular flow of exposition continues.

Future extensions

Authors may end with one or more new conjectures, open questions, or suggestions for future extensions or generalizations.

9.2.4   Final sections

Acknowledgments and funding

In the Acknowledgments section, the authors acknowledge all persons who contributed to the article in any way, by providing, for example, expert advice, insightful ideas, corrections, proofreading services, data, or software tools. Furthermore, if they received financial support from a private organization, business, or government grant, the authors must acknowledge the funding source(s) as well as any conflict of interest that could have influenced their research, such as having a financial, commercial, legal, or professional relationship with certain organizations.

Bibliography

The Bibliography is a complete listing of all the works cited through-out the article. The format of this section depends on the citation style used by the journal (Subsection 10.4.3 contains more information about bibliographies and citation styles).

Appendices and supplementary material

Appendices appear at the end of an article, after the Bibliography, and contain supplementary material that may relate to any part of the article. Authors and/or journal editors often choose to move some of the article's content to appendices to prevent interrupting the flow of the narrative with details that are not essential to the overall understanding of the article, such as lengthy mathematical calculations, proofs, tables, figures, data, or computer codes. In some cases, the supplementary material does not appear in print and is instead further relegated to an external website.

Editors of nonmathematical journals often judge that all math-ematical details mentioned in an article are outside the scope of the readers' interest and relegate them to the supplementary material. Therefore, readers interested in the mathematical models themselves must often search for them in the supplementary material.

EXERCISES FOR SECTION 9.2

1. Choose a peer-reviewed original research article about a topic in a natural or social science (physics, biology, ecology, epidemiology, economics, etc.) that uses a mathematical modeling approach. Then answer the following questions in a few sentences.

   (a) **Authors.** Who are they? What are their research interests? How did they contribute to the article? Is there a "first author" or did all co-authors contribute equally?

   (b) **Journal.** What is the scope, frequency, and impact factor of the journal?

   (c) **Publication date and impact.** When was the article published? How many times has it been cited to date?

   (d) **Motivation and education of reader.** To what field(s) outside of mathematics does this research apply? Briefly describe the main issue, its importance, and the benefits of studying it. Give the full reference of at least one outside source that a reader can consult to gain more background on the topic.

   (e) **Goal of project.** What is the main research question and/or the main goal of the research project?

   (f) **Literature search and originality.** Give the full reference of at least three peer-reviewed, original research articles cited in your article, and explain how the authors use these previous results in their research.

(g) **Originality.** What aspect of this research is original? What research gap are the authors setting out to fill?

(h) **Data.** What quantitative or qualitative data about the real life system do the authors use to build their model? If the authors collected their own data, describe the experimental procedure. Otherwise, provide the full reference(s) of the data source(s).

(i) **Model.** What type of model is it? What are the variables, parameters, equations? What simplifying assumptions do the authors make?

(j) **Solution methods.** Do the authors solve the model analytically or numerically? If they use a computer, what numerical method, software, and hardware do they use?

(k) **Results and discussion.** What type of results do the authors obtain? Which ones are particularly important, surprising, or counterintuitive? How do the results support or reject the main research question? What are the strengths and weaknesses of the model?

(l) **Conclusion.** What future steps do the authors suggest to improve their model in the future?

(m) **Acknowledgments and funding.** Whom do the authors acknowledge and who funded the research?

2. Choose a peer-reviewed original research article in pure mathematics. Then answer the following questions in a few sentences.

(a) **Authors.** Who are they? What are their research interests? How did they contribute to the article? Is there a "first author" or did all co-authors contribute equally?

(b) **Journal.** What is the scope, frequency, and impact factor of the journal?

(c) **Publication date and impact.** When was the article published? How many times has it been cited to date?

(d) **Situating the topic.** What field or subfield of mathematics does this research fall into? How did this field originate? What is its history?

(e) **Definitions.** What definitions are provided in the article? Clearly indicate which definitions are original (i.e., created by the authors), if any.

(f) **Structure of the main body.** Describe the structure of the article, i.e., the sequence and hierarchy of the results. Clearly indicate which theorems in the article are original (i.e., proved by the authors), and provide the full reference for an outside source containing a past result mentioned in the article.

(g) **Corollaries and applications.** Are there any interesting corollaries or applications?

(h) **Conclusion.** What open questions or new conjectures do the authors suggest as future research directions?

(i) **Acknowledgments and funding.** Whom do the authors acknowledge and who funded the research?

9.3 Reading and Expositing Original Research Articles

Whereas the previous section described the external "anatomy" of a mathematical article, the present section offers specific strategies for tackling the content of the article itself.

Reading an original research article is an exciting adventure into a new territory that only a handful of people may have explored. A reader may even be the first to read the article aside from the reviewers. Reading primary literature in mathematics is also an excellent way to become familiar with the style and syntax of formal mathematical writing, which we present in more detail in Section 10.1.

However, reading a journal article is a challenging task that one must approach strategically. Research articles can seem intimidating and unwelcoming at first glance; their formal mathematical style, purposely cold and dispassionate, often differs from the more student-centered approach encountered in mathematics textbooks. Moreover, English may not be the authors' native language. Although articles undergo extensive peer review, this process does not necessarily involve much line or copyediting. Furthermore, authors often assume that readers have a certain background in their field and may refer to theoretical concepts, technical jargon, recent advances, acronyms, abbreviations, or notations that are unfamiliar to a nonspecialist reader.

The goal for a first-time reader is not to become a specialist himself or herself, but rather to extract the general idea of an article while building up the necessary background on the fly. To this aim, we suggest several strategies in Subsection 9.3.1: *intensive reading, backward learning, preliminary research,* and *citation chains.* Subsection 9.3.2 describes the process of *expositing* a research article, i.e., explaining its contents to a broader audience. This exercise requires that the reader simultaneously engage the original article at a deep enough level to understand its content and that he or she then synthesizes the material to clearly explain it to others.

9.3.1 Reading an original research article

Intensive Reading

As we discussed in Section 7.3.1, the most effective way to approach a mathematical text is by a combination of *extensive reading* and *intensive reading.*

Also called "reading for gist," extensive reading consists in moving through the text relatively quickly, in order to apprehend the main ideas, the flow of the narrative, and the general structure of the paper. This type of reading is, by its nature, superficial. Although it exposes the reader to general concepts, it does not promote mathematical learning. Therefore, it must be followed by intensive reading [7, 12, 16, 22].

Intensive reading involves a complete analysis of the text in detail to absorb as much information from it as possible. This requires fully engaging with the text, atteding to every detail, analyzing every word in every sentence, and pausing to look up every concept that is not completely understood. It also involves frequently consulting external resources such as dictionaries and encyclopedias. Depending on the reader's background and complexity of the text, the pace of an intensive reading can be extremely slow and mentally taxing, as the reader may have to pause to look up almost every word and learn the background of certain concepts to fully understand them.

Backward Learning

In the course of an intensive reading, it may become apparent that to fully understand a concept described in the article, the reader may need to consult more than just dictionaries and encyclopedias. For example, he or she may need to find and read one of the articles cited in the introduction, delve deeper to learn about a particular mathematical concept or solution method mentioned in the article. In these cases, the reader is working backward to build his or her background as needed to fully understand a specific article. This process is called *backward learning.*

Backward learning differs from the traditional learning approach of a typical "forward" learning that occurs in a traditional mathematics course, which progresses sequentially according to an organized curriculum, often following a textbook, progressing through its sections one by one, in a prescribed order. In forward learning, the student is exposed to theoretical concepts and methods from the ground up, from the general to the specific, starting with general concepts and definitions, progressively moving toward more complex and specialized branches, and finally studying specific examples and applications. Backward learning reverses the process: the student begins with identifying a specific application or extension as the goal, then building knowledge backward to achieve that goal, from the specific to the general, collecting information as needed, on the fly, to bolster the foundation needed to approach a specific result.

Backward learning is an essential skill when reading primary literature. Because a research article is typically written from the point of view of an expert in a very narrow, specialized, possibly multidisciplinary field, even readers with extensive experience and knowledge in related fields may have to engage in some backward learning to fully understand the article. Backward learning is a common way to acquiré specialized knowledge when forward learning is impossible or impractical.

Backward learning does not suddenly make a first-time reader a specialist, but it does allow him or her to gain understanding of a specific article. It has several other advantages:

- it provides an entry point into a new field, which the reader can then continue exploring beyond the scope of the article,

- it allows the reader to more easily understand other articles in the same field,

- it allows someone new to a field to follow conversations between specialists and engage them by asking relevant questions and using newly acquired specialized jargon.

As a reader engages in backward learning via dictionaries, encyclopedias, websites, other research articles, books, and other resources, he or she may quickly end up collecting a large number of references (articles, webpages, etc.). For this reason, it is helpful to work with a *reference manager* (see Subsection 9.4.3) to easily find them later, and if necessary, compile them into a Bibliography.

The following sections can help a reader organize the backward-learning process.

Preliminary research

Before reading an article, it is essential to do some preliminary research on the authors, journal, and date of publication.

- **Authors.** Researching the authors may reveal other articles on the same topic. Furthermore, learning about the authors' background and research interests helps to understand their point of view. If their contact information is available, it may even be useful to contact them directly with questions about their article. Authors are usually delighted to discover that someone is interested in their research and often answer enthusiastically.

- **Journal.** Researching the journal's scope indicates what type of research it prioritizes and sheds light on why certain aspects of the research are highlighted in the article. The journal's metrics also give an idea of its importance.

- **Date.** The publication date should be noted, especially in fast-moving fields of research, where results may no longer be considered new after a few years.

The field

A nonspecialist reader may need to invest a considerable amount of time learning about various aspects of a new or unfamiliar field of research. One may begin by asking the questions below to put the article in context and understand the authors' point of view and their intended audience.

- What is the history of this field? When did it begin? Who were the first research pioneers in this field?

- Is there any specific reason (other than mere curiosity) that caused this field to emerge?

- What past results in this field that have become important milestones?

- What open problems are this field's researchers working on?

If the article features a mathematical model, one may ask the following additional questions regarding the field of application:

- What is the main motivation for doing research in this field?
  - Is it mere discovery/curiosity, i.e., is the goal merely to understand how some aspect of the real world works?
  - Do the results provide an economic or logistical benefit?
  - Do the results provide another, less tangible, benefit to the world or to humanity, such as protecting the environment, healing a disease, increasing the quality of life, reducing poverty or human suffering?

- When did this field become mathematized? What are the advantages and challenges of using a mathematical approach in this field?

- Are there other applications (other than the one mentioned in the article) for the methods described in the article?

- Are other types of mathematical methods (other than the one(s) described in the article) used in this field of application?

- Who is funding research in this field? Why?

Terminology

Every field of study develops its own *jargon*, which specialists use when communicating with each other with precision. Though perfectly natural to specialists, jargon-loaded conversations may sound like a secret and incomprehensibe language to outsiders. Because original research articles are often written to an audience of specialists, they usually contain a large amount of jargon, which can make even the title sound intimidating. It is essential that nonspecialist readers familiarize themselves with the jargon before attempting to decipher the sentences.

Learning jargon is an important, albeit time-consuming, element of backward learning. The process requires stopping to look up and take notes about every unfamiliar word, phrase, and acronym. The first step when researching the meaning of an unfamiliar term is to look it up in a dictionary or encyclopedia. For mathematical terms, specialized references such as *Wolfram MathWorld*[4] or *Springer's Encyclopedia of Mathematics*[5] can yield more detailed information than an all-purpose search engine. To fully understand a term, it may be necessary to look up other terms in its definition, working backward until one reaches familiar ground. Section 7.3.2 discusses the critical importance of learning precise mathematical definitions and the additional, deceptive difficulty when a mathematical concept borrows a term from everyday English.

Citation chains

Research articles typically build upon the ideas contained in past articles, which provide inspiration and launching points for future research. Therefore, articles in the same research topic can be arranged into a chronological chain, in which each article builds upon the ideas in a previous article. An interested reader may find it helpful to study other articles in a citation chain: past articles help understand the context, and future articles help measure the impact. A reader may also look further down the citation chain to find similar information more clearly presented.

The *literature review* in the article's Introduction contains references to past articles, which provide context, background, and motivation for the current research, while highlighting the research gap that supports the authors' claim of originality. On the other hand, a reader may research the future sources that cite an article, whether the are original research articles, review articles, expository articles, or books. Authors often conclude their article by suggesting future extensions to their research; studying future work may reveal whether anyone extended the research along one of those suggested avenues.

Some articles become seminal articles with thousands of citations, eventually becoming landmark publications in their field. Others may have very few or no citations at all. This does not necessarily mean that the article is poor; it may simply be too recent for anyone to have had the time to produce any further original results.

A reader will need the engine of an indexing service to find and access the articles of a citation chain. These tools can also order the references chronologically or by their own impact. We discuss scholarly search engines and indexing services in Subsection 9.4.3.

---

[4] www.mathworld.wolfram.com
[5] www.encyclopediaofmath.org

9.3.2   Expositing an article

An excellent exercise that promotes comprehension of a research article consists in *expositing* its content. As teachers and tutors can attest, explaining a difficult concept to another person ends up solidifying one's own grasp of the material. Furthermore, mathematicians frequently play an expository role, often being called upon to explain their work to people who do not share their specialized background.

Expository papers and presentations

As secondary literature, expository papers refer to results stated in primary sources, but do not contain any novel results themselves. An expository paper may provide an alternate proof for an already proved theorem or suggest an alternate solution method to approach a problem. Another motivation for writing an expository paper is making the contents of a primary article more accessible to a broader audience. Because primary sources are intimidating and difficult to penetrate, nonspecialists often find it easier to approach novel results via expository papers and presentations aimed specifically at a broader audience. Expository authors explain the main ideas contained in an original research article in layman's terms and will often include additional background information, definitions, and comments, in an effort to make their article more accessible. Similarly, for the sake of clarity and conciseness, an expository paper or presentation often omits certain very technical details of the original article.

An expository author decides how much detail to include depending on the mathematical background, interest level, and goals of the target audience, which may be the general public, undergraduate students, graduate students, or specialists in a different field. In interdisciplinary research, it is common to write expository pieces for people who are unfamiliar with one or more aspects of an interdisciplinary field, such as mathematicians who are not familiar with the field of application, or vice versa.

Original research can be exposited as a written article, an oral presentation, or both. Expository articles appear in peer-reviewed journals but often also in non peer-reviewed magazines, newsletters, and websites.[6]

---

[6]A good example of an expository resource is the free online magazine *Quanta Magazine* (www.quantamagazine.org), which publishes expository articles in mathematics, computer science, physics, and biology. Each article is written for a general audience with minimal background, but also provides links to the original articles to which it refers.

Goals

An expository author plays the dual role of learner and teacher. On one hand, he or she must be sufficiently comfortable with the process of scholarly communication to research and acquire knowledge directly from a primary source through backward learning and intensive reading. On the other hand, he or she must be capable of explaining the main concepts clearly to their target audience. To this effect, an expository author faces three main tasks: (1) background teaching, (2) highlighting, and (3) omitting.

1) **Background teaching.** Expositing research may require a significant amount of teaching the background, motivation, terminology, and previous results, before delving into the content of the original article itself.

2) **Highlighting.** An expository author judges which parts of the original article are particularly relevant, and which are not. He or she extracts and highlights the main idea(s), important key steps, and most significant results from the original article.

3) **Omitting.** An expository author may decide to omit certain nonessential portions of the original research (such as proofs, corollaries, or solution methods), if they do not compromise the overall understanding of the main points.

Depending on the additional background information and omitted material, an expository paper may be longer or shorter than the original article to which it refers. The following examples show how a short sentence in a primary research article may require a longer explanation in an expository paper.

Example 9.3.1. Sentence in original article:

"We use RKF45 to solve the ODE."

Expository article:

"The researchers use a method known as the *Runge-Kutta-Fehlberg-4-5* method (abbreviated RK45) to solve the ordinary differential equation (ODE). This numerical method consists in discretizing the continuous time variable t into discrete time steps (Δt), leading to an approximate solution." $\triangle$

Example 9.3.2. Sentence in original article:

"We use our model to study the interactions of proteins in the ER of pancreatic beta cells."

Expository article:

"The researchers use their model to study cells of the human body located in the pancreas. These cells, called *pancreatic beta cells*, store and produce insulin, a hormone which plays a role in regulating blood sugar levels. In particular, the model focuses on protein interactions in the *endoplasmic reticulum* (ER), which is the part of these cells where insulin is synthesized."   △

Style and syntax

With the exceptions noted below, expository papers follow the syntax and style of formal mathematical writing outlined in Section 10.1, and expository oral presentations follow the guidelines for mathematical presentations presented in Section 10.5.

- Whereas primary literature is often written in the first person ("we"), expository papers use the third person to refer to the authors of the original paper (as in the examples above).

- Whereas primary literature uses the present tense, expository papers often refer to the original authors' work in the past tense, especially when referring to data collection or background research that the original researchers conducted prior to establishing their results.

EXERCISES FOR SECTION 9.3

1. Below is the Abstract of an original research article titled, *Buckling and Collapse Analysis of Embedded Carbon Nanotubes*. Read the title and Abstract intensively, looking up every unfamiliar or unclear word and phrase (you may need to do some backward learning to properly understand some of the concepts). Then write a short paper that exposits this Abstract to an audience of undergraduate mathematics students.

> We present a mathematical model to analyze the deformation, buckling, and fracture modes under compression of carbon nanotubes embedded in a polymeric membrane. The model, which was calibrated by experimental measurements, consists of a second order ordinary differential equation based on Euler's theory of elastic instability. Results show that the compressive strengths of thin- and thick-walled nanotubes are found to be about 2 orders of magnitude higher than the compressive strength of any other known fiber.

2. Below is the Abstract of an original research article titled, *Finite element modeling of the mitral valve*. Read the title and abstract

intensively, looking up every unfamiliar or unclear word and phrase (you may need to do some backward learning to properly understand some of the concepts). Then write a short paper that exposits this abstract to an audience of undergraduate mathematics students.

Finite element modeling represents an established method for the comprehension of the mitral valve function and for the simulation of interesting clinical scenarios. However, current models still do not include all the key aspects of the real system. We implement a new three-dimensional structural finite element model to examine deformation and stress patterns in the mitral valve under systolic loading conditions. Our model incorporates all essential anatomic components, regional tissue thickness, collagen fiber orientation and related anisotropic material properties. Computational results agree to a great extent with experimental data from the literature. The results provide insight into some of the features characterizing normal mitral function, such as contraction of the mitral annulus and tissue anisotropy. Some of the computed results may be useful in the design of surgical devices and techniques. The model can thus aid both the surgeon and the biomedical engineer in improving the materials and techniques available for the repair and/or replacement of the mitral valve.

9.4 Researching Primary and Secondary Sources

The previous three sections attempt to demystify the world of primary literature and offer guidance on how to approach deciphering a journal article. Learning to read and learn from this type of literature is a skill developed over time, called *scientific literacy*. This skill fits within the broader framework of *information literacy*, which includes knowing not only how to read articles, but also how and where to look for specific information that is relevant, reliable, and up to date.

Information literacy has undergone a fundamental shift in recent decades with the coming of the Information Age. The advent of computing has revolutionized the way people collect, share, and store information, causing an exponential increase in the quantity of available information. The Information Age has also transformed the role of *libraries*, which have taken the central role of facilitating the flow of information, a complex task that involves managing, organizing, and storing vast amounts of information and making it accessible to users.

Subsection 9.4.1 describes the characteristics and unique challenges brought forth by the Information Age. Subsection 9.4.2 describes the essential roles of libraries in the process of scholarly communication and the services they offer to users. Subsection 9.4.3 describes the essential role of the internet and online tools for information management. Finally, Subsection 9.4.4 discusses the challenges associated with the cost of access to scholarly material.

9.4.1   The Information Age

Starting in the mid-20th century, the *Information Age* (or *Digital Age*) is the historical era in which we live, characterized by a shift to an economy based on collecting, storing, sharing, and exploiting information rather than material products. The world's technological capacity to store, transmit, and compute data has increased exponentially since the mid-1980s. Scholarly literature has followed this trend, with more than 30,000 active journals in the world today, publishing more than 1 million articles every year. In addition to several hundred journals devoted exclusively to mathematics, there are several thousands more devoted to the natural and social sciences, many of which contain original results in applied mathematics. The quantity of existing mathematical knowledge contained in primary literature and the rate at which new mathematical knowledge is being generated are mind-boggling.

The sheer amount of mathematical knowledge currently available poses a serious challenge for anyone trying to find anything. Locating a specific piece of mathematical information amounts to finding a needle in a haystack. Moreover, conducting a search for a more general mathematical topic is likely to return thousands of relevant articles, all published in peer-reviewed journals in recent years. How does one choose where to start? In the Information Age, the challenge has shifted from being able to find any information at all, to being able to sift through vast amounts of information and discern its quality and reliability.

Fortunately, the Information Age has also brought forth a wide range of specialized tools to help search for information efficiently and systematically. Various search engines and online databases help researchers search for specific information and organize their research materials.

The Information Age has also shifted the role of libraries and the profession of librarians from managing a repository of knowledge contained in physical books to managing information available on the internet. Academic libraries are equipped with specialized search tools to access and locate specific information in primary and secondary sources. As highly trained professionals, academic librarians partner with researchers by leveraging powerful information-managing tools to facilitate the dissemination and preservation of knowledge.

9.4.2 Libraries

Libraries have always played a central role in organizing, disseminating, and storing information. Historically, libraries were centered around a collection of physical books. With the rise of means for digital storage and transmission of information, libraries have become the gateways to information available in digital form. One can classify libraries into several types according to their primary purpose:

- **Public libraries.** The primary purpose of public libraries is serving the needs of the general public. Their mission is to educate and promote self-learning with a special emphasis on programs for children and youth. Public libraries are *lending* (or *circulating*) libraries, meaning that most of their materials are available for certain patrons to check out for a period of time.

- **Academic libraries.** Housed at an institution of higher education (college or university), these libraries serve the research and education needs of the students and faculty. We discuss these in greater detail below.

- **National libraries.** These reference-only libraries are established by the nation's government as a central repository of information. Some national libraries are vested with the right of *legal deposit*, meaning that all publishers in that country are legally required to deposit a copy of every publication with the library. For example, the *Library of Congress* serves as the national library of the United States. Similarly, the *United States National Library of Medicine*, managed by the U.S. *National Institutes of Health*, holds copies of every medical publication in the world. National libraries are among the largest libraries in existence, some housing millions of physical books, and many times more information in digital format.

- **Private libraries.** Private institutions, such as businesses, law firms, churches, hospitals, museums, and research laboratories, may establish a library to serve their own specialized research and educational needs.

Academic libraries

The primary purpose of an academic library is to support the educational and research mission of their institution. Academic libraries manage facilities for storing information in physical and digital format and provide the necessary hardware and software for users to access digitally stored information. In addition, academic libraries facilitate users' interaction with their materials by providing space for study and research, such as reading rooms for individual study and consultation of library materials and private meeting rooms for collaborative work. Unlike public libraries, academic libraries do not make their materials accessible to the general public and only a portion of their holdings is circulating.

An academic library may be divided into departments, sometimes across several buildings, each one with its own staff of specialized librarians. In addition to the standard academic divisions, some libraries boast one or more *special collections* of rare and valuable books or reference materials on a specialized topic.

In addition to providing educational support and customer service, the main tasks of academic libraries consist in managing their (1) book collection, (2) journal subscriptions, and (3) database subscriptions.

1) **Book collection.** Libraries organize and maintain the collection (both physical books and digital e-books) up to date.

2) **Journal subscriptions.** Libraries manage journal subscriptions and access to journal articles. The majority of journals only exist in digital format nowadays, but a library may hold some print issues.

3) **Databases.** Libraries invest in subscription-based reference databases, journal indexes, and specialized search engines. Such systems enable users to find specific information and organize a systematic and comprehensive topical search.

Managing library holdings, whether books, journals, or databases, encompasses the following aspects:

- **Development.** A library is constantly developing its collection of books, journals, and databases to best serve the needs of their institution. This includes deciding which materials to keep and in what format (physical or digital), which to purchase, and which to discard.

- **Organizing.** Another important aspect of a library's work consists in classifying and cataloging the books, journals, and databases so they are discoverable to users.

- **Assessment.** Academic libraries constantly assess their resources and make adjustments to ensure they use their budget in a fiscally responsible way. To this end, libraries gather usage statistics and calculate cost per use of their resources to guide their decisions on collection development.

In the past, libraries operated largely independently from one another. Now, libraries take advantage of their ability to share digital information by partnering with other libraries worldwide. Academic libraries subscribe to several *union catalogs* that itemize the collections of other libraries (such as *WorldCat*, with over 17,000 libraries in 123 countries), and they make items from other libraries accessible to their users via an *interlibrary loan*.

Academic librarians

Academic librarians are tasked with developing, organizing, and assessing the collection of books, journals, and databases, as well as providing educational resources and customer service to students and faculty. In particular, academic librarians facilitate the dissemination of scholarly information as it becomes available in scholarly journals, providing expertise to search, find, and access specific and up-to-date information. As experts in all aspects of scholarly communication, they are knowledgeable in journal metrics, publishing policies, copyright, intellectual property, author rights, and other issues pertaining to online digital scholarship.

Academic librarians typically hold at least a graduate degree in *library and information science*, and many hold additional credentials in one or more areas of specialty. At many institutions, academic librarians enjoy faculty status, therefore having the same privileges and expectations as professors in terms of producing scholarly work, publishing, developing curricula, and playing an active role in shaping the long-term academic direction of their institution.

9.4.3 Online tools

Conducting an internet search effectively requires the right set of tools, especially when searching for specialized information in a narrow field that it is relevant, reliable, and up-to-date. All-purpose search engines like *Google* tend to dredge up large amounts of irrelevant and low-quality material. Such a tool can provide preliminary information on a general topic, but a more pointed search requires a specialized scholarly search engine.

Scholarly search engines

A *scholarly search engine* functions like a standard online search engine, but it narrows down search results to scholarly material only, such as peer-reviewed journal articles and academic books. Users can also specify other search parameters and filters, such as the publication language or date range. The following scholarly search engines are all free to any user:

- **Google Scholar.** This user-friendly interface managed by Goog
  can be linked to a specific academic library to show which resources are accessible through the library's journal and index
  subscriptions.

- **BASE.** Operated by the Bielefeld University Library in Bielefeld, Germany and managed by academic librarians, Bielefeld
  Academic Search Engine (BASE) is continually curating its system in response to academic needs. It also allows verified authors to interact with the system to point out mistakes or updates in specific search items.

- **ArXiv.** This is a worldwide central repository for *preprints*
  in mathematics, physics, and computer science. See Subsection
  9.1.5 for more information on preprints.

Although scholarly search engines filter out most nonscholarly material, not all of them are successful at filtering out *predatory journals*,
i.e., fraudulent publications that pose as *bona fide* peer-reviewed journals, but whose articles are not submitted to any standard of quality
or accuracy whatsoever.

Indexes and Databases

An *indexing service* (or *index*) manages a *database* of scholarly material (mostly journal articles and books), provides specialized tools for
locating specific information in the database, and guarantees a level
of information quality and reliability. Some databases are free, but
most can only be accessed by paying a yearly subscription. Academic
libraries typically subscribe to several databases.

Hundreds of different databases exist and reputable journals opt
to be indexed in several of them to increase their prestige and visibility. Each indexing service operates a sophisticated search engine
that allows subscribers to search the database by a variety of criteria related to every article's *metadata* (title, author(s), affiliation(s),
journal, issue/volume/page numbers, keywords, and abstract). Indexes also keep up-to-date the citation count of each article in their
database and use this information to compute various citation-based
metrics for each journal, article, and author. (See Subsection 9.1.3
for more information on indexes and citation-based journal metrics).

The following are the most common indexes for mathematics and
science:

- **MathSciNet.** Managed by the American Mathematical Society, this database contains the largest collection of mathematics
  articles in the world. MathSciNet reviews every article to verify
  its correct classification according to the *MSC2020 Mathematics
  Subject Classification System* (see Subsection 7.1.3).

- **Web of Science.** Operated by Clairvate Analytics, Web of Science specializes in scholarly sources in science and mathematics. Web of Science's uses its citation data to calculate the *impact factor* metric for each journal in the database.

- **Scopus.** Owned and operated by Elsevier (one of the largest publishers of scientific journals), Scopus specializes in science and mathematics. It provides the CiteScore (CS), SCImago Journal Rank (SJR), and Source Normalized Impact per Paper (SNIP) citation-based metrics for each journal in the database.

- **ScienceDirect.** Like Scopus, ScienceDirect is operated by Elsevier and focuses on science and mathematics.

- **PubMed.** Managed by the United States Library of Medicine, PubMed features the largest collection of medical publications in the world, many of which contain original results in mathematics applied to medical-related fields.

Working with an index can simplify the work of a research in the following ways:

- **Topical search.** A user can search for articles in a specific topic by specifying words that appear in the title or abstract, or author-chosen keywords pertaining to the article.

- **Ranking search results.** Database search engines can rank the results of a topical search by publication date, citation count, or citation-based journal metric, thereby allowing a user to organize articles chronologically or by relative impact.

- **Citation chain search.** Indexes link every article in the database to all past articles it cites and to all future articles that cite it, thereby allowing a user to trace the chronological development of ideas by recreating a citation chain.

- **Alerts.** Some indexes allow users to sign up for email alerts to be notified of new items in a topic of interest. This feature allows subscribers to stay on top of new advances in a rapidly evolving field of research.

Reference managers

When researching a topic with search engines and databases, one can quickly end up with a large number of articles, books, websites, and other references, all of which must be stored and organized in a way that makes it possible to quickly retrieve the information. *Reference managers*, such as *EndNote*, *Mendeley*, and *Zotero*, are computer applications designed for this task. These systems collect the *metadata* provided by online databases for each of their items (i.e., title, author, journal, date, etc.) and organize each reference in a personal library searchable by metadata criteria, much like a personalized indexing service. Reference managers can also compile the references into a Bibliography in various citation styles or into BibTeX format.

Most reference managers also function as *scholarly collaboration networks*, social media platforms that allow users to collaborate and share their research with each other (see Subsection 7.2.2).

9.4.4 Access

Accessing new publications, especially articles in peer-reviewed journals, can be expensive. Journals typically maintain a listing of article titles and abstracts, but the full texts are hidden behind a *paywall*. Users must pay a *pay-per-view* (PPV) fee, typically $20-$100 per article. Academic libraries may have agreements in place with certain journal publishers for a discounted PPV fee, or may purchase yearly subscriptions for unlimited access to commonly accessed journals. These subscriptions may cost over $10,000 for a single journal.

The cost to access scholarly material nowadays is at an all-time high and continues to rise, a phenomenon called the *serials crisis*. Even though most journals have become fully online, subscription costs continue to escalate at many times the rate of inflation. The reasons for this are complex; one of them stems from the peer-review model itself, by which a journal publisher who publishes an original result holds a temporary monopoly over the original idea (until the idea is cited in another publication), and is therefore free to set its price unrestrained by any competition. Furthermore, journals often force authors to surrender the copyright to their work, preventing them from freely sharing their own research with colleagues.

Academic libraries struggle to keep up with this cost increase caused by the serials crisis. They are often forced to cancel journal subscriptions or to pass on a portion of access costs to users, effectively limiting the audience for research. Professional researchers, who rely on accessing the very latest results published in their field for their work, often have to allocate a substantial amount of their research budget to access articles.

Open Access

Created in the 1990s as a reaction against the serials crisis, *Open Access* (OA) is a model where the journal charges the publication costs to the author(s) instead of the reader(s). Under OA, the authors (or their institutions) pay a $100-$3000 *article processing charge* (APC) to the journal publisher for the article to become permanently accessible to the public for free. Academic libraries and researchers heavily promote the OA model; OA journals are gaining in prestige, and OA publishing overall has seen rapid growth in the past few years. See Section 9.1.4 for more information on open access.

EXERCISES FOR SECTION 9.4

1. Do some research on your institution's library:

   (a) Is there a separate library for mathematics?

   (b) Does the library have a dedicated librarian for mathematics?

   (c) Does your library subscribe to any of the following databases: MathSciNet, Web of Science, Scopus, ScienceDirect? Does it subscribe to other databases that focus on science and mathematics?

2. Choose a topic of your interest in applied or pure mathematics. Use a database to find the following references related to your chosen topic:

   (a) Five peer-reviewed journal articles published in the last 12 months. Then rank them by citation count.

   (b) Five peer-reviewed journal articles cited at least 100 times. Then order them by publication date.

   (c) Three peer-reviewed journal articles whose full text is accessible through your institutional library's journal subscriptions.

   (d) Three peer-reviewed journal articles that are open access.

3. Choose a peer-reviewed journal article in a topic of your interest in applied or pure mathematics. Use a database to find the following references:

   (a) Five peer-reviewed journal articles that are cited in your chosen article.

   (b) Five peer-reviewed journal articles that cite your chosen article.

CHAPTER 10

Writing and Presenting Mathematics

In a 2019 podcast entitled, *Algorithms, Complexity, Life, and The Art of Computer Programming*, original programmer of LATEX Donald Knuth quipped, "A good technical writer, trying not to be obvious about it, says everything twice: formally and informally. Or maybe three times." Just like every field of human inquiry, mathematics possesses its own unique methods of and culture around communication within the discipline.

This chapter focuses on writing formal documents with proper mathematical style and structure. Like all other academic fields, mathematics uses its own lexicon of specialized terminology, and mathematical prose conforms to a specific style. Whether the problem involves building a model or proving a theorem, it is essential that the student communicate the steps to others with clarity and precision. Section 10.1 outlines the features of formal mathematical writing. Section 10.2 offers guidelines to write and structure a project report for a mathematical project.

Writing mathematics on a computer presents a special challenge, as standard word processors are ill-suited for embedding mathematical symbols into regular text. For this reason, instead of a word processor, mathematicians universally use a markup language called LATEX, a system designed by mathematicians for mathematicians that offers great flexibility for structuring and typesetting mathematical documents. Section 10.3.1 introduces the basic features of LATEX, and Section 10.4 presents its more advanced capabilities, such as embedding arrays, tables, figures, and bibliographies. Finally, Section 10.5 focuses on designing a visual support (presentation slides or poster) for an oral presentation.

10.1   Mathematical Writing

Elementary mathematics classes do not focus much on writing. Instead, the main activity often reduces to solving problems by executing calculations that lead to some predefined "right answer." In upper-level mathematics courses, the focus shifts away from merely *doing* the math to *explaining* how it is done. As the mathematics student gathers specialized knowledge and increases in mathematical maturity, he or she will be required to communicate in writing how their mind proceeds through the solution of a mathematical problem or the proof of a theorem.

Learning how to communicate ideas effectively is a central part of doing mathematics for several reasons. First, explaining the steps of a problem sequentially is beneficial for learning, as it solidifies knowledge of problem-solving methods and sequences of deductive reasoning. Second, mathematical writing opens the door to collaboration and interdisciplinary work. Problem solving is becoming increasingly collaborative and interdisciplinary, requiring collaborators to communicate their problem-solving approaches effectively to each other. Finally, mathematical writing sharpens the ability to express oneself with clarity, precision, and succinctness, three hallmarks for which mathematics graduates are prized in any profession, as these features often transpire into the way a mathematically adept professional writes business letters, emails, and informal memos.

Undergraduate mathematics students have certainly been exposed to a great deal of informal mathematical writing, such as notes taken in class during a lecture or scratch paper used to write assignments. However, students at this level often have much less exposure to the formal mathematical style that appears in formal papers, assignments, reports, journals, and textbooks. A good first step toward learning correct mathematical writing is to familiarize oneself with the proper style and syntax by reading these types of written works.

Some aspects of formal mathematical writing are similar to other types of academic writing. Like formal academic documents written in the sciences and humanities, the student must write mathematical documents with flowing prose, in full sentences, observing proper grammar and punctuation, and following a formal structure. However, mathematical writing differs from other types of formal writing in its specific use of tense, person, mathematical symbols, and the unique structure of its documents.

Subsection 10.1.1 outlines main features of mathematical writing. Then, Subsection 10.1.2 describes the elements of mathematical style and syntax. Finally, Subsection 10.1.3 describes how to apply these guidelines to the types of documents required in upper-level mathematics courses.

10.1.1 Features of mathematical writing

Rhetoric vs. symbolic mathematical writing

Modern mathematics is a unique field in its use of specialized symbols to communicate ideas. Symbolic mathematical writing has the advantage of making ideas clearer and more concise. Furthermore, by transcending language barriers, standardized symbols enable mathematicians of all languages and cultures to communicate and understand each other's thoughts.

Before the 15th century, mathematics was entirely written in words (with occasional abbreviations), in what is called *rhetorical* (as opposed to *symbolic*) syntax. After the first symbolic notations appeared, it took about 200 more years for the more common mathematical symbols to become sufficiently standardized so that everyone would understand their meaning. For example the "=" symbol, which everyone now understands as meaning "equals", was only popularized in the late 17th century by Newton and Leibniz [10].

Example 10.1.1. The following two paragraphs illustrate the difference between rhetorical and symbolic syntax.

"To find the roots of a quadratic equation, first we find the discriminant by subtracting four times the product of the coefficients of the quadratic and constant terms from the square of the coefficient of the linear term. The two roots are computed by alternately adding or subtracting the square root of the discriminant thus obtained to the opposite of the linear term's coefficient, and finally dividing the resulting expressions by twice the coefficient of the quadratic term."

and

"The values of x that satisfy the equation $ax^2 + bx + c = 0$ are $\dfrac{1}{2a}\left(-b \pm \sqrt{b^2 - 4ac}\right)$."

The two paragraphs convey the same mathematical calculation, but the second benefits from more clarity and succinctness thanks to symbolic notation.   △

Nowadays, students are very familiar with symbolic mathematics, to the point of sometimes losing sight that the essence of mathematics lies in the abstract concepts represented by the notation, not in the notation itself. In elementary classes, students often submit assignments written entirely in symbolic form, without a single English word. In upper-level mathematics, however, mathematical symbols are not able to fully express all the complex ideas and concepts behind a mathematical computation or proof. It is necessary to use English words, in conjunction with mathematical symbols, to convey a mathematical explanation in the clearest way possible.

The following paragraph is a poor attempt at explaining how to find the points on the line $y = x$ that are at a distance of 5 from the point $(2, 1)$.

Example 10.1.2.

$$(x - 2)^2 + (x - 1)^2 = 5 \cdot 5 = 25$$

$$(x - 2)^2 + (x - 1)^2 = x^2 - 4x + 4 + x^2 - 2x + 1 = 25$$

$$2x^2 - 6x - 20 = 2(x + 2)(x - 5)$$

$$x = -2, 5 \Rightarrow (-2, -2), (5, 5) \qquad \triangle$$

Although the last line seems to conclude (correctly) that there are two points $(-2, -2)$ and $(5, 5)$, the paragraph is by no means a clear explanation. It is, at best, a disorganized set of "scratch paper" notes used to arrive at the final solution.

In upper-level mathematics, conveying a clear explanation of how the solution is obtained is more important than merely "getting the right answer." In most cases, this cannot be accomplished entirely in symbols; some English words are required to guide a reader through the writer's train of thought. On the other hand, it may be possible (and even preferable) to omit steps of routine algebra and calculus that a reader can accomplish without guidance, as too much detail can muddle an explanation and make it tedious and unclear. Keeping these two guidelines in mind, we may rewrite the previous example as follows:

Example 10.1.3. To find the points on the line $y = x$ that lie at a distance of 5 from the point $(2, 1)$, we substitute the given information in the distance formula, which yields the equation

$$(x - 2)^2 + (y - 1)^2 = 5^2.$$

Since the points we seek are on the line $y = x$, we substitute y for x in this equation. After some algebra, the equation becomes

$$2(x + 2)(x - 5) = 0.$$

The solutions of this equation are $x = -2$ and 5. Therefore, the two points we seek are $(-2, -2)$ and $(5, 5)$. $\qquad \triangle$

Language

In the mid-20th century, English gradually became the *lingua franca* of international communication, scholarship, and science. Therefore, all rhetorical (nonsymbolic) portions of mathematical writing must be in English. It is important to keep in mind that English is not the first language of most readers and writers of mathematics. Thus, a mathematical document written in English is likely to have been written by a person whose first language is not English and primarily read by people whose first language is not English, either.

English has not always been the *lingua franca* of mathematics; in fact, the trend is quite recent. In Western mathematics, French, German, and Russian were the most frequently used languages of mathematical writing throughout the 19th and early 20th century, and Latin was the undisputed *lingua franca* of mathematics and science for many centuries before then.[1]

Mathematical style

Good mathematical writing is direct, concise, and clear. Its distinctive style is best described as dispassionate, nonemotional, and nonargumentative. Unlike the scholarly style encountered in the humanities, which relies on convincing argumentation, in mathematics, the facts and logic speak for themselves. It is not necessary to argue for the correctness of a result or the truth of a statement.

The absolute lack of emotion in mathematical style can seem very unwelcoming and aloof. At first, the student may be taken aback by the unfriendly style of formal mathematical documents like journal articles and technical reports. Yet this style has a beauty of its own, which mathematicians come to appreciate in time. British mathematician Bertrand Russell says,

> Mathematics, rightly viewed, possesses not only truth, but supreme beauty—a beauty cold and austere, like that of sculpture, without appeal to any part of our weaker nature, without the gorgeous trappings of painting or music, yet sublimely pure, and capable of a stern perfection such as only the greatest art can show [68].

In mathematical writing, clarity and beauty go hand-in-hand. Clarity is the main objective, so anything that interferes with the clarity of an explanation must be reworded or eliminated. Mathematical style prizes simple words and grammar, which not only increases its clarity, but also its accessibility to nonnative English speakers.

---

[1]Interestingly, the written form of a *lingua franca* often outlives its spoken form. Thus, Newton, Leibniz, Euler, and Gauss all published their work in Latin, the written *lingua franca*, but a dead spoken language for over a thousand years.

Semantic density

Good mathematical writing is characterized by *semantic density* (also known as *economy of expression*), which means that it expresses a large amount of meaning with a small number of words. Let us compare the following two sentences:

1) "We decided to make use of the method of Gaussian elimination in order to solve the system of linear equations."

2) "We use Gaussian elimination to solve the linear system."

Both sentences express the same content, but the second one is more semantically dense. It uses a mere nine words to express what the first one says in twenty.

Expressing oneself with semantic density is not natural to anyone. Therefore, one often approaches mathematical writing as a two-step process: we write a first draft in a way similar to oral speech, laying down the content without concern for density. Then, we condense the writing, reducing the word count without reducing its content. For example, the first draft will likely use phrases reminiscent of oral speech, such as "for the purpose of" and "has been shown to be," which can later be pared down to simply "to" and "is." Similarly, the first draft may contain "empty" phrases that do not add any substance to the meaning, and instead obscure the clarity of the explanation:

- it is interesting to note that...

- it may be reasonable to suppose that...

- the fact is that ...

- it should be noted that ...

The process of condensing the writing involves streamlining the syntax, avoiding repetitions, and eliminating empty words and phrases. These empty words include most adverbs (such as "actually," "basically," "clearly," "essentially," "extremely," "obviously," "really," "very") and adjectives (such as "important," "interesting," "clear," "relevant," etc.). With some practice, it is usually possible to drastically reduce the word count significantly without losing any of the content.

Here is a first draft of an Abstract for an article describing a mathematical model for the spread of the bacteria *E. coli* in Lake Superior.

"In this paper, we will describe the method by which we were able to model the spread of *E. coli* bacteria in Lake Superior using a differential equation in which one of the terms is nonlinear. Basically what we did is that we included a limited carrying capacity K to model the growth of the bacteria, coupled with a linear decaying rate D. We were able to solve the differential equation analytically by using the method of separation of variables to obtain a solution S. It is particularly interesting to note that the solution that we found predicts that the concentration of *E. coli* will oscillate in time, and that the oscillation will decrease in amplitude, and eventually will stabilize below the toxic level of approximately two-hundred and twenty-five parts per billion by the end of the year."

By streamlining the writing, we can condense this Abstract to almost half its length without eliminating any of the content:

"We describe our method of modeling the spread of *E. coli* bacteria in Lake Superior by a nonlinear differential equation, in which the growth term includes a limited carrying capacity, and the decay term is linear. The method of separation of variables allows us to solve the differential equation analytically. The solution predicts that the *E. coli* concentration oscillates in time with decaying amplitude, eventually stabilizing below the toxic level of approximately 0.225 ppm by the end of the year."

Audience

In any written document (not just in mathematics), the target audience determines the tone, content, and level of explanatory detail.

In elementary mathematics courses, the intended audience of an assignment is typically an instructor or teaching assistant, a professional mathematician who does not require a full explanation, but who often asks that a student "show work" to verify that he or she is indeed following the correct steps. Instructors often accept submitted assignments written as "scratch paper," i.e., in a format that consists of handwritten mathematical computations in symbolic form devoid of explanations, often containing crossed-out mistakes and wrong turns, eventually ending with a boxed expression, which symbolizes the arrival at some "final answer."

In upper-level mathematics, assignments must go beyond merely "showing work" to an instructor. Their main purpose is to learn to properly describe a progression through a mathematical problem, as if explaining it to a fellow student who has the same level of mathematical knowledge. Some assignments may even require writing for a reader who knows less mathematics than the writer, an excellent exercise to prepare for any professional career where one is required to clearly and succinctly explain a problem-solving process to a non-mathematician.

10.1.2   Mathematical style and syntax

Sequential writing

A mathematical explanation proceeds like a story, with a beginning, a series of events that take place in a prescribed order, and an end. For example, it may begin with a problem set before the reader; then it proceeds with the solution, explaining and illustrating each step in sequence by a symbolic expression or mathematical computation. The writer should introduce each step by a transition word or phrase, such as "first," "then," "thus," "therefore," "since," "because," "on one hand," "on the other hand," "besides," "furthermore," "similarly," "likewise," "however," "moreover," and "finally."

Example 10.1.4. To solve the initial-value problem

$$\frac{dx}{dt} = kx \qquad x(0) = x_0,$$

we use the method of separation of variables. First, we separate the variables:

$$\frac{dx}{x} = kdt,$$

then, we integrate:

$$\ln x = t + c,$$

where c is an arbitrary constant. Solving for x gives

$$x(t) = ce^{kt}.$$

Finally, the initial condition allows us to determine that $c = x_0$. Therefore, the solution is

$$x(t) = x_0 e^{kt}. \qquad\qquad \triangle$$

A good mathematical explanation should introduce every mathematical computation by an explanation in words. The last sentence should clearly state the final answer to the question posed at the beginning.

Throughout the narrative, one must guide the reader through the steps of the problem, providing clarifying comments as needed. Thus, the writer plays the role of a tour guide pointing out the important features, or a sports newscaster giving a running commentary of a game. For example, the author should justify every assertion with a phrase like "because $x > 0$, ..." or "as can be seen on the diagram,...".

Turning back to Examples 10.1.3 and 10.1.4, we observe that every symbolic expression or equation is introduced by an English sentence. We also observe that several steps of simple algebra are omitted. If writing for an audience of peers, it is not necessary to muddle the narrative by including simple routine calculations that a fellow student would be able to do without assistance.

Symbolic expressions and numbers

Mathematical symbols are simply placeholders for English words (for example, $\geq$ and $\in$ stand for "is greater than or equal to" and "is an element of," respectively). As such, they must follow the standard rules of English grammar and syntax, just like the English words they represent. For example, we may say,

To find the eigenvalues of matrix $A = \begin{bmatrix} -3 & 12 \\ -2 & 7 \end{bmatrix}$, we solve the characteristic equation $\begin{vmatrix} -3 - \lambda & 12 \\ -2 & 7 - \lambda \end{vmatrix} = 0$, which simplifies to $(\lambda - 1)(\lambda - 3) = 0$, yielding the eigenvalues $\lambda = 1$ and 3.

There are two notation formats for embedding symbolic expressions and equations in a sentence: *inline* (on the same line as the words) or *display* (centered on their own line). The writer should only use inline format when the expressions are short and do not contain large symbols. For all other cases, display format is preferable. Thus, a good writer would write the mathematical expressions in the example above in display (instead of inline) format:

Example 10.1.5. To find the eigenvalues of matrix

$$A = \begin{bmatrix} -3 & 12 \\ -2 & 7 \end{bmatrix},$$

we solve the characteristic equation

$$\begin{vmatrix} -3 - \lambda & 12 \\ -2 & 7 - \lambda \end{vmatrix} = 0, \tag{10.1}$$

which simplifies to

$$(\lambda - 1)(\lambda - 3) = 0, \tag{10.2}$$

yielding the eigenvalues $\lambda = 1$ and 3. △

Whether in inline or display format, the writer must follow all punctuation and capitalization rules, as if the expressions appeared in words. In the previous examples, a comma or period follows every expression or equation, even in display format.

The writer may also use display format to emphasize or number an expression or equation (as Equations 10.1 and 10.2 in the previous example). It is not necessary to number every single displayed expression, only those that will be referred to later in the text.

The writer should strive to use mathematical and numerical symbols when available, as they are usually more concise and clearer than a rhetorical explanation, especially to a reader versed in mathematics. However, as a matter of style, the writer should avoid starting a sentence with a symbol.

Unclear	Better
The theorem holds for all values of x that are greater than or equal to zero.	The theorem holds for all $x \geq 0$.
The population stabilizes between two hundred and fifty and three hundred.	The population stabilizes between 250 and 300.
z is a complex number with positive real part.	The complex number z has positive real part.

Finally, the writer should avoid informal mathematical symbols in a formal document. The following table lists some of the informal symbols that one may use for scratch paper, for taking notes, or for chalkboard and oral presentations slides. In formal writing, an equivalent English phrase is more appropriate.

Informal symbol	Formal English phrase
$\Rightarrow$, $\therefore$	therefore
$\Leftrightarrow$	is equivalent to
iff	if and only if
st	such that
ae	almost everywhere
$\forall$	for all
$\exists$	there exists
WLOG	without loss of generality
TFAE	the following are equivalent

Person and tense

Mathematical writing uses either the third person or the first person plural (the *academic* or *authorial* "we"). Even a single author should use the plural, as he or she plays the role of a guide on a mathematical excursion coming alongside the reader, explaining and pointing out important steps along the way. Here are some examples:

"The variable x represents the distance in meters."
"This allows us to find the eigenvalues of the matrix."
"We separate the variables to solve the equation."

It is also acceptable to use the passive voice instead of the authorial "we," although this often makes the writing more cumbersome.

"The equation is solved by the method of separation of variables."
"The expression can be simplified by a trigonometric identity."

Although instructors commonly use the second person imperative on worksheets, quizzes, and exams, the writer should avoid it in formal writing.

"Solve the equation by separation of variables." ✗
"Find the distance between the two points." ✗

The writer should preferably write a formal mathematical paper exclusively in the present tense. Again, this is because he or she plays the role of an active guide, progressing through an explanation alongside the reader in real time, rather than merely reporting steps that he or she carried out in the past. Every example in this section is in the present tense.

There are, however, a few rare cases when a writer may choose to use the past tense, as in the following example, where he or she provides numerical data or reports an event that took place in the past and cannot be replicated in the present.

"We determine the model's parameters using data from a survey that we gave to 325 students in 2015."

Aside from these exceptional cases, one should avoid the past tense (and all verb tenses other than the present). The use of the present tense is a strong distinctive of formal and professional scientific syntax, which goes a long way in lending clarity to the content.

Terminology

Writing a mathematical explanation is particularly challenging in that one has just one chance to convey the correct mathematics with no ambiguity and perfect accuracy. Unlike other types of writing, mathematical explanations cannot leave any room for doubt or misunderstanding. Merely providing a general idea or capturing the "gist" is not sufficient. A reader must understand every step of a calculation or proof, down to the last details, and be able to reproduce them. Explaining mathematical steps clearly and correctly takes practice and goes beyond merely being able to execute them. It is a valuable skill, as it contributes to clarify and solidify mathematical knowledge in the writer's own mind as well as the reader's.

Semantic field. Mathematics is its own *semantic field*, meaning that it is a language of its own, with its own specialized vocabulary. To write effectively, one must learn the correct use of mathematical terms. Following are some examples of common mathematical terms that are frequently confused.

Example 10.1.6. The terms *expression*, *equation*, *identity*, and *formula* do not mean the same thing.

- An *expression* is a mathematical word in symbolic form, such as

$$\sum_{n=1}^{\infty} \frac{1}{n^2}.$$

- An *equation* is a mathematical statement that two things are equal, so it usually contains the "equals" symbol. An equation may contain unknown values, and may or may not be true, depending on the values of the unknowns. For example,

$$x^2 + 3xy - y^2 = 5.$$

- An *identity* is an equation that is always true, regardless of the values of the unknowns. For example,

$$\sin^2 x + \cos^2 x = 1.$$

- A *formula* is an equation that assigns a mathematical expression to a well-defined quantity. For example, the formula for the volume of a sphere is

$$V = \frac{4}{3}\pi r^3. \qquad \triangle$$

Example 10.1.7. *Solution* and *answer* are another pair of common mathematical terms that students often use incorrectly.

- A *solution* of an equation (or system of equations) is an assignment to the unknown variables that *satisfies* the equation, i.e., that makes the equation true. A solution can also apply to a mathematical problem, since many problems require finding a mathematical expression that satisfies an equation.

  "The equation $x^2 - x - 1 = 0$ has two real solutions."
  "The differential equation $y'(x) = 2x$ has an infinite number of solutions."

- An *answer* is a statement used in response to a question.

  The correct answer to the question, "Are the rows of matrix A linearly independent?" is "Yes". $\triangle$

Example 10.1.8. The terms *positive* and *negative* do not include zero, while in contrast *nonnegative* and *nonpositive* do.

"The expression $\sqrt{x^2}$ is nonnegative for all $x \in \mathbb{R}$."
"The derivative of $f(x) = x^3 + x^2$ is positive for all $x \in \mathbb{R}$." $\triangle$

Borrowed terms. Mathematics often borrows words from everyday English, assigning them a mathematical definition that may be very different from their common, everyday meaning. For example, the adjective "similar" has a precise definition in geometry, and another precise (but different) definition in linear algebra. Similarly, modern algebra borrows the English words "group," "ring," "field," and "ideal" for its own purposes, while calculus uses "limit," "derivative," and "continuous." These double-meanings can be particularly treacherous, as the misuse of a mathematical term not only makes a statement unclear, but incorrect. See Subsection 8.3.4 for more potential sources of confusion when using mathematical terms.

Names of common methods. Many common mathematical methods, techniques, algorithms, lemmas, and theorems have their own proper name. When writing a mathematical explanation, one must not only be able to execute the steps, but also describe them precisely. Providing the proper name of a method, algorithm, or theorem greatly enhances clarity and conciseness, and it dispenses the writer from having to explain every step. For example,

"The roots of the *characteristic equation* are the eigenvalues."
"We use *L'Hospital's Rule* to find the limit."
"We use the *quadratic formula* to find roots of the polynomial."
"We solve the linear system by *Gaussian elimination*."
"We use the *Simplex algorithm* to find the maximum."

If...then, if and only if. A mathematical writer should be careful with the conjunction "if," especially when it indicates a *conditional* ("if...then") or *biconditional* (equivalence) statement ("if and only if"). Conditional statements must clearly separate the hypothesis and conclusion by stating both "if" and "then." Similarly, biconditional statements (including definitions) should always use the full phrase "if and only if."

Unclear	Better
If $x > 0$, $x^2 > 0$.	If $x > 0$, then $x^2 > 0$.
Definition: The matrix A is *singular* if $\det A = 0$.	Definition: The matrix A is *singular* if and only if $\det A = 0$.

Informal phrases. In keeping with the formal and impersonal style of mathematical writing, the writer should replace informal terms by more formal equivalents and avoid contractions.

Informal phrase	Equivalent formal phrase
We *plug in* the given values for x and y.	We *substitute* the given values for x and y.
We *pull out* the constant.	We *factor out* the constant.
The equation *can't* be solved analytically.	The equation *cannot* be solved analytically.

Letters and fonts

Letters. In addition to symbols, mathematical writing frequently uses letters to represent mathematical objects and quantities.

Letters	Common use
x, y, z	real numbers, variables
i, j, k, m, n	integers, indices
f, g, h	functions
c, k	constants
s, t	variable parameters

Case and fonts. Mathematical writing uses both upper- and lowercase letters, as well as various fonts for different mathematical objects (vectors, matrices, sets, etc.).

Font	Examples
bold	**ABCDEFGHIJKLMNOPQRSTUVWXYZ**
blackboard	$\mathbb{ABCDEFGHIJKLMNOPQRSTUVWXYZ}$
calligraphic	$\mathcal{ABCDEFGHIJKLMNOPQRSTUVWXYZ}$
script	$\mathscr{ABCDEFGHIJKLMNOPQRSTUVWXYZ}$
fraktur	$\mathfrak{ABCDEFGHIJKLMNOPQRSTUVWXYZ}$

Greek letters. Greek letters are very common in mathematical writing to represent geometrical angles, angular parameters, or constant parameters in a mathematical model.

Upper case	Lower case	Name	Upper case	Lower case	Name
A	α	alpha	N	ν	nu
B	β	beta	Ξ	ξ	xi
Γ	γ	gamma	O	o	omicron
Δ	δ	delta	Π	π	pi
E	ϵ or ε	epsilon	P	ρ	rho
Z	ζ	zeta	Σ	σ	sigma
H	η	eta	T	τ	tau
Θ	θ or ϑ	theta	Y	υ	upsilon
I	ι	iota	Φ	ϕ or φ	phi
K	κ	kappa	X	χ	chi
Λ	λ	lambda	Ψ	ψ	psi
M	μ	mu	Ω	ω	omega

10.1.3 Mathematical documents

Many upper-level mathematics classes require students to write formal mathematical papers. Some are short, one-page homework assignments that describe the solution of a problem or proof of a theorem. Other papers may be longer and contain figures, tables, or a bibliography.

Short papers

The two most common types of standard homework assignment submissions consist in either (1) the solution to a mathematical problem (*problem & solution*), or (2) the proof of a theorem (*theorem & proof*). Before writing a paper of this type, the student should solve the problem or prove the theorem beforehand on scratch paper. Then, he or she can organize the work into a formal explanation, highlighting the key steps while omitting the simple and obvious ones. We suggest the following structure: preliminary items, given problem, and solution/proof.

- **Preliminary items.** At the very least, short formal papers must include at the top a title, the date, and the name(s) of the students(s)[2]. The instructor may also require additional information here, such as the course number.

- **Given problem.** Formal documents must be self-contained, i.e., they must not require that the reader refer to another book, worksheet, etc. Therefore, the first section of the paper consists in simply rewriting the given statement of the problem to solve along with all given information.

---

[2]For a group project, it is customary for team members to list their names alphabetically by *last name*.

- **Solution/Proof.** To structure the solution, the writer should follow these guidelines:

  - divide the solution or proof into key steps,
  - introduce each step with an English sentence and illustrate each one with a symbolic expression or equation as needed,
  - write the paper in the present tense only, using the third person or the academic "we,"
  - aim for an appropriate balance of rhetoric (English words) and symbolic (mathematical symbols) syntax. Too much of either makes the explanation unclear.
  - write short symbolic expressions in inline format but place longer ones in display format,
  - pay close attention to punctuation, especially in sentences that include mathematical expressions in display format,
  - replace any informal terms and "shorthand" symbols by more formal equivalents,
  - define all variables, parameters, or unknowns. For example, if the solution introduces the symbol c, one must clarify whether it represents a variable, constant, integer, real number, vector, etc.

Other types of mathematical papers

Instructors in upper-level mathematics courses may sometimes require that students write longer papers. These fall into the following types:

- **Project reports.** Similar to a short *problem & solution* paper (described above), a project report is a formal document that describes the process of approaching and solving a larger, longer, or more open-ended problem. We discuss project reports in Section 10.2.

- **Expository papers.** We discuss this type of paper in Section 9.3.2.

- **Research papers.** These formal documents typically follow the structure of a journal article, which we describe in Section 9.2.

- **Oral presentations.** Section 10.5 contains guidelines for generating the visual support (slides or poster) of an oral presentation.

EXERCISES FOR SECTION 10.1

1. For each of the following, indicate if it is an expression, equation, identity, or formula.

   (a) $e^{i\pi} + 1 = 0$

   (b) $I_z = \iiint_E \left(x^2 + y^2\right) \rho(x, y, z) \, dV$

   (c) $y'' + 3y' - 4y = \sin 3x$

   (d) $\sqrt{x^2 + y^2 + z^2}$

   (e) $x = \frac{1}{2a}\left(-b \pm \sqrt{b^2 - 4ac}\right)$

   (f) $\cos\theta = \frac{1}{2}(1 + \cos 2\theta)$

   (g) $\int_0^x e^{-t^2} \, dt$

   (h) $A\vec{x} = \vec{b}$

2. Evaluate the following integral

$$\int_0^1 \sqrt{1 - x^2} \, dx.$$

   Write your answer as a formal *problem & solution* paper (as outlined in Subsection 10.1.3). Below are some "scratch paper" notes to guide through the required steps.

$$x = \sin t, \ dx = \cos t \, dt \Rightarrow \int_0^{\frac{\pi}{2}} \sqrt{1 - \sin^2 t} \, \cos t \, dt$$

$$\sin^2 t + \cos^2 t = 1 \Rightarrow \int_0^{\frac{\pi}{2}} \cos^2 t \, dt$$

$$\cos^2 t = \frac{1 + \cos 2t}{2} \Rightarrow \frac{1}{2} \int_0^{\frac{\pi}{2}} (1 + \cos 2t) \, dt$$

$$= \frac{1}{2}\left(t + \frac{1}{2}\sin 2t\right)\Big|_0^{\frac{\pi}{2}} = \frac{1}{2}\left(\frac{\pi}{2} + 0 - 0 - 0\right) = \frac{\pi}{4}.$$

3. Find the arclength of the parametric curve

$$x = 1 + 3t^2 \qquad y = 4 + 2t^3$$

   for $0 \leq t \leq 1$. Write your answer as a formal *problem & solution* paper (as outlined in Subsection 10.1.3). Below are some "scratch paper" notes to guide through the required steps.

$$S = \int_0^1 \sqrt{\left(\frac{dx}{dt}\right)^2 + \left(\frac{dy}{dt}\right)^2} \, dt = \int_0^1 \sqrt{(6t)^2 + (6t^2)^2} \, dt =$$

$$= \int_0^1 \sqrt{36t^2 + 36t^4} \, dt = \int_0^1 6t\sqrt{1 + t^2} \, dt$$

$$\text{u-sub} \quad \Rightarrow \quad du = 2t \, dt$$

$$\Rightarrow \quad t = 0 \to 1 \quad \Rightarrow \quad u = 1 \to 2$$

$$= 3\int_1^2 \sqrt{u} \, du = 2u^{\frac{3}{2}}\Big|_1^2 = 2\left(2^{\frac{3}{2}} - 1\right) \approx 3.657$$

4. Find the moment of inertia of a disc of radius R centered at the origin, of constant density k, around the x-axis. Write your answer as a formal *problem & solution* paper (as outlined in Subsection 10.1.3). Below are some "scratch paper" notes to guide through the required steps.

$$I_x = \iint_{disc} y^2 k\, dA$$

polar coordinates $\quad \Rightarrow \quad x = r\cos\theta,\ y = r\sin\theta\ dA = r\, dr\, d\theta$

$$\text{disc}\quad 0 \le r \le R\quad 0 \le \theta \le 2\pi$$

$$= k \int_0^{2\pi} \int_0^R r^2 \sin^2\theta\, r\, dr\, d\theta = k \int_0^{2\pi} \sin^2\theta\, d\theta \int_0^R r^3\, dr$$

$$\text{Trig identity}\quad \sin^2\theta = \frac{1}{2}(1 - \cos 2\theta)$$

$$= k \int_0^{2\pi} \frac{1 - \cos 2\theta}{2}\, d\theta \int_0^R r^3\, dr = \frac{k}{2}\left(\theta - \frac{1}{2}\sin 2\theta\right)\Big|_0^{2\pi} \frac{r^4}{4}\Big|_0^R =$$

$$= \frac{k}{2}(2\pi - 0 - 0 + 0)\frac{R^4}{4} = \frac{k\pi R^4}{4}$$

5. Solve the following linear system and write the solution as a set of vectors.

$$\begin{cases} x_1 - x_2 + x_3 &= 3 \\ 2x_1 - x_2 + 4x_3 &= 7 \\ 3x_1 - 5x_2 - x_3 &= 7 \end{cases}$$

Write your answer as a formal *problem & solution* paper (as outlined in Subsection 10.1.3). Below are some "scratch paper" notes to guide through the required steps.

$$\begin{bmatrix} 1 & -1 & 1 & 3 \\ 2 & -1 & 4 & 7 \\ 3 & -5 & -1 & 7 \end{bmatrix} \rightarrow \begin{bmatrix} 1 & -1 & 1 & 3 \\ 0 & 1 & 2 & 1 \\ 0 & -2 & -4 & -2 \end{bmatrix} \rightarrow \begin{bmatrix} 1 & -1 & 1 & 3 \\ 0 & 1 & 2 & 1 \\ 0 & 0 & 0 & 0 \end{bmatrix}$$

$$\Rightarrow \begin{cases} x_1 - x_2 + x_3 &= 3 \\ x_2 + 2x_3 &= 1 \end{cases} \quad \text{Let } x_3 = r \in \mathbb{R}.$$

$$\Rightarrow x_2 = 1 - 2r \quad x + 3 = 3 + (1 - 2r) - r = 4 - 3r$$

$$\Rightarrow \begin{bmatrix} x_1 \\ x_2 \\ x_3 \end{bmatrix} = \begin{bmatrix} 4 - 3r \\ 1 - 2r \\ r \end{bmatrix} \quad \Rightarrow \quad \left\{ \begin{bmatrix} 4 - 3r \\ 1 - 2r \\ r \end{bmatrix} \middle|\, r \in \mathbb{R} \right\}$$

6. Prove that an angle inscribed in a half-circle is a right angle. Referring to the diagram, write your answer as a formal *theorem & proof* paper (as outlined in Subsection 10.1.3). Below are some "scratch paper" notes to guide through the required steps.

$$\overline{AO} = \overline{OC} = r \quad \Rightarrow \quad AOC \text{ is isoceles} \quad \Rightarrow \quad \alpha = \gamma$$

$$\overline{BO} = \overline{OC} = r \quad \Rightarrow \quad BOC \text{ is isoceles} \quad \Rightarrow \quad \beta = \delta$$

$$\sum \text{angles} = 180° \ (\text{triangles})$$

$$\Rightarrow (AOC): \alpha + \gamma + \lambda = 2\gamma + \lambda = 180°$$

$$\Rightarrow (BOC): \beta + \delta + \mu = 2\beta + \mu = 180°$$

$$\lambda + \mu = 180° \quad \Rightarrow \quad (180° - 2\gamma) + (180° - 2\delta) = 180°$$

$$\Rightarrow \gamma + \delta = 90°$$

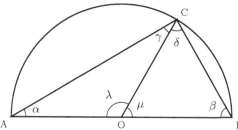

7. Prove the Law of Cosines for an acute triangle. Referring to the diagram, write your answer as a formal *theorem & proof* paper (as outlined in Subsection 10.1.3). Below are some "scratch paper" notes to guide through the required steps.

$$a^2 = x^2 + h^2$$

$$c^2 = (b - x)^2 + h^2$$

$$\Rightarrow h^2 = a^2 - x^2 = c^2 - (b - x)^2$$

$$\Rightarrow a^2 = c^2 - b^2 + 2bx$$

$$\therefore c^2 = a^2 + b^2 - 2ab\cos\gamma$$

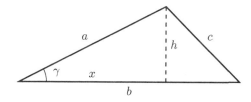

8. Prove the Law of Sines for an acute triangle. Create a diagram similar to the one for Exercise 10, then write your answer as a formal *theorem & proof* paper (as outlined in Subsection 10.1.3).

9. Write the statement of Proposition 2.2.9 entirely in rhetorical syntax (no symbols, numerals, or operations).

10. Write the statement of Theorem 2.2.12 entirely in rhetorical syntax (no symbols, numerals, or operations).

11. Replace the following expressions by a single equivalent word. (a) based on the fact that, (b) for the purpose of, (c) a greater number of, (d) longer period of time, (e) has been shown to be, (f) by means of, (g) is possible that, (h) in regards to, (i) a great number of times, (j) in order that, (k) despite the fact that, (l) during the course of, (m) by virtue of the fact that, (n) because of all these things.

12. Rewrite the following first draft of the abstract (200 words) for a facility location project. Follow all the guidelines for proper formal mathematical writing described in this section, and condense the abstract to 100 words or fewer without losing any of the content.

> In this paper, we will be presenting the model that was designed by our team of collaborators to determine the best location for a distribution center of perishable seafood for the four hundred and thirty-eight restaurants located in the city of Los Angeles, in southern California. After we collected the relevant data provided to us by a large number of restaurants in the Los Angeles area, we were able to formulate an objective function that took into account not only the distance from the distribution center to the various restaurants, but also the typical, horrible traffic delays that the various delivery vehicles would probably have to deal with. To this end, we used the relevant data graciously provided to us by the Los Angeles Department of Transportation. We also would like to point out the fact that we considered the very high cost of real estate in the Los Angeles area in our model. We then proceeded to solve the optimization model using the method of Lagrange multipliers. Based on the results we obtained, we were finally able to conclude that the best location for the distribution center would have to be around the Westlake area, north of Olympic Boulevard.

10.2 Project Reports

Similar to the *problem & solution* document we described in the previous section, a *project report* describes a larger, more complex or open-ended problem and its solution.

The concept and format of a project report is similar to the type of reports that project managers write in many industries to communicate the status, progress, and results of the projects they manage. Thus, writing a good project report is an important skill to develop for mathematics students destined for a career in industry.

A project report is not to be confused with a *scientific report*, which one would write to describe a research project. These reports typically follow the IMRAD (Introduction, Methods, Results and Discussion) format we describe in Section 9.2.2. Although a project report bears some similarities with a scientific report, it serves a different purpose and addresses a different audience. A project report typically addresses two different types of readers: stakeholders, usually nonmathematicians primarily interested in the overal motivation, solution(s) and practical implications; and technical experts interested in the mathematical details.

A good report is organized, easy to follow, and concise; its readers are often busy people who highly value a well-written report that conveys relevant information without having to hunt through several pages to find it. In Subsection 10.2.1, we suggest a report structure that the student may apply to any project that consists in solving a practical problem using a mathematical approach. It divides the report into four standard sections: Introduction, Methods, Solution, and Summary of Results; we discuss each one in Subsections 10.2.2-10.2.5. Finally, Subsection 10.2.6 discusses three other common types of reports for longer, multistage projects: project proposals, progress reports, and executive summaries.

10.2.1   Structure and style of a project report

Required items and overall structure

Every project report must begin with the title, date, and name(s) of the authors.

A good title is descriptive with no abbreviations or acronyms, providing key words so that a reader can uniquely identify the problem it addresses. It should also mention the type of document (i.e., a project report), for example: "Traffic Analysis for Unionville, Missouri - Project Report."

In a group project, team members should list their names alphabetically by last name.

The rest of the report is divided into four standard sections:

1) Introduction (or Problem and Data),

2) Methods (or Model),

3) Solution,

4) Summary of results (and/or **Recommendations**).

The contents of each section are explained below.

Audience and level of detail

Project reports have two different target audiences:

1) **Stakeholders** may be superiors, clients, sponsors, investors, or nontechnical team members primarily interested in the main question, results, and practical implications. The writer should address the Introduction and Summary of Results to stakeholders, aiming to be concise with a minimum of technical jargon.

2) **Technical experts** will be interested in the Methods and Solution. These sections should contain enough detail for an interested reader to understand the approach and replicate the
results. The writer may assume these readers to be well versed
in basic mathematics, although not necessarily familiar with
specialized mathematical methods.

Style

Project reports follow the style and syntax guidelines for formal mathematical writing we outline in Subsections 10.1.1 and 10.1.2. In particular, the writer should take care to:

- use an appropriate balance of rhetoric and symbolic writing,

- write in the present tense, using the first or third person,

- divide the explanations into steps and proceed through them
  sequentially,

- aim for conciseness and semantic density to avoid information
  overload,

- avoid informal phrases and shorthand notation,

- properly number, caption, and cross-reference tables and figures, as we explain in Subsection 10.4.2.

10.2.2   Introduction

The *Introduction* (or *Problem and Data*) section describes the problem and provides all given data. A report must stand on its own without referring to other documents, so the writer must provide here all
given background information and numerical data, which may require
reproducing tables or diagrams.

10.2.3   Methods

If the project involves using a mathematical modeling approach to
solve a problem, the *Methods* section describes the model, so it may
be titled *Model*. The contents of this section vary by type of model; we
describe three types hereafter: (1) empirical models, (2) mechanistic
models, and (3) operations research models.

Empirical model

In a data-driven empirical model, the writers must explain how they
use the given data to assign variables and build the model's equations.
We suggest organizing this description into separate subsections:

1) **Variables.** Here, the writer defines every variable by name with its symbol and unit, as well as the type of measurement scale (nominal, ordinal, interval, or ratio). The writer must then explain how to determine the numerical values from the data.

2) **Equations.** The writer provides the equations and describes the role of each variable (independent, dependent, control, moderator, mediator, etc.).

3) **Modeling assumptions and justification.** Here, the writer explains and justifies the assumptions behind the form of the equations, for example, the assumed distribution of the variables and error terms, absence of multicollinearity, etc.

Mechanistic model

A mechanistic model makes assumptions about the underlying mechanisms or laws that govern a system's behavior, typically written as differential equations (if the independent variables are continuous) or recurrence relations (if they are discrete). Thus, the description of a mechanistic model must provide and justify the modeling assumptions and provide the equations. We suggest organizing the model's description into separate subsections:

1) **Variables.** In a mechanistic model, the dependent variables may be time-dependent, space-dependent, or both. Here, the writer defines every variable (dependent and independent) by name with its symbol and unit. In the case of a dynamical (i.e., time-dependent) model, it may be necessary to define the origin of time (when is $t = 0$?).

2) **Parameters.** Parameters include all quantities other than variables. The writer must define each one with its symbol and unit. Parameters may include constants in the equations, initial and auxiliary conditions, and specific variable values, such as P_{max}, P_{min}, P_{final}, $P_{steadystate}$, etc. The writer should provide here the numerical value of parameters that can be easily determined from the given data. If determining parameter values requires solving the model, the writer should provide those values in the *Solution* section instead.

3) **Equations.** The equations (differential or recurrence relations) appear here. If it is a discrete time step model, the writer must provide the time step size here as well.

4) **Modeling assumptions and justification.** Here, the writer explains and justifies which aspects of the real-life problem one is simplifying or ignoring, in particular:

(a) which variables and/or parameters one is ignoring and why it is reasonable to do so,

(b) which laws and mechanisms one is assuming are at play and why it is reasonable to assume them,

(c) what the equations assume to be linear, constant, or negligible, and why it is reasonable to assume this.

Operations research model

The description of an operations research model must include:

1) **Variables.** Here, the writer defines every variable by name with its symbol and unit. Some variables may be *decision* (or *control*) variables, over which one has control to determine the objective function's optimum value(s).

2) **Parameters.** Parameters include all quantities other than variables. The writer must define each one with its symbol and unit. Parameters may include constants in the equations and specific variable values, such as optimum values, shadow prices, etc. The writer should provide here the numerical value of parameters that can be easily determined from the given data.

3) **Objective function and constraints.**

4) **Modeling assumptions and justification.** Here, the writer explains and justifies which aspects of the real-life problem one is simplifying or ignoring, in particular:

(a) which variables and/or parameters one is ignoring and why it is reasonable to do so,

(b) what the equations assume to be linear, constant, or negligible, and why it is reasonable to assume this.

10.2.4   Solution

In the *Solution* section, the writer explains how he or she uses the model to answer the questions posed in the Introduction. This section must clearly convey the steps from model to solution(s), providing sufficient mathematical detail for readers to understand the approach and replicate the results on their own. In particular, the writer must explain whether he or she uses an *analytic* or *numerical* method.[3]

---

[3]The terms *analytic* and *numerical* are particularly confusing, because both methods are analytical in some sense and both generate numerical values. Furthermore, the adjective *analytic* has a different definition when speaking of an *analytic function* or *analytic continuation*.

- An *analytic method* leads to an exact solution by symbolic manipulation. A solution of this type can be written in *closed form*, i.e., as a formula in terms of parameters. One may then obtain a numerical solution by simply substituting the parameters' numerical values into the formula. One advantage of this approach is that if the parameter values change, one may easily recalculate the solution by substituting the new values. Therefore, one should carry out all calculations symbolically, only substituting numerical values at the end.

- A *numerical method* leads to an approximate solution, usually by an algorithm or other method of numerical analysis. If implemented by a computer, one should mention here the software used, for example.

Depending on the type of questions posed in the Introduction, a writer may divide the Solution section into several subsections addressing different questions or analyzing different scenarios. For example, one may need to solve a model multiple times using different variables or initial conditions, perform a sensitivity analysis, or perform an error analysis.

10.2.5 Summary of Results

As its name implies, the *Summary of Results* summarizes the results. As a mere summary, this section should not contain any new information that has not already been stated in previous sections. The writer should assume that readers of this section did not read the Methods or Solution sections and are looking for succinct answers to the questions posed in the Introduction without the details pertaining to their derivation. Here, one merely restates the final answers in numerical form, spelling out any acronyms, abbreviations, or symbols that one may have introduced in the Methods or Solution.[4]

If requested, the Summary of Results may include *recommendations*, also called *expert opinions*, which are practical comments or suggestions that inform stakeholder decisions based on the results.

10.2.6 Other types of reports

A project may be the subject of multiple reports throughout its lifetime if it is long, involves many people, or is completed in stages. Large, multistage projects are often launched with a *project proposal* that introduces the problem, sets the budget, and defines a timeline for completing various stages. Project managers then generate

---

[4]It is acceptable to cross-reference figures and tables that appear in the Methods and Solution.

a series of *progress reports* at various stages until the project's completion. At any stage, a project manager may generate an *executive summary*, a short report for stakeholders interested in the scope and desired outcome of a project without the technical details. We discuss these reports below.

Project proposal

A *project proposal* sets the vision for a project and the requirements in time, resources, and manpower necessary for its completion. In particular, it:

1) explains the project's topic, scope, relevance, motivation, related work, and desired outcomes,

2) provides any available data and explains who will be collecting additional data and how,

3) proposes solution approaches, as well as the requirements in terms of equipment, software, hardware, manpower, and monetary resources to implement them,

4) proposes a budget,

5) introduces the project manager(s), the team members, their respective roles, and why these people are suited to work on this problem,

6) provides a timeline for the project's progress, including dates of progress reports and the estimated completion date.

A project proposal is often made public, as an announcement for the launch of a project and as a plea for feedback and resources. Thus, a project proposal often seeks to "sell" the idea of a project to prospective sponsors. In academic settings, researchers regularly generate similar reports to solicit grants to fund their research.

The content of a project proposal is similar to the Introduction of a stand-alone project report, but may also include general ideas about the methods and preliminary results. Project proposals follow the standard guidelines for mathematical writing, but they commonly use the past tense to describe the background of the project and future tense for what they propose to do.

Progress report / Status report

A *progress report* (or *status report*) gives an update on a project's progress. In particular, it

1) describes any changes to the project's title, topic, or scope, and provides reasons for the changes,

2) explains the solution methods and describes any difficulties, hurdles, or roadblocks encountered, as well as the strategies implemented to overcome them,

3) provides the results (or partial results) obtained thus far,

4) documents any unexpected expenses and delays and proposes updates to the budget and timeline accordingly,

5) explains what remains to be done, and what additional resources are needed for the remaining stages of the project.

Progress reports officially communicate the status of a project to all concerned parties and document the progress and changes since the last report. Outside stakeholders often request periodic status reports at predefined dates to monitor a project's progress and verify that the timing and budget are on track. Project managers use these reports to highlight important accomplishments and milestones and to solicit suggestions for upcoming phases. Finally, the report may help team members assigned to various phases of the project to help anticipate future workload.

Progress reports follow the standard guidelines for mathematical writing, except that they use the past tense to describe the portions of the project that have been completed, and the future tense for what remains to be done.

Executive summary

Similar to the *abstract* of an academic project, an *executive summary* is a short document (often limited to a single page or paragraph) that describes a project's motivation, scope, methods, results, and relevance. It communicates the general idea and highlights the main points without the technical details. A writer usually writes an executive summary after writing the final project report.

10.3 Mathematical Typesetting

Nowadays, people write everything on a computer, even mathematics. Although mathematicians often generate first drafts of mathematical documents by hand, at some point they need to write them on a computer. Unfortunately, most common document-writing computer tools are not well suited for mathematics.

A *word processor* is a software application for writing and storing text on a computer. These programs create documents containing basic text and simultaneously define how the text appears on the page by providing a variety of formatting options, such as font sizes, margin widths, paragraph spacing, heading styles, etc. Many can also embed figures and tables into a document. Outside of mathematics, the

great majority of professional academics and business people prepare
their documents on word processors such as *Microsoft Word* or *Google
Docs*. Although versatile and user-friendly, word processors are not
capable of typesetting mathematics in an acceptable fashion. They
have other shortcomings as well:

- Word processors are designed mostly for text-only documents.
  Although some offer the option of inserting a few mathematical
  symbols (for example, Microsoft Word's *equation editor*), their
  ability to typeset mathematics is very limited.

- A word processor processes the content and appearance of the
  document simultaneously, forcing the writer to constantly interrupt his or her train of thought to attend to the formatting
  of the document.

- Word processors are not universal. There is no guarantee that a
  document generated on a word processor will appear the same
  (or even be viewable at all) on a different word processor, or
  even a different version of the same word processor. This makes
  collaboration problematic. Furthermore, at the pace at which
  computer applications are replaced by newer ones, there is no
  guarantee that a document generated on a commonly available
  word processor today will be easily viewable in a decade.

- The printed quality of word-processor-generated documents is
  poor. Word processors follow the "What You See Is What You
  Get" (WYSIWYG) system, meaning that the printed document
  looks exactly like what appears on the computer screen. Even
  a high-quality screen has no more than 400 pixels per inch,
  whereas a printer typically has a resolution of 2400 or more
  pixels per inch, so a WYSIWYG program usually does not utilize the full resolution of a printer [54].

- Documents generated on a word processor take up a lot of computer memory because of all the hidden formatting commands
  that are part of the computer file. For example, a document
  containing the single sentence, "Hello World." (10 characters,
  one space, and one period), when written in a plain text editor,
  occupies 12 bytes. On a word processor, the file size for the
  same document is over 10,000 bytes [54].

Instead of a word processor, mathematicians prefer a markup language called LaTeX, a system developed by mathematicians specifically for mathematical documents. Although LaTeX is not as user-
friendly as a word processor, the advantages it affords are well worth
investing the time and effort to learn it.

Subsection 10.3.1 introduces the main features of LaTeX. Subsection 10.3.2 presents the different elements that make up the LaTeX

source code contained in a TₑX file, and Subsection 10.3.3 contains specific codes and instructions for inserting mathematical expressions in a LaTeX-generated document. Finally, Subsection 10.3.4 discusses front end applications that can be used in conjunction with LaTeX, and Subsection 10.3.5 presents the LaTeX-based online collaboration platform *Overleaf*.

10.3.1 LaTeX

LaTeX is a descriptive *markup language*, a typesetting system that embeds formatting instructions (called *markup commands*) into the text. These commands describe how the content is to appear on the page but do not reveal the document's final appearance until the document is *compiled*. The *Hypertext Markup Language* (HTML), for generating webpages, is an example of a similar descriptive markup language.

History

LaTeX is itself a derivative of a similar system called TₑX developed in the early 1970s by the American mathematician and computer scientist Donald Knuth. One of the primary strengths of TₑX is its precise knowledge of the sizes of all characters and symbols (down to 10^{-6} of an inch) and its ability to use this information to compute the optimal arrangement of letters per line and lines per page. This results in beautiful and professional-looking documents whose quality is far superior to those generated by word processors.

In the mid-1980s, American computer scientist Leslie Lamport created LaTeX, which complements the basic TₑX system with several macro packages to make the system more user-friendly. The idea behind LaTeX is to separate the document's structure from its presentation, so the task of writing (producing content) is cleanly separated from the task of formatting (presenting the content in an aesthetically pleasing way).

LaTeX has emerged as one of the most sophisticated digital typographical systems, especially prevalent in technical fields, such as mathematics, computer science, physics, and engineering, where it has become the universal document-writing system of choice. It is free, portable, platform-independent, and has global reach, making it ideal for collaboration. One may compile, view, and print a LaTeX-generated document on any computer in the world, regardless of type or operating system. Finally, unlike many word processors, LaTeX is extremely stable, has no bugs, and does not crash [54].

LaTeX is also popular in nontechnical fields, such as law, journalism, or any field that requires generating documents with non-Latin scripts (Greek, Russian, Chinese, Arabic, etc.) or specialized symbols.

LaTeX is pronounced LAY-tek, LAH-tek, or la-TEK. The final "X" in LaTeX and TeX is the Greek letter "chi" in the Greek word $\tau\varepsilon\chi\nu\eta$, which means "art" (or "craft,") and is the root of the English words "technical" and "technique."

LaTeX distributions

To use LaTeX on a computer, one must first download and install a LaTeX *distribution*, which includes the TeX engine, the LaTeX macros, and several additional programs, packages, and files required for LaTeX to run properly. Many distributions also include a plain text editor.

Several LaTeX distributions can be downloaded for free from the internet; the most common ones are TeX Live, MikTeX, and MacTeX.

The LaTeX source code

The LaTeX source code, which contains all the information needed to generate a document (content and markup commands), is a plain text file with a `.tex` extension, called a "TeX file." Plain text files are the most versatile of file types, containing only characters. Any computer, regardless of its type or operating system, is equipped with a *plain text editor* to create, view and edit plain text files.[5] Although one can create a TeX file with any plain text editor, most LaTeX distributions come with a TeX-specific plain text editor (such as TeXworks, TeXMaker, and TeXstudio) with additional features, like a spellchecker, drop-down menus, auto-completion of LaTeX commands, and auto-compilation.

Compiling a LaTeX document

To view the final form of a LaTeX-generated document, one must first *compile* it. Using the content and formatting instructions contained in the source code, a LaTeX *compiler* assembles the document and creates an *output file*, which one may view and print.

The *Portable Document Format* (PDF), a file type created in the 1990s by Adobe Inc., is the most common output file type today. Independent of application software, hardware, and operating systems, PDF files can be viewed and printed on any computer equipped with one of the many freely available PDF viewers (such as *Adobe Reader*). PDF has risen to become the leading global document-sharing format across operating system platforms.

---

[5]For example, *Notepad* and *TextEdit* are the native plain text editors for the Windows and macOS operating systems, respectively.

10.3.2 The TEX file

The TEX file contains the LATEX *source code*, which consists of the document's content, the *markup commands*, and comments.

Markup commands

Every markup command begins with a backslash (\). Commands that apply to the entire document appear in the *preamble*, while those that only apply to a specific part of the document are embedded into the content itself. For example, writing

```
\geometry{margin=2in}
```

in the preamble creates 2-inch margins for the entire document. Conversely, in the following line of text,

```
The function is \textbf{continuous}.
```

the command \textbf{...} only applies to the word "continuous," making it appear in bold font.

Comments

The LATEX ignores anything on a line after a "%" symbol.[6] A writer often uses this feature to insert personal notes and comments in the code, which will not appear in the compiled document.

Preamble

The *preamble* of the LATEX code contains a list of commands that affect the document as a whole. These include:

- **The document class.** The writer may choose one of several different LATEX document types (called *classes*): article, book, letter, report, slides, and many others. Short documents use the article document class, so in this case, the first line of the LATEX code is \documentclass{article}. The writer may insert optional formatting instructions in brackets. For example, the following imposes 12 point font in an article:

  ```
  \documentclass[12pt]{article}
  ```

- **Packages.** LATEX commands come in bundles called *packages*. For example, the package amsmath contains commands for common mathematical symbols, geometry has commands for page dimensions and margins, and fancyhdr contains those for headers and footers. Although a standard LATEX distribution comes

---

[6]To type %, use \%.

with several commonly used packages, occasionally a user may need to install an additional package containing a command for a specific format or newly created symbol.[7] The preamble invokes the packages containing all the commands that the writer will use throughout the document. For example, the preamble may invoke the three packages mentioned above as follows:

```
\usepackage{amsmath}
\usepackage{geometry}
\usepackage{fancyhdr}
```

- **Title, date, author.** These items appear in the preamble as follows:

```
\title{The title goes here}
\author{The author goes here}
\date{The date goes here}
```

Environments

After the preamble, the writer enters the content of the document between the commands \begin{document} and \end{document}. All content between these commands is said to be in the "document" *environment.*

Within the document, one may nest other environment types, such as quote (for quotes) or center (for a centered paragraph). Every nested environment begins and ends with \begin{...} and \end{...}. For example, to include a quote in a document:

```
\begin{document}
.

.

.

  \begin{quote}
    The quote goes here.
  \end{quote}
.

.

.

\end{document}
```

Lists. The commands enumerate and itemize create a numbered list and a bulleted list, respectively. Within these lists, the command \item instructs LaTeX to switch to the next item on the list. For example:

---

[7]The Comprehensive TeX Archive Network (ctan.org) is the world's central repository for LaTeX packages. Any package can be downloaded and installed from ctan.org for free.

```
\begin{enumerate}
  \item First item goes here.
  \item Second item goes here.
  \item Third item goes here.
  .
  .
  .
\end{enumerate}
```

Enumerations and itemizations can be nested within one another, up to four levels deep.

Mathematics. LaTeX uses two mathematics-specific environments, *inline math* and *display math*, which we discuss in Subsection 10.3.3 below.

Arrays, tables, figures. The `array`, `tabular`, `table`, and `figure` environments are presented in Subsection 10.4.1.

Sectioning commands

Within the `document` environment, the command `\maketitle` creates the *top matter* (title, author, and date) with the information entered in the preamble. One may divide the rest of the document into sections (`\section{...}`), which can be further divided into subsections (`\subsection{...}`), and subsubsections (`\subsubsection{...}`). For example:

```
\section{Title of the first section}
  \subsection{Title of the first subsection}
    \subsubsection(Title of the first subsubsection}
    .
    .
    .
    \subsubsection{Title of second subsubsection}
    .
    .
    .
  \subsection{Title of second subsection}
  .
  .
  .
```

By default, LaTeX assigns numbers to chapters, sections, subsections, and subsubsections when compiling the document. To suppress the numbering, one may instead use `\chapter*`, `\section*` and so on. One may also modify the default format of the headings (font and spacing) by appropriate commands in the preamble.

Text, spacing, fonts

When entering text, the amount of space between words is unimportant, because LaTeX will adjust the spacing automatically when compiling the document. The writer may use the ~ symbol to insert a "protected" space that LaTeX will not break between lines. A blank line begins a new paragraph. The command \vspace{*length*} inserts a vertical space between lines, where *length* can be specified in inches or centimeters, for example \vspace{0.25in}.

To italicize, boldface, or underline text, the writer can utilize the commands \textit{...}, \textbf{...}, and \underline{...}, respectively.

10.3.3   Mathematics in LaTeX

Mathematics-specific environments

LaTeX uses two environments for writing mathematics: *inline* math and *display* math.

Inline math, for short expressions and equations, uses dollar signs $...$ as delimiters embedded in a line of text.

Display math, for longer expressions and equations, displays expressions centered on their own line. The delimiters are \[and \], and LaTeX writers usually put them on separate lines for clarity.

Example 10.3.1. LaTeX code:

```
The equation of the unit circle in the $xy$-plane is
$x^2+y^2=1$.
```

Compiled output:
    The equation of the unit circle in the xy-plane is $x^2 + y^2 = 1$. $\triangle$

Example 10.3.2. LaTeX code:

```
The values of $x$ that satisfy the equation
\[
ax^2+bx+c=0
\]
are
\[
x=\frac{1}{2a}\left(-b\pm\sqrt{b^2-4ac}\right).
\]
```

Compiled output:
    The values of x that satisfy the equation

$$ax^2 + bx + c = 0$$

are

$$x = \frac{1}{2a}\left(-b \pm \sqrt{b^2 - 4ac}\right).$$ △

One may add a *label* (\label...) to an equation in display environment. When the document is compiled, LaTeX will assign a number to labeled equations. One may also use these labels to refer to specific equations in the text. Section 10.4.2 contains more information about labels in LaTeX.

Spacing and fonts

LaTeX ignores spaces in mathematics environments, so for example $x   y$ appears as xy in the compiled document. Therefore, the writer must insert spaces in the text on either side of the inline environment to prevent words from sticking to a mathematical expression. One may also enter spaces of different sizes manually inside a mathematics environment with the following commands:

thin space	\,
medium space	\:
thick space	\;
quad space (width of a capital M)	\quad
double quad space	\qquad

Because LaTeX uses a math-specific font in mathematics environments, the writer should always enter mathematical expressions in a mathematics environment. Thus, for example, the variable x should be entered as x.

Mathematical symbols

The amsmath LaTeX package contains commands for hundreds of common mathematical symbols, and thousands of less-common ones are available in specialized packages. The tables below list the commands for some of the most common symbols; for others, one may find several searchable lists of symbols on the internet.[8]

- **Subscripts and superscripts** are entered by _ and ^ respectively. For example a_{ij}x^2 gives $a_{ij}x^2$.

- **Functions.** LaTeX uses a math-specific font for functions in mathematics environments, so the writer should always enter them in a mathematics environment. Thus, for example, $\sin x$ should be entered as $\sin x$. The following table contains a list of common functions. One may write others manually

---

[8] *Detexify* (detexify.kirelabs.org) allows users to draw a symbol by hand and suggests LaTeX commands for possible matches.

with the command `\text{...}`, for example `$\text{Col}A$` for Col$A$ (the column space of A).

sin	`\sin`	sec	`\sec`	ln	`\ln`	min	`\min`
cos	`\cos`	csc	`\csc`	log	`\log`	max	`\max`
tan	`\tan`	cot	`\cot`	arg	`\arg`	det	`\det`

- **Greek letters.** The following table lists the lowercase letters. Capitalizing the letter names produces uppercase letters, for example `\Gamma` for Γ.

α	`\alpha`	η	`\eta`	ν	`\nu`	τ	`\tau`
β	`\beta`	θ	`\theta`	ξ	`\xi`	υ	`\upsilon`
γ	`\gamma`	ϑ	`\vartheta`	o	o	ϕ	`\phi`
δ	`\delta`	ι	`\iota`	π	`\pi`	φ	`\varphi`
ϵ	`\epsilon`	κ	`\kappa`	ρ	`\rho`	χ	`\chi`
ε	`\varepsilon`	λ	`\lambda`	ϱ	`\varrho`	ψ	`\psi`
ζ	`\zeta`	μ	`\mu`	σ	`\sigma`	ω	`\omega`

- **Variable-sized delimiters.** Delimiter commands always appear in pairs (left and right).[9] The writer should always use the proper variable-size delimiter commands in mathematical environments, so that their size matches the height of their contents.

(...)	`\left(...\right)`	⟨...⟩	`\left\langle...\right\rangle`
[...]	`\left[...\right]`	|...|	`\left\vert...\right\vert`
{...}	`\left\{...\right\}`	‖...‖	`\left\Vert...\right\Vert`

The following tables show the commands for variable-size constructs and operators, operations and other symbols, arrows, accents, and fonts.

- **Variable-sized constructs and operators**

$\frac{abc}{xyz}$	`\frac{abc}{xyz}`	$\sum_{i=1}^{n}$	`\sum_{i=1}^{n}`
$\sqrt{abc}$	`\sqrt{abc}`	$\int_{a}^{b}$	`\int_{a}^{b}`
$\sqrt[n]{abc}$	`\sqrt{n}{abc}`	$\lim_{x\to a}$	`\lim_{x \rightarrow a}`
$\iint_{D}$	`\iint_{D}`	$\iiint_{E}$	`\iiint_{E}`

- **Operations and other symbols**

---

[9] For a single delimiter on one side only, use `\left.` or `\right.` to omit the delimiter on the other side.

·	\cdot	≠	\neq	⪇	\nleq	∀	\forall
*	\ast	≡	\equiv	⪈	\ngeq	∃	\exists
⋆	\star	≈	\approx	⊂	\subset	∄	\nexists
∘	\circ	∼	\sim	⊆	\subseteq	′	\prime
ℓ	\ell	≤	\leq	⊈	\nsubseteq	∂	\partial
±	\pm	≥	\geq	∈	\in	∇	\nabla
×	\times	≶	\lessgtr	∉	\notin	∞	\infty
∪	\cup	≮	\nless	∅	\emptyset	⋯	\cdots
∩	\cap	≯	\ngtr	∅	\varnothing	⋮	\vdots

- **Arrows**

←	\leftarrow	→	\rightarrow	↔	\leftrightarrow
⇐	\Leftarrow	⇒	\Rightarrow	⇔	\Leftrightarrow

- **Accents**

$\hat{q}$	\hat{a}	$\bar{a}$	\bar{a}	$\dot{a}$	\dot{a}
$\tilde{a}$	\tilde{a}	$\vec{a}$	\vec{a}	$\ddot{a}$	\ddot{a}

- **Fonts**

blackboard $\mathbb{ABCDEFGHIJ}$	\mathbb{ABCDEFGHIJ}
bold **ABCDEFGHIJ**	\mathbf{ABCDEFGHIJ}
calligraphic $\mathcal{ABCDEFGHIJ}$	\mathcal{ABCDEFGHIJ}
script $\mathscr{ABCDEFGHIJ}$	\mathscr{ABCDEFGHIJ}
fraktur $\mathfrak{ABCDEFGHIJ}$	\mathfrak{ABCDEFGHIJ}

The following example shows the code and output for a paragraph containing several symbols, constructs, and variable-size delimiters in inline and display environments.

Example 10.3.3. LaTeX code:

```
\begin{document}
To evaluate the integral
\[
\int_{0}^{1}x\sqrt{1+3x^{2}}\,dx,
\]
we use the substitution $u=1+3x^{2}$, which implies that
$du=6x\,dx$.  The integral becomes
\[
\frac{1}{6}\int_{1}^{4}\sqrt{u}\,du=
\frac{1}{9}\left(4^{\frac{3}{2}}-1\right)=\frac{7}{9}.
\]
\end{document}
```

Compiled output:

To evaluate the integral

$$\int_0^1 x\sqrt{1+3x^2}\,dx,$$

we use the substitution $u = 1 + 3x^2$, which implies that $du = 6x\,dx$. The integral becomes

$$\frac{1}{6}\int_1^4 \sqrt{u}\,du = \frac{1}{9}\left(4^{\frac{3}{2}} - 1\right) = \frac{7}{9}. \qquad\qquad \triangle$$

10.3.4   Front end applications for LaTeX

Because the learning curve for learning LaTeX can be steep, various *front end applications* provide a user-friendly interface to write a LaTeX document. The open-source program *LyX* (www.lyx.org) is a popular add-on tool for beginners, offering drop-down menus and icons to help with math symbols, formatting, graphics placement, compiling, and viewing, while simultaneously showing the corresponding LaTeX commands to help users learn them gradually. LyX uses the WYSIWYM concept ("What You See Is What You Mean"); it tries to interpret what the user intends to write and renders on the screen an approximation of the compiled document. LyX creates a .lyx file as well as the .tex file with the LaTeX code. It comes with a tutorial, spellchecker, thesaurus, and other advanced editing features.

Other popular front end applications for LaTeX include *TeXshop*, *Writer2LaTeX*, and *Scientific WorkPlace*.

10.3.5   Overleaf

With a user base of over 2,000,000 users in 180 countries, *Overleaf*, (found at www.overleaf.com), is the most popular online collaboration platform that runs on LaTeX. Overleaf allows users to generate, share, compile, and display LaTeX documents online and offers other advantages:

- **Accessibility**. Anyone with an internet connection can access and use Overleaf. There is no need to install a LaTeX distribution onto a personal computer.

- **Immediate Output**. Overleaf boasts an *auto-compile* feature, i.e., it can compile and display the output document in real time, as the writer types the LaTeX code.

- **Storage**. Users can create a free Overleaf account to store documents in the cloud, making them readily accessible from any computer with an internet connection.

- **Collaboration.** Multiple collaborators can work on the same document at the same time and see changes to the document in real time.

Overleaf also comes with a tutorial and helps catch coding errors by displaying warning messages. The only major drawback to Overleaf is the need for a relatively fast and reliable internet connection.

EXERCISES FOR SECTION 10.3

1. The proofs in Sections 2.2 and 2.3, and the problems in Section 10.1 are excellent exercises to practice LaTeX.

10.4 Advanced Typesetting

This section builds on the previous one with guidelines for more advanced typesetting features: arrays (Subsection 10.4.1), tables and figures (Subsection 10.4.2), and bibliographies (Subsection 10.4.3).

10.4.1 Arrays

Arrays and matrices

The **array** environment creates a rectangular grid (called an *array*) nested inside a mathematics environment (either inline or display). The user can specify the number of rows and columns, and whether the content of each cell is to be centered, left-aligned, or right-aligned. The writer may use an array to display a matrix, a system of equations, or a bracketed function.

- The number of columns is defined in curly brackets at the beginning, along with the alignment of the content in each column (c for "centered", l for "left-aligned" and r for "right-aligned"). For example, {ccccc} means five centered columns, and {rcl} means three columns, where the content appears right-aligned, centered, and left-aligned, respectively.
- The content of each cell is entered row by row, using an ampersand (&) symbol to switch to the next cell in the same row, and a double backslash (\\) to switch to the next row.
- The entire array can be framed by variable-size delimiters, such as \left(and \right), whose size automatically adjusts to the size of the array.[10]

Example 10.4.1. LaTeX code:

---

[10]Alternatively, one may use the LaTeX environments pmatrix or bmatrix for matrices in parentheses or brackets, respectively.

```
\[
\left(
\begin{array}{ccc}
 a & b & c \\
 d & e & f \\
 g & h & i \\
\end{array}
\right)
\]
```

Compiled output:

$$\left(\begin{array}{ccc} a & b & c \\ d & e & f \\ g & h & i \end{array} \right).$$

△

Bracketed functions

For systems of equations and bracketed functions (such as piecewise-defined functions), one may use the \text{...} command to enter text (in text font) inside cells.

Example 10.4.2. LaTeX code:

```
\[
|x|=\left\{
\begin{array}{ccl}
 -x & \text{if} & x<0,\\
  x & \text{if} & x \geq 0.
\end{array}
\right.
\]
```

Compiled output:

$$|x| = \left\{ \begin{array}{ccl} -x & \text{if} & x < 0, \\ x & \text{if} & x \geq 0. \end{array} \right.$$

△

Here, the \left{ ... \right. pair produces the single variable-size bracket on the left, and the words "if" are entered as text in their own center-aligned column.

10.4.2   Tables and figures

Tables and figures play an important role in mathematical documents. Tables organize large amounts of quantitative information and figures convey graphs, diagrams, images, and other visual information that is difficult to describe otherwise. The clarity of a mathematical paper depends heavily on the clarity of its tables and figures. For this reason, a writer will often spend more much time on them than any other part of the paper. Furthermore, readers often study the tables and figures of a paper before reading the text, so a reader's first impression of a paper is largely dependent on the quality of its tables and figures.

Tables and figures must stand on their own, i.e., the writer should not assume that a reader has access to other parts of the paper when reading them. The reader should not have to hunt for definitions of symbols, variables, abbreviations, or acronyms; in tables and figures, these items must be spelled out in full.

Formal mathematical documents only use the terms *table* and *figure* (not "chart," "diagram," "graph," etc., although a figure may contain a diagram or a graph). Furthermore, the writer must make sure that every table and figure is (1) numbered, (2) captioned, and (3) cross-referenced, as we explain below.

Tables

In LATEX, the `tabular` environment creates a table in much the same way as the `array` environment.[11]

- The number of columns is defined in curly brackets at the beginning, along with the alignment of the content in each column (c, l, or r). If desired, a vertical separator line (|) or double line (||) may be specified between columns. For example, {||c|c|c||} means three centered columns separated by vertical lines, with double lines at the left and right ends.

- The content of each cell is entered as in an array, using the ampersand and double backslash symbols to switch to the next cell and row, respectively.

- The command \hline inserts a horizontal line across all of the columns.

Example 10.4.3. LATEX code:

---

[11]But unlike an *array*, a table cannot be nested inside a mathematics environment.

```
\begin{center}
 \begin{tabular}{|c||c|c|}
 \hline
 Age (mo) & Average weight (lbs) & Average
 height (in) \\
 \hline
 \hline
 0 & 7.3 & 19.4 \\
 \hline
 6 & 16.6 & 25.9 \\
 \hline
 12 & 20.4 & 29.2 \\
 \hline
 18 & 23.4 & 31.8 \\
 \hline
 24 & 26.4 & 33.5 \\
 \hline
 \end{tabular}
\end{center}
```

Compiled output:

Age (mo)	Average weight (lbs)	Average height (in)
0	7.3	19.4
6	16.6	25.9
12	20.4	29.2
18	23.4	31.8
24	26.5	33.5

The writer usually chooses to nest the **tabular** environment inside a **center** environment, so that the table appears centered on the page.   △

Figures

Although certain LaTeX packages (e.g., *TikZ*) allow writers to create graphic elements in the LaTeX code itself, in some cases it may be necessary (or easier) to import an image (such as a graph or diagram) created on another application, such as Microsoft Excel, PowerPoint, Adobe Illustrator, MATLAB, or Mathematica. In this case, one must

1) invoke the "graphicx" package with \usepackage{graphicx} in the preamble,

2) import the image by typing

\includegraphics[*options*]{*path/to/filename*}

in the code. Here, *path/to/filename* is the name and location of the graphics file (which may have the extension .png, .jpg, .eps, etc.), and *options* lists optional specifications regarding the size and placement of the image. For example, the option scale=0.5 scales the graphic to 50% of its original size.[12]

When importing graphics created on other applications, the user may need to scale and label axes (with unabbreviated quantities and units), adjust the thickness and colors of curves, and increase the font size of the text. Finally, in some cases, it may be preferable to combine various graphics into a single *multipaneled figure* with a common caption, as this reduces the total number of figures and helps the reader compare similar graphs or diagrams.

Floats and numbering

The preferred method of positioning a table or figure in a LaTeX document consists in putting it (along with its caption) inside a box called a *float*, which is free to float around the document. This allows LaTeX to move it when compiling the output document, so it does not get cut between pages or leave unappealing blank spaces before or after it. One is not required to place a table or figure in a float, but it is strongly recommended to do so, so that LaTeX can determine its optimum positioning on the page.[13]

The environments

table   (\begin{table}...\end{table})

and

figure   (\begin{figure}...\end{figure})

are for table and figure floats, respectively. LaTeX will number each table and figure in sequence when it compiles the output document.

Captioning

In a formal paper, every table and figure must be accompanied by a short paragraph placed below it (called a *caption*) that explains and/or clarifies its content. In keeping with the self-contained nature

---

[12]If using Overleaf (or any other online LaTeX platform), one must upload the image file to the Overleaf server.

[13]At first, it may be off-putting to see the tables and figures move to a different placement than originally planned. However, seasoned readers of mathematical papers are not fazed by this; they are accustomed to skipping over tables and figures if they appear in the middle of a paragraph and going back to them only when they are specifically cross-referenced in the text. For this reason, the writer must always properly number and cross-reference every table and figure, or else a reader may skip it and never come back to it.

of tables and figures, captions must be sufficiently detailed so that a reader can completely understand the information that the table or figure is intended to convey without referring to other parts of the paper. For this reason, a caption may be several lines long and repeat word for word explanations that also appear elsewhere in the paper. If the table or figure contains data or images taken from someone else's work, the caption must include a citation for their source.

In LaTeX, captions are entered with \caption{...} placed immediately after the table or figure, but still inside the environment.

Labeling and cross-referencing

In a formal paper, the writer must *cross-reference* every table and figure in the text by its number. To do this, one must

1) create a label for it \label{tab:*name*} or \label{fig:*name*} (for a table or figure, respectively) in the float, where *name* is a user-defined identifier.

2) cross-reference the table or figure later in the text with the code \ref{tab:name} or \ref{fig:name}.

The identifier name does not appear in the compiled document; LaTeX replaces it by its appropriate number in the sequence of tables and figures.

If the table or figure is placed in a float, the writer must take care to cross-reference by number, not by position (e.g., "the table below"), because in the compiled document, the float may have moved to a different position.

In the same way, one may also label and cross-reference equations, sections, and subsections, using respectively \label{eq:name}, \label{sec:name}, and \label{subsec:name}.

Finally, the writer must take care to capitalize the words table, figure, equation, section, and subsection when they are referring to a specific item by its number, for example, "the data in Figure 2 shows...".

In the following example, a float contains a table, caption, and label.

Example 10.4.4. LaTeX code:

```
The average weight and height of girls between 0 and 2
years old born in the United States between 2000 and
2010 appear in Table~\ref{tab:weight_height_girls}.
```

```
\begin{table}[ht]
\begin{center}
\begin{tabular}{|c||c||c|c|}
```

```
\hline
Age (mo) & Average weight (lbs) & Average height~(in) \\
\hline
\hline
0 & 7.3 & 19.4 \\
\hline
6 & 16.6 & 25.9 \\
\hline
12 & 20.4 & 29.2 \\
\hline
18 & 23.4 & 31.8 \\
\hline
24 & 26.4 & 33.5 \\
\hline
\end{tabular}
\end{center}
\caption\{Average weight and height of girls between
0 and 2 years old born in the United States between
2000 and 2010. Source: Center for Disease Control
and Prevention (CDC), www.cdc.gov.}
\label{tab:weight_height_girls}
\end{table}
```

Compiled output:

The average weight and height of girls between 0 and 2 years old born in the United States between 2000 and 2010 appear in Table 1.

Age (mo.)	Average weight (lbs.)	Average height (in.)
0	7.3	19.4
6	16.6	25.9
12	20.4	29.2
18	23.4	31.8
24	26.5	33.5

Table 1: Average weight and height of girls between 0 and 2 years old born in the United States between 2000 and 2010. Source: Center for Disease Control and Prevention (CDC), www.cdc.gov.

The option [ht] to \begin{table}[ht] tells LATEX to attempt to put the table [h] "here" (where the code for the table appears) and, if that does not work well, to place it at the [t] "top" of the current page or more likely the next.   △

Using data or images from another source

To avoid plagiarism, if any graphic, data, or result presented in a table or figure comes from another source, then its caption must credit the source, either by stating the full reference or by providing an in-text citation that refers to a Bibliography entry. Furthermore, if a figure contains an *image* created by another person, the writer must provide the name of the copyright owner and certify that he or she was granted permission to legally use the image. Three cases are possible:

- **Copyrighted Images.** If an image is protected by copyright, it is not enough to merely cite its source. Before using the image, one must obtain permission from the copyright owner, who may be the creator of the original image or a third party who later obtained the copyright. Permission is not automatically granted, and the user may need to pay royalties to the copyright owner.

- **Creative Commons.** Creative Commons is a type of license made available by the copyright owner that allows others to copy, distribute, and display an image for noncommercial purposes as long as they give credit to its creator. In most cases, images under a Creative Commons license can be freely used if they are properly cited.

- **Public Domain.** Public domain images are images whose creator is unknown, or has relinquished ownership and control over the image. These images may be freely used and it is not necessary to provide a citation.

10.4.3 Bibliographies

Citations

As in all academic fields, the author of a document must provide a *citation* to acknowledge the source of any portion of the content that contains or is based on the work of another person. Failure to do so constitutes *plagiarism*, which is unethical, unlawful, and carries heavy legal consequences. In a mathematical paper, several things may come from external sources:

- previously proved theorems,

- concepts, conjectures, or ideas developed by others,

- data and results, including tables and graphs,

- methods, techniques, algorithms, and software tools,

- images, diagrams, and figures.

A citation may be in the form of a direct quote or paraphrase.[14] In either case, a citation is signaled by a short note or alphanumeric code in the body of the text, called an *in-text citation*, usually enclosed in parentheses or brackets. For example,

"According to John Mackey, the logistic model does not accurately describe the growth of this population [3]."

Every in-text citation refers to an entry in the *Bibliography* section, usually at the end of the paper. In-text citations can appear almost anywhere in a paper, including in captions of tables and figures. However, the Abstract must not contain any citations, as is it often published by itself, without the Bibliography.

Bibliography

The *Bibliography* (also called *references* or *works cited*, depending on the citation style) contains the full source information for all the works cited in a paper. All in-text citations refer to a Bibliography entry, and every item in the Bibliography must cross-reference at least one in-text citation.

A Bibliography entry must provide sufficient data to uniquely define a source and enable a reader to find it easily. The items required for a valid reference vary by source type. For journal articles, books, and webpages, at least the following items are required:

- **Journal articles.** Author(s), title of article, publication date, journal name, volume and issue, page number(s).

- **Book.** Author(s), title of book, publication date, publisher, page number(s).

- **Webpage.** Author(s), title of webpage, URL, date accessed.[15]

Similar lists of requirements exist for other types of sources, such as conference proceedings, dissertations, and working papers.

Bibliography and citation styles

A *citation style* refers to the specific format of a Bibliography entry and its corresponding in-text citation. There are hundreds of different citation styles; they vary in terms of capitalization, punctuation, font styles, formatting of names, order of the items in each entry, and order of the entries in the Bibliography. Therefore, the writer must take care to remain consistent in sticking to one style and providing

---

[14]In mathematics, it is usually not necessary to quote word for word using quotation marks; in fact, it is preferred not to do so.

[15]Because the information on a webpage can change, the reference for a webpage must include the "date accessed."

complete references in accordance to that style. Here are five different citation styles for the same journal article:

- Peter Adams, *Local Lyapunov exponent and dimension of strange attractors*, Journal of Nonlinear Dynamics **23** (2015), no. 2, 145-156. (AMS style)

- Adams, P. (2015). Local Lyapunov exponent and dimension of strange attractors. *Journal of Nonlinear Dynamics*, 23(2):145-156. (APA style)

- P. Adams, Local Lyapunov exponent and dimension of strange attractors, *Journal of Nonlinear Dynamics*, vol. 23, no. 2, pp. 145-156, 2015. (IEEE style)

- P. ADAMS, *Local Lyapunov exponent and dimension of strange attractors*, Journal of Nonlinear Dynamics, 23:2 (2015), pp. 145-156. (SIAM style)

- Peter Adams. Local Lyapunov exponent and dimension of strange attractors. *Journal of Nonlinear Dynamics*, 23(2):145-156, 2015. (Plain style)

Similarly, depending on the citation style, the in-text citation may list the author, publication year and/or page number, an alphanumeric code, or just a number. Here are some examples:

- [Ada15] (AMS style)

- [Adams, 2015] (APA style)

- [3] (Plain style)

Certain academic fields have their own citation style, while others borrow one of the standard styles listed in the table below.

Citation style	Academic fields
APA	Psychology, business, economics, education, social sciences.
CMS	History, fine arts, anthropology, philosophy.
CSE	Biology, chemistry, geology, physics.
IEEE	Engineering, computer science, information science.
MLA	English, literature, foreign languages, communications, religious studies.

There is no standard citation style for mathematics. Some mathematicians use the style employed by the American Mathematical Society (AMS) or the Society for Industrial and Applied Mathematics (SIAM); others simply prefer the all-purpose "Plain" style.

BibTex

Writing a Bibliography in the correct style can be tedious work.[16] For this reason, LaTeX users have created a system called *BibTeX* that manages all aspects of the Bibliography and corresponding in-text citations.

A *BibTeX file* is a plain text file that contains all the bibliographical data in "raw" format. The writer can attach a BibTeX file to a LaTeX code, specify a citation style, and LaTeX will render the bibliography and all in-text citations in the correct format. Thus, one can easily change the citation style without having to re-type the Bibliography and go through the document to correct every in-text citation. Also, as a plain text file, the BibTeX file is portable and can be attached repeatedly to any LaTeX document.

BibTeX file

The BibTeX file is a plain text file that contains a list of *records*, one for each Bibliography entry. Each record begins by identifying the source type by either @article{...}, @book{...}, @manual{...}, @misc{...}, @phdthesis{...}, @proceedings{...}, etc. The information contained in each record starts with the identifier label, followed by a list of *fields* in the format

field = value

where each *field* may be the author, title, date, etc., as required by the source type.

Example 10.4.5. Reference for a journal article in Plain style:

Peter Adams. Local Lyapunov exponent and dimension of strange attractors. *Journal of Nonlinear Dynamics,* 23(2):145-156, 2015. △

The same reference as a BibTeX record:

```
@article{adams2015,
    title={Local {L}yapunov exponent and dimension of
    strange attractors},
    author={Peter Adams},
    journal={{Journal of Nonlinear Dynamics}},
    volume=23,
    number=2,
    pages={145-156},
    year=2015,
}
```

---

[16]It is equally tedious to have to re-type an existing Bibliography in a different style.

In this example, the record begins with @article{...}. This identifies the entry as a journal article. The code adams2015 is a user-defined identifier label for the record. The title, author, journal, volume, number, pages, and year appear as separate fields. The order in which the fields appear, formatting of names, and capitalization of titles are unimportant, because LaTeX will adjust the format to the specified citation style. However, because some styles do not capitalize any words in the title, it may be necessary to "protect" a capital letter by framing it in curly brackets, as the "L" of "Lyapunov."

Example 10.4.6. The following BibTeX file contains three records: a journal article, a book, and a website.

```
@article{adams2015,
  title={{Title of Article}},
  author={Peter Adams},
  journal={{Title of Journal}},
  volume=23,
  number=2,
  pages={145-156},
  year=2015,
}
@book{babbington2015,
  title={{Title of Book}},
  author={Peter Babbington},
  publisher={name of publisher},
  year=2015,
}
@misc{cidi2015,
  title={Center for International Disaster Information},
  howpublished={www.cidi.org},
  year=2015,
  note={Online, accessed 2020-09-30},
}
```

The writer may type a BibTeX file manually, one field at a time. However, there are a few ways to simplify the task:

1) Indexing services such as *Google Scholar* and *Web of Science* provide ready-made BibTeX records that a user may copy and paste into a BibTeX file.

2) Reference manager tools, such as *Mendeley Desktop*, can generate the entire BibTeX file by simply selecting the relevant references in the application's library.

Attaching a BibTeX file to a LaTeX code

To attach a BibTeX file to a LaTeX file, one must

1) change the extension of the BibTeX file to .bib.

2) invoke the BibTeX file in the code at the place in the document where the Bibliography is to appear (usually at the end) by typing

\bibliography{*path/to/bibfile*}
\bibliographystyle{*stylename*}

where *path/to/bibfile* is the name and location of the Bib-TeX file (without the .bib extension) and *stylename* is the citation style (such as apa or plain).[17]

To insert an in-text citation, the writer enters the command

\cite{*label*}

in the appropriate location in the text, where *label* is the identifier label of the Bibliography record. For example, to cite the journal in the previous example, one would type \cite{adams2015}.

EXERCISES FOR SECTION 10.4

1. The problems in Section 10.1 illustrated by diagrams are excellent exercises to practice LaTeX by typetetting a solution that includes a figure.

2. Find the minimum and maximum values of the following function and sketch its graph.

$$f(x) = 2x^3 - 12x^2 + 18x.$$

Write your answer as a formal *problem & solution* paper (as outlined in Subsection 10.1.3).

3. The table below shows the percentages of women graduating with a bachelor's degree in four different STEM fields in the U.S.

Year	Natural Sciences	Social Sciences	Mathematics	Engineering
1970	24.1	42.7	37.6	1.4
1975	30.7	42.2	38.4	3.0
1980	39.5	49.2	38.1	10.6
1985	36.3	40.7	30.3	14.2
1990	48.9	52.4	36.5	16.2
1995	50.3	54.1	35.8	18.5
2000	55.6	59.0	33.7	20.8
2005	60.2	60.4	29.0	21.1
2010	58.9	59.3	26.5	19.6

The source of these data is the following article:

---

[17]If using Overleaf (or any other online LaTeX platform), one must upload the BibTeX file to the Overleaf server.

Kahn, S. and Ginther, D. (2015), Are recent cohorts of women with engineering bachelors less likely to stay in engineering?, *Frontiers in Psychology*, 6:1144, 1-15.

Create a short L#TEX document that contains the following items.

(a) A short paragraph that introduces the data and discusses what is noteworthy or surprising about the data.

(b) The data presented in a table. Recreate the table above in L#TEX.

(c) The data illustrated graphically. Create a figure containing a graph that shows the percentages for the four STEM fields versus time on the same graph, interpolated linearly (i.e., connecting the data points by straight lines). Use the software of your choice to generate the graph.

(d) A one-entry Bibliography in the Plain citation style.

Make sure that the table and figure are clear, legible, centered, scaled, numbered, captioned, cross-referenced, and cite the source of the data in the captions.

4. Choose a topic of your interest in applied or pure mathematics and search for information or data sources related to your topic. Select at least (a) three peer-reviewed articles, (b) two websites, and (c) one book. Create a BibTeX file for all the references. Write a short L#TEX document that cites all the sources and include a Bibliography in the Plain citation style.

10.5   Oral Presentations

Giving a good oral presentation is a necessary skill in many professions. In the mathematics community, oral presentations are ubiquitous; they play a vital role in the transmission of mathematical knowledge and constitute the central activity at academic conferences. Businesses that employ mathematicians expect them to be able to communicate the essence of their work orally, in public, and to a variety of audiences.

Preparing and delivering a good mathematical presentation takes skill and practice. Unlike oral presentations at academic events in the humanities, where presenters merely read a recently published paper, mathematical presentations require an active teaching component, rely heavily on a visual support, and must exploit the interplay between explanations and visuals effectively. Furthermore, a presentation is different than a lecture. It is usually short (10-15 minutes) and must get across the essence of a body of work that may have taken months to complete. Although the presentation itself is only a few minutes long, the presenter may spend several hours planning the content, preparing the visual supports, and rehearsing the delivery.

One of the distinctive aspects of an oral presentation in mathematics and science is the heavy reliance on a visual support. Mathematical calculations are difficult to follow without seeing them, and abstract mathematical concepts that are difficult to describe in words may be apprehended more readily by a diagram or graph. Nowadays, the two most common forms of visual support are (1) presentation slides projected onto a screen and (2) a poster. We describe the advantages and disadvantages of each format in Subsection 10.5.1.

The presenter should divide the process of preparing an oral presentation into three phases: (1) planning, (2) designing, and (3) rehearsing. The planning phase consists in clarifying the goal, requirements, expectations, guidelines, and constraints, all of which will guide decisions regarding the content. The presenter should determine these things first to prevent the stress of having to redo the presentation and visual support later, on short notice. After the planning phase, the second and third phases consist in designing the visual supports (either slides or poster) and rehearsing the delivery. We discuss each of these phases in Subsections 10.5.2 (planning the content), 10.5.3 (designing the visual support), and 10.5.4 (rehearsing the delivery).

10.5.1 Presentation formats

Slides and posters

The most common type of presentation format is a *slide presentation* (or simply, a *talk*), where a presenter talks to a room full of people (usually for 10-20 minutes) before inviting questions from the audience. The setting and size of the audience vary, ranging from very intimate (a handful of people sitting around a conference table) to a large lecture hall with hundreds of people. The visual support, which complements the presenter's words, is a set of presentation slides generated on a computer application like *Microsoft PowerPoint*, *Google Slides*, *Prezi*, or *Beamer*.

A *poster presentation* is an alternate presentation format common in academic circles. Instead of slides, the presenter prints the visual information onto a single, large poster (typically 3 feet high by 4 feet wide), which is mounted onto a display board in a large room or hallway, alongside other posters, for a public two- to three-hour-long poster session. During this time, the presenters stand by their posters while participants attending the session browse through the posters, interacting with the presenters one by one.

A poster presentation is variable in its content and length, as the presenter can ask questions before presenting to customize the talk to their background level, interests, and time availability. Thus, the presentation takes the form of a two-way conversation, where the listener is free to interrupt with questions and the presenter can provide clarifications on the fly. During the course of a two-hour session, a presenter may give multiple slightly different presentations to different listeners.

A small poster session may feature a handful of posters and presenters, while poster sessions at national conferences can boast hundreds of presenters and several thousands in attendance. A session lasts two to three hours, but the posters themselves are often left on display for the duration of the conference.

Poster presentations are a more recent phenomenon (compared to traditional talks with slides), but they are gaining in popularity and are increasingly replacing oral sessions at academic conferences.

Differences between slide and poster presentations

Although a professor or event organizer may sometimes dictate the presentation format, presenters often have the freedom to opt for either a slide or a poster presentation, especially at large conferences. Each format has its advantages and disadvantages, some of which are outlined below. Seasoned presenters may be partial to one or the other, but ultimately the choice comes down to personal preference or personality type.

- Conference organizers schedule contributed slide presentations at precise time slots within a session running in parallel with other sessions. Thus, interested persons may miss the talk if the time slot is inconvenient or conflicts with conference events happening at the same time. In contrast, at major large-attendance events, poster sessions are scheduled at prime times that rarely conflict with other scheduled activities. Furthermore, even if one misses the poster session, one may still view the posters later, as posters often remain on display for the duration of the conference.

- Conference organizers group presentations into topic-specific sessions attended mainly by specialists already interested in the topic. In contrast, poster sessions attract many walk-by curious attendees who may become interested in a poster that caught their eye, even though they may be unfamiliar with the topic.

- Slide presentations are linear and irreversible. The audience is locked into the presenter's rate of delivery and must adjust to it. Furthermore, the presenter receives no feedback from the audience (except for a few questions at the end). At a poster presentation, the presenter and listener are engaged in a two-way conversation, so the presenter receives constant feedback. Listeners can ask the presenter to back up or spend more time on parts that are more interesting to them. Similarly, the presenter can easily notice when their interlocutor is looking puzzled, address them directly, and adjust his or her explanations.

- Designing a slide presentation that caters to people from different backgrounds (for example, those in the same subdiscipline, those further afield, and nonmathematicians) is difficult. In contrast, poster presentations are customized to their audience, so multiple audiences can be reached easily.

- Designing a poster requires more work than a set of slides. The limited real estate on a poster of fixed dimensions makes it challenging to fit all the relevant information in a balanced, sequential manner that is visually appealing without being overwhelming. Therefore, posters require extensive attention to layout and formatting. Slides offer more flexibility, as the presenter has the freedom to add more slides to distribute the content. Unlike a poster, inserting or modifying one slide does not require reformatting the entire layout.

- Because a poster must be printed, the presenter must complete it several days before the presentation. In contrast, one may modify and adjust a slide presentation right up until the last minute.

- Slides can be easily stored in multiple copies (on a USB stick, cloud server, and email) to minimize risk of loss. A poster is large and awkward to carry, especially during travel, and if it gets lost, or damaged, it is difficult to replace on short notice.

Other formats

Although slides and posters represent the great majority of oral presentations in mathematics, other formats exist. In a "chalk talk," the presenter uses only a chalkboard (or whiteboard). This method is common for lecturing, as it allows revealing sequential steps of a calculation or reasoning in real time. But the limited time of an oral presentation makes this method ineffective.

Older types of visual support include transparencies projected by an overhead projector and 35-millimeter photographic slides viewed with a mechanical light projector.[18] Although only a few decades old, these technologies are now considered obsolete and rarely used.

---

[18]The term *slide* in modern presentations comes from these small transparencies that "slide" in front of a light beam.

10.5.2 Planning the content

Audience

The audience is a central decision factor when planning a presentation, as it dictates what content to include or omit and the level of the explanations. The presenter must know the background and interests of the listeners, what they expect to get out of the presentation, and what the presenter expects the audience to walk away with. In a class, the professor may provide information about the background and expectations of the audience. At a conference, the audience depends on the theme of the session. Conference organizers use the previously submitted abstracts to group presentations into sessions revolving around a common theme. Therefore, a presenter can anticipate the audience's interests and background level by inquiring about the session's theme ahead of time. In interdisciplinary and applied mathematics, a topic may attract listeners well versed in mathematics but not in the area of application, or vice versa. Even when experienced presenters give the same presentation multiple times, they often re-purpose it for different audiences.

Poster sessions also typically revolve around specific themes but are more likely to attract curious attendees with a more diverse range of expertise and interests. However, posters offer more flexibility, as the presentation itself can be customized to each listener.

Format

With slides, the time limit is the primary factor that dictates the content and design, as the presenter must capitalize on every second available without going over the time limit. A good rule of thumb is to plan for 1 to 2 slides per minute; this allows for a good pacing without feeling rushed. The presenter may also want to inquire about the estimated size of the room and audience, and how much time is allotted to questions at the end of the talk.

A poster presentation is more flexible in terms of timing. Instead, the most important factor is the size of the poster, which is usually dictated by the size of the display boards available at the event venue.

Sectioning and content

The sections of a presentation are the same as in a written paper, whether a short problem, project report, research paper. For example, the presenter would divide an oral presentation of a project report into four sections: problem and data, model, solution, and summary of results (as we suggest in Section 10.2). Similarly, a presentation for a science project that features a mathematical model follows the

standard IMRAD sequence (introduction, methods, results, and discussion). In pure mathematics, oral presentations follow a logical
sequence of definitions, lemmas, theorems, and corollaries.

After defining the sections of the presentation, the presenter must
decide how many slides to allocate to each one (for a slide presentation), keeping in mind the pacing (1-2 slides per minute) and total
time available. For a poster, one must decide on the overall layout
and how much space to allocate to each section. Posters are most
often displayed in landscape orientation, divided into 3 to 4 columns
to be read down and across.

The content of each section depends on the background and expectations of the audience. The goal of an oral presentation is not
to teach all the details with the expectation that the audience will
emerge with perfect understanding. Instead, oral presentations convey an overall theme, providing the main concepts and steps, while
highlighting a few details and specific examples. The goal is to elicit
interest and curiosity so that listeners feel inspired to ask questions
and seek out more information for themselves. Presenters expect interested attendees to walk away from their presentation with their
contact information, a business card, or an appointment to meet at
a future date to ask more questions and even possibly initiate a collaboration.

10.5.3   Designing slides and posters

The single most important thing to remember when designing slides
or posters is that they are intended to *support* the oral presentation,
not replace it. In an oral presentation, unlike a written paper, one
delivers most of the content orally. Therefore, the primary purpose of
the visual support is to complement the speaker's words by organizing
the content, highlighting certain points, and displaying visually complex information. One uses the visual support primarily to convey
concepts that are difficult to describe with words, such as diagrams,
pictures, and mathematical calculations. Slides and posters need not
be intelligible on their own, disconnected from the presenter's explanations. Keeping this in mind avoids the common mistake of wasting
valuable space with too many words.

Software

A large variety of software applications are available on the market
for designing slides and posters.

- **Microsoft PowerPoint** is the industry's standard slide presentation platform. *Keynote* is a similar program released by
  Apple Inc. for Macintosh computers. Both PowerPoint and

Keynote are versatile tools, universally available, and relatively user-friendly, but as word processor extentions, both suffer from the disadvantages of word processors (see Section 10.3). For example, different versions may alter the formatting.[19] Both PowerPoint and Keynote feature a rudimentary equation editor for mathematics (and Keynote's editor can process basic mathematics entered in LaTeX), but remain limited compared to the mathematical capabilities of LaTeX.

- **Google Slides.** As a web-based system, Google Slides has the advantage of allowing multiple collaborators to access and work on the same slides simultaneously. Like PowerPoint, Google Slides features an equation editor for typesetting simple mathematical expressions.

- **Beamer** is the method of choice for designing presentations with a large amount of mathematical content. A German product, Beamer is a LaTeX document class for producing presentation slides, so it offers all the advantages of using LaTeX. As any LaTeX-produced document, Beamer slides are compiled into PDF format, viewable on any computer equipped with any PDF viewer. However, formatting slides and posters that contain images requires attaching external image files to the LaTeX source code and writing code to position and scale each one. This is a labor-intensive process compared to the simple click-and-drag action of other programs.

Posters are created as a single large slide using any of the programs mentioned above. Microsoft PowerPoint is a common choice for designing posters, and there are many templates available on the internet to get started.

Style and syntax

The guidelines for style and syntax of mathematical documents outlined in Subsection 10.1.2 apply to slides and posters. However, because a visual support is considered less formal than a written paper, some rules are more flexible:

- It is not necessary to write in full sentences. In fact, full sentences are discouraged, as they contain nonessential words that take up valuable space. Instead, one should opt for bullet lists and short phrases that are easy for the audience to apprehend while simultaneously listening to the accompanying oral explanations.

---

[19]This can be especially problematic if one must present on a different computer than the one used to create the slides. For this reason, presentations should be exported to PDFs, which locks the formatting.

- Informal mathematical symbols ($\Rightarrow$, $\therefore$, $\Leftrightarrow$, $\forall$, $\exists$, etc.) and acronyms (iff, st, ae, WLOG, TFAE, etc.) are allowed.

- Tables and figures need not be numbered, captioned, or cross-referenced (see below).

Because space is at a premium, one should prioritize diagrams, images, tables, mathematical calculations, and other visual elements that are difficult to convey orally. Also, because reading and listening to oral explanations simultaneously can be taxing for the audience, text should be kept at a minimum. As a general rule of thumb, a slide should not exceed 8 lines and 40 words. This requires maximum semantic density, condensing phrases, eliminating semantically "empty" words (the, is, are, etc.), and omitting anything that can be explained orally.

Unlike reading a paper, the audience must follow the presentation sequentially and does not have the possibility to back up to review previous slides. To this effect, slide titles and bullet lists play the role of section headings in a paper, helping the audience understand the overall organization of the presentation. Because the audience will not necessarily remember everything on every slide, the speaker should repeat important information on several slides, such as central concepts, definitions, or notations that appear frequently. For mathematical calculations, the meaning of variables and unfamiliar symbols should be reminded on every slide where they appear.

Format, colors, and fonts

The presenter should choose an overall style that is sober and professional, with a simple background, and a color scheme that enhances visual appeal without being too busy or distracting from the content (2-3 colors is ideal).

One should aim to fill the available space in a balanced way, avoiding dense text, as it causes information overload and makes for an exhausting presentation. On the other hand, large empty spaces are also to be avoided.

The presenter should use only sans-serif fonts (such as Arial, Calibri, Tahoma, or Verdana).[20] On slides, the font size can range from 30 to 40 for titles and headings, and 16 to 20 for body text. For posters, the title should use font size 80-100, so that it is visible from several feet away, and 40-50 for the text, so that readers standing 3-4 feet away can read comfortably.

---

[20]Sans-serif fonts refer to fonts that lack *serifs*, the small extensions at the end of strokes. Serif fonts, such as Times New Roman, are easier to read at small sizes on printed documents, but more difficult to read on slides and posters.

Mathematics

The guidelines for mathematical style and symbolic expressions are the same as for any mathematical document (see Subsection 10.1.2). Only the key steps of mathematical calculations are necessary, as the goal of an oral presentation is to convey the general approach to solving a problem or proving a theorem, referring the audience to a complete paper for the full details if necessary. Too much mathematical detail makes for overcrowded slides and causes information overload.

One should define unfamiliar terms, notation, and symbolic variables on every slide where they appear. Unlike a paper where a reader can flip back to a previous page to be reminded of definitions, the audience is locked into the pre-determined slide sequence and must be reminded of these details frequently (on more than one slide) throughout the presentation.

Inserting mathematical expressions on slides is straightforward when using Beamer. On all other platforms, a common roundabout method consists in generating a dummy LaTeX document with the mathematical portions of the presentation, and copy-pasting snapshots from the output PDF file as images into the slides or poster.

Tables and figures

Tables should have no more than 5 to 7 rows and columns, to prevent information overload, and ideally contain visual cues that draw attention to specific rows, columns, or cells that are worth noticing. When presenting large amounts of data or results, one should provide only a representative sample in the presentation and to refer to an external paper for the complete details.

The guidelines for tables and figures are the same as in Subsection 10.4.2, except that they need not be numbered, captioned, or cross-referenced. However, if they contain any data, result, diagram, or image taken from another source, the source must be cited on the same slide (see below).

Citations and bibliographies

As in a paper, the presenter must cite the source of any part of the presentation that contains the work of another person. There are two ways to handle citations in a slide presentation:

1) One may insert in-text citations on the slides throughout the presentation (similar to a paper). At the end, we include a *bibliography slide* containing all works cited throughout the presentation.

2) Alternatively, one may provide the bibliographical information on the same slide as the cited content, as a footnote. In this case, the presenter may use a smaller font for this information, so that it does not overcrowd the slide and distract from the more important elements.

Audiences usually prefer the second method, so they can view the relevant bibliographical source information immediately, without having to wait for a bibliography slide at the end. The second method also avoids having to end with a usually overcrowded and overwhelming bibliography slide, a poor way to end a presentation.

In a poster, one may insert in-text citations as in a paper, cross-referencing a bibliography section at the end.

As mentioned in Subsection 10.4.2, for any image created by another person, the presenter must, in addition to the citation, provide the name of the copyright owner, if any, and certify that he or she was granted permission to legally use the image.

Title and outline

The first slide of a presentation, called the *title slide*, must contain the following items:

- title of the presentation,

- name and affiliation of presenter(s),

- name, date, and place of the event hosting the presentation.

In a poster, these items appear in the *title section*, a broad band that runs across the entire width at the top of the poster.

After the title slide, the presentation may begin with an optional *outline slide* that listing the sections of the talk, to give the listeners the overall structure of the presentation upfront.

Acknowledgments and contact information

Presenters often end a presentation with an *acknowledgments slide*, which acknowledges everyone who contributed to the content by providing guidance, advice, ideas, data, or software. Additionally, the presenters may be required to acknowledge any source of funding.

Because the goal of an oral presentation is to pique the audience's interest, presenters often choose to end a talk with a final slide showing their *contact information* and an invitation to contact them later with questions or to further discuss the content. Interested listeners often need some time to process what they heard before asking questions. They can write down the presenters' contact information during the few minutes of public Q&A time that follow the presentation and privately reach out to the presenters later.

10.5.4   Delivery

A presenter should rehearse the delivery of a presentation several times to make sure it is clear, concise, and relaxed. Rehearsing improves the flow, as the presenter gradually memorizes key phrases and anticipates transitions. The presenter should also carefully gauge the timing, especially in a slide presentation, where one is not allowed to go over the allotted time.

Slide presentations

Time is a central factor when rehearsing a slide presentation; the goal is to optimize the use of allotted time without going over the time limit. Ideally, the presentation should end with at most 1 or 2 minutes remaining. Ending too soon is the sign of an unprepared talk and a poor use of the given time. At a conference, every oral session is managed by a *moderator* in charge of introducing the presentations and staying on schedule.[21] This person moderator typically gives a hand signal to the presenters when there is 1 or 2 minutes remaining and will interrupt speakers when the time is up.

Delivering a good oral presentation takes practice. The following suggestions will enhance the talk for the speakers and the audience:

- A presenter should begin by introducing himself or herself and thank the audience for their time and interest.

- The pacing should flow at about 1 or 2 slides per minute on average. Some slides may only take a few seconds, while others may take a full minute. One should avoid dwelling more than a minute on a single slide.

- The presenter should refrain from reading the slides out loud. The purpose of slides is to complement oral explanations, not duplicate them. Reading slides out loud is a poor use of presentation time and (because the audience can read them faster) makes the presentation seem slow.

- One should keep in mind that the audience has to read the slides and listen to oral explanations simultaneously. This type of multitasking can be mentally taxing, especially if a slide is dense with information, so the presenter should slow down to give the audience time to read and apprehend complex visual information while listening.

- Because the audience must follow the presentation sequentially and cannot back up to review previous parts, the presenter should repeat new concepts and important ideas frequently, reminding essential elements (such as definitions and symbols)

---

[21]In a class, the professor plays the role of the moderator.

periodically in case the audience forgets them or missed them the first time.

- If possible, the presenter should make eye contact with the audience and move around to make the presentation more engaging.

- One should avoid paper notes, as they increase the temptation to read them, divert eye contact from the audience, and give the talk the impression of being unrehearsed. If absolutely necessary, one should only glance at them infrequently and quickly.

The time of questions from the audience at the end of a presentation is extremely valuable for both the presenter and the audience. It is the only feedback presenters receive from the audience, the only gauge of whether the pace and level of explanations was appropriate, and the only opportunity to clarify misunderstandings. Providing complete answers to questions on the fly is also an opportunity for the presenter to display command of knowledge about the topic in areas not specifically covered in the presentation itself.

Poster presentations

When presenting a poster, one gives a similar talk multiple times with no strict time limit to a revolving audience of 1 to 5 people. Over the course of a 2 to 3 hour poster session, the presenter may repeat the same talk with slight variations dozens of times, so by the end it is well rehearsed and flows smoothly. Furthermore, as the presenter interacts with more listeners, he or she can anticipate their questions and incorporate the answers into subsequent presentations, enhancing the clarity with each repetition.

The following suggestions will enhance a poster presentation for the presenter and the audience:

- Like slides, a poster should complement the oral explanations, not duplicate them. The presenter should refrain from reading the poster out loud word for word but instead provide explanations that complement the poster's content. Listeners must read the poster and listen to oral explanations simultaneously, so one should give them a few seconds to read through the poster before engaging them with the presentation.

- One of the most important advantages of the poster presentation format is the possibility to customize each repetition of the talk to the audience. To do this effectively, the presenter should engage interested listeners in conversation to determine their background, interests, and time constraints, and then use this information to customize the presentation. For example, a poster featuring a mathematical model for studying bird migration patterns using partial differential equations may attract a mathematician interested in the differential equations aspect, and later a nonmathematician interested in the application to birds. Each one will receive a different presentation that emphasizes their respective areas of interest.

• The presenter should avoid launching into a 10-minute talk with no interruption (as if it were a slide presentation). Instead, he or she should engage the listeners periodically throughout the talk to ensure that they are following and allow them to ask questions during the presentation.

• One must be flexible, prepared to adjust each repetition to the particular characteristics of the audience. For this, one should prepare multiple different versions of the same talk ahead of time, aimed at people with different interests, background knowledge, and time availability (for example, a 1-minute, 3-minute, and 5-minute talk).

Equipment and logistics

Even a well-prepared, well-designed, and well-rehearsed presentation can go terribly wrong at the last minute because of technical issues or misunderstandings regarding the presentation equipment. Wasting valuable time to troubleshoot technical issues can easily derail the timing of an entire presentation. Talks given at conferences or in a class run on a tight schedule and are not permitted to run late if a presenter is underprepared.

A well-prepared presenter should store a slide presentation in multiple places (on a laptop, cloud drive, email, USB stick) to minimize the risk of loss or corruption of the file containing the slides. A presenter who plans to connect a laptop to the projector should make sure the battery is charged, the screen saver is off, and bring the necessary cables and adapters to connect, or make sure ahead of time that they are available at the venue. A presenter who plans to use a different computer (one already connected to the projector), must make sure it is equipped with software capable of reading and displaying their slides. Even the right software installed on a different computer may have difficulty displaying slides correctly if they contain complex formatting, graphics, or embedded videos. Finally, one should bring a laser pointer and remote control device, as these items are usually not supplied.

Posters require a tubelike carrying case for transportation. The presenter should bring tacks or tape to affix the poster to the display board on site, or make sure that these items will be provided.

Finally, presenters often bring business cards or page-size summaries of their presentation to hand out to interested attendees who may approach them during or after the event.

APPENDIX A

Rubric for Assessing Proofs

Every faculty that teaches proof writing skills develops his or her own habits of assessment and feedback. This appendix provides a rubric and a list of feedback codes the authors use in assessing proofs. The core of this rubric comes from Chuck Collins [20].

The following rubric involves four categories: logic, understanding, creativity, and communication. Each section provides a measure of quality in each category along with a list of common flaws. These categories cannot be comprehensive. In particular, the rubric does not address errors that arise from "simple" mistakes, namely incorrect algebra or errors in calculations, differentiation algorithms, integration algorithms, and so on.

A.1 Logic

Proper logic stands as the central quality of a good proof. Here is a metric to measure the quality of a proof in terms of its logic.

Logic Rubric Score

0) Proof shows no logic or is too incomplete to evaluate.

1) Proof reflects a 1-step solution; it has no middle argument.

2) Individual steps are logically correct mostly, but the overall argument lacks logical order or the steps are unsupported.

3) Proof has good logic and overall reasoning, but either several small steps or one big step is wrong or missing.

4) Proof is logical and complete but is too mechanical in the details or makes some small mistakes.

5) Proof is correct, efficient, and shows proper detail in all parts.

Errors in logic are called fallacies. By a fallacy we mean affirming a statement that does not follow from the validity of a given hypothesis or previous statements in a proof.

The following gives a list of common fallacies or responses to common problems or errors in a proof.

AC Affirming the Conclusion. Also known as Affirming the Consequent. This is a logical fallacy. This is when one knows that the proposition $p \to q$ is true and that the proposition q is true and then concludes that p is true. (Example: Consider this true theorem: If $2^n - 1$ is prime, then n is prime. The fallacy of affirming the conclusion would be to suppose that n is prime and then conclude that $2^n - 1$ must be prime. As a counter-example, when 11 is prime, $2^{11} - 1 = 23 \times 89$.)

CE Counter-Example. This is not so much a fallacy as a reminder that the easiest way to disprove a universally quantified statement is to find one counter-example. It is good practice to find a simple counter-example that exhibits a basic understanding of why the hypothesized universal statement is not true.

CR Circular Reasoning, also called Begging the Question. This is a fallacy in which the writer assumes something (without proof) that is equivalent to the result that he/she wants to prove. (See Section 2.1.5.)

DH Denying the Hypothesis. This is also known as Denying the Antecedent. This is a logical fallacy. This is when one knows that the proposition $p \to q$ is true and that the proposition $\neg p$ is true and then concludes that $\neg q$. This fallacy resembles modus tollens, but it is inverted in a way that does not give a valid rule of inference.

DNF Does Not Follow. This means that the stated conclusion does not follow logically without amendments or additional hypotheses. This is usually where a student makes an incorrect leap of reasoning.

DNFRA Does Not Follow Right Away. This is less egregious than DNF but simply means that the writer must do more work to establish the stated conclusion from the given reasons.

FG Faulty Generalization. To prove a general result, we must use arbitrary objects (e.g., elements, sets, functions) and reason with the arbitrary objects. We cannot pick a few examples, show that the desired property holds for those, and then claim "so this property always holds." This is an example of a quantification fallacy. (Consider the Goldbach Conjecture; computations confirm it up to 400,000,000,000,000 even integers. Yet this does not mean that the conjecture is true.)

PINC Pattern Is Not Convincing. Sometimes, a student might illustrate something by showing a pattern. This might be helpful, but it is not always fully convincing. This shortcoming in explanation arises when the writer does not provide a solid explanation for a conclusion that they deduce from the pattern.

UC Unproven Claim. An assertion without proof or citation and used as a key part of the proof. (This is an egregious error.)

UCC Unproven Claim that is Correct. Any UC in a proof is bad. At times, students learning to write proofs state without proof a claim that happens to be correct. This does not at all establish the proof. It only means that by some accident or observing a pattern, the student guessed a useful stepping stone.

UCI Unproven Claim that is Incorrect. This is the worst form of a UC. It is essentially an incorrect guess.

A.2 Understanding / Terminology

In some academic disciplines, there is not always a clear consensus on definition of terms. These differences become important conversation points in those disciplines. In contrast, because of the logical precision of mathematics, people cannot have conversations unless they use the same precise definitions. Even when discrepancies occur in definitions or symbols, it is essential for people to be aware of these discrepancies.

Understanding / Terminology Rubric Score

0) No understanding or improper use of terminology.

1) Misused terminology or definitions.

2) Proper terminology but incomplete understanding.

3) Shows understanding but uses intuitive or imprecise language.

4) Shows understanding and uses proper terminology, but misses some finer points.

5) Shows understanding of all parts and uses terminology properly.

The following issues and associated codes describe some common problems that fall under this rubric.

CC Confusing Categories. Operations and reasoning with certain operations only happen in a certain context. For example, it does not mean anything to divide a number by a vector, to intersect a line with a number, to take the square root of a set. (For example, saying that "$\vec{u} \cdot \vec{v} = 1$ implies $\vec{v} = 1/\vec{u}$" does not make any sense because nowhere in linear algebra do we ever divide by a vector in $\mathbb{R}^n$.)

IUOT Improper Use Of Terms. Mathematical terms refer to specific classes of objects. For example, we never talk about a differentiable set or the bisector between two functions. Those examples sound ridiculous, but here are a few other examples of improper use of terms: a vector that is independent from another vector (the adjective "independent" is only used about a set of vectors); a set that is bijective (a set can be in bijection with another set, but we never say that a set is just bijective).

NSO Not So Obvious. It is not uncommon in mathematical writing to come across expressions like "it is obvious that..." or "it is clear that..." Professional mathematicians may say this when they are very familiar with a certain area, especially when they feel their readers should also be equally familiar. For math students, the claim that something is "clear" or "obvious" sometimes hides precisely where the subtlety is. If one sentence would make a claim clear, then a writer should say that one sentence; if it would take more than one sentence, then maybe the truth of the statement is not so obvious after all.

TI Too Intuitive. It is not uncommon to hold an intuitive understanding of certain concepts. This intuition can be helpful to lead us to conjectures, but it is never a substitute for the precise definition. (For example, recall that a set of vectors $\{\vec{v}_1, \vec{v}_2, \ldots, \vec{v}_n\}$ in a vector space V is called *linearly independent* if for any scalars $c_1, c_2, \ldots, c_n$

$$c_1\vec{v}_1 + c_2\vec{v}_2 + \cdots + c_n\vec{v}_n = \vec{0} \implies c_1 = c_2 = \cdots = c_n = 0.$$

An intuitive perspective on the concept of linear independence, might be to say that "they all point in different directions." This is far too imprecise for any proof. On the other hand, consider a third way of thinking of linear independence: "A set of vectors is linearly independent if no vector in the set is a linear combination of the others." This statement might feel more intuitive, but it is in fact precise because there exists a theorem that affirms that this statement is equivalent to the definition of linear independence.)

WDCNCA   Without Definition Can Not Conclude Anything.
When working with mathematical objects, if we do not explicitly define the object (an element, a set, a function, a subset, a subspace, and so on), we cannot conclude anything specific about that object. (For example, if we do not state what function we are working with, we cannot possibly conclude that it is a bijection. For example, consider this reasoning. "Let $f : \mathbb{R} \to \mathbb{R}^2$ be a function. Since $f(x_1) = f(x_2)$ implies that $x_1 = x_2$, then f is one-to-one." This is partly circular reasoning and partly nonsensical.)

A.3   Creativity

Many exercises, especially the simpler ones in high school mathematics, and even calculus or linear algebra, simply require the application of an established algorithm. Because advanced mathematics courses aim to introduce students to the more open-ended nature of mathematical investigation, proof problems move away from algorithmic thinking. Consequently, it is up to the student/mathematician to establish the steps of reasoning behind a proof and this aspect of proof writing requires a certain amount of creativity.

Some proof problems are straightforward and some require considerable creativity to find a strategy. Some deep theorems in mathematics emerge from finding a new proof strategy. Even as we propose a Creativity Rubric, we are aware that it is impossible to quantify creativity.

Creativity Rubric Score

  0) No attempt at finding a solution.

  1) Properly uses the definitions to set up the problem.

  2) Properly models after an example.

  3) Identifies an effective proof strategy but fails to implement it.

  4) Identifies all parts of an effective strategy and implements it correctly.

  5) Exhibits a deep and concise insight.

The following issues and associated codes describe some common problems that fall under this rubric.

CPS Correct Proof Strategy. Even if not carried through fully, this indicates that the stated strategy should work if carried out properly. (This is a positive code.)

FETC Following Example Too Closely. When proving some hypothesis, it is sometimes possible to get a useful idea from an example or the proof of some related result. This is a good learning strategy. However, sometimes students follow the example too closely and do not recognize the subtle differences between the example they are working from and their current proof problem.

IPS Incorrect Proof Strategy. Often, with proofs, there can be more than one way to establish a claim or conjecture. However, sometimes a selected proof strategy might not be likely to help.

RD Restate Definition. Just to get started with some proofs, it is necessary and useful to restate the definitions of objects involved in the proof. (This may also be an issue of style.)

A.4 Communication

Though the quality of these rubrics often is correlated, a proof also requires good communication.

At the worst, some students in early math courses act as if all that matters is the answer. However, any answer is irrelevant if it is not supported by a proof. Furthermore, the majority of interesting results in mathematics are theorems, which only can be called a theorem once there is a proof of the claim. Not quite as bad but still poor, are proofs that might have the pieces but are not organized; or even still proofs where the writer did not bother to use good English.

Here is a common rubric used to measure the quality of communication in proof writing. Note that in this rubric, the requirements after a score of 0 are cumulative. In other words, to get a score of n for $1 \leq n \leq 5$, the proof must satisfy all the requirements between 1 and n.

Communication Rubric Score

0) Has structure or is unreadable.

1) Follows proper basic structure for type of proof.

2) Provides proper support or reasons for important steps.

3) Has proper use of notation.

4) Uses complete sentences and makes no spelling or grammar mistakes.

5) Has good flow.

The following categories and associated codes are just some issues that may arise in the communication rubric for proof writing.

CT Cite Theorem. In mathematical reasoning, we use theorems all the time in proofs. Sometimes, the theorems we use become so commonplace that to cite them all would become tedious, especially when the reader should be familiar with them. However, before the writer or the reader are at that same stage of shared context, the writer should cite the theorems that he or she uses. Even when the writer can anticipate that his or her readers share the same context, a proof should cite most named-theorems or theorems that provide key ideas in the proof. It is important for a proof writer to communicate that they are confident of their building blocks.

LC Logical Connectors. In a sequence of calculations, if we want to say that one line logically implies the next, we should write $\implies$ at the beginning of the next line. If we want to say that one line is logically equivalent (if and only if), we should write $\iff$ at the beginning of the next line. Some writers also use $\therefore$ to read "therefore," which is the same as "implies." Without logical connectors, the writer is simply listing statements and not making claims about whether they are connected or not.

MN Mathematical Notation. Mathematical notation is well-defined enough to say and write everything with precision. Everyone writing mathematics should use the (generally) accepted notation. Furthermore, unless a student is on the cutting edge of research, the mathematics community has likely already settled on notation for most concepts.

NN Not Necessary. This code does not mean that something is incorrect. It simply means that the work and statement is not necessary for the proof in question. Especially when writing a proof, you want to be concise.

PS Poor Structure. This is a serious problem. A good proof must have a structure. A good proof does not involve a list of ideas and expect the reader to discern the proper order to create a valid argument. Also, it is not good form to start a proof by listing a number of definitions and theorems that seem relevant; these might be helpful for scratch work or gathering thoughts, but the proof should only cite theorems or definitions following a natural flow.

U=C Use the = Symbol Correctly. It is very bad math style to put an equal symbol between things that are not equal. If a student does this, even if they are trying to use short-hand, they will likely confuse themselves and their readers. (Quite a few first-year calculus students use random symbols like $\to$ instead of equal. This should not be done. In upper-level courses, when someone needs to calculate various powers of a number or matrix or element in a group, a lazy writer might write something like: $2^2 = 4 \cdot 2 = 8 \cdot 2 = 16 \cdot 2 = 32 = 2^5$. This is ugly and incorrect: 2^2 is not equal to 2^5.)

APPENDIX B

Index of Theorems and Definitions from Calculus and Linear Algebra

Aimed at mathematics students, this book assumes the reader has taken calculus and linear algebra. Some students transitioning to advanced mathematics have not yet developed the habit of retaining the definitions and theorems with precision. With this in mind, this appendix provides a list of some of the key definitions and theorems encountered in a first course in calculus and a course in linear algebra.

In the following lists, the statement is a definition unless it carries the name of "Theorem." We do not at all intend the following lists of definitions and theorems as comprehensive or fully representing those that a student should remember. We give these lists to support the examples and discussion in other parts of this book.

B.1   Calculus

Increasing; Decreasing. Let I be an interval of $\mathbb{R}$. A function $f : I \to \mathbb{R}$ is called

- *increasing* if $x < y$ implies $f(x) \leq f(y)$;
- *strictly increasing* if $x < y$ implies $f(x) < f(y)$;
- *decreasing* if $x < y$ implies $f(x) \geq f(y)$;
- *strictly decreasing* if $x < y$ implies $f(x) > f(y)$.

Convex; Concave. Let I be an interval of $\mathbb{R}$. A function $f : I \to \mathbb{R}$ is called *convex* if for all $x_1, x_2 \in I$ and for all $t \in [0, 1]$,

$$f((1 - t)x_1 + tx_2) \leq (1 - t)f(x_1) + tf(x_2).$$

A function $f : I \to \mathbb{R}$ is called *concave* if for all $x_1, x_2 \in I$ and for all $t \in [0, 1]$,

$$f((1 - t)x_1 + tx_2) \geq (1 - t)f(x_1) + tf(x_2).$$

Limit at a point. Let $c \in \mathbb{R}$ and let f be a function defined on an interval containing c, except perhaps at c itself. The *limit* of f at c is L, and we write

$$\lim_{x \to c} f(x) = L$$

if for all $\varepsilon > 0$ there exists $\delta > 0$ such that $0 < |x - c| < \delta$ implies $|f(x) - L| < \varepsilon$.

With quantifiers, we can write this definition as

$$\lim_{x \to c} f(x) = L$$
$$\iff \forall \varepsilon > 0 \, \exists \delta > 0 \, \forall x \in \mathbb{R} \, (0 < |x - c| < \delta \longrightarrow |f(x) - L| < \varepsilon).$$

Left-Hand Limit. $\lim_{x \to c^-} f(x) = L$ if for all $\varepsilon > 0$ there exists $\delta > 0$ such that $0 < x - c < \delta$ implies $|f(x) - L| < \varepsilon$.

Right-Hand Limit. $\lim_{x \to c^+} f(x) = L$ if for all $\varepsilon > 0$ there exists $\delta > 0$ such that $0 < -(x - c) < \delta$ implies $|f(x) - L| < \varepsilon$.

Continuity at a Point. A function f is *continuous* at c if f is defined at c and $\lim_{x \to c} f(x) = f(c)$.

One-sided Continuity. A function is continuous from the left (resp. right) at c if $\lim_{x \to c^-} f(x) = f(c)$ (resp. $\lim_{x \to c^+} f(x) = f(c)$.

Continuity on an Interval. A function f is *continuous on an interval I* if it is continuous at every element of the interval. If I has endpoints, then we understand f to be continuous from the right at the left endpoint, and continuous from the left at the right endpoint.

Differentiability at a Point. Let $a \in \mathbb{R}$. A function f defined on an interval containing a is called *differentiable* at a if the limit

$$\lim_{h \to 0} \frac{f(a + h) - f(a)}{h}$$

exists. If this limit exists, we call it the *derivative* of f at a and denote it by $f'(a)$.

Differentiability on an Interval. Let I be an open interval. A function f is differentiable on I if it is differentiable at every real number of I.

Intermediate Value Theorem (IVT). Let f be a continuous function over an interval $[a, b]$. If y_0 is any number between $f(a)$ and $f(b)$, where $f(a) \neq f(b)$, then there exists $c \in (a, b)$ such that $f(c) = y_0$.

Extreme Value Theorem (EVT). Let f be a continuous function over an interval $[a, b]$. Then f attains an absolute maximum value $f(c)$ and an absolute minimum value $f(d)$ for some $c, d \in [a, b]$.

Mean Value Theorem (MVT). Let f be a continuous function over an interval $[a, b]$ that is differentiable over (a, b). Then there exists $c \in (a, b)$ such that

$$\frac{f(b) - f(a)}{b - a} = f'(c).$$

Fundamental Theorem of Calculus. Suppose that f is continuous on $[a, b]$. Then

1) The function $g : [a, b] \to \mathbb{R}$ defined by $g(x) = \int_a^x f(u)\, du$ is continuous over $[a, b]$, is differentiable over (a, b), and satisfies $g'(x) = f(x)$.

2) $\int_a^b f(x)\, dx = F(b) - F(a)$, where the function $F : [a, b] \to \mathbb{R}$ is any antiderivative of f.

B.2 Linear Algebra

The student in linear algebra learns that the theory of vectors in $\mathbb{R}^n$ extends to the theory of elements in a general vector space. After we give the definition of a general vector space (over $\mathbb{R}$), we state many definitions both in terms of elements in a vector space, knowing that $\mathbb{R}^n$ is one example of a vector space.

Vector space. A vector space (over $\mathbb{R}$) is a set V equipped with a binary operation $+ : V \times V \to V$ and a function $\cdot : \mathbb{R} \times V \to V$ that satisfy the following conditions:

1) $+$ is commutative: $\forall \mathbf{u}, \mathbf{v} \in V$, $\mathbf{u} + \mathbf{v} = \mathbf{v} + \mathbf{u}$.

2) $+$ is associative: $\forall \mathbf{u}, \mathbf{v}, \mathbf{w} \in V$, $(\mathbf{u} + \mathbf{v}) + \mathbf{w} = \mathbf{u} + (\mathbf{v} + \mathbf{w})$.

3) $+$ has an identity: There exists an element of V called the *zero vector* and denoted $\mathbf{0}$ such that $\forall \mathbf{u} \in V$, $\mathbf{u} + \mathbf{0} = \mathbf{u}$.

4) $+$ has inverses: For every $\mathbf{u} \in V$, there exists an element called the *negative* and denoted $-\mathbf{u}$ such that $\mathbf{u} + (-\mathbf{u}) = \mathbf{0}$.

5) $\forall c \in \mathbb{R}\, \forall \mathbf{u}, \mathbf{v} \in V$, $c(\mathbf{u} + \mathbf{v}) = c\mathbf{u} + c\mathbf{v}$.

6) $\forall c, d \in \mathbb{R}\, \forall \mathbf{u} \in V$, $(c + d)\mathbf{u} = c\mathbf{u} + d\mathbf{u}$.

7) $\forall c, d \in \mathbb{R}\, \forall \mathbf{u} \in V$, $c(d\mathbf{u}) = (cd)\mathbf{u}$.

8) $\forall \mathbf{u} \in V$, $1\mathbf{u} = \mathbf{u}$.

Subspace. A *subspace* of a vector space V is a subset $W \subseteq V$ that satisfies the following conditions:

1) W is closed under addition: $\forall \mathbf{u}, \mathbf{v} \in W, \mathbf{u} + \mathbf{v} \in W$.

2) W is closed under scalar multiplication: $\forall c \in \mathbb{R} \,\forall \mathbf{u} \in W, c\mathbf{u} \in W$.

Linear combination. Let $S = \{\mathbf{u}_1, \mathbf{u}_2, \ldots, \mathbf{u}_k\}$ be a finite subset of a vector space V. A *linear combination* of S is any vector of the form $c_1\mathbf{u}_1 + c_2\mathbf{u}_2 + \cdots + c_k\mathbf{u}_k$, where $c_1, c_2, \ldots, c_k \in \mathbb{R}$.

Linear independence. A subset $\{\mathbf{u}_1, \mathbf{u}_2, \ldots, \mathbf{u}_k\}$ of a vector space V is called *linearly independent* if

$$c_1\mathbf{u}_1 + c_2\mathbf{u}_2 + \cdots + c_k\mathbf{u}_k = \mathbf{0} \implies c_1 = c_2 = \cdots = c_k = 0.$$

Otherwise, the set of vectors is called *linearly dependent.*

Span. The *span* of a finite set of vectors $\{\mathbf{u}_1, \mathbf{u}_2, \ldots, \mathbf{u}_k\}$ in a vector space V is the set of all linear combinations of $\{\mathbf{u}_1, \mathbf{u}_2, \ldots, \mathbf{u}_k\}$; in other words

$$\mathrm{Span}(\mathbf{u}_1, \mathbf{u}_2, \ldots, \mathbf{u}_k) = \{c_1\mathbf{u}_1 + c_2\mathbf{u}_2 + \cdots + c_k\mathbf{u}_k \mid c_1, c_2, \ldots, c_k \in \mathbb{R}\}.$$

Linear transformation. Let V and W be two vector spaces over $\mathbb{R}$. A function $T : V \to W$ is called a *linear transformation* if:

1) $\forall \mathbf{u}, \mathbf{v} \in V, T(\mathbf{u} + \mathbf{v}) = T(\mathbf{u}) + T(\mathbf{v})$; and

2) $\forall c \in \mathbb{R} \,\forall \mathbf{u} \in V, T(c\mathbf{u}) = cT(\mathbf{u})$.

Kernel and image. Let $T : V \to W$ be a linear transformation between vector spaces.

1) The *kernel* of T is $\mathrm{Ker}\,T = \{\mathbf{u} \in V \mid T(\mathbf{u}) = \mathbf{0}\}$.

2) The *image* of T is $\mathrm{Im}\,T = \{\mathbf{w} \in W \mid \exists \mathbf{u} \in V \; T(\mathbf{u}) = \mathbf{w}\}$.

Eigenvalue and eigenvector of a matrix. Let A be an $n \times n$ matrix. A scalar $\lambda \in \mathbb{R}$ is called an *eigenvalue* of A if there exists a nonzero vector $\mathbf{v} \in \mathbb{R}^n$ such that $A\mathbf{v} = \lambda\mathbf{v}$. The vector $\mathbf{v}$ is called an *eigenvector* of A.

Eigenvalue and eigenvector of a linear transformation. Let V be a vector space and let $T : V \to V$ be a linear transformation of V to itself. A scalar $\lambda \in \mathbb{R}$ is called an *eigenvalue* of T if there exists a nonzero vector $\mathbf{v} \in V$ such that $T(\mathbf{v}) = \lambda\mathbf{v}$. The vector $\mathbf{v}$ is called an *eigenvector* of T.

Orthogonal. Two vectors $\mathbf{u}, \mathbf{v} \in \mathbb{R}^n$ are called *orthogonal* (to each other) if $\mathbf{u} \cdot \mathbf{u} = 0$. A square matrix $A \in M_{n \times n}(\mathbb{R})$ is called *orthogonal* when $A^\top A = I_n$.

Bibliography

[1] CareerCast. careercast.com, 2021 (accessed January 7, 2021).

[2] Georgetown University's Center on Education and the Workforce, 2021 (accessed January 7, 2021).

[3] National Association of Colleges and Employers. www.naceweb.org, 2021 (accessed January 7, 2021).

[4] National Institute of Education. services.math.duke.edu/major/whyMajor.html, 2021 (accessed January 7, 2021).

[5] U.S. Bureau of Labor Statistics. stats.bls.gov/ooh, 2021 (accessed January 7, 2021).

[6] Michael J. Barany. The World War II origins of mathematics awareness. *Notices of the AMS*, 64(4), 2018.

[7] M.L. Barton and C. Heidema. *Teaching Reading in Mathematics*. Association for Supervision and Curriculum, 2nd edition, 2002.

[8] Michael Beaney. *The Frege Reader*. Blackwell Oxford, 1997.

[9] Eric Temple Bell. *Men of Mathematics*. Simon and Schuster, 2014.

[10] William P. Berlinghoff and Fernando Q. Gouvea. *Math Through the Ages*. Oxton House Publishers, 2002.

[11] Mary Blocksma. *Reading the Numbers: A Survival Guide to the Measurements, Numbers, and Sizes Encountered in Everyday Life*. Viking, 1989.

[12] Matt Boelkins and Tommy Ratliff. How we get our students to read the text before class. https://www.maa.org/how-we-get-our-students-to-read-the-text-before-class, 2000 (accessed January 7, 2021).

[13] N. N. Bogolyubov, G. K. Mikhailov, and A. P. Yushkevich, editors. *Euler and Modern Science.* The MAA Tercentenary Euler Celebration. Mathematical Association of America, Washington, DC, 2007.

[14] James Bradley and Russell W. Howell. *Mathematics Through the Eyes of Faith.* Harper Collins, 2011.

[15] David M. Burton. *A History of Mathematics.* McGraw-Hill, 4th edition, 1999.

[16] Melanie Butler. Preparing Our Students to Read and Understand Mathematics. *Journal of Humanistic Mathematics,* 9(1):158–177, 2019.

[17] Judith N. Cederberg. *A Course in Modern Geometries.* Undergraduate Texts in Mathematics. Springer-Verlag, New York, 2nd edition, 2001.

[18] Larry Cochran. *The Sense of Vocation: A Study of Career and Life Development.* State University of New York Press, 1990.

[19] Paul Cohen. *The Independence of the Axiom of Choice,* 1963 (accessed January 11, 2021).

[20] Chuck Collins. *Guide to the Presentation of a Proof (Rubric/Best Practices),* 2008 (accessed December 19, 2019).

[21] Mark Colyvan. The miracle of applied mathematics. *Synthese,* 127(3):265–277, 2001.

[22] C. Cowen. Teaching and testing mathematics reading. *The American Mathematical Monthly,* 98(1):50–53, 1991.

[23] Lokenath Debnath. A short biography of Joseph Fourier and historical development of Fourier series and Fourier transforms. *International Journal of Mathematical Education in Science and Technology,* 43(5):589–612, 2012.

[24] Keith Devlin. *The Language of Mathematics: Making the Invisible Visible.* Macmillan, 2000.

[25] Keith Devlin. *Mathematics: The New Golden Age.* Columbia University Press, 2001.

[26] Stillman Drake et al. *Discoveries and Opinions of Galileo.* Doubleday New York, 1957.

[27] William Dunham. *Journey Through Genius: The Great Theorems of Mathematics.* Wiley, 1990.

[28] William Dunham. *Euler: The Master of Us All*, volume 22. American Mathematical Society, 2020.

[29] D. A. Edge. Oliver Heaviside (1850-1925) – Physical Mathematician. *Teaching mathematics and its applications*, 2(2):55–61, 1983.

[30] Albert Einstein. *Sidelights on Relativity*. Good Press, 2019.

[31] Murray Eisenberg. *Axiomatic Theory of Sets and Classes*. Holt, Rinehart and Winston, New York, 1971.

[32] Richard P. Feynman. New textbooks for the "new" mathematics. *Engineering and science*, 28(6):9–15, 1965.

[33] Abraham A. Fraenkel and Yehoshua Bar-Hillel. *Foundations of Set Theory*. North-Holland Publishing Company, Amsterdam, 1958.

[34] Torkel Franzén. Gödel's theorem, an incomplete guide to its use and abuse. *AK Peters, Wellesley*, 2005.

[35] Carl Friedrich Gauss. *Disquisitiones Arithmeticae*. Springer-Verlag, New York, 1986. Translated and with a preface by Arthur A. Clarke, Revised by William C. Waterhouse, Cornelius Greither, and A. W. Grootendorst and with a preface by Waterhouse.

[36] Walter Gautschi. Leonhard Euler: His life, the man, and his works. *SIAM Review*, 50(1):3–33, 2008.

[37] Kurt Gödel. The consistency of the axiom of choice and of the generalized continuum-hypothesis. *Proceedings of the National Academy of Sciences of the United States of America*, 24(12):556–557, 1938.

[38] Stephen J. Gould. *The Value of Science: Essential Writings of Henri Poincare*. Modern Library, 2012.

[39] Marvin Greenberg. *Euclidean and Non-Euclidean Geometries: Development and History*. W. H. Freeman, New York, 4th edition, 2007.

[40] S. Gunderson, R. Jones, and K. Scanland. *The Jobs Revolution: Changing How America Works*. Copywriters, Inc., 2004.

[41] Margie Hale. *Essentials of Mathematics*. The Mathematical Association of America, 2003.

[42] Paul R. Halmos. *Naive Set Theory*. D. Van Nostrand, Princeton, NJ, 1960.

[43] Godfrey Harold Hardy. *A Mathematician's Apology*. Cambridge University Press, 1992.

[44] Oliver Heaviside. On operators in pysical mathematics, part II. *Proceedings of the Royal Society of London*, 54:105–143, 1893.

[45] Horst Herrlich. *Axiom of Choice*. Springer-Verlag, Berlin, 2006.

[46] Reuben Hersh and Phillip J. Davis. *The Mathematical Experience*. Birkhäuser, Boston, MA, 1981.

[47] Kevin Houston. *How to Think Like a Mathematician: A Companion to Undergraduate Mathematics*. Cambridge University Press, 2009.

[48] Edward V. Huntington. Complete sets of postulates for the theory of real quantities. *Transactions of the American Mathematical Society*, 4(3):358–370, 1903.

[49] George Ifrah. *The Universal History of Numbers*. Wiley, Hoboken, NJ, 2000.

[50] Henry Ketcham. *The Life of Abraham Lincoln*. Perkins Book Company, New York, 1901.

[51] Israel Kleiner. *Excursions in the History of Mathematics*. Birkhauser Boston, 2011.

[52] Morris Kline. *Mathematics in Western Culture*. Oxford University Press, 1964.

[53] Morris Kline. *Mathematics, the Loss of Certainty*. Oxford University Press, 1982.

[54] Steven Krantz. *A Primer of Mathematical Writing*. American Mathematical Society, 2017.

[55] Thomas S. Kuhn. *The Structure of Scientific Revolutions*. University of Chicago Press, 2012.

[56] Stephen Lambert and Ruth J. DeCotis. *Great Jobs for Math Majors*. VGM Career Horizons, 1999.

[57] Joseph M. Landsberg. *Geometry and Complexity Theory*, volume 169. Cambridge University Press, 2017.

[58] Pierre-Simon Laplace. *Pierre-Simon Laplace Philosophical Essay on Probabilities: Translated from the fifth French edition of 1825 with notes by the translator*, volume 13. Springer Science & Business Media, 1998.

[59] Glenn Ledder. *Mathematics for the Life Sciences: Calculus, Modeling, Probability, and Dynamical Systems.* Springer Science & Business Media, 2013.

[60] G.W. Leibniz. *New Essays on Human Understanding.* Cambridge University Press, 2nd edition, 2008.

[61] J. Donald Monk. *Introduction to Set Theory.* McGraw-Hill, New York, 1969.

[62] Giuseppe Peano. *Arithmetices Principia: Novo Methoda Exposita.* Fratres Bocca, Rome, Italy, 1889.

[63] Roger Penrose. *The Emperor's New Mind.* Penguin Books, 1991.

[64] Henri Poincaré. *Science et Méthode.* Ed. Flammarion, 1908.

[65] Ahmed Renima, Habib Tiliouine, and Richard J. Estes. *The Islamic Golden Age: A Story of the Triumph of the Islamic Civilization.* Springer, 2016.

[66] Gian-Carlo Rota and David Sharp. Mathematics, philosophy, and artificial intelligence. https://sgp.fas.org/othergov/doe/lanl/pubs/00326965.pdf, 1985 (accessed September 17, 2021).

[67] Herman Rubin and Jean E. Rubin. *Equivalents of the Axiom of Choice.* North-Holland Publishing, Amsterdam, 1963.

[68] Bertrand Russell. *Mysticism and Logic.* New York, Longmans, Green and co., 1919.

[69] C. Edward Sandifer. *How Euler Did It.* The MAA Tercentenary Euler Celebration. The Mathematical Association of America, Washington, DC, 2007.

[70] Thomas Q. Sibley. *Thinking Geometrically: A Survey of Geometries.* The Mathematical Association of America, Washington, DC, 2015.

[71] George F. Simmons. *Calculus Gems.* McGraw-Hill, 1992.

[72] Robin Smith. Aristotle's logic. In Edward N. Zalta, editor, *The Stanford Encyclopedia of Philosophy.* Fall 2020 edition.

[73] Mark Steiner. The Applicabilities of Mathematics. *Philosophia Mathematica*, 3(3):129–156, 1995.

[74] Max Tegmark. *Our Mathematical Universe.* Alfred A. Knopf, 2014.

[75] Steven Weinberg. *Dreams of a Final Theory*. Vintage, London, 1993.

[76] Charles W.A. Whitaker et al. *Aristotle's De Interpretatione: Contradiction and Dialectic*. Oxford University Press, 1996.

[77] Eugene Wigner. The unreasonable effectiveness of mathematics in the natural sciences. *Comm. Pure Applied Math.*, 13(1):1–14, 1960.

[78] Elizabeth Woodward-Smith. *Cultural awareness and foreign language teaching*. Universidade da Coruña, 1997.

[79] Martin M. Zuckerman. *Sets and Transfinite Numbers*. Macmillan Publishing, New York, 1974.

Index

Printed in the United States
by Baker & Taylor Publisher Services